Edited by
Alexander Gromov and Ulrich Teipel

Metal Nanopowders

Related Titles

Koch, E.

Metal-Fluorocarbon Based Energetic Materials

2012
ISBN: 978-3-527-32920-5 (Also available in digital formats)

Agrawal, J.P.

High Energy Materials

Propellants, Explosives and Pyrotechnics

2010
ISBN: 978-3-527-32610-5 (Also available in digital formats)

Tjong, S.C.

Polymer Composites with Carbonaceous Nanofillers

Properties and Applications

2012
ISBN: 978-3-527-41080-4 (Also available in digital formats)

Teipel, U. (ed.)

Energetic Materials

Particle Processing and Characterization

2005
Print ISBN: 978-3-527-30240-6 (Also available in digital formats)

Edited by
Alexander Gromov and Ulrich Teipel

Metal Nanopowders

Production, Characterization, and Energetic Applications

WILEY-VCH
Verlag GmbH & Co. KGaA

The Editors

Prof. Dr. Alexander Gromov
Tomsk Polytechnic University
30 Lenin Prospekt
634050 Tomsk
Russia

and

Technical University Nürnberg
Georg-Simon-Ohm
Process Engineering Department
Wassertorstr. 10
90489 Nürnberg
Germany

Prof. Dr. Ulrich Teipel
Technical University Nürnberg
Georg-Simon-Ohm
Process Engineering Department
Wassertorstr. 10
90489 Nürnberg
Germany

Library of Congress Card No.: applied for

British Library Cataloguing-in-Publication Data
A catalogue record for this book is available from the British Library.

Bibliographic information published by the Deutsche Nationalbibliothek
The Deutsche Nationalbibliothek lists this publication in the Deutsche Nationalbibliografie; detailed bibliographic data are available on the Internet at <http://dnb.d-nb.de>.

Print ISBN: 978-3-527-33361-5
ePDF ISBN: 978-3-527-68072-6
ePub ISBN: 978-3-527-68071-9
mobi ISBN: 978-3-527-68070-2
oBook ISBN: 978-3-527-68069-6

Cover-Design Grafik-Design Schulz, Fußgönheim, Germany
Typesetting Laserwords Private Limited, Chennai, India
Printing and Binding Markono Print Media Pte Ltd, Singapore

Printed on acid-free paper

Contents

Foreword

Interest in studying the combustion of metal powders dramatically raised since Russian scientists Kondratyuk and Tsander suggested the use of metals as energetic additives to rocket fuels at the beginning of the twentieth century. Since that time, it is obvious that an increase in the dispersion of flammable substances participating in heterogeneous combustion processes leads to an increase in rate and heat of combustion. The major energy contribution belongs to the process of oxidation, which is also bound up with powder dispersion and purity. Burning of metal nanopowders is accompanied by new physical and chemical laws (such as high reactivity under heating, threshold phenomena, formation of nitrides in air), which allow to fully appreciate the advantages and disadvantages of nanopowders when used in fuel systems.

Widespread use of metal nanopowders is currently hampered by the lack of enough advanced technology for their preparation, certification, and standardization procedures, instability during storage, and subjective factors: the possible toxicity of nanopowders, investment risks, cost of nanotechnologies, and so on. Therefore, the main objective for the authors is to inform a wide readership of fundamental and applied studies on the processes of oxidation and combustion of metal nanopowders.

Prof. Dr.-Ing. George Manelis,
Institute of Problems of Chemical Physics,
Russian Academy of Science, Chernogolovka,
Russia

Prof. Dr.-Ing. Hiltmar Schubert,
Fraunhofer Institute of
Chemical Technology,
Pfinztal, Germany

List of Contributors

Giovanni Colombo
Dipartimento di Scienze e
Tecnologie Aerospaziali, SPLab
Politecnico di Milano Campus
Bovisa Sud
34 Via La Masa
I-20156 Milan
Italy

Luigi T. De Luca
Dipartimento di Scienze e
Tecnologie Aerospaziali, SPLab
Politecnico di Milano Campus
Bovisa Sud
34 Via La Masa
I-20156 Milan
Italy

Stefano Dossi
Dipartimento di Scienze e
Tecnologie Aerospaziali, SPLab
Politecnico di Milano Campus
Bovisa Sud
34 Via La Masa
I-20156 Milan
Italy

Edward L. Dreizin
University Heights
Otto H. York Department of
Chemical, Biological, and
Pharmaceutical Engineering
New Jersey Institute of
Technology
138 Warren St
Newark
NJ 07102-1982
USA

N. Eisenreich
Institute of Problems of Chemical
and Energetic Technologies
Russian Academy of Science
Socialisticheskaya str., 1
659322 Byisk
Russia

Cory Farley
Texas Tech University
Mechanical Engineering
Department
Corner of 7th and Boston Ave.
Lubbock
TX 79409-1021
USA

Marco Fassina
Dipartimento di Scienze e
Tecnologie Aerospaziali, SPLab
Politecnico di Milano Campus
Bovisa Sud
34 Via La Masa
I-20156 Milan
Italy

Franz Dieter Fischer
Montanuniversität Leoben
Institute of Mechanics
Franz-Josef-Straße 18
A-8700 Leoben
Austria

Luciano Galfetti
Dipartimento di Scienze e
Tecnologie Aerospaziali, SPLab
Politecnico di Milano Campus
Bovisa Sud
34 Via La Masa
I-20156 Milan
Italy

Alexander Gromov
Tomsk Polytechnic University
Lenin prospekt, 30
634050 Tomsk
Russia

and

George Simon Ohm University
of Applied Sciences
Processing Department
Wassertorstr. 10
90489 Nürnberg
Germany

Sh. Guseinov
State Research institute for
Chemistry and Technology of
Organoelement Compounds
(GNIIChTEOS)
Shosse Entuziastov str., 38
105118 Moscow
Russia

Alexander Il'in
Tomsk Polytechnic University
Lenin prospekt, 30
634050 Tomsk
Russia

Keerti Kappagantula
Texas Tech University
Mechanical Engineering
Department
Corner of 7th and Boston Ave.
Lubbock
TX 79409-1021
USA

Oksana V. Komova
Boreskov Institute of Catalysis SB
RAS
Pr. Akademika Lavrentieva 5
630090 Novosibirsk
Russia

Larichev Mikhail Nikolaevich
V.L. Talrose Institute for Energy
Problems for Chemical Physics
Russian Academy of Science
Leninsky prospect
bl. 38/2:, 119334 Moscow
Russia

M. Lerner
Institute of Strength Physics and Material Science
Russian Academy of Science
8/2 Academicheskiy St.
634021 Tomsk
Russia

Filippo Maggi
Dipartimento di Scienze e Tecnologie Aerospaziali, SPLab
Politecnico di Milano Campus
Bovisa Sud
34 Via La Masa
I-20156 Milan
Italy

Olga Nazarenko
Tomsk Polytechnic University
Lenin prospekt, 30
634050 Tomsk
Russia

Olga V. Netskina
Boreskov Institute of Catalysis SB RAS
Pr. Akademika Lavrentieva 5
630090 Novosibirsk
Russia

Michelle Pantoya
Texas Tech University
Mechanical Engineering Department
Corner of 7th and Boston Ave.
Lubbock
TX 79409-1021
USA

Christian Paravan
Dipartimento di Scienze e Tecnologie Aerospaziali, SPLab
Politecnico di Milano Campus
Bovisa Sud
34 Via La Masa
I-20156 Milan
Italy

Julia Pautova
Tomsk Polytechnic University
Lenin prospekt, 30
634050 Tomsk
Russia

Alice Reina
Dipartimento di Scienze e Tecnologie Aerospaziali, SPLab
Politecnico di Milano Campus
Bovisa Sud
34 Via La Masa
I-20156 Milan
Italy

Mirko Schoenitz
University Heights
Otto H. York Department of Chemical, Biological, and Pharmaceutical Engineering
New Jersey Institute of Technology
138 Warren St
Newark
NJ, 07102-1982
USA

Valentina I. Simagina
Boreskov Institute of Catalysis SB RAS
Pr. Akademika Lavrentieva 5
630090 Novosibirsk
Russia

P.X. Song
Institute of Particle Science and Engineering
University of Leeds
Leeds, LS2 9JU UK
National Institute of Clean-and-Low-Carbon Energy
Future Science & Technology Park
Changping District
Beijing 102209
China

Andrea Sossi
Dipartimento di Scienze e Tecnologie Aerospaziali, SPLab
Politecnico di Milano Campus Bovisa Sud
34 Via La Masa
I-20156 Milan
Italy

P. Storozhenko
State Research institute for Chemistry and Technology of Organoelement Compounds (GNIIChTEOS)
Shosse Entuziastov str., 38
105118 Moscow
Russia

Ulrich Teipel
George Simon Ohm University of Applied Sciences
Processing Department
Wassertorstr. 10
90489 Nürnberg
Germany

Dmitry Tikhonov
Tomsk Polytechnic University
Lenin prospekt, 30
634050 Tomsk
Russia

Dieter Vollath
Nano Consulting
Primelweg 3
D-76297 Stutensee
Germany

A. Vorozhtsov
Tomsk State University
Lenin str., 36
634050 Tomsk
Russia

and

Institute of Problems of Chemical and Energetic Technologies
Russian Academy of Science
Socialisticheskaya str., 1
659322 Byisk
Russia

Alfred P. Weber
Technical University of Clausthal
Institute of Particle Technology
Leibnizstrasse 19
D-38678 Clausthal-Zellerfeld
Germany

D.S. Wen
Institute of Particle Science and Engineering
University of Leeds
Leeds, LS2 9JU UK
National Institute of Clean-and-Low-Carbon Energy
Future Science & Technology Park
Changping District
Beijing 102209
China

Introduction

Stabilization of low-dimension structures, especially nanosized ones, and their use in the heterogeneous chemical reactions as nanopowders allow considering *high specific surface* as an independent thermodynamic parameter along with the temperature, pressure, concentration of reactants, and so on. New characteristics of 2D nanomaterials are well known – the thermal conductivity of graphene ($5000\,W\,(m\,K)^{-1}$) with $1000\,m^2\,g^{-1}$ specific surface exceeds those for metals in a factor 10 [1]. The use of the advantages of high specific surface of 3D nanostructures – nanopowders in catalysis, oxidation, and combustion results in high rates of heterogeneous reactions and reduction in activation energies of ignition due to the small size of solid reactants. The laws of classical chemistry and physics are little applicable to the analysis of processes with metal nanopowders. An example of such a system is the burning of the composition nanoAl/nanoMoO_3 at the rate of about $1\,km\,s^{-1}$ [2].

In USSR, metal ultrafine (in fact, nano-) powders with reproducible properties were first obtained during World War II. In the 1960s and 1970s, numerous works were carried out on metal nanopowder production by electrical explosion of wires [3], evaporation-and-condensation method [4], and the technologies of metal nanopowder application for nuclear synthesis in the USSR and the US. In 1977, the result of these works was published in Morokhov's book [5], where the methods for metal nanopowder production by thermal decomposition of salts were viewed. In Western Europe and the US, the term *nanocrystalline material* appeared and spinned off after the Gleiter's publication in 1980 [6].

Since the discovery by Yu. Kondratyuk and F. Tsander in 1910 [7], the possibilities of powdery metal being used as an additive in energetic materials and as the reagents for self-propagating high-temperature synthesis [8] were intensively studied. Several books (e.g., the work of Pokhil *et al.* [9] and Sammerfield [10]) were published, where the laws of combustion of micron-sized metal powders ($5\div500\,\mu m$) in high-temperature oxidizing environments were discussed. The study of the laws of combustion of powdered metals was done mainly for Al, Be, Mg, Ti, Zr, and B. The lack of micron-sized metal powders were detected during the first test of metallized fuels in the 1940s: an agglomeration of particles (especially for aluminum and magnesium) in the heating zone of energetic material, a low degree of metal reaction in the vapor phase (incomplete combustion), significant biphasic loss of a specific impulse (15% or more for the compositions containing 20–25 wt% Al) [9].

In the 1970s, Zeldovich and Leipunsky *et al.* [11] showed one of the approaches to reduce this lack by using low-sized metallic particles for fuels and combustion catalysts, in particular, metal nanopowders. This book summarizes the efforts of several teams over the world to realize those ideas.

The revitalization of the use of metal nanopowders in materials science and engineering became further possible in the 1990s, when the technologies for the large-scale production of those materials became available. Nowadays, tons of rather inexpensive metal nanopowders are produced in several countries for different technological applications, while the problems of their standardization, storage, handling, toxicity, correct application, and so on, are still unsolved.

The idea of this book is also to show the true picture of the properties of metal nanopowders and, correspondingly, their application avenues. The "romantic atmosphere" around nanomaterials and metal nanopowders accordingly should be left in the twentieth century forever. Nanoparticles and, especially, metal nanoparticles are very "capricious" technological raw materials with metastable physical and chemical properties in many cases, because nanometals (in addition to small particle size) show very high reducing properties: nanoCu react similarly to bulk Zn – release the hydrogen from acids, nanoAl show the properties of bulky alkali metals – react with water under room temperature, and so on.

Special scientific and engineering interests represent the new fundamental laws of combustion for the metal nanopowders, analysis of the combustion regimes, and intermediate and final burning products reported in this book. Excited by the experimental works of Ivanov and Tepper [12], scientists worked in the direction of nanometals application in energetic materials intensively during the past decade and the most valuable results are presented in this book.

In conclusion, we want to underline that the study of industrially available metal nanopowders allowed opening previously unknown laws and they will open the significant application prospects in science and technology of the twenty-first century.

Alexander Gromov
Ulrich Teipel

References

1. Seol, J.H., Jo, I., Moore, A.L., Lindsay, L., Aitken, Z.H., Pettes, M.T., Li, X., Yao, Z., Huang, R., Broido, D., Mingo, N., Ruoff, R.S., and Shi, L. (2010) Two-dimensional phonon transport in supported graphene. *Science*, **328** (5975), 213–216.
2. Bockmon, B.S., Pantoya, M.L., Son, S.F., and Asay, B.W. (2003) Burn rate measurements in nanocomposite thermites. Proceedings of the American Institute of Aeronautics and Astronautics Aerospace Sciences Meeting, Paper No. AIAA-2003-0241.
3. Chase, W.G. and Moore, H.K. (eds) (1962) *Exploding Wires*, Plenum Press, New York.
4. Gen, M.Ya. and Miller, A. (1981) A method of metal aerosols production. USSR Patent 814432. No. 11. p. 25.

5. Morokhov, I.D., Trusov, L.I., and Chizhik, S.P. (1977) *Ultradispersed Metal Medium*, Atomizdat, Moscow.
6. Gleiter, H. (1989) Nanocrystalline materials. *Prog. Mater. Sci.*, **33** (4), 223–315.
7. Gilzin, K.A. (1950) Rocket Engines, Moscow.
8. Merzhanov, A.G., Yukhvid, V.I., and Borovinskaya, I.P. (1980) Self-propagating high-temperature synthesis of cast refractory inorganic compounds. *Dokl. Akad. Nauk USSR*, **255**, 120.
9. Pokhil, P.F., Belyaev, A.F., Frolov, Y.V., Logachev, V.S., and Korotkov, A.I. (1972) *Combustion of Powdered Metals in Active Media*, Nauka, Moscow.
10. Sammerfield, M. (ed.) (1960) *Solid Propellant Rocket Research*, Academic Press, New York.
11. Zeldovich, Y.B., Leipunsky, O.I., and Librovich, V.B. (1975) *Theory of Non-Stationary Combustion of Powders*, Nauka, Moscow.
12. Ivanov, G.V. and Tepper, F. (1997) "Activated" aluminum as a stored energy source for propellants. *Int. J. Energetic Mater. Chem. Propul.*, **4** (1–6), 636–645.

1
Estimation of Thermodynamic Data of Metallic Nanoparticles Based on Bulk Values

Dieter Vollath and Franz Dieter Fischer

1.1
Introduction

It is a well-accepted fact: the temperature of phase transformation is particle-size dependent. In general, this dependency is described as

$$T_{\text{trans-nano}} = T_{\text{trans-bulk}} - \frac{\alpha}{d} \tag{1.1}$$

In Equation 1.1, the quantities $T_{\text{trans-nano}}$ and $T_{\text{trans-bulk}}$ stand for the transformation temperature of nanoparticles and the bulk material, respectively, d for the particle diameter, and α is a constant value depending on the entropy of transformation and the difference of the surface energy in both phases [1]. The same description, as proved for phase transformations, was found to be valid for the enthalpy of phase transformations. As typical examples, experimental results obtained for aluminum particles were given by Eckert [2], for tin particles by Lai *et al.* [3], or by Suresh and Mayo [4, 5] on yttrium-doped zirconia particles.

The range of particle sizes where Equation 1.1 is valid is limited. In the case of larger particles, Coombes [6] has shown that these have a surface layer of about 3 nm, where melting starts. As long as this surface layer dominates the behavior of the particles, Equation 1.1 cannot be applied. The existence of such a surface layer was also shown by Chang and Johnson [7] by theoretical considerations, concluding that this surface layer is less ordered than the center of the particles. As it was shown by Kaptay [8], the thickness of this premelting layer can be estimated by the rules of classical thermodynamics. Therefore, the assumption of a surface layer where melting starts is well justified. Now, one may ask if there is also a lower limit of particle sizes, below which Equation 1.1 is not applicable. Experimental results suggest this. Figure 1.1 displays the melting temperature of gold nanoparticles according to Castro *et al.* [9]. In this graph, the melting temperature is plotted versus the inverse particle size. According to Equation 1.1, one has to expect a linear relation.

The experimental data of Castro *et al.* may be separated into two ranges: Range I, which follows Equation 1.1 and a separated Range II, which is far off from the expected value. A linear fit of the experimental data in both ranges delivers an

Metal Nanopowders: Production, Characterization, and Energetic Applications, First Edition.
Edited by Alexander Gromov and Ulrich Teipel.

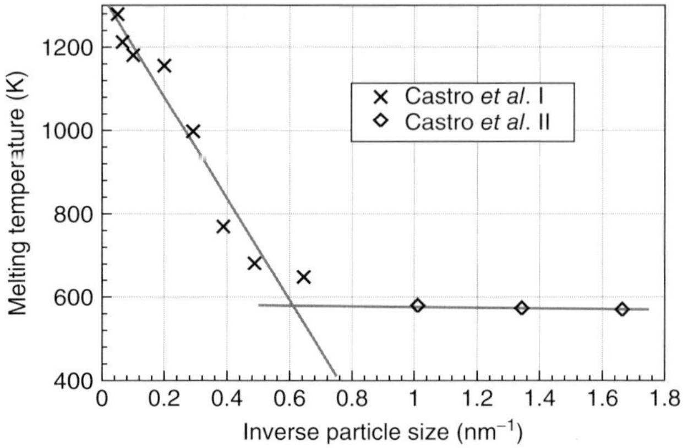

Figure 1.1 Experimental data for the melting temperature of gold nanoparticles, according to Castro *et al.* [9], together with linear fits plotted versus the inverse particle size. This graph shows clearly two separated ranges of the melting temperature: at larger particles, a range following Equation 1.1 (Range I) and a second range with particle-size-independent temperature (Range II).

intersection at an inverse particle size of 0.62 nm^{-1}, which is equivalent to a particle size of 1.6 nm. Obviously, for particle sizes below this intersection, Equation 1.1 is no longer valid. Such a phenomenon or similar ones are quite often described; as, for example, in the case of sodium particles [10]. Well in line with the above-described phenomenon, found for gold and sodium particles, are experimental results of Shvartsburg and Jarrold [11] on small tin particles consisting of 19–31 atoms, exhibiting melting points significantly higher than those of the bulk material. Besides a reduction of the melting temperature, close to melting or crystallization, additional phenomena are observed. Oshima and Takayanagi [12] found in 6 nm tin particles crystalline embryos with sizes around 1.5 nm. It is remarkable that this size is in the range of the limitation of Equation 1.1, as was found in the case of the melting of gold particles.

1.2
Thermodynamic Background

A general analysis of these phenomena needs detailed quantum mechanical studies. However, in most cases, one is interested in just a first approach using thermodynamic data of metallic nanoparticles. It is the aim of this chapter to show a simplified approach in this direction. The most important tool for any analysis of this kind is classical thermodynamics. Certainly, as this tool describes continuous systems, such an approximation cannot deliver phenomena depending on the quantum nature of the cohesion energy of small particles, or, in other words, magic particle sizes, superatoms, or jellium shell concepts cannot be expected as the result. These phenomena are excluded.

To analyze phase transformations, a detailed knowledge of the thermodynamic data of the materials in question is necessary. In addition, in the case of nanoparticles, knowledge of the surface energy in both phases is of great importance. As typical and well-studied examples for a phase transformation, melting, and crystallization were selected. In the following considerations, for reasons of simplicity, the minor changes of geometry and density are neglected. Generally, in the proximity of the melting point, the difference of the free enthalpy

$$G_{\text{m-nano}} = G_{\text{liquid-nano}} - G_{\text{solid-nano}}$$
$$G_{\text{m-bulk}} = G_{\text{liquid-bulk}} - G_{\text{solid-bulk}} \tag{1.2a}$$

at the temperature T are

$$G_{\text{m-nano}} = H_{\text{m-nano}} - TS_{\text{m-nano}} + \Delta\gamma A$$
$$G_{\text{m-bulk}} = H_{\text{m-bulk}} - TS_{\text{m-bulk}} + \Delta\gamma A \tag{1.2b}$$

The quantity H_{m} is the enthalpy and S_{m} the entropy of melting, both with the additional subscript "nano" or "bulk." The term $\Delta\gamma$ stands for the difference in the surface energy in the liquid and solid states. The quantity A represents the surface area per mol of nanoparticles. It is important to note that the quantities in Equations 1.2 are the differences of the thermodynamic quantities observed during the melting process.

$$H_{\text{m}} = H_{\text{liquid}} - H_{\text{solid}} > 0$$
$$S_{\text{m}} = S_{\text{liquid}} - S_{\text{solid}} > 0$$
$$\Delta\gamma = \gamma_{\text{liquid}} - \gamma_{\text{solid}} < 0 \tag{1.3}$$

In the case of bulk materials, the surface energy term $\Delta\gamma A$ of Equation 1.2b is generally neglected but it is of relevance in the case of nanoparticles.

For lack of better data, in most cases, the material data of the bulk material $H_{\text{m-bulk}}$ and $S_{\text{m-bulk}}$ are used for nanoparticles, too, yielding

$$G_{\text{m-nano}} = H_{\text{m-bulk}} - TS_{\text{m-bulk}} + \Delta\gamma A \tag{1.4}$$

Setting $G_{\text{m-nano}} = 0$ leads to the well-known reduction in the melting point of nanoparticles $T_{\text{m-nano}}$ in comparison with the one of the bulk material, $T_{\text{m-bulk}}$, as was described for the first time more than a hundred years ago by Pawlow [13] using $G_{\text{m-bulk}} = 0$, $H_{\text{m-bulk}} = T_{\text{m-bulk}}S_{\text{m-bulk}}$, and neglecting $\Delta\gamma A$ for the bulk material, and more recently in [14–16] as

$$\frac{T_{\text{m-nano}}}{T_{\text{m-bulk}}} = 1 + \frac{\Delta\gamma A}{T_{\text{m-bulk}}S_{\text{m-bulk}}} = 1 - \left|\frac{\Delta\gamma A}{T_{\text{m-bulk}}S_{\text{m-bulk}}}\right|$$
$$= 1 - \left|\frac{\Delta\gamma}{T_{\text{m-bulk}}S_{\text{m-bulk}}}\right|\frac{6M}{\rho d} = 1 - \left|\frac{\Delta\gamma}{H_{\text{m-bulk}}}\right|\frac{6M}{\rho d} \tag{1.5}$$

In Equation 1.5, M stands for the molecular weight and ρ for the density of the particles. To visualize the general trend in the reduction of the melting temperature with decreasing particle size, the use of the absolute value of the fraction $|\Delta\gamma/H_{\text{m-bulk}}|$ is the only correct way in the case of melting and crystallization.

As already mentioned, the derivations leading to Equation 1.5 assume that it is allowed to use bulk data for nanomaterials: however, this is problematic in the case of nanoparticles. The following gives a series of indications.

In a review on the melting of solids, Mei and Lu [17] devote a whole chapter to abnormal size effects on melting. Most interesting in this context are experimental findings of Shvartsburg and Jarrold [11] reporting that tin clusters consisting of 19–31 atoms exhibit melting points significantly above that of the bulk material. Molecular dynamic (MD) simulations for clusters of C_n, Si_n, Ge_n, and Sn_n clusters for $n \leq 13$ by Lu *et al.* [18] also revealed melting points significantly above that of the bulk materials.

- The material data of nanoparticles differ from those in bulk materials. For example, Vollath *et al.* [19] found drastic changes in the thermodynamic data of phase transformations in nanoparticulate zirconia, leading in the case of small particles to a change in the phase sequence with respect to temperature being reversed. An analogous phenomenon was found by Ushakov *et al.* [20] for pure and La-doped zirconia and hafnia nanoparticles with a diameter of 5–6 nm, where, at room temperature, the amorphous phase was more stable than the tetragonal one, in contrast to the general opinion that amorphous particles are the least stable ones.
- Even more dramatic is the influence on surface energy. There are experimental indications for a six times larger surface energy for nanoparticles of gold [21] and silver [22] compared to the bulk values. However, the model leading to such an evaluation of these experiments is seriously questioned [23, 24]. An increase in the surface energy by a factor of roughly two for aluminum nanoparticles was predicted in a theoretical study by Medasani and Vasiliev [25]. Both results contradict theoretical estimates that find a reduction in the surface energy with decreasing particle size [23, 24].

To estimate the thermal behavior of nanoparticles, knowledge of thermodynamic quantities and surface energy is essential. Therefore, it is the goal of this contribution to present proper and reliable approaches to estimate the thermodynamics of nanoparticles based on bulk data.

1.3 Size-Dependent Materials Data of Nanoparticles

For quite some time, there have been approaches to estimate the thermodynamic data of nanoparticles as a function of their size. Tolman [26, 27] presented such a relation for the surface energy as

$$\gamma_{\text{nano}} = \frac{\gamma_{\text{bulk}}}{(1 + (4\delta/d))} \tag{1.6}$$

The quantity δ is the so-called Tolman length. Das and Binder [28] generalized the Tolman relation to a wider applicable equation of type

$$\gamma_{\text{nano}} = \frac{\gamma_{\text{bulk}}}{(1 + 8(\tilde{l}/d)^2)} \tag{1.7}$$

In Equation 1.7, the quantity $\tilde{l}$ is again a characteristic length. However, Equation 1.7 was developed not for a free surface but an interface between two coexisting phases. A further relation, appraised as a "universal" relation, was reported by Guisbiers [29] in the form of

$$\xi_{\text{nano}} = \xi_{\text{bulk}}\left(1 - \frac{\alpha}{d}\right)^{\frac{1}{2s}} \tag{1.8}$$

Guisbiers argues that this relation is valid for the material property ξ, which may be the melting temperature, Debye temperature, superconducting temperature, Curie temperature, cohesive energy, activation energy of diffusion, or vacancy formation energy. The quantity α is a material constant with the dimension of a length, and s is a positive number depending on the material property. For $s = 1/2$ and $\alpha/d \ll 1$ Equation 1.6 and Equation 1.8 are practically equivalent. Furthermore, Equation 1.5 for the melting temperature obeys relation 1.8 with $s = 1/2$.

A more sophisticated relation for γ_{nano} compared to Equation 1.8, based on the cohesive energy of nanocrystals, was reported by Lu and Jiang [23, 24] and Ouyang *et al.* [30], as

$$\gamma_{\text{nano}} = \gamma_{\text{bulk}}\left(1 - \frac{d_0}{d - d_0}\right)\exp\left(-\frac{d_0}{d - d_0}\right) \tag{1.9a}$$

The quantity d_0 is the "smallest size" of d if this equation is valid. For $d/d_0 \gg 1$, Equation 1.9a can be rewritten as

$$\gamma_{\text{nano}} \approx \gamma_{\text{bulk}}\left(1 - \frac{d_0}{d - d_0}\right)^2 \tag{1.9b}$$

and as a further approximation as

$$\gamma_{\text{nano}} \approx \gamma_{\text{bulk}}\left(1 - \frac{d_0}{d}\right)^2 \tag{1.9c}$$

Equation 1.9c agrees again with Equation 1.8 for $s = 1/4$.

A different physical approach was reported by Li [31], using a layer-by-layer structure of the reference crystal from which the nanoparticle is cut out. This concept leads to an extremely complicated relation, which yields for $d_0/d \ll 1$ the same approximation as Equation 1.9c.

With respect to the melting point of nanomaterials $T_{\text{m-nano}}$, the thermodynamic approach of Letellier *et al.* [15, 16] should be noted. These authors have concluded the same tendency as shown in this chapter (Equation 1.5)

$$\frac{T_{\text{m-nano}}}{T_{\text{m-bulk}}} = 1 - c\left(\frac{d_0}{d}\right)^s \tag{1.9d}$$

However, they introduce also as a conceptual extension a positive exponent s to the quantity $1/d$ as $(d_0/d)^s$ with d_0 being a reference quantity and c, a positive constant factor. They report a value of $s = 0.79$ for lead nanoparticles and 1.20 for tin nanoparticles.

Safaei and Attarian Shandiz [32] published a model for the melting entropy of metallic nanoparticles, based on the methods of statistical physics. It is important to emphasize that their final formulae for the entropy of melting and the melting temperature confirm the earlier work by Jiang and Shi [33, 34], who developed, on the basis of Mott's equation for the melting entropy, a model for size-dependent melting temperature and entropy. Using an earlier approach to calculate the melting temperature of nanoparticles [35], Attarian Shandiz and Safaei [36] derived the same relations as Jiang and Shi [33, 34] for the melting temperature of metallic nanoparticles. The special feature of these derivations may be found in the fact that the final formulae depend on the thermodynamics and the crystal structure of the bulk material only. Furthermore, this approach inherently incorporates the influence of the surface energy. Therefore, this term no longer appears explicitly in the further equations.

Neglecting the electronic contribution, according to Attarian Shandiz and Safaei [32, 36], the ratio of melting temperatures is given by

$$\frac{T_{\text{m-nano}}}{T_{\text{m-bulk}}} = 1 - (1 - q)\frac{2d_0}{d + d_0} \tag{1.10}$$

In Equation 1.10, d_0 stands for a critical particle size, where the particle consists of surface atoms only. It is important to point out that, from its derivation, Equation 1.10 already contains the influence of the difference of the surface energy. From its definition, d_0 depends on the crystal structure of the particle. Attarian Shandiz and Safaei [36] give a table of this quantity for different lattices, for example, for the fcc structure $d_0 = \delta\sqrt{6} = 2.45 \cdot \delta$, where δ stands for the diameter of one atom in the metallic environment. It is obvious that a table as in [36] for d_0 is applicable only in exceptional cases where the structure of the smallest particle is identical to the bulk structure. This assumption is not necessarily correct; for example, for gold, see Tian *et al.* [37]. Therefore, in many cases, it may be necessary to fit the parameter d_0 with experimental data. The quantity $q = Z_S/Z_V$ is the ratio of the coordination numbers at the surface, Z_S, and in the volume, Z_V, of the bulk material. For bulk materials and larger particles, $q = 0.5$ is valid. Comparing calculated values with different results from the literature led Attarian Shandiz and Safaei [35, 36] to the conclusion that in the case of very small particles, a value $q = 0.25$ is more appropriate. This finding is well in line with a study on coordination numbers as a function of particle size and structure by Montejano *et al.* [38]. Therefore, a fit function was developed, which gives for the bulk material a value 0.5 and which decreases to 0.25 for the particle size d_0. Hence, it may be appropriate to select an expression such as

$$q = 0.5\frac{d}{d + d_0} \tag{1.11}$$

to fit experimental data of melting temperatures of nanoparticles. Equation 1.10 and Equation 1.11 contain the same and only one fit parameter d_0. Obviously, d_0 is the lower limit for d when this approximation may be applied. Certainly, this fitting process does not fulfill the claim of the original intention with respect to Equation 1.10, namely, to describe the melting properties of nanoparticles free of fitting parameters, depending on well-known bulk properties only. However, it will be shown later on in this contribution that the parameter d_0 can be expressed by bulk values, making fitting processes unnecessary.

For the entropy and enthalpy of melting, Attarian Shandiz *et al.* [32, 35, 36] and Jiang *et al.* [33, 34] derived the following equation:

$$S_{\text{m-nano}} - S_{\text{m-bulk}} = \frac{3R}{2} \ln\left(\frac{T_{\text{m-nano}}}{T_{\text{m-bulk}}}\right) \Rightarrow \frac{T_{\text{m-nano}}}{T_{\text{m-bulk}}} = \exp\left(\frac{2}{3R}\left(S_{\text{m-nano}} - S_{\text{m-bulk}}\right)\right) \tag{1.12a}$$

Equation 1.12a does not take into account the electronic contribution to the entropy of melting, for which Safaei and Attarian Shandiz [32] proposed the term

$$S_{\text{m-nano}} - S_{\text{m-bulk}} = \frac{3R}{2} \ln\left(\frac{T_{\text{m-nano}}}{T_{\text{m-bulk}}}\right) + S_{\text{E}}\left(1 - \frac{T_{\text{m-nano}}}{T_{\text{m-bulk}}}\right) \tag{1.12b}$$

including the electronic contribution to the enthalpy of melting, described by the second term using the parameter S_{E}. However, Safaei and Attarian Shandiz [32] pointed out that this term is zero in cases where the numbers of free electrons per ion in the solid and liquid phase are equal and not size dependent. According to Safaei and Attarian Shandiz [32], this may be correct in the case of metals, but certainly not for semiconductors.

The enthalpy of melting as a function of the particle size is given by

$$H_{\text{m-nano}} = H_{\text{m-bulk}} \frac{T_{\text{m-nano}}}{T_{\text{m-bulk}}} \frac{S_{\text{m-nano}}}{S_{\text{m-bulk}}} \tag{1.13}$$

Equation 1.13 stems from the relation $G_{\text{m}} = H_{\text{m}} - T_{\text{m}} S_{\text{m}} = 0$, valid both for the bulk and nanoparticles (see the additional subscripts "bulk" and "nano"); recall that $\Delta\gamma A$ is incorporated in the current version of G_{m}.

From Equation 1.12 and Equation 1.13, the free enthalpy of melting is calculated as

$$\begin{aligned} G_{\text{m-nano}} = {} & H_{\text{m-bulk}} \frac{T_{\text{m-nano}}}{T_{\text{m-bulk}}} \frac{S_{\text{m-nano}}}{S_{\text{m-bulk}}} \\ & - T\left[S_{\text{m-bulk}} + \frac{3R}{2} \ln\left(\frac{T_{\text{m-nano}}}{T_{\text{m-bulk}}}\right) + S_{\text{E}}\left(1 - \frac{T_{\text{m-nano}}}{T_{\text{m-bulk}}}\right)\right] \end{aligned} \tag{1.14}$$

Lastly, Equation 1.2b, containing the surface energy explicitly, and Equation 1.10, incorporating the surface energy, must lead to identical results. Therefore, these equations are well suited to calculate $\Delta\gamma$ as a function of the particle diameter, yielding

$$\frac{T_{\text{m-nano}}}{T_{\text{m-bulk}}} = 1 - (1 - q)\frac{2d_0}{d + d_0} = 1 - \left|\frac{\Delta\gamma}{T_{\text{m-bulk}} S_{\text{m-bulk}}}\right| \frac{6M}{\rho d} = 1 - \left|\frac{\Delta\gamma}{H_{\text{m-bulk}}}\right| \frac{6M}{\rho d} \tag{1.15a}$$

and, consequently

$$(1-q)\frac{2d_0}{d+d_0} = \left|\frac{\Delta\gamma}{H_{\text{m-bulk}}}\right| \frac{6M}{\rho d} \tag{1.15b}$$

This leads to $\Delta\gamma$ as a function of d, using Equation 1.11, as

$$\begin{aligned} |\Delta\gamma| &= \left(1 - \frac{T_{\text{m-nano}}}{T_{\text{m-bulk}}}\right) |H_{\text{m-bulk}}| \frac{\rho d}{6M} = (1-q)\frac{2d_0}{d+d_0} |H_{\text{m-bulk}}| \frac{\rho d}{6M} \\ &= \frac{d_0(d+2d_0)}{(d+d_0)^2} |H_{\text{m-bulk}}| \frac{\rho d}{6M} \end{aligned} \tag{1.15c}$$

Certainly, one may criticize that this approach does not take care of the influence of the possible faceting of the surface, especially for the smallest particles. However, this may be justified in a first approximation and by the fact that, experimentally, faceting is rarely observed in the case of very small metallic particles.

Next, one may ask about the existence and size of stable nuclei for crystallization within the melted particles. To calculate the critical diameter of the stable nuclei, Θ_{crit}, one needs to determine the extremum of the free enthalpy of melting yielding

$$\frac{\partial G_{\text{m-nano}}}{\partial d} = 0 \tag{1.16}$$

with $d = \Theta_{\text{crit}}$ as solution. (The lengthy derivation with its details is given in Appendix 1.A.)

Equation 1.16 leads to a transcendent expression; the roots can be found numerically only. Nuclei or particles with diameter d below Θ_{crit} are fluctuating; they develop and disappear. Nuclei or particles with larger diameter are stable; they do not fluctuate, and, if possible, have a tendency to grow (see also [39, 40]). Looking at the critical diameter Θ_{crit} as a function of temperature, one may expect the following cases:

- $\Theta_{\text{crit}} < d$: In this stable case, one may expect a crystallization process of the particles starting from the interior. This process was already assumed to describe experimental results at the melting point of nanosized lead particles [6]. Furthermore, in this case, one expects the formation and decay of nuclei, which has been proved experimentally [12].
- $\Theta_{\text{crit}} > d$: This is the unstable, fluctuating case. As the formation of nuclei is spontaneous, instantaneous melting or crystallization of the whole particle may be expected. Therefore, one has to expect perpetual fluctuations between the solid and the liquid state.

1.4
Comparison of Experimental and Calculated Melting Temperatures

It is quite difficult to find reliable experimental or calculated data for the melting process of nanoparticles. In the first example, melting data of silver calculated by MD methods [41] were selected. Figure 1.2 shows the calculated and the fitted

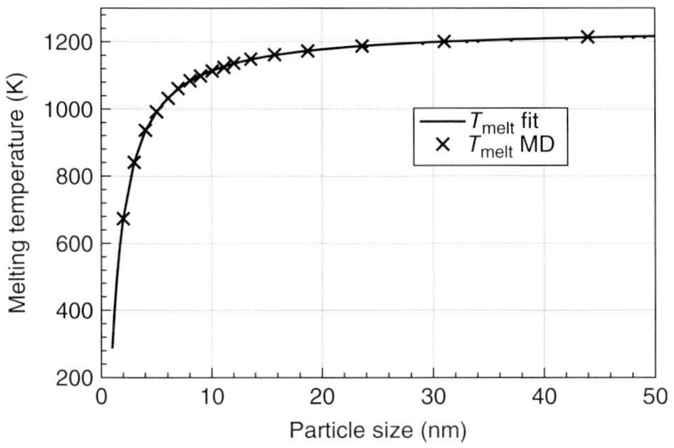

Figure 1.2 Melting temperature of silver nanoparticles. The MD data were taken from Luo *et al.* [41]. Fitting was performed using Equation 1.10 and Equation 1.11.

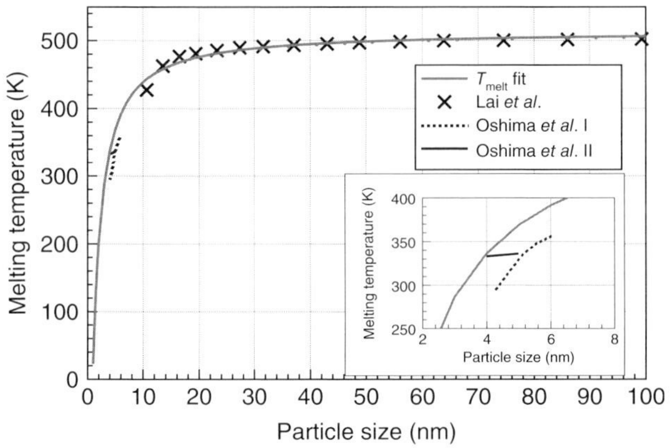

Figure 1.3 Melting temperature of tin nanoparticles. The experimental data used for fitting were taken from Lai *et al.* [3]. In addition, experimental data of Oshima and Takayanagi [12] are plotted. The inset shows, enlarged, the particle size–temperature range, where the experimental data are found. Fitting was performed using Equation 1.10 and Equation 1.11.

data by application of Equation 1.10 in connection with Equation 1.11. The fitting procedure led to a value of $d_0 = 1.04 \pm 0.002$ nm.

In the second example, experimental data for melting of tin nanoparticles obtained by Lai *et al.* [3] were fitted. As shown in Figure 1.3, in this case also, the fit is nearly perfect, resulting in the fitting parameter $d_0 = 1.5 \pm 0.1$ nm. In this figure, experimental data of Oshima and Takayanagi [12] were plotted as well, but not used in the fitting procedure. These data are of special interest because they split into two different paths at the particle size of about 5 nm. This phenomenon is better visible in Figure 1.4, where the data shown in Figure 1.3 are plotted versus the inverse particle size.

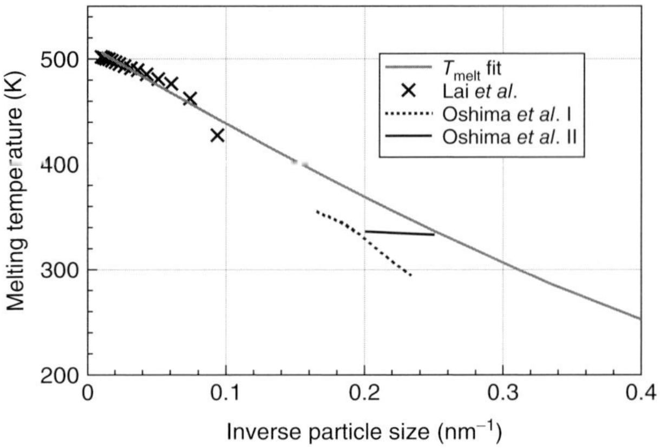

Figure 1.4 Melting temperature of tin nanoparticles plotted versus the inverse particle size. In this graph, the same data as in Figure 1.2 were used [3, 12].

In Figure 1.4, in the field limited by the two different paths in the experimental data of Oshima and Takayanagi [12], liquid particles with crystallized inclusions can be observed in the electron micrographs. As the inclusions are unstable, this field was denominated as "pseudocrystalline."

As a last example, the melting of gold nanoparticles is discussed. For gold, the experimental results of Castro *et al.* [9], which were already applied for demonstration in Section 1.1, are used. Figure 1.5 displays the melting data for gold nanoparticles, as discussed. For reasons of clarity, similar to Figure 1.4, the melting temperature of gold nanoparticles is plotted versus the inverse particle size.

Fitting of the experimental data for gold, denoted as Castro *et al.* I [9] in Figure 1.5, resulted in the fit parameter $d_0 = 1.08 \pm 0.07$ nm. In addition, the

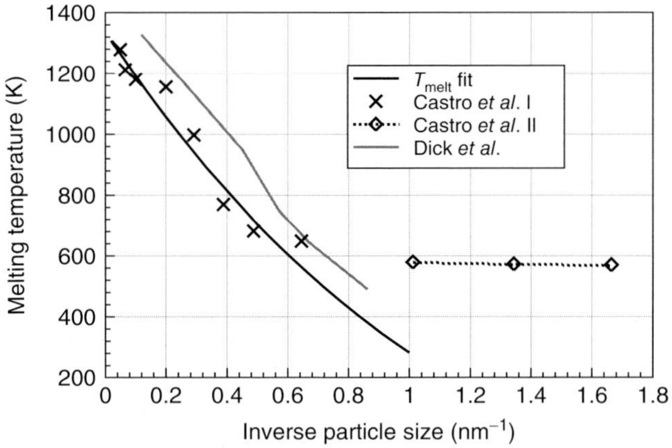

Figure 1.5 Experimental and fitted data for the melting temperature of gold nanoparticles versus the inverse particle size. In this graph, experimental results of Castro *et al.* [9] and Dick *et al.* [42] are plotted. The results of these authors are different because Castro *et al.* used free and Dick *et al.* particles enclosed in silica.

experimental results of Dick *et al.* [42] are plotted in this graph. Unfortunately, the experimental conditions of these two groups were so different that a common fitting is impossible. This is, because in the experiments of Dick *et al.* [42], the gold particles were mechanically confined in a silica matrix. Therefore, the experimental results of Dick *et al.* lead, in extrapolation, to a significantly higher melting temperature for the bulk material, $T_{m\text{-bulk}} = 1522 \pm 41$ K. Interestingly, the value $d_0 = 1.0 \pm 0.04$ nm is practically identical in both experiments. Similar to the experimental results of Oshima and Takayanagi [12] for tin, the results obtained by Castro *et al.* [9] show, in the case of small particles (marked as Castro *et al.* II in Figure 1.5), a remarkable deviation from the fitted results.

Besides the data used earlier, there are more data obtained by MD simulations published in the literature. However, these data were not used because they predict melting points for the bulk material that are far off from the well-known ones.

Calculating the free enthalpy of melting as a function of temperature and particle size by application of Equation 1.14, with respect to Equation 1.10, Equation 1.11, and Equations 1.12 and $S_E = 0$, sheds more light on the melting behavior of metallic nanoparticles. On the basis of data calculated from the experimental results shown in Figure 1.1, Figure 1.2, Figure 1.3, Figure 1.4, and Figure 1.5, the diagrams in Figure 1.6a–c demonstrates the free enthalpy of melting for the

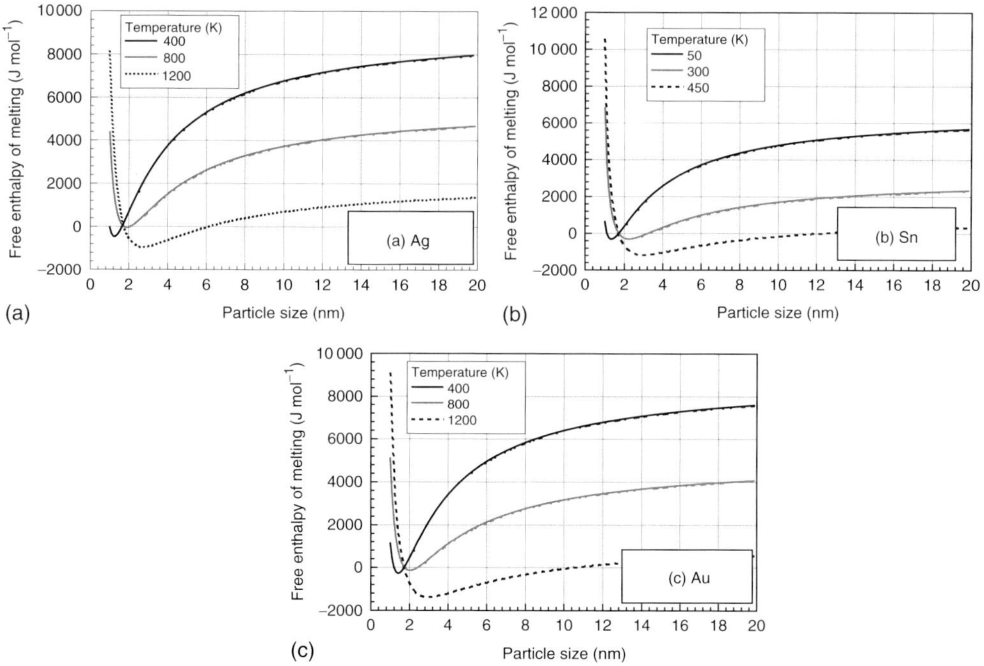

Figure 1.6 Free enthalpy of melting for (a) silver, (b) tin, and (c) gold versus particle size and temperature as parameter.

Table 1.1 Characteristic parameters calculated from the experimental data.

Metal and literature data used for fitting	d_0 (nm) Equation 1.10 and Equation 1.11	d_{char} (nm) Equation 1.18b	T_{char} (K) Equation 1.19
Ag [41]	1.04 ± 0.002	1.76	607
Sn [4]	1.5 ± 0.1	1.71	164
Au [9]	1.08 ± 0.07	1.73	626

examples explained. In the case of gold, the data obtained from the experiments by Castro *et al.* [9] were applied.

The graphs in Figure 1.6a–c have a few features in common: each of the curves shows two "zero-crossings" (roots with respect to Equation 1.14 using Equation 1.10 and Equation 1.11, one at a particle size showing a temperature-dependent root and the second a temperature-independent root.

The temperature-independent root is possible only if the condition

$$S_{\text{m-nano}} = 0 \tag{1.17}$$

is fulfilled; for mathematical details, see Appendix 1.A. By combining Equation 1.10, Equation 1.11, and Equation 1.12, the condition in Equation 1.17 leads to the equation

$$d^2(1-\alpha) + dd_0(1-2\alpha) - d_0^2(1+\alpha) = 0 \tag{1.18a}$$

with the abbreviation $\alpha = \exp(-2S_{\text{m-bulk}}/3R)$. Equation 1.18a, a quadratic one, has two roots; however, only the positive root is physically reasonable, which is denominated as the "characteristic particle size" d_{char} and Equation (1.10) with Equation (1.11) as

$$d_{char} = \frac{d_0}{2(1-\alpha)}\{2\alpha - 1 + (5-4\alpha)^{0.5}\} \tag{1.18b}$$

This characteristic particle size is related to a characteristic temperature T_{char} by combining Equation 1.12a with Equation 1.17 as

$$T_{char} = T_{\text{m-bulk}} \exp\left(-\frac{2S_{\text{m-bulk}}}{3R}\right) = T_{\text{m-bulk}}\left[\frac{d_{char}^2 + d_{char}d_0 - d_0^2}{(d_{char} + d_0)^2}\right] \tag{1.19}$$

Table 1.1 displays the characteristic parameters evaluated for the three examples shown in Figure 1.6a–c.

It is not too astonishing that the d_0 values for gold and silver are nearly identical. More striking is the fact that d_{char} is nearly identical for the three metals in discussion.

Table 1.1 teaches that the characteristic particle size d_{char} is larger than d_0, which is the smallest one, where these considerations are valid. In the interval of particle sizes between d_{char} and d_0, one may expect special phenomena, as the free enthalpy of melting (Figure 1.6a–c) changes sign at d_{char}.

The graphs in Figure 1.6a–c exhibits the remarkable fact that, because of the change in the sign of the free enthalpy of melting, particles with sizes below d_{char} are liquid if their temperature is lower than T_{char} and crystallized at higher temperatures.

Obviously, this is a case of reverse behavior of the phases, a rather strange phenomenon, which is quite often described. Very early, this phenomenon was predicted by Tammann [43]. Experimentally, the existence of this strange behavior was proved by findings of pressure-induced amorphization at constant temperature by Mishima *et al.* [44] for ice and by Hemley *et al.* [45] for silica. Crystallization at constant pressure with increasing temperature was found by Rastogi *et al.* [46, 47] for organic compounds. Recently, a theoretical basis for Tammann's generalized phase diagram was given by Stillinger and Debenetti [48] and Schupper and Shnerb [49].

With reference to the examples selected for this chapter, the phenomenon of reverse melting is better visible in a graph, where the free enthalpy of melting is plotted versus the temperature. Using silver as an example, this is shown for four different particle sizes in Figure 1.7, where the phenomenon of reverse melting and crystallization is perfectly visible. Particles with sizes of 1.5 nm are liquid at lower temperatures and solid at higher ones. In contrast, particles with 2 and 3 nm diameters behave in the usual way. In addition, the zero-valued, temperature-independent free enthalpy of melting for the characteristic particle size, 1.76 nm, is also plotted.

From Figure 1.6a–c, one learns that at each temperature level, there are two particle sizes, where the solid and the liquid state are in equilibrium, $G_{m\text{-}nano} = 0$. These equilibrium sizes, calculated by $G_{m\text{-}nano} = 0$, according to Equation 1.14, are plotted in Figure 1.8a–c (See also Appendix 1.A.) for the three examples – silver,

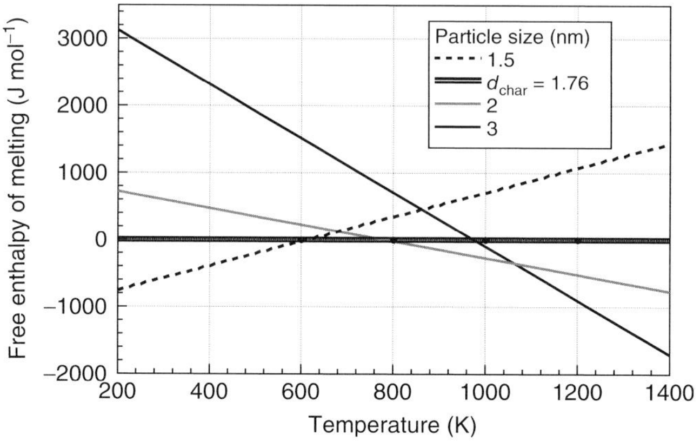

Figure 1.7 Free enthalpy of melting for silver particles versus temperature and particle size as parameter. Particles are solid in the case of positive values of the free enthalpy of melting and liquid in the case of negative values. Comparing the courses for the particle sizes, one realizes the reverse behavior of melting and crystallization for the particle size 1.5 nm.

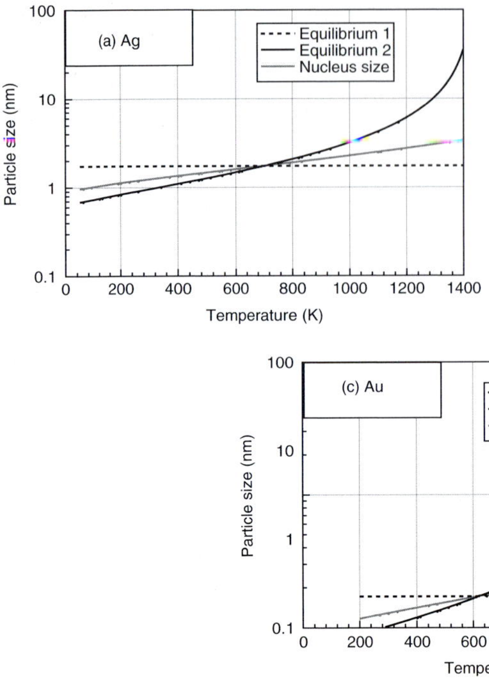

Figure 1.8 Equilibrium particle sizes (two roots with respect to Equation 1.14 using Equation 1.10) versus temperature for (a) silver, (b) tin, and (c) gold. In addition, the sizes of the stable nuclei (nucleus size) for crystallization according to Equation 1.16 are plotted. Note that these three curves have one point in common.

tin, and gold. In these graphs, the sizes of the nuclei, fulfilling Equation 1.16, are plotted, too. The graphs in Figure 1.8 show a few remarkable features. There is, as it is also visible in the earlier figures, one temperature-independent equilibrium particle size (Equilibrium 1), the "characteristic particle size," d_{char}. The plot of the second temperature-dependent equilibrium particle size (Equilibrium 2) intersects the temperature independent equilibrium size at the "characteristic temperature." The plot of the size of the stable nuclei Θ_{crit}, according to Equation 1.16, intersects at the same point; see the proof in Appendix 1.A. Obviously, d_{char} and Θ_{crit} are identical. The existence of the intersection of these three sizes at one point is a strong indication that this particle size characterizes the material.

The consequences of the relative position of the sizes of the stable nuclei and the equilibrium particle sizes are obvious: At a particle size larger than the critical one, d_{char}, the size of the particles is larger than the size of the critical nuclei and the temperature-independent equilibrium particle size, which is smaller than the size of the critical nuclei Θ_{crit}. Therefore, at any temperature, these small particles repeatedly crystallize and melt. This is a continuing process of fluctuation, especially as these particles are smaller than the stable nuclei. Possibly, these nuclei are responsible for the crystallization process in melted metals. For a long

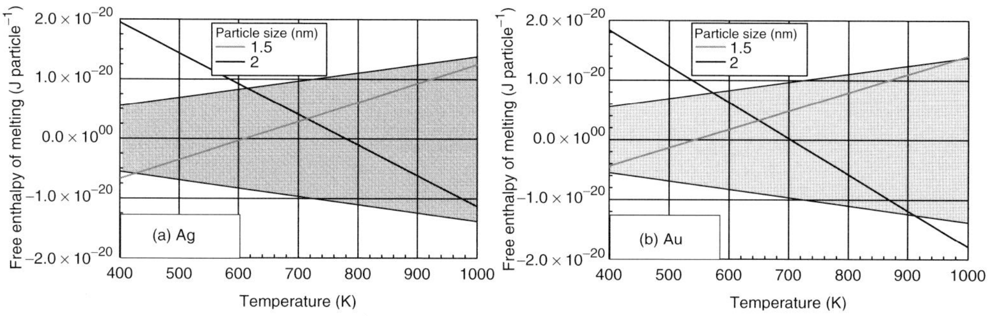

Figure 1.9 Free enthalpy of melting per particle for (a) silver and (b) gold versus particle size as parameter. Within the shaded area, limited by the thermal energy $\pm kT$, fluctuation is possible.

time, these nuclei were postulated (see, e.g., [39, 40, 50]) and finally experimentally verified [51]. The situation is different for particle sizes smaller than Θ_{crit}. In this case, the size of the nuclei is larger than the one of the temperature-dependent equilibrium particles. Therefore, a priori, one cannot expect that these particles are stable, whereas particles with the characteristic particle size d_{char} may be considered, with respect to their phase, as stable ones.

To analyze the stability of the particles against fluctuation, Figure 1.9a,b displays the free enthalpy of melting per particle as a function of the temperature for different particle sizes. In addition, the range where fluctuation is possible, limited by the lines $\pm kT$ (see, e.g., [52, 53]), is marked as shaded areas in Figure 1.9a,b.

Particles with size and temperature in this range may always alternate between the solid and the liquid state. Most interesting are the particles with the size d_{char} because for these particles the free enthalpy of melting is zero, independent of the temperature. Therefore, these particles are never stable: they always fluctuate between the two possible phases. Furthermore, looking at Figure 1.5, one realizes the strange phenomenon that melting of particles with sizes below 1 nm occurs at a constant temperature around 600 K. Analyzing Figure 1.6c and Figure 1.9b carefully makes it clear that for crystalline particles in this range of size and temperature inverse melting may be expected.

Equation 1.15c gives the difference in the surface energy between the solid and the liquid state for the bulk material and yields immediately for $d \to \infty$

$$
\begin{aligned}
|\Delta\gamma_{\text{bulk}}| &= \lim_{d\to\infty}\left[\left(1-\frac{T_{\text{m-nano}}}{T_{\text{m-bulk}}}\right)|H_{\text{m-bulk}}|\frac{\rho d}{6M}\right] \\
&= |H_{\text{m-bulk}}|\frac{\rho}{6M}d_0\lim_{d\to\infty}\left(1-\frac{d_0^2}{(d+d_0)^2}\right) = |H_{\text{m-bulk}}|\frac{\rho}{6M}d_0 \qquad (1.20)
\end{aligned}
$$

An evaluation of this formula for the three examples – silver, tin, and gold – is shown in Figure 1.10.

Looking at Figure 1.10, one realizes the striking phenomenon that, especially for particles with sizes below 5 nm, the difference in the surface energy between the

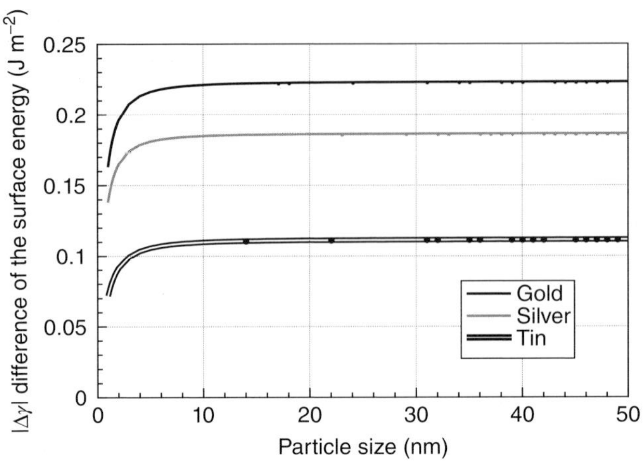

Figure 1.10 Difference in the surface energy between the solid and the liquid state $|\Delta\gamma| = |\gamma_{\text{liquid}} - \gamma_{\text{solid}}|$ for silver, tin, and gold at the melting temperature versus particle size.

solid and the liquid state decreases with decreasing particle size. This result is in agreement with calculations by Chang and Johnson [7], leading to the fact that the surface of nanoparticles gets less ordered (i.e., liquid-like) with decreasing particle size and supports similar theoretical findings by Lu and Jiang [23, 24].

Equation 1.20 may now be used to calculate d_0 using $|\Delta\gamma_{\text{bulk}}|$

$$d_0 = \frac{6M}{H_{\text{m-bulk}}\rho}|\Delta\gamma_{\text{bulk}}| \tag{1.21}$$

As d_0 is now expressed by bulk values, it is possible to describe the thermodynamics of the phase transformation from solid–liquid for nanoparticles by using bulk data only. Now, there exist no longer any fit parameters in the system of equations.

1.5 Comparison with Data for the Entropy of Melting

In the examples discussed, the melting temperature was used to check the validity of Equation 1.10 and Equation 1.11 to calculate the melting temperature of metallic nanoparticles using bulk data only. To check this approach for the thermodynamic data, the values obtained by MD modeling for the entropy of melting for silver nanoparticles [41] were applied. Figure 1.11 displays the values of the entropy of melting $S_{\text{m-nano}}$ for silver given by Luo *et al.* [41] together with the results obtained by application of Equation 1.12a in combination with Equation 1.10 based on bulk data only. According to Sheng *et al.* [54], who concluded that the entropy of melting for metallic particles is primarily vibrational in nature, it is not necessary to consider any influence of the electronic structure.

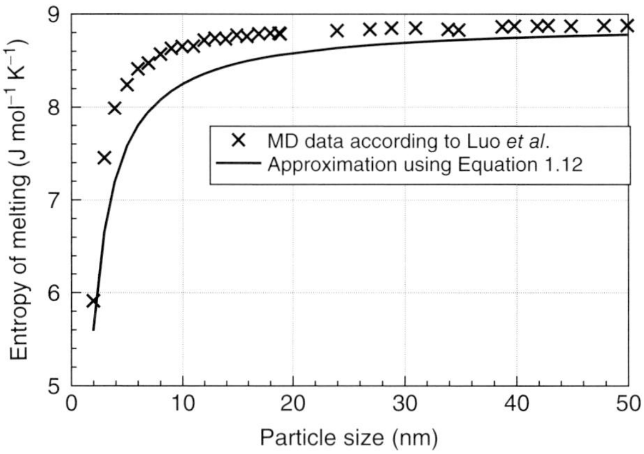

Figure 1.11 Entropy of melting for silver versus particle size particles according to Luo *et al.* [41] together with the results using Equation 1.12a in combination with Equation 1.10 based on bulk data.

Analyzing Figure 1.11, one realizes that the approximation based on Equation 1.12a delivers systematically too low values. For this approximation, the thermodynamic parameters obtained from the bulk values, which led in the case of the melting temperatures to a perfect fit, are applied. However, looking at the paper by Luo *et al.* [41], one realizes similar deviations between the values obtained by MD calculations and experimental values referred to in their paper.

As visible in Figure 1.11, the MD-derived values for the smallest particles are significantly higher as compared to the trend of the data. This systematic deviation does not occur for the melting temperature. In this case, the deviation of the values calculated from the bulk data with the ones from the MD modeling is, as it is visible in Figure 1.2, less than 0.5%.

1.6 Discussion of the Results

Starting with the relations for the ratio of the melting temperature $T_{\text{m-nano}}/T_{\text{m-bulk}}$ and the associated equation for excess entropy of transformation of nanoparticles $S_{\text{m-nano}} - S_{\text{m-bulk}}$ derived by Attarian Shandiz *et al.* [32, 35] and Jiang *et al.* [33, 34], an improved system to calculate the thermodynamic data of nanoparticles was developed. Most important is the introduction of a minimum particle size d_0 for the validity of the derivation that can be related to materials data of the bulk material. The significant progress achieved in this paper was made possible by the development of an empirical approximation formula for the ratio of the coordination numbers at the surface, Z_S, and that of the volume in the bulk material, Z_V, $q = Z_S/Z_V$. In all analyzed cases, this approximation gives better results when using a constant value of q.

This refined approach was applied to three examples to calculate the melting temperature:

1) numerical data for silver (this case applied results of MD simulations);
2) experimental data for tin;
3) experimental data for gold stemming from two different groups.

In all three cases, the fit is perfect. However, the most important finding is that in all cases, a particle size exists, where the entropy and enthalpy of melting change sign. This "characteristic particle size" is connected to a "characteristic temperature." Therefore, for particle sizes below the characteristic one, inverse melting has to be assumed. It is important to mention that in all three analyzed cases, the "characteristic particle size" is very similar; maybe d_{char} can be related to a universal constant.

The particle size where $S_{\text{m-nano}} = 0$ and $\partial G_{\text{m-nano}}/\partial d = 0$ are valid (in this chapter called *characteristic particle size*) is often interpreted as the minimum particle size, where crystallization is observable (see, for example, [55, 56]). However, this interpretation is not in agreement with the experimental results determined for gold [9]. The reason for this misinterpretation may be found in the neglection of the phenomenon of inverse melting. On the other hand, particles with sizes smaller than d_{char} may never be assumed as stable ones, as is visible in Figure 1.8a–c. The size of the critical nuclei Θ_{crit} is always larger than the one of the particle. Insofar, in view of an experiment, the phenomenon of inverse melting is somewhat meaningless. However, the experimental data marked in Figure 1.1 as Castro *et al.* II and in Figure 1.3 and Figure 1.4 as Oshima *et al.* II, characterized by nearly horizontal lines, may now be interpreted as the upper limits of the fluctuation range.

Well in line with the above-described phenomenon, resembling inverse melting found for gold particles, are the experimental results of Shvartsburg and Jarrold [11] on small tin particles consisting of 19–31 atoms exhibiting melting points significantly above those of the bulk material. Similar results were obtained by MD simulations for clusters of C_n, Si_n, Ge_n, and Sn_n for the number of atoms $n \leq 13$ by Lu *et al.* [18]. Certainly, the actual behavior of such small clusters is more complex than that described by the equations in this chapter; however, the general tendency is well reproduced.

The system of equations, developed in this chapter, allows calculating the difference between the surface energy in the solid and liquid state. Extrapolating these results to large particle sizes, one obtains this difference for the bulk material with surprising precision. This result gives confidence in calculating the melting data of nanoparticles using bulk values for melting temperature, melting entropy, and the difference in surface energy at the melting point of the bulk material.

A further interesting finding stems from the analysis of the free enthalpy of melting. In any temperature, the calculation delivers two particle sizes with zero free enthalpy of melting. One of these sizes is temperature independent and the other temperature dependent. In the case of both particle sizes being identical, particles with such a diameter fluctuate between the crystallized and the liquid state.

Looking at the size of the crystallization nuclei, as they are depicted in Figure 1.8a–c, one realizes that for particle sizes larger than the characteristic one, the size of the crystallization nuclei is smaller than the size of the particles, but larger than the characteristic particle size. Therefore, besides the thermodynamic reasons, as shown in Figure 1.9a,b, one cannot expect that the temperature-independent particle shows a stable phase at temperatures above the characteristic one. This situation may be the reason for the fluctuating crystallized particles found by Oshima and Takayanagi [12] inside of the particles. The size limit given by these authors is, possibly, not based on the thermodynamics of the system: it may be connected to the contrast conditions in the electron microscope. The inversion of the phase sequence for particles sizes below the characteristic one is the reason for the observation of crystallized gold particles with sizes of less than 1 nm in diameter at temperatures around 600 K by Castro *et al.* [9]. Without the phenomenon of the inversion of the phase sequence, this observation would be impossible. The temperature-independent equilibrium particles may be the well-known nuclei for crystallization, which were assumed to exist for a while and indirectly proved by experiments. The situation is reversed for particles with $d > d_{char}$. In this case, one may expect stable crystallized particles.

Finally, it is important to point out that these derivations are valid only in the vicinity of the equilibria in discussion. Looking at the entire possible range of temperatures, in the case of inverse phase transformations, the phase sequence is significantly more complex [57, 58].

1.7 Conclusions

On the basis of the equations published by Attarian Shandiz *et al.* [32, 35] and Jiang *et al.* [33, 34], an improved system to calculate the thermodynamic behavior of metallic nanoparticles was developed. The special attractiveness of this improved set of equations is that it allows estimations based on bulk data only. The following set of equations gives a good approximation for the thermodynamic behavior of metallic nanoparticles.

The ratio of the melting temperature of nanoparticles and the one of the bulk material is given by

$$\frac{T_{\text{m-nano}}}{T_{\text{m-bulk}}} = 1 - (1 - q)\frac{2d_0}{d + d_0} \tag{S1}$$

using the abbreviations

$$q = 0.5\frac{d}{d + d_0} \quad \text{and} \quad d_0 = \frac{6M}{\rho}\left|\frac{\Delta\gamma_{\text{bulk}}}{H_{\text{m-bulk}}}\right| \tag{S2}$$

The entropy of melting is calculated using

$$S_{\text{m-nano}} = S_{\text{m-bulk}} + \frac{3R}{2}\ln\left(\frac{T_{\text{m-nano}}}{T_{\text{m-bulk}}}\right) + S_{\text{E}}\left(1 - \frac{T_{\text{m-nano}}}{T_{\text{m-bulk}}}\right) \tag{S3}$$

The newly introduced parameter d_{char} is given by

$$d_{char} = \frac{d_0}{2(1-\alpha)}\{2\alpha - 1 + (5-4\alpha)^{0.5}\} \tag{S4}$$

using the abbreviation $\alpha = \exp(-2S_{m\text{-}bulk}/3R)$.

The only problematic point in the systems of these equations is the quantity S_E, the contribution of the electrons to the entropy of melting, which is unknown. However, in the case of metallic particles, the electronic influence may be neglected.

A detailed analysis of Equation S3 leads to a characteristic particle size where entropy and enthalpy of melting are zero. In the case of smaller particle sizes, this equation leads to an inversion of the phase sequence. This phenomenon is well known in many other systems and proved in the case of nanoparticles by a series of experimental results and theory. However, as these particles are smaller than the critical size of the crystallization nuclei, one must not expect any stable phase at particle sizes below d_{crit}. Furthermore, in the range of temperatures where this extrapolation may be allowed, the particles will, in any case, thermally fluctuate between the phases (Figure 1.9a–c).

The sizes of the stable nuclei for crystallization, which are larger than Θ_{crit}, are in accordance with a series of experimental results. The system of Equation S1, Equation S2, and Equation S3 gives the thermodynamic background for crystallization and melting phenomena and Equation S4 the material specific constant observed in the connection to nanoparticles and bulk systems.

1.A
Appendix: Zeros and Extrema of the Free Enthalpy of Melting $G_{m\text{-}nano}$

Applying Equation 1.13 and keeping in mind that the influence of the surface energy is incorporated, the free entropy of melting can be written as

$$G_{m\text{-}nano} = H_{m\text{-}bulk}\frac{T_{m\text{-}nano}}{T_{m\text{-}bulk}}\frac{S_{m\text{-}nano}}{S_{m\text{-}bulk}} - TS_{m\text{-}nano} \tag{1.A1}$$

with the notations in Section 1.1. With the abbreviations $\widetilde{T} = T_{m\text{-}nano}/T_{m\text{-}bulk}$ and $\widetilde{S} = S_{m\text{-}nano}/S_{m\text{-}bulk}$, Equation 1.A1 is rewritten as

$$G_{m\text{-}nano} = (H_{m\text{-}bulk} \cdot \widetilde{T} - TS_{m\text{-}bulk})\widetilde{S} \tag{1.A2}$$

The quantities $\widetilde{T}$ and $\widetilde{S}$ can be expressed as functions of the particle size d; see Equation 1.10, Equation 1.11, and Equation 1.12b with $S_E = 0$, respectively, as

$$\widetilde{T} = \frac{d^2 + dd_0 - d_0^2}{(d+d_0)^2} \tag{1.A3}$$

$$\widetilde{S} = 1 + \frac{3R}{2S_{m\text{-}bulk}}\ln(\widetilde{T}) \tag{1.A4}$$

One can immediately see from Equation 1.A2 that $G_{\text{m-nano}} = 0$ has two related root equations, namely

$$\widetilde{S} = 0 \quad \text{and} \quad H_{\text{m-bulk}} \cdot \widetilde{T} - TS_{\text{m-bulk}} = 0 \tag{1.A5}$$

Consequently, Equation $1.\text{A5}_1$ yields a zero (root) independent of T (denoted in Figure 1.7 as Equilibrium 1) and Equation $1.\text{A5}_2$ dependent on T (denoted in Figure 1.7 as Equilibrium 2).

To study now the extrema of $G_{\text{m-nano}}$ and the according necessary condition $\partial G_{\text{m-nano}}/\partial d = 0$, for sake of completeness the derivatives of $\widetilde{T}$ and $\widetilde{S}$ with respect to d are listed as

$$\frac{\partial \widetilde{T}}{\partial d} = \frac{(d + 3d_0)d_0}{(d + d_0)^3} \tag{1.A6}$$

$$\frac{\partial \widetilde{S}}{\partial d} = \frac{3R}{2S_{\text{m-bulk}}} \frac{1}{\widetilde{T}} \cdot \frac{\partial \widetilde{T}}{\partial d} \tag{1.A7}$$

It should be noted that both derivatives are always positive!

The derivative of $G_{\text{m-nano}}$ follows as

$$\frac{\partial G_{\text{m-nano}}}{\partial d} = \left[H_{\text{m-bulk}} \cdot \widetilde{S} + \left(H_{\text{m-bulk}} \cdot \widetilde{T} - TS_{\text{m-bulk}} \right) \cdot \frac{3R}{2S_{\text{m-bulk}}} \frac{1}{\widetilde{T}} \right] \frac{\partial \widetilde{T}}{\partial d} \tag{1.A8}$$

Because $\partial \widetilde{T}/\partial d > 0$, zeros (roots) of $\partial G_{\text{m-nano}}/\partial d = 0$ are those of the expression in brackets in Equation 1.A8 and allow formulating a relation $d(T)$, which is denoted as "nucleus size" in Figure 1.7a–c.

A specific situation occurs if the particle size according to Equilibrium 1, Equation 1.A5, coincides with the particle size according to Equilibrium 2, Equation 1.A5 (see the crossing point of the curves in Figure 1.7a–c). Then the curve $d(T)$ runs also through this crossing point, as Equation 1.A8 includes Equations 1.A5 and 1.A5 making Equation 1.A8 to zero.

References

1. Vollath, D. (2008) *Nanomaterials – An Introduction to Synthesis, Properties, and Applications*, Wiley-VCH Verlag GmbH, Weinheim.
2. Eckert, J., Holzer, J.C., Ahn, C.C., Fu, Z., and Johnson, W.L. (1993) Melting behavior of nanocrystalline aluminum powders. *Nanostruct. Mater.*, **2**, 407–413.
3. Lai, S.L., Guo, J.Y., Petrova, V., Ramanath, G., and Allen, L.H. (1996) Size-dependent melting properties of small thin particles: nanocalorimetric measurements. *Phys. Rev. Lett.*, **77**, 99–102.
4. Suresh, A., Mayo, M.J., Porter, W.D., and Rawn, C.J. (2004) Crystallite and grain-size-dependent phase transformations in yttria-doped zirconia. *J. Am. Ceram. Soc.*, **86**, 360–362.
5. Suresh, A., Mayo, M.J., and Porter, W.D. (2003) Thermodynamics of the tetragonal-to-monoclinic phase transformation in fine and nanocrystalline yttria-stabilized zirconia powders. *J. Mater. Sci.*, **18**, 2913–2921.
6. Coombes, C.J. (1972) The melting of small particles of lead and indium. *J. Phys. F: Met. Phys.*, **2**, 441–449.

7. Chang, J. and Johnson, E. (2005) Surface and bulk melting of small metal clusters. *Philos. Mag.*, **85**, 3617–3627.
8. Kaptay, G. (2012) Nano-Calphad: extension of the Calphad method to systems with nano-phases and complexions. *J. Mater. Sci.*, **47**, 8320–8335.
9. Castro, T., Reifenberger, R., Choi, E., and Andres, R.P. (1990) Size-dependent melting temperature of individual nanometer-sized metallic clusters. *Phys. Rev. B*, **42**, 8548–8556.
10. Schmidt, M., Kusche, R., von Issendorff, B., and Haberland, H. (1998) Irregular variations in the melting point of size-selected atomic clusters. *Nature*, **393**, 238–240.
11. Shvartsburg, A.A. and Jarrold, M.F. (2000) Solid clusters above the bulk melting point. *Phys. Rev. Lett.*, **85**, 2530–2532.
12. Oshima, Y. and Takayanagi, K. (1993) Solid–liquid phase transition of tin particles observed by UHV high resolution transmission electron microscopy: pseudo-crystalline phase. *Z. Phys. D*, **27**, 287–294.
13. Pawlow, P. (1909) Über die Abhängigkeit des Schmelzpunktes von der Oberflächenenergie eines festen Körpers. *Z. Phys. Chem.*, **65**, 1–35.
14. Guisbiers, G., Kazan, M., Van Overschelde, O., Wautelet, M., and Pereira, S. (2008) Mechanical and thermal properties of metallic and semiconductive nanostructures. *J. Phys. Chem. C*, **112**, 4097–4103.
15. Letellier, P., Mayaffre, A., and Turmine, M. (2011) in *Handbook of Nanophysics: Clusters and Fullerenes*, vol. **2** (ed K.D. Sattler), CRC Press, Boca Raton, FL, pp. 19.1–19.11.
16. Letellier, P., Mayaffre, A., and Turmine, M. (2007) Melting point depression of nanosolids: nanoextensive thermodynamic approach. *Phys. Rev. B*, **76**, 045428-1–045428-8.
17. Mei, Q.S. and Lu, K. (2007) Melting and superheating of crystalline solids: from bulk to nanocrystals. *Prog. Mater. Sci.*, **52**, 1175–1262.
18. Lu, Z.Y., Wang, C.Z., and Ho, K.M. (2000) Structures and dynamical properties of C_n, Si_n, Ge_n and Sn_n clusters with n up to 13. *Phys. Rev. B*, **61**, 2329–2334.
19. Vollath, D., Fischer, F.D., Hagelstein, M., and Szabó, D.V. (2006) Phases and phase transformations in nanocrystalline ZrO_2. *J. Nanopart. Res.*, **8**, 1003–1016.
20. Ushakov, S.V., Brown, C.E., and Navrotsky, A. (2004) Effect of La and Y on crystallization temperatures of Hafnia and Zirconia. *J. Mater. Res.*, **19**, 693–696.
21. Nanda, K.K., Maisels, A., and Kruis, F.E. (2008) Surface tension and sintering of free gold nanoparticles. *J. Phys. Chem. C*, **112**, 13488–13491.
22. Nanda, K.K., Maisels, A., Kruis, F.E., Fissan, H., and Stappert, S. (2003) Higher surface energy of free nanoparticles. *Phys. Rev. Lett.*, **91**, 106102-1–106102-5.
23. Lu, H.M. and Jiang, Q. (2004) Size-dependent surface energies of nanocrystals. *J. Phys. Chem. B*, **108**, 5617–5619.
24. Lu, H.M. and Jiang, Q. (2006) Size dependent surface energy and surface tension. Emerging Technologies – Nanoelectronics, IEEE Conference, January 10–13, pp. 21–24.
25. Medasani, B. and Vasiliev, I. (2009) Computational study of the surface properties of aluminum nanoparticles. *Surf. Sci.*, **603**, 2042–2046.
26. Tolman, R.C. (1948) Consideration of the Gibbs theory of surface tension. *J. Chem. Phys.*, **16**, 758–774.
27. Tolman, R.C. (1949) Effect of droplet size on surface tension. *J. Chem. Phys.*, **17**, 333–337.
28. Das, S.K. and Binder, K. (2011) Universal critical behaviour of curvature-dependent interfacial tension. *Phys. Rev. Lett.*, **107**, 235702-1–235702-4.
29. Guisbiers, G. (2011) Schottky defect in nanoparticles. *J. Phys. Chem. C*, **115**, 2616–2621.
30. Ouyang, G., Tan, X., and Yang, G. (2006) Thermodynamic model of the surface energy of nanocrystals. *Phys.Rev. B*, **74**, 195408-1–195408-5.
31. Li, X. (2012) The influence of the atomic interactions in out-of-plane on surface energy and its applications in nanostructures. *J. Appl. Phys.*, **112**, 013524-1–013524-7.

32. Safaei, A. and Attarian Shandiz, M. (2010) Melting entropy of nanocrystals: an approach from statistical physics. *Phys. Chem. Chem. Phys.*, **12**, 15372–15381.
33. Jiang, Q. and Shi, F.G. (1998) Entropy for solid–liquid transition in nanocrystals. *Mater. Lett.*, **37**, 79–82.
34. Jiang, Q., Yang, C., and Li, J.C. (2002) Melting enthalpy depression of nanocrystals. *Mater. Lett.*, **56**, 1019–1021.
35. Attarian Shandiz, M., Safaei, A., Sanjabi, S., and Barber, Z.H. (2007) Modeling size dependence of melting temperature of metallic nanoparticles. *J. Phys. Chem. Solid*, **68**, 1396–1399.
36. Attarian Shandiz, M. and Safaei, A. (2008) Melting entropy and enthalpy of metallic nanoparticles. *Mater. Lett.*, **62**, 3954–3956.
37. Tian, D., Li, J., Zhao, Y., Zhao, J., and Guo, X. (2011) Structural evolution of Au*n* ($n = 29–35$) clusters: a transition from hollow cage to amorphous packing. *Comput. Mater. Sci.*, **50**, 2359–2362.
38. Montejano, J.M., Aguilera-Granja, F., and Moran-Lopez, J.L. (1997) Direct enumeration of the geometrical characteristics of clusters. *Nanostruct. Mater.*, **8**, 269–297.
39. Turnbull, D. (1950) Formation of crystal nuclei in liquid metals. *J. Appl. Phys.*, **21**, 1022–1028.
40. Hollomon, J. and Turnbull, D. (1953) Nucleation. *Prog. Met. Phys.*, **4**, 333–388.
41. Luo, W., Hu, W., and Xiao, S. (2008) Size effect on the thermodynamic properties of silver nanoparticles. *J. Phys. Chem. C*, **112**, 2359–2369.
42. Dick, K., Dhanasekaran, T., Zhang, Z., and Meisel, D.J. (2002) Size-dependent melting of silica-encapsulated gold nanoparticles. *J. Am. Chem. Soc.*, **124**, 2312–2317.
43. Tammann, G. (1903) *Kristallisieren und Schmelzen. Ein Beitrag zur Lehre der Änderungen des Aggregatzustandes*, Johann Ambrosius Barth, Leipzig.
44. Mishima, O., Calvert, L.D., and Whally, E. (1985) An apparently first-order transition between two amorphous phase of ice induced by pressure. *Nature*, **314**, 76–78.
45. Hemley, R.J., Jephcoat, A.P., Mao, H.K., Ming, L.C., and Manghnani, M.H. (1988) Pressure-induced amorphisation of crystalline silica. *Nature*, **334**, 52–54.
46. Rastogi, S., Höhne, G.W.H., and Keller, A. (2000) Unusual pressure-induced phase behavior in crystalline poly (4-methylpentene-1): calorimetric and spectroscopic results and further implications. *Macromolecules*, **32**, 8897–8909.
47. Rastogi, S., Newman, M., and Keller, A. (1993) Unusual pressure-induced phase behavior in crystalline poly-4-methylpentene-1. *J. Polym. Sci., Part B: Polym. Phys.*, **31**, 125–139.
48. Stillinger, F.H. and Debenetti, G.P. (2003) Phase transitions, Kauzmann curves, and inverse melting. *Biophys. Chem.*, **105**, 211–220.
49. Schupper, N. and Shnerb, N.M. (2004) Spin model for inverse melting and inverse glass transition. *Phys. Rev. Lett.*, **93**, 037202-1–037202-4.
50. Predel, B. (1963) Post-melting effect of pure metals. *Z. Metallkd.*, **54**, 206–212.
51. Vollath, D. (1966) Kinetics of the melting process of alloys. *Z. Metallkd.*, **57**, 772–775.
52. Vollath, D. and Fischer, F.D. (2009) Bistablility, hysteresis and fluctuations in adiabatic ensembles of nanoparticles. *J. Nanopart. Res.*, **11**, 1485–1500.
53. Vollath, D. and Fischer, F.D. (2011) Fluctuations, bistability and hysteresis connected to phase transformations of nanoparticles. *Prog. Mater. Sci.*, **56**, 1030–1076.
54. Sheng, H.W., Ren, G., Peng, L.M., Hu, Z.Q., and Lu, K. (1997) Epitaxial dependence of the melting behaviour of In nanoparticles embedded in Al matrices. *J. Mater. Res.*, **12**, 119–123.
55. Safaei, A. and AttarianShandiz, M. (2009) Size-dependent thermal stability and the smallest nanocrystal. *Physica E*, **41**, 359–364.
56. Jiang, Q., Tong, H.Y., Hsu, D.T., Okuyama, K., and Shi, F.G. (1998) Thermal stability of crystalline thin films. *Thin Solid Films*, **312**, 357–361.

57. Ferro, R., Saccone, A., Delfino, S., Cardinale, A.M., and Maccio, D. (1996) Inverse melting in binary systems: morphology and microscopy of catatectic alloys. *Metall. Mater. Trans. B*, **27B**, 379–386.

58. Bormann, R. (1994) Thermodynamic and kinetic requirements for inverse melting. *Mater. Sci. Eng., A*, **A179/A180**, 31–35.

2
Numerical Simulation of Individual Metallic Nanoparticles

D.S. Wen and P.X. Song

2.1
Introduction

Traditionally, the study of oxidation, ignition, and combustion of reactive particles is conducted at the macroscale, where the experiments can be performed and numerical simulation built on conventional laws of momentum and energy transports. Such experiments and simulation can predict the bulk behavior, but cannot reveal detailed mechanisms at the individual nanoparticle level. In the end, it is the interaction of nanoparticles that decides the properties at the bulk scale. How to bridge the individual particle property with the bulk property has always been an intensive research topic. Owing to the increasing influence of surface atoms at the nanoscale, the properties of nanomaterials exhibit a unique size-dependent phenomenon. A classical example is the melting temperature depression at the nanoscale. The established properties and transport laws at the bulk scale cannot be extended to the nanoscale. Experimental probing into individual nanoparticles is also difficult and will inevitably encounter various technical problems even regardless of the cost. It is therefore expected that atomistic simulation can play a key role in understanding the transport and reaction mechanisms at the individual nanoparticle level.

With the rapid development of computing technologies, many simulation techniques have been developed to understand different kinds of physical and chemical phenomena ranging from the electron structure of the atom to the system scale. The multiscale simulation approach has been widely used, as shown schematically in Figure 2.1. At the bottom of the simulation is the quantum mechanics simulation (based on the First Principles method), where the electronic structure is determined by approximately solving the quantum mechanical Schrödinger equation, which yields the wave function, energy, and gradients. Currently, the *ab initio* method is accurate but is time consuming and expensive for large systems in the nanometer scale. Mesoscale simulation (such as the lattice Boltzmann method) aims to bridge the discrete particle level to the continuum simulation. Molecular mechanics simulation refers to theoretical methods and computational techniques to model or mimic the kinetic behavior of molecules, which have been widely used in the fields

Metal Nanopowders: Production, Characterization, and Energetic Applications, First Edition.
Edited by Alexander Gromov and Ulrich Teipel.

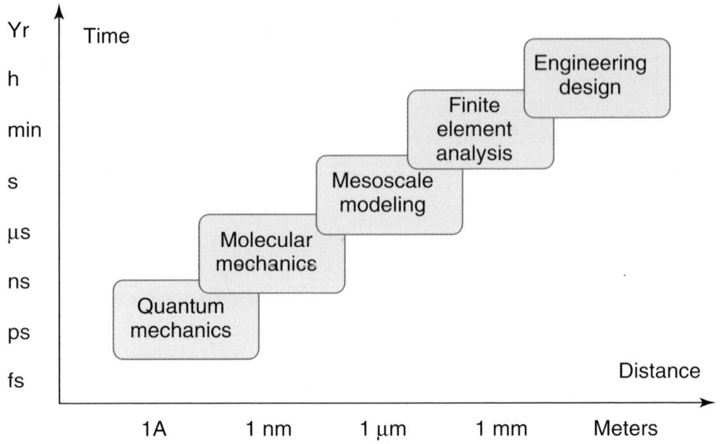

Figure 2.1 Overview of the multiscale simulation.

of computational chemistry, computational biology, and materials science to study molecular systems ranging from a few atoms to millions of molecules.

To simulate thermal–physical properties of individual nanoparticles, molecular simulation would be the first choice. There are essentially two approaches to performing molecular simulations: the stochastic and the deterministic approach. The stochastic approach, called *Monte Carlo (MC)*, is based on exploring the energy surface by randomly probing the configuration space of a molecular system. The deterministic approach, called *molecular dynamics (MD)*, simulates the time evolution of the molecular system, which provides the actual trajectory of the system. The information generated from simulation methods can be used to fully characterize the thermodynamic state of the system. The choice between MC and MD is largely determined by the phenomenon under investigation. Generally, MC simulations are preferable for the simulation of gas or other low-density systems, and MD is favorable for simulating liquid or solid system, where molecular collisions exchange energy between molecules and improve the ability of MD to sample conformations. Hence, MD is used in this chapter throughout to simulate energetic nanoparticles related to their ignition and combustion behavior.

Reactive metallic particles involve many complicated physical and chemical processes. Taking combustion of nanoparticles as an example, it involves heating of individual particles, which may lead to particle melting, the interaction, agglomeration, and possibly sintering of different particles, accompanied by the initial slow oxidation interacted with different phase transitions, until it reaches the required temperature that leads to particle combustion. Such complicated processes cannot be revealed from bulk experiments. Using MD simulation as an example, this chapter explores the fundamental heating and reaction at the individual nanoparticle level. A brief overview of MD simulation is first introduced, and then the chapter focuses on the simulation of individual nanoparticle behavior related to ignition and combustion of energetic particles, which includes (i) size-dependent properties of individual nanoparticles, (ii) sintering and agglomeration of metallic nanoparticles, (iii) oxidation of nanoparticle in the presence of oxygen, and (iv) heating and cooling of core–shell-structured nanoparticles. Such information will be very useful in understanding the mechanism of reactive particles at the nanoscale, being complementary to bulk scale experiments and simulation.

2.2 Molecular Dynamics Simulation

MD simulations have contributed significantly to enhance the fundamental understanding of physical and chemical mechanisms at the nanoscale. They can be used to study many important unanswered questions relating to reactive nanoparticles that cannot be directly addressed by continuum approaches. For completion, a quick summary of MD simulation is presented here.

2.2.1 Motion of Atoms

MD simulations are modeled as ensembles of interacting particles under specific conditions. Given the initial coordinates and velocities of an ensemble of particles that interact in a force field (FF) whose total potential is denoted as U, based on Newton's second law states,

$$m_i \frac{d^2 r_i}{dt^2} = F_i = -\frac{\partial U}{\partial r_i} \tag{2.1}$$

This method integrates the equation of motion numerically, and updates coordinates and velocities at each integration time step. From the obtained particle trajectories, one can calculate various global system properties as statistical averages. The MD used conventional Newton's second law to describe the interaction between atoms. For N-body systems, the equation of motion based on it is as

$$\frac{d^2 r_i}{dt^2} = F_i(r_1, r_2, \dots, r_N) \tag{2.2}$$

where

$$F_i(r_1, r_2, \dots, r_N) = -\nabla_{r_i} U_i(r_1, r_2, \dots, r_N) \tag{2.3}$$

Here, r_i is the position vector of the particle i, F_i is the total force acting on particle i, and U_i is the total potential energy from which the force is derived.

Considering computational efficiency and accuracy, only several algorithms are suitable for MD simulation such as "leap frog" and "predictor-corrector." The velocity Verlet algorithm is among them and used in this work because it does not alter time reversal symmetry of the integration, which is crucial to preserve the accuracy of locations and velocities [1]. The form of the velocity Verlet integration scheme is given in the following equations.

$$x_i(t+\Delta t) = x_i(t) + \Delta t v_i(t) + \frac{(\Delta t)^2}{2m_i} F_i(t)$$
$$v_i(t+\Delta t) = v_i(t) + \frac{\Delta t}{2m_i}[F_i(t) + F_i(t+\Delta t)] \tag{2.4}$$

In Equation 2.4, x is the position of atom i, v is the velocity, and F_i is the net force on the given atom i. Positions and velocities are obtained and updated at the next time step, $t+\Delta t$, based on the values at the current time step t. In order to do the very first iteration, initial values of x, v, and F are needed, so a simulation is always initiated before a run can start. For example, particles may be placed in an array or on a grid for simplicity and velocities can be randomly assigned from an appropriate distribution. The initial configuration of a metallic solid is convenient to be its perfect crystal structure. The first set of forces can then be obtained from the initial positions.

2.2.2 Temperature and Potential Energy

Temperature is a state variable that specifies the thermodynamic state of the system. It is also an important concept in MD simulations. This macroscopic quantity is related to the microscopic description of simulations through the kinetic energy, which is calculated from the sum of atomic velocities. Temperature is a thermodynamic quantity that is meaningful only at the equilibrium. It is related to the average kinetic energy of the system via the equipartition principle of the statistic thermodynamics. This principle states that each degree of freedom either in momentum or in coordinates, which appears as a squared term in the Hamiltonian, has an average energy of $k_B T/2$ associated with it. This is true for the momenta p_i which appear as $p_i^2/2m$ in the Hamiltonian, where the bracket denotes the ensemble average,

$$\left\langle \sum_i^N \frac{p_i^2}{2m} \right\rangle = \langle K \rangle = \frac{N_f k_B T}{2} \tag{2.5}$$

In an unrestricted system with N atoms, the number of degrees of freedom N_f is $3N$ because each atom has three velocity components (v_x, v_y, and v_z). It is convenient to define an instantaneous kinetic temperature function:

$$T_{\text{instant}} = \frac{2K}{N_f k_B} \tag{2.6}$$

The average of the instantaneous temperature T_{instant} is the thermodynamic temperature T in MD simulation. Temperature is calculated from the total kinetic energy and the total number of degrees of freedom. For a nonperiodic system:

$$\frac{(3N-6)k_B T}{2} = \sum_{i=1}^{N} \frac{m_i v_i^2}{2} \tag{2.7}$$

Six degrees of freedom are subtracted because both the translation and rotation of the center of mass are ignored. And for a periodic system, Equation 2.7 becomes

$$\frac{(3N-3)k_{\mathrm{B}}T}{2} = \sum_{i=1}^{N} \frac{m_i v_i^2}{2} \tag{2.8}$$

where only the three degrees of freedom corresponding to translational motion can be ignored, as the rotation of a central cell imposes a torque on its neighboring cells. The sum on the right side of Equation 2.8 is the kinetic potential and the rest from the entire energy is potential energy, which is calculated by the various forms of potentials defined and parameter in the FF. In most cases, the potential energy is the sum of the pair, bond, angle, dihedral, improper, and *k*-space long-range energy. During dynamics, kinetic and potential energy exchange with each other and the temperature changes as a consequence. To maintain the correct temperature, the computed velocities have to be adjusted appropriately. To keep the simulation at a constant T or P, so-called "thermostat," or "barostat" are used as common control methods. Several different algorithms have been developed such as the Andersen [2] and the Parrinello–Rahman barostats [3], and the Nosé–Hoover [4, 5] and the Berendsen thermostats [6]. The Nosé–Hoover thermostat is used throughout this study whenever it is referred to. In addition to maintaining the desired temperature and pressure, the "thermostat" and "barostat" must produce the correct statistical ensemble, which demands that the probability of occurrence of a certain configuration obeys the laws of statistical mechanics.

2.2.3 Ensembles

In addition to maintaining the desired temperature, the temperature-control mechanism must produce the correct statistical ensemble. This means that the probability of occurrence of a certain configuration obeys the laws of statistical mechanics. Restricted by computer speed, the simulation system can only contain some limited numbers of particles from which there may be big differences compared with real materials. Ensemble is a concept in statistical thermodynamics. For an equilibrium system, macroscopic properties can be calculated from the average of every possible microscopic state. The equilibrium ensembles usually include microcanonical ensemble (N, V, E), canonical ensemble (N, V, T), isothermal–isobaric ensemble (N, P, T), isobaric–isoenthalpic ensemble (N, P, H), grand canonical ensemble (μ, V, T), Gibbs ensemble, and semi-grand ensemble. When the number of particles within consideration is up to quite a large value, these various equilibrium ensembles are equivalent to each other for treating real materials. When the number of particles in MD is far less than the one considered in statistical physics, the statistical average will be different for various ensembles. Generally speaking, the NVE ensemble is not practical when comparing with experimental or theoretical results. So, usually other ensembles are preferred depending on what thermodynamic quantity is kept constant including the number of particles, volume, pressure, temperature, or chemical potential: either canonical ensemble, that is, constant-NVT, isothermal–isobaric ensemble, that is, constant-NPT, or grand canonical ensemble, that is, constant-μVT.

2.2.4 Energy Minimization

According to the initial temperature, the initial velocities are generated so as to produce a Maxwell–Boltzmann distribution at the desired temperature; the distribution does not remain constant as the simulation continues. This is especially true when the system does not start at a minimum-energy configuration of the structure. This situation often occurs because structures are commonly minimized only enough to eliminate any hot spots. The minimization of configuration is another tricky topic of MD. A long-term stable configuration usually corresponds to that at global energy minima. Unfortunately, there is no rigorous method to find general global minima. This makes searching global minima all over the phase-space to be the Holy Grail of multidimensional optimization. Despite the theoretical and practical difficulties, there are two mathematical techniques guiding the minima search, that is, steepest descents (SDs) and conjugate gradient (CG), widely used to find local minima that might be the global minimum by some chance. SD is difficult to be converged. However, owing to its robustness, SD can be always performed no matter how far the current position deviates from the minimum energy location. Furthermore, global optimization is done by more advanced methods such as simulated annealing and genetic algorithm, which usually take longer time. A general strategy of minimization is that SD is conducted at the very first steps, and followed by other optimization logarithms, usually CG. Even though all of these methods do not guarantee the capture of global minima in any finite time, it is still necessary to perform minimization/optimization on the system before MD simulation to avoid the *N*-body system falling into a metastable stage associated with a local minimum. The metastable situation basically means that the thermodynamic states are not genuinely equilibrium stable, but have much longer lifetime during which their values appear unchanged.

2.2.5 Force Field

Similar to the Newtonian gravitational field, the term *force field* refers to a theoretical construction of all contributing interactions posed on the system, although sometimes the term *potential* holds the same implication. Along with the development of some reliable FF such as tight binding (TB) method, embedded-atom method (EAM), and Finnis–Sinclair (FS), MD is proved to be suited for the study of metallic systems [7–9]. These theories describe the energy contribution of an atom in an N-atom interacting system as a function of local electron density at the atom's site.

FF in MD is the explicit function form of comprehensive interaction potentials acting on atoms in a system, which is usually an analytical function of total energy to atom coordinates. The selection of FF is crucial to the validation of MD simulation.

The well-known Lennard-Jones potential (L-J potential) was proposed as early as in 1924, which is a pioneering molecular potential expressed as given here:

$$U_{\mathrm{LJ}}(r_{ij}) = 4\varepsilon \left[\left(\frac{\sigma}{r_{ij}} \right)^{12} - \left(\frac{\sigma}{r_{ij}} \right)^{6} \right] \tag{2.9}$$

where ε is the depth of the potential well, σ is the finite distance at which the interatomic potential is zero, and r_{ij} is the distance between the particles i and j. This is a typical pairwise potential and consists of a strong repulsive term $(1/r_{ij}^{12})$ describing the Pauli repulsion at short ranges and a weak attraction term $(1/r_{ij}^{6})$ describing the long-range van der Waals force. Although very few materials except argon dimer can be accurately modeled by L-J potential, it and its derived forms have been implemented in many of the present-day FFs as a pairwise fractional potential.

Metal system is widely investigated under the framework of EAM potential in MD. It is based on the effective medium theory and derived from the energy functional theory, where all atoms are viewed as embedded in the host system consisting of all other atoms [8]. The embedding energy is electron-density dependent. EAM FF is expressed as

$$U_{\mathrm{EAM}} = \frac{1}{2} \sum_{ij} \Phi(r_{ij}) + \sum_{i} \Theta(\vec{\rho}_i) \tag{2.10}$$

The potential function consists of the pair-interaction term $\Phi(r_{ij})$ and density-dependent potential $\Theta(\vec{\rho}_i)$. In the approximation of EAM, $\Phi(r_{ij})$ is the pair interaction between atoms whose separation is given by r_{ij}. The pair-interaction potential can drop smoothly by introducing the cubic spline function where the sums are over the atoms i and j. The embedding function Θ is the energy required to embed an atom into the background electron density $\vec{\rho}_i$ at site i, where $\vec{\rho}_i$ is given by a linear supposition of spherically averaged atomic electron densities, whereas in the modified embedded-atom method (MEAM), the embedding function is augmented by angularly dependent terms. EAM is easily extended to alloys in which the pair-interaction term $\Phi(r_{ij})$ is designed to be an arithmetic or geometric average of each atom's individual pairwise potential, which has been improved as a feasible approach [10].

2.2.6
Potential Truncation and Neighbor List

For an N-body system, if the interactions are not truncated, $N(N-1)/2$ pair interactions must be counted in. Consequently, the $N \times N$ scale computations are needed for the evaluation of energy. The computational cost of a large system will be tremendous if all these pairwise interactions are calculated. To reduce the computation cost and preserve sufficient accuracy, the potential is truncated from a given distance beyond which the interactions between atoms are relatively weak. It is to avoid the singular point in potential function by inserting a function to have a smooth decrease to zero at cut-off distance. The idea of potential truncation is to exclude the interaction beyond a cut-off distance, r_c. Only interactions of particles located within the cut-off distance in each cell box are evaluated. The potential can be truncated at the cut-off radius in two ways: the first is to "simple cut," as shown in Equation 2.11 and the second one is to impose a smooth truncation at cut-off radius as in Equation 2.12. The two methods make a negligible difference without drift of results for MD simulation if the numerical algorithm is carefully designed.

$$U^S(r_{ij}) = \begin{cases} U(r_{ij}) - U_c & (r_{ij} < r_c) \\ 0 & (r_{ij} > r_c) \end{cases} \tag{2.11}$$

$$U^S(r_{ij}) = \begin{cases} U(r_{ij}) - U_c - \left(\frac{dU(r_{ij})}{dr_{ij}}\right)_{r_{ij}=r_c} & (r_{ij} < r_c) \\ 0 & (r_{ij} > r_c) \end{cases} \tag{2.12}$$

The other method that is able to save CPU time is to conditionally renew the particles within the cut-off distance. Each such update is expensive, as it involves $N \times N$ operations for an N-particle system. A common technique is called a *neighbor list*, in which the particle environment is book-kept. Verlet list is a commonly used method, illustrated in Figure 2.2. A second cut-off distance r_v is introduced, within which a list of particles is made. Only if the particle's displacement is bigger than "the shell", $(r_v - r_c)$, is the neighbor list renewed. Otherwise, only the particles in the current list are considered. This design in order to save computational time avoids checking every particle in the system every time a force calculation is made. In low-temperature solids where the particles do not move very fast, it is possible to perform an entire simulation without or with only a few updates, whereas in simulation of liquids, updating every 5 or 10 steps is quite common.

2.2.7 Simulation Program and Platform

This chapter focuses on the simulation of individual nanoparticle behavior related to the ignition and combustion properties. A number of energetic nanoparticles have been studied, and this chapter focuses on three commonly used materials:

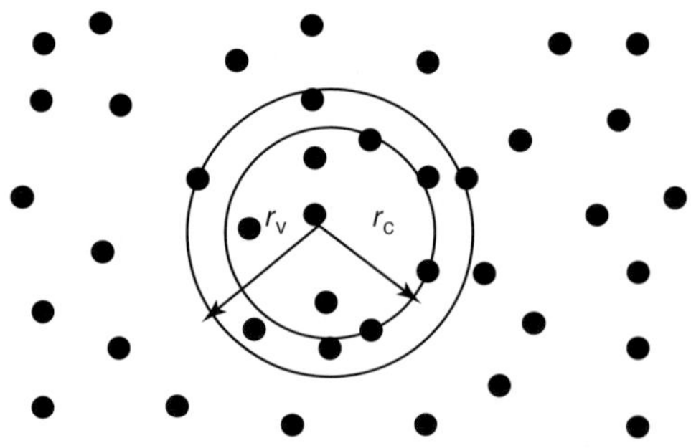

Figure 2.2 Scheme of neighbor list.

aluminum, silicon, and nickel. The size-dependent properties will be simulated on aluminum and nickel particles. Different-sized nickel particles will be used to simulate the sintering and agglomeration process, and the exothermic oxidation process will be simulated by silicon particles. A nickel passivation layer will be added on an aluminum particle to form a core–shell structure, aiming to control the ignition temperature and reduce the sintering process before and after combustion. Detailed simulation setting will be described in the corresponding sections.

The large-scale atomic/molecular massively parallel simulator (LAMMPS) is used throughout this study (*http://lammps.sandia.gov/*). It is an MD program developed by Sandia National Laboratories, USA. The visualization for the MD trajectory is through visual molecular dynamics (VMD), which is developed by the University of Illinois at Urbana-Champaign. Another MD simulation package, MD++, developed by Stanford University, USA, is used to generate a starting configuration of the silicon cluster. An optimization tool, Packmol, developed by State University of Campinas, Brazil, is used to optimize the starting configuration of the silicon and oxygen system. LAMMPS, VMD, MD++, and Packmol are all open source codes distributed under the terms of the GNU General Public License. A FORTRAN-version of reactive force field (ReaxFF) and a C++ version of function file are used to process ReaxFF bonding.

2.3 Size-Dependent Properties

2.3.1 Introduction

Using melting as an example, we reveal size-dependent properties using MD simulation. Melting is a phase transformation from a crystalline solid state that is rigid and topologically long-range ordered into a liquid state that is topologically long-range disordered. Melting is crucial to thermodynamic properties of nanoparticles/clusters. Under equilibrium conditions, melting of a bulk material usually occurs in a very narrow temperature and pressure range. A small change in temperature or pressure could largely change the phase from one status to another. When the size of the material is reduced dramatically, the surface-to-volume ratio is increased, which makes the melting of the nanoparticles different from that of their bulk counterpart. For instance, reduced particles have complicated size dependence and premelting phenomena. There are a few empirical equations that describe the size-melting relation based on either experiment and/or theory.

Because of the large surface/volume ratio, nanoparticles are more prone to early melting. Such an early melting phenomenon can significantly affect subsequent thermal–chemical behavior such as particle sintering and oxidation. In the framework of classical thermodynamics, the melting of a solid is known as a *first-order discontinuous phase transformation* occurring at a critical temperature at which Gibbs free energies of the solid and the liquid states are equal, that is, T_0 of the solid.

The melting process of a solid involves an abrupt change in latent heat and volume at T_0, which is a convenient indicator in terms of numerical simulation. Melting kinetics of low-dimensional materials is much deviated from that for the conventional bulk solids. For example, the melting point of free-standing nanometer-sized particles is remarkably depressed relative to T_0 [11]. It is also observed that when the nanoparticles are properly coated by (or embedded in) a high-melting-point matrix, the melting point of the particles can be elevated above T_0 [11], which is also found in our following simulations; this rarely happens in bulk solids [12, 13]. Here, MD simulations are conducted for individual particles. These will be used to benchmark the methodology and establish a baseline for the following simulations.

Conventionally, a melting transition is determined by measuring the caloric curve of the material, that is, heat change as a function of temperature. The heat capacity is the derivative of the caloric curve. The melting temperature of a nanoparticle has been widely investigated by caloric curve. When the heat capacity curve displays sharp peaks, the phase transition is first order, which is sensitively related to the volume change of the system. For MD simulations, a useful method of evaluating the melting temperature of a nanocluster is to evaluate the point of discontinuity in the potential energy. This is based on the fact that the potential energy is a function of volume of the system [11]. For particles of a few nanometers, the first-order transition from the solid to the liquid phase is of a fast transient nature and typically occurring in the order of 10^{-14} s [14, 11], which can be well reproduced by MD studies. There have been many studies in which the melting of nanostructured materials is evaluated by this approach [11, 12].

2.3.2
Simulation Setting

Both nickel and aluminum particles are simulated in this work. The classical EAM potential developed by Foiles [15] was used in this MD study of two nickel particles. Three quantities are required for the EAM total energy in the form of Equation 2.13, that is, the pair potentials $\Phi(r)$, the embedding functions $\Theta(\rho)$, and the electron densities ρ_i of atom i, which is a sum of the local atomic electron densities, ρ^{a}

$$\rho_i = \sum_{j(\neq i)} \rho_j^{\mathrm{a}}(r_{ij}) \tag{2.13}$$

The atomic electron densities ρ^{a} obtained from the Hartree–Fock calculations and the form of atomic density is then written as

$$\rho^{\mathrm{a}}(r) = (N - N_{\mathrm{s}})\rho_{\mathrm{d}}^{\mathrm{a}}(r) + N_{\mathrm{s}}\rho_{\mathrm{s}}^{\mathrm{a}}(r) \tag{2.14}$$

where N is the total number of outer electrons, N_{s} is a measure of the s electron content of atomic density, $\rho_{\mathrm{s}}^{\mathrm{a}}$ is the density of the outer s orbit, and $\rho_{\mathrm{d}}^{\mathrm{a}}$ is the density of the outer d orbital. The pairwise potential is expressed in form of the simple polynomial of an effective charge $z(r)$,

Table 2.1 Fitting parameters of EAM potential (nickel).

P	$\Theta(\rho)$
0.0	0.0
0.01446	−3.5847
0.02891	−5.1449
0.05783	−3.4041
0.06650	0.0

$$\Phi(r) = \frac{Z^2(r)}{r}$$
$$Z^2(r) = a_1(r_0 - r)^3 + a_2(r_0 - r)^4 \tag{2.15}$$

The values for nickel are $N_s = 0.86$, $a_1 = 0.070937$, $a_2 = 0.146031$, $r_0 = 3.0045$. The electron densities are in units of inverse cubic angstrom and the energies are in electron volts (eV). The embedding energies are expressed in terms of natural splines. The knots used for the spline and the corresponding value of $\Theta(\rho)$ are listed in Table 2.1.

The MD simulations were carried out in the NVT canonical ensemble and the constant temperature was maintained by the Nosé–Hoover thermostat. Free-forming nickel nanoparticles were investigated in a vacuum without imposing periodic boundary conditions to simulate isolated nanostructures. The possible oscillation induced by the Nosé–Hoover thermostat in the simulation was avoided by placing a dragging force periodically by equilibrating the system at the imposed constant temperature for 10 000 steps with the time step fixed at 1.0 fs. The potential cut-off distance was defined as 4.5 Å. The Verlet neighbor list was employed in all simulations.

2.3.3 Size-Dependent Melting Phenomenon

Both constant rates of heating (CRH) and isothermal methods were investigated in the melting of individual nickel and aluminum particles. Both methods revealed similar phenomenon with slightly different values probably caused by different relaxation times. The potential energy per atom for the nickel and aluminum particle is shown in Figures 2.3 and 2.4. A similar pattern is observed. At low temperatures, the average potential energy per atom increases linearly with increasing temperatures. The melting phenomenon is identified by the distinct rise in the average potential energy due to the release of latent heat, which occurs at ~1100 and ~1500 K for the small and large nickel clusters, respectively, as revealed. For the aluminum particle, the melting temperature is determined as ~620 K, which is significantly lower than its bulk value, 933 K.

Figures 2.3 and 2.4 also reveal that the discontinuity in the average potential energy has a wide temperature range for the melting point due to the large fraction of

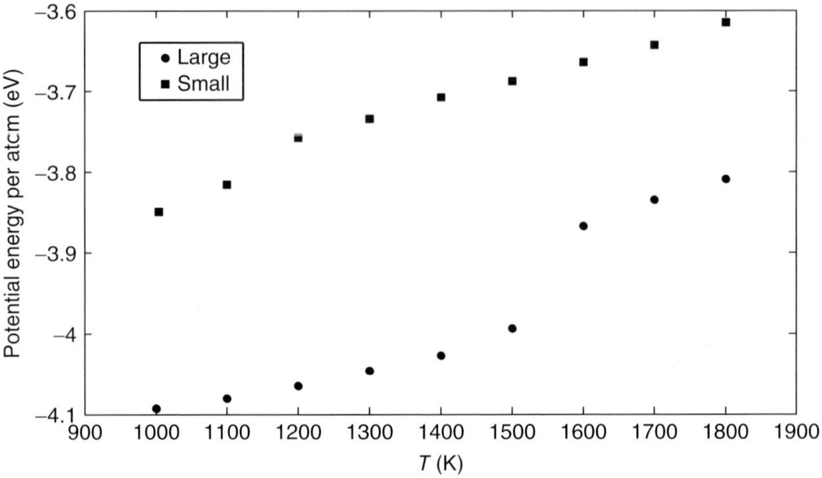

Figure 2.3 Potential energy per atom under isothermal heating for two nickel particles.

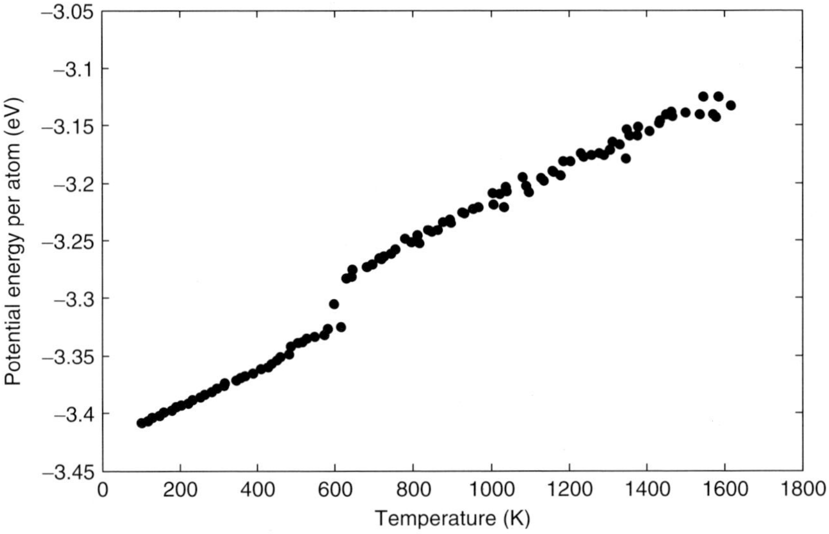

Figure 2.4 Potential energy per atom under isothermal heating for a 4 nm aluminum particle.

the surface atoms. It has been demonstrated both experimentally and theoretically that the melting process of a crystalline substance starts from the surface layer and propagates into the interior, resulting in a surface melting temperature significantly smaller than the bulk melting points. One would expect a similar phenomenon to occur for nanoscale nickel clusters. To reveal such a phenomenon, the large nickel particle was partitioned into different layers, numbered from 1 to 9 outward from the cluster center. The average potential energy per atom of each layer

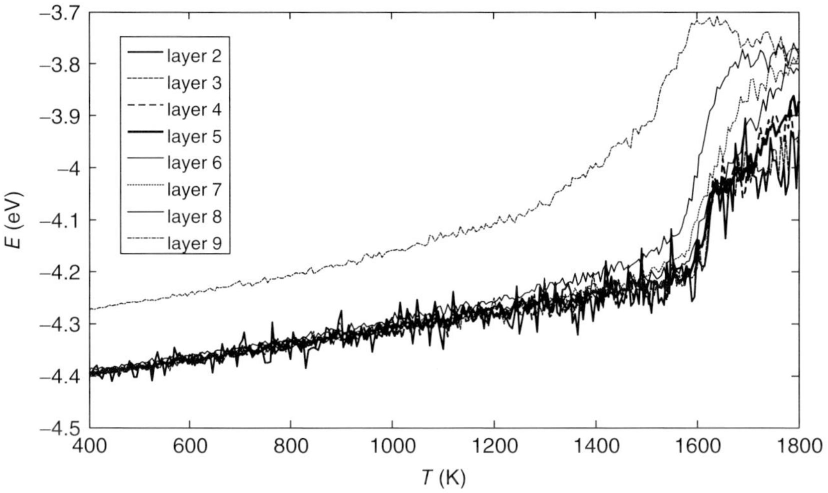

Figure 2.5 Layered potential energy per atom of large cluster.

was calculated as shown in Figure 2.5 under CRH conditions. It shows that the energy per atom of different regimes increases linearly with temperature at low temperatures but starts to jump at a characteristic temperature. However, the value of the characteristic temperature is different, increasing from the surface to the core atoms, which implies a stepwise melting from the surface to the inner cluster. Such a result is plausible as the core atoms of the cluster have higher coordination numbers than the surface atoms. As the cluster heats up, disorder appears first at the surface because of the broken bonds associated with surface atoms, resulting in surface melting, which then propagates inward. As the surface melting occurs, the core still remains rigid and solid-like. The surface melting temperature is ~1200 K, while the overall melting occurs at ~1500 K for the large cluster. Similarly, a surface temperature of ~900 K is identified for the small cluster.

The surface melting and formation of the liquid-like surface layer will likely induce particle agglomeration, even sintering, which affects the subsequent thermal–chemistry conversion process. It has been revealed by thermodynamic analysis and experiments that the coalescence process begins with nanoparticle contact, followed by the alignment of the coalescing planes at the interface between the nanoparticles, where the liquid-like mobility of the nanoparticle surface layers is essential to achieve this. When a small particle coalesces with a larger one, the smaller one rotates to orient its places to those of the larger one [16].

To reveal the influence of particle size, different-sized aluminum nanoparticles were simulated. The result shows a strong size-dependent melting temperature, which is due to the increased fraction of loosely bounded surface atoms at reduced dimension. The size-dependent MD-simulated melting temperatures are compared with a few other experimental and simulation results [17, 18]. One example is shown in Figure 2.6, where consistent results are obtained, with a maximum difference of 25%. For nickel particles, the predicted melting temperature depression, defined

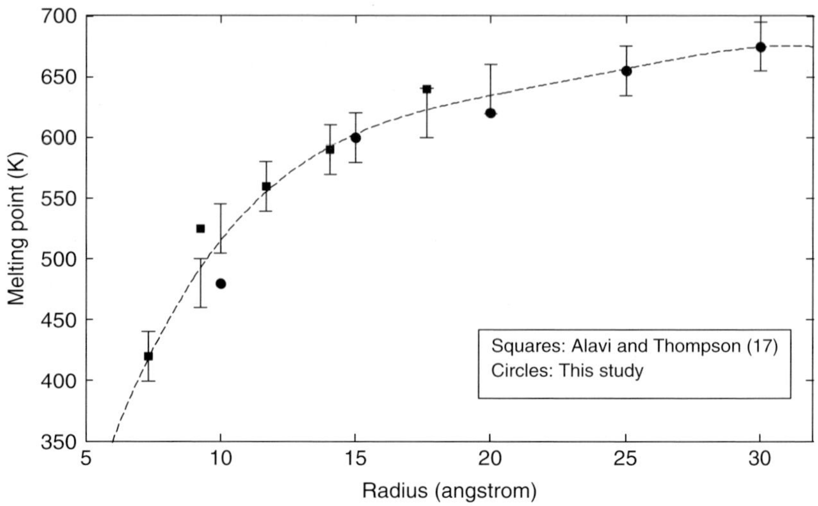

Figure 2.6 Comparison of the melting point of Al particle.

as the melting temperature of particles over that of the bulk value, are 0.7 and 0.5 for 3.52 and 1.76 nm nickel particles from Cao *et al.* [19], while the MD-simulated results are 0.75 and 0.55 based on the surface melting temperature. Such a difference is believed to be associated with the accuracy of the interaction potential. It has been suggested that the values of melting temperature for aluminum using this EAM potential would be higher than the experimental value, while the values for nickel are lower than expected based on experiment [20].

Such a size-dependent melting depression shows the robustness of the current MD simulation and also serves as a benchmark study for the following simulations.

2.4 Sintering Study of Two Nanoparticles

2.4.1 Introduction

From the structure–property point of view, particle size and morphology are critical to the oxidation behaviors. Particles synthesized from different routes may have a wide range of particle size distribution. Because of the melting-depression phenomenon, particles tend to melt first during the heating and oxidation process. Smaller particles will become molten first, which may accelerate particle–particle interaction and the sintering process.

For large particles, there are six different contributing mechanisms including surface diffusion, lattice diffusion from the surface, vapor transport, grain boundary diffusion, lattice diffusion from grain boundary, and lattice diffusion through dislocations. Several studies reveal that the sintering process increases and will have distinct features as particle size decreases [21]. In order to characterize the sintering process quantitatively, the Koch–Friedlander (K–F) model was developed to explain particle growth in a process of coagulation and sintering. Described by the K–F model, the sintering rate depends mainly on the excess surface of the particle over a spherical particle with the same mass. A particle tends to reduce its free surface to minimize the free energy. This causes the rounding of the particle and finally the formation of a sphere. The asymptotic equation in Equation 2.16 describes the sintering process [21]

$$\frac{da}{dt} = -\frac{1}{t_s}(a - a_{\text{final}}) \tag{2.16}$$

where a is the surface area, a_{final} is the spherical surface, and t_s is the sintering time. From macroscopic continuum theories of sintering via surface diffusion, the sintering time t_s is predicted to vary with the fourth power of the particle size

$$t_s = \frac{k_B T_S}{25 D \gamma d_A^4} R^4 \tag{2.17}$$

where R, D, γ, T_S, and d_A are the radius of the primary particles within the agglomerates, the diffusion constant, the surface free energy, sintering temperature, and the atomic diameter. The K–F model is also a foundation of the surface volume model [22] and the primary particle list model [23]. The volume and the surface area of each particle as independent variables are tracked in the surface–volume models, which are difficult to be determined accurately at the nanoscale.

Many experimental and theoretical studies have been conducted on the coalescence and sintering of different types of metal and metal oxide nanoparticles including gold, copper, nickel, silicon, iridium, and titanium oxide [24–32]. The general trend resulting from these studies reveals that the sintering process increases as particle size decreases, and it has been generally reported that surface and grain boundary diffusion are the two contributing transport processes for the sintering of copper and gold nanoparticles [24, 25, 33]. There are, however, still many inconsistencies and contradictions even. Most of these sintering studies employed two equal-sized spherical particles, although, in reality, the engineering process involves coalescence and sintering a number of particles with uneven size distributions. Actually, the macroscopic theory is mainly developed on the basis of the two-particle model. The study of the initial stage of sintering is basically established on the two-particle model. It is because the distinction of this stage is the neck shrinkage, considering more particle does not make a theoretical difference. As the sintering proceeds, the interconnected pores form a spatial structure between multiple particles. The quantitative description of sintering is very difficult and the pores, instead of particles, become the object of sintering study.

Only a few limited studies for unequal-sized particles revealed that different phenomena were expected [24, 30, 34]. From the materials aspect, it is a big challenge to produce nanoparticles with controlled size and shape. Polydispersity is always the case for all commercial particles. In the previous section, we demonstrated a strong particle size effect on the melting temperature. Consequently, for any engineering application of energetic nanomaterials, it would be expected that early melting can occur first for smaller particles, which actively intensify the particle–particle interactions. The induced particle agglomeration or even sintering could significantly affect the following thermochemical energy conversion process. A proper sintering study needs to consider nonuniform particle size distribution and these size-dependent properties

2.4.2
Simulation Setting

The EAM potential described in Section 2.3 was used in this part. Two uneven-sized nickel particles were simulated in this study. The large nickel cluster consists of 2112 atoms, measuring 3.52 nm in diameter, which are arranged spatially in a face-centered cubic (fcc) lattice in a spherical shape. Similarly, a small nickel cluster consisting of 276 atoms, or 1.76 nm in diameter, is placed in the close vicinity of the large particle. The two crystallographic axes of the two nickel particles are consistent with the coordinate's axis to have the same space orientation without tilting. This makes the possible grain boundary naturally coherent and does not cause reorientation of particles. Initially, the atom clusters were constructed. In order to investigate the kinetic of sintering of particles in different regimes, the nickel crystals were partitioned into different layers with equal distance of $dR = 1.25$ Å each, the half of unit cell constant, the interatomic distance in pure nickel, which results in 10 and 5 radial layers of nickel atoms for the large and small nickel particles. The outmost layer of nickel atoms was defined as "surface layer," the first three layers from the surface, including the surface layer, was defined as "outer layer" and further inward layers were defined as "core." Tracing atoms are arranged to study the dynamics of each regime during the sintering process.

Both isothermal heating and CRH methods were investigated by MD simulations. For the isothermal study, simulations were performed at constant temperatures between 800 and 1500 K at a temperature interval of 100 K. Before each run, the two clusters were equilibrated separately for 5×10^4 steps. After removing the angular momentum, they were positioned in a single simulation box and the small cluster only has the translational displacement along the abscissa of the large cluster. The distance between the two surface layers was set as 4 Å, just within the EAM cut-off distance, to enable the surface atomic interactions. The sintering simulation was then initiated for 200 ps. The same temperature span was used for the CRH method for 40 ps, which brings the nominal heating rate of 17.5 K ps^{-1}.

2.4.3
Sintering Process Characterization

A typical sintering scenario is shown in Figure 2.7 at a constant temperature of 1300 K. Two clusters were positioned within the cut-off distance of the interaction potential and sintering was initiated at $t = 0$ ps after the initial equilibrium of the

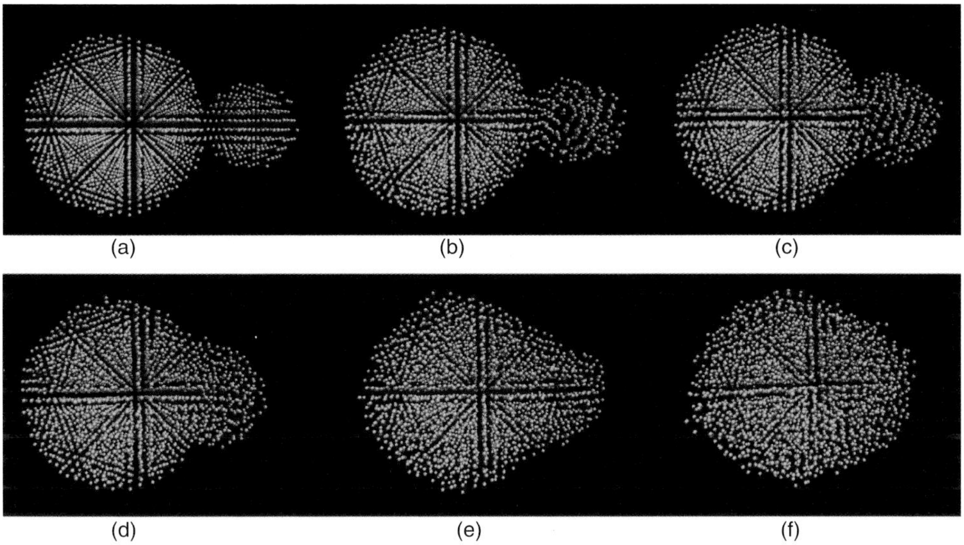

Figure 2.7 Nickel melting temperature comparisons at 1300 K (simulation time point: 0, 0.01, 0.05, 0.5, 2, and 200 ps for snapshots from (a) to (f)).

two clusters (Figure 2.7). The sintering process occurs rapidly in the first few picoseconds. A neck forms at $t = 0.01$ ps (Figure 2.7b), and reaches a diameter close to the size of the small cluster at $t = 0.5$ ps (Figure 2.7c,d). This was then followed by a slow process (Figure 2.7e,f), reaching a full sintering at 200 ps where the small cluster becomes undistinguishable.

The sintering process can be quantified by measuring the shrinkage, $\zeta = \Delta L / L_0$, which is defined as the change in the center-to-center distance of the particles ΔL ($\Delta L = L_0 - L$) over the initial distance, L_0. For all sintering under different temperatures, the shrinkage curve can be characterized by the three stages illustrated in Figure 2.8. Stage I is the neck formation and is marked by a sharp decrease in the center-to-center distance between the particles. The shrinkage appears to be independent of temperature in this initial stage, as evidenced by the similar shrinkage gradients, $d\Delta L/dt$, among different sintering temperatures. Such an observation is similar to that of Zeng *et al.* [24] and Arcidiacono *et al.* [25], which is presumably because the dominant mechanism for early-stage nanoscale sintering does not require thermal activation. The simulation also shows that in stage I the shrinkage ratio and duration increases with increasing temperature, implying an easier sintering process at elevated temperatures. Stage II comes after the rapid neck formation and is marked by a temperature-dependent characteristic with lower sintering rates, which is seen by the slope change after 5 ps. In this stage, the shrinkage is sensitive to thermal activation as the surface diffusion and grain boundary diffusion become dominant here; such behavior becomes important after some degree of shrinkage is formed and the neck curvature is significantly reduced. It is noteworthy that for sintering at low temperatures, that is, $T < 900$ K, there is

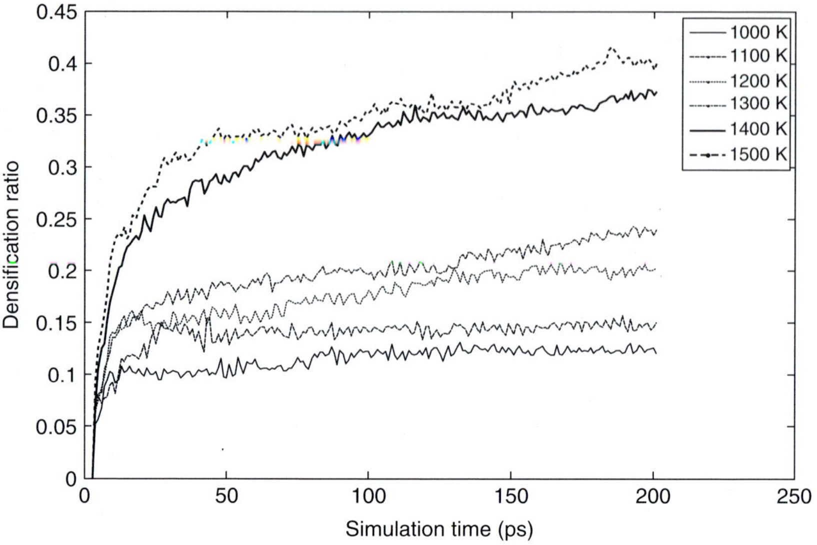

Figure 2.8 The shrinkage ratio under different temperatures during sintering of two nickel particles.

a small recovering process in the second stage and the center-to-center distance increases slightly after the initial rapid neck formation. This is presumably due to the solid status of both particles, resembling behaviors of soft collision. Stage III marks the end of the sintering, as evidenced by the nearly constant shrinkage values, varied between 0.1 and 0.4 depending on the sintering temperature from partial to full sintering. The increasing shrinkage ratio with temperature is associated with the increased diffusion process at higher temperatures.

In the nanoscale regime, the diffusion pathways are mixed and overlapped. The melting of particles makes it harder to adopt the continuum model. Instead of identifying the six classical diffusions, the diffusivity of different geometric regions was investigated. The mean square displacement (MSD) is a measure of the average distance a given particle in a system travels and can provide a good approximation of diffusion. The MSD is defined as

$$\text{MSD} = \langle r^2(t) \rangle = \left\langle \frac{1}{N} \sum_{i=0}^{N} \left(r_i(t) - r_i(0) \right)^2 \right\rangle \tag{2.18}$$

Here, N is the number of particles, t corresponds to time, and $r_i(t) - r_i(0)$ is the vector distance traveled by a given particle over the time interval. MSD shows how far a group of atoms travel in a given time interval.

In order to reveal the sintering mechanism, the MSD of different regimes during the sintering process at two typical temperatures are plotted in Figure 2.9. It is evident that the largest magnitude of MSD is observed on the surface layer, and the magnitude decreases toward the cluster center. The result also illustrates that the MSD behaves quite differently between the large and small clusters. For the

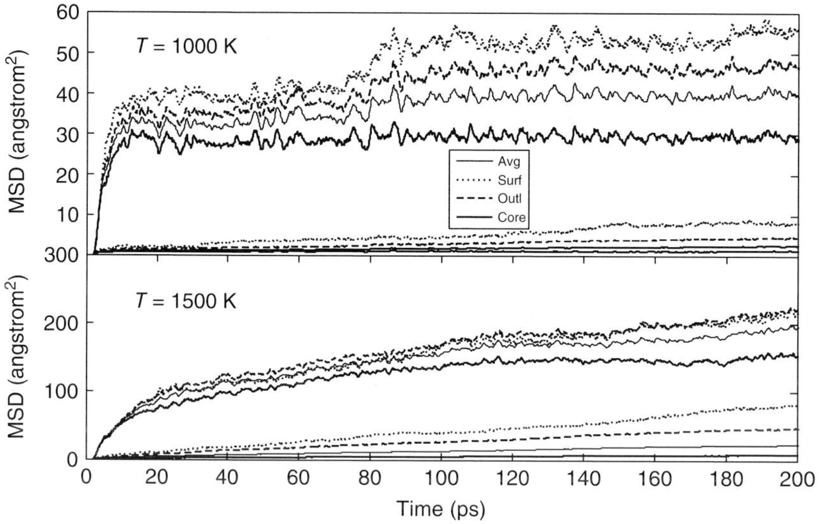

Figure 2.9 Comparison of MSD of two clusters of different layers (thick solid: core; dash: outlayer; dot: surface; and thin solid: average.)

large cluster at low temperature, that is, ∼1000 K (see Figure 2.9), the MSD of all regions increases slowly with time with small MSD values, 10 Å^2 in 200 ps for the surface layer. The MSD difference between the core and surface regime is small, 6 Å^2, which implies that there is no remarkable diffusion from this region at this temperature, and atoms tend to oscillate around the balanced lattice position. When the isothermal temperature was elevated to 1200 K or higher (Figure 2.9), large temporal evolution of MSD was observed in all selected regions of the large cluster. The MSD of the surface layer increases nearly linearly, reaching ∼80 Å^2 in 200 ps and the difference between the surface layer and the core becomes large, owing to the surface melting of the large cluster.

One phenomenon of interest is that the small cluster had a large initial MSD increase, from 0 to 20–30 Å^2 at $T = 1000$ K and 50 Å^2 at $T = 1500$ K, at the beginning stage independent of temperature. This is due to the displacement in the first stage of the sintering, caused by the "attraction" from the large cluster, that is, the surface and outer layer atoms, as the MSDs of these temperature-independent behaviors are larger than the separation distance. A new equilibrium position is reestablished for the small cluster as the repulsion and attraction upon the small cluster is roughly balanced. It should be noted that as there is no long-range interaction similar to the "gravity" for the EAM potential with cutting off distance, it is impossible for the small cluster to move entirely like a rigid particle. There is possibly some plastic deformation during this stage, as reported by Zeng *et al.* [24]. Some results presented by Hawa and Zachariah [34] are thought to be associated with this phenomenon. The ratio of the MSD at the surface to that of the core reflects the surface melting phenomenon. There is a smooth decrease in MSD values from the surface layer toward the inside at low temperature ∼1000 K, as the surface melting occurs for the small cluster while the core is still solid-like. Such a trend disappears at temperatures over ~1200 K, where the MSD value of the surface layer and the outmost layer are similar, which is believed to be associated with the global melting of the small cluster.

One advantage of MD is that the properties of interest are derived from a well-established statistical method without mechanical assumption, typically referring to the continuum model. The significance of MSD is that it contains information on the atomic diffusivity in relation to diffusion coefficient. The slope of the MSD is related to the diffusion coefficient $D(T)$, which can be calculated using the Einstein equation,

$$D(T) = \lim_{t\to\infty} \frac{\langle r^2 \rangle}{2dt} \tag{2.19}$$

where $\langle r^2 \rangle$ is the MSD of the atoms in the time t and d is the dimensionality of the system, which equals 3 in this work. As shown in Figure 2.5, it is clear that surface properties play an important role in the coalescence event. Different regimes have different diffusivities, being larger for atoms closer to the surface. The result of the diffusivity of different regimes calculated from MSD is shown in Figure 2.10. It clearly shows that the diffusivity is a regime-dependent and size-dependent behavior, that is, being larger in the surface regime and being smaller for the large particle. Nearly all these regimes displaced a linear relationship between ln (D) and $1/T$, typical Arrhenius type, which can be expressed as

$$D(T) = D_0 \exp\left[-\frac{Q_v}{k_B T}\right] \tag{2.20}$$

where D_0 is the pre-exponential factor that is independent of temperature, Q_v is the activation energy, T is the absolute temperature, and k_B is the Boltzmann

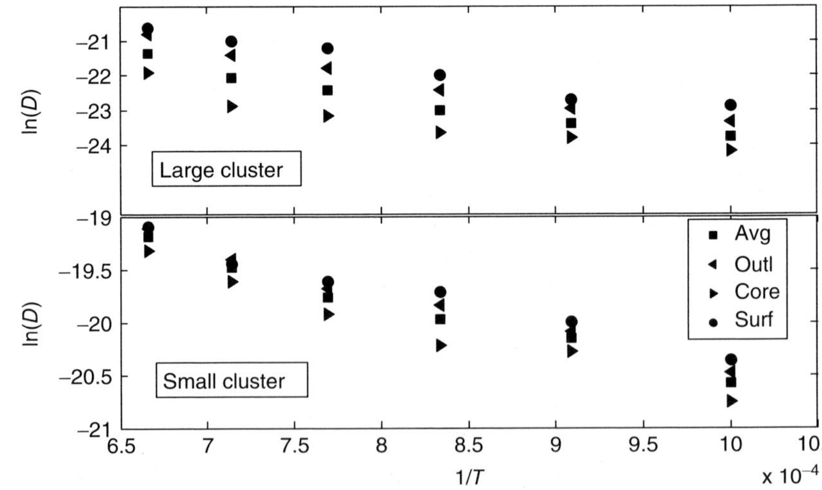

Figure 2.10 Diffusivity calculation for the sintering of two nickel clusters.

constant. The average activation energies of the two particles are calculated as 0.497 and 0.581 eV, respectively, for the large and small clusters; such values are much smaller than the bulk values, which are in the range of 2.0–2.9 eV, but in the similar values compared with other nanoscale nickel sintering experiments. For instance, the average activation energy for sintering of nickel particles with $d = 6.2$ nm was determined as 0.62 eV from both the direct *in situ* differential mobility analyzer (DMA) and the *ex situ* transmission electron microscopy (TEM) analysis [35]. Panigrahi [27] also obtained experimentally the activation energy of sintering as 0.68 eV for 30 nm ball-milled nickel particles.

For large particles, it has been widely accepted that the activation energy of surface diffusion is proportional to the melting temperature of crystalline solids, $E_{sp}/E_{sb} = T_{mp}/T_{mb}$, where E_{sp} and E_{sb} refer to the surface diffusion activation energy at the particle and bulk level, and T_{mp} and T_{mb} refer to the melting temperature at the particle and bulk level. If such proportionality is extendable to nanoparticles, the activation energy ratio should be around 0.5–0.7 as revealed from the melting study. However, the MD study reveals a different phenomenon, the activation energy reduction ratio, which is in the range of 0.2–0.3, is much smaller than the melting temperature depression ratio. Other studies were also conducted, including the modeling of the necking growth and coalescence time, t_s, based on Equation 2.16 in the K–F model. The results show that surface diffusion is not the sole mechanism for sintering. It is apparent that beside the surface diffusion, there are other mechanisms that contribute to the sintering process that may include the grain boundary diffusion, viscous flow, plastic deformation, and surface tension gradient resulting from a nonperfect sphericity in the solid–liquid interface.

2.5 Oxidation of Nanoparticles in the Presence of Oxygen

2.5.1 Introduction

ReaxFF, developed in the California Institute of Technology by Dr van Duin and his colleagues [36–39] provides an atomic level, full-chemistry, and full-physics description of energetic processes of materials without any assumptions or simplifications other than those intrinsic to the FF description of the interactions, in a way similar to that of classical nonreactive FF. Such a molecular level character is essential for the development of predictive, physics-based materials models. ReaxFF has covered broad elements in the periodic table including Al, B, Ba, Bi, C, Cl, Co, Cu, F, Fe, H, Li, Mg, Mo, N, Na, Ni, O, Pt, Ru, S, Sb, Se, Si, Ti, W, Y, Zn, and Zr. It aims to include more elements in future development. The development of ReaxFF in recent years has enabled, for example, simulations with quantum chemistry (QC)-comparable accuracy of the thermal- or shock-induced decomposition of energetic materials under realistic loading conditions, oxygen ion transport in fuel cell, grain boundary proton diffusion, oxidative dehydrogenation, melting and crystallization of aluminum, catalysis of nickel, dynamic transition in the shock structure of an energetic crystal, catalytic-selective oxidation processes, self-assembly of silica nanocages, and dynamical crack propagation in silicon [40, 38, 41–43, 62]. The ReaxFF also displays its profound capability of simulating mesoscale phenomena.

The total energy of ReaxFF is divided by different potential contributions [37].

$$U_{system} = E_{bond} + E_{vdWaals} + E_{Coulomb} + E_{val} + E_{tors} + E_{over} + E_{under} \tag{2.21}$$

Allowing bonds to break and form during simulations, ReaxFF does not assign chemical bonds or predefine connectivity to atoms before dynamic moves are deployed. The partial potential function is designed to be atomic-coordinate dependent. The key parameter is bond order (BO) between atom i and j, BO'_{ij}, that is, directly calculated from the sum of σ ("single bond"), π ("double bond"), and ππ ("triple bond") BOs from the given equation

$$\begin{aligned} BO'_{ij} &= BO'^{\sigma}_{ij} + BO'^{\pi}_{ij} + BO'^{\pi\pi}_{ij} \\ &= \exp\left[p_{bo,1}\left(\frac{r_{ij}}{r_0^{\pi}}\right)^{p_{bo,2}}\right] + \exp\left[p_{bo,3}\left(\frac{r_{ij}}{r_0^{\pi}}\right)^{p_{bo,4}}\right] + \exp\left[p_{bo,5}\left(\frac{r_{ij}}{r_0^{\pi\pi}}\right)^{p_{bo,6}}\right] \end{aligned} \tag{2.22}$$

where r_{ij} is the atomic distance, r_0 is the equilibrium bond length, and p with subscript is the adjusting unitless parameters whose value changes with atomic distance.

Because it is directly associated with bond break and formation, the bond energy potential is presented as

$$E_{bond} = -D_e^{\sigma} BO_{ij}^{\sigma} \exp[p_{be,1}(1 - (BO_{ij}^{\sigma})^{p_{be,2}}] - D_e^{\pi} BO_{ij}^{\pi} - D_e^{\pi\pi} BO_{ij}^{\pi\pi} \tag{2.23}$$

All parameters in Equation 2.22 and Equation 2.23 are listed in Table 2.2, Table 2.3, and Table 2.4, in which of note is that there is no existence of ππ ("triple bond") in the Si–O system.

All of the covalent interactions including E_{val} and E_{tors} dealing with bond angles and torsion angles are expressed in similar analytical forms of BOs. The BOs are corrected by E_{over} and E_{under} once the bonds are overcoordinated or undercoordinated. Totally, there are 148 parameters in ReaxFF for the Si–O system describing the total energy potential, and the entire related equations and full parameters were described in detail among Refs [36–38].

Table 2.2 Bond radius parameters of the Si–O system.

Atoms	r_0^{σ} (Å)	r_0^{π} (Å)
Si–Si	2.013	1.563
Si–O	1.610	1.294
O–O	1.169	1.020

Table 2.3 Bond order parameters of the Si–O system.

Bond	$p_{bo,1}$	$p_{bo,2}$	$p_{bo,3}$	$p_{bo,4}$	$p_{bo,5}$	$p_{bo,6}$
Si–Si	−0.069	7.94	−0.20	7.54	n/a	n/a
Si–O	−0.519	4.45	−0.37	4.26	n/a	n/a
O–O	−0.161	5.66	−0.25	6.52	−0.16	10.51

Table 2.4 Bond energy parameters.

Bond	D_e^{σ} (kcal mol^{-1})	D_e^{π} (kcal mol^{-1})	$p_{be,1}$	$p_{be,2}$	$p_{be,3}$
Si–Si	113.8	54.0	0.25	0.26	0.07
Si–O	193.1	41.1	−0.21	0.92	0.77
O–O	118.9	42.8	0.91	0.86	0.86

Using the silicon nanoparticle as an example, this section investigates the oxidation process of nanoparticles in the presence of oxygen. The oxidation process will be characterized in terms of bonding configuration based on the atomic coordinates and the bonding list. The study is used to justify the feasibility of ReaxFF FF as a benchmark for further oxidation investigation at a scaled-up dimension. This study does not tend to go deeper into chemical details than QC, but to present the FF's ability of reproducing thermal and chemical behaviors in the interest of engineering aspects.

2.5.2 Simulation Setting

The simulation was performed in a three-dimensional cell box. Each dimension was measured as 54.31 Å in length. The silicon cluster was placed in the center of the box containing 184 atoms. The same amount of oxygen molecules, 184 O_2, was located in the rest space of the simulation box. Initially, the silicon cluster was trimmed from the silicon crystal's face-centered diamond cubic structure with a lattice spacing of 5.43071 Å. The oxygen molecules were arranged in a simply spatial order in which a minimum displacement between molecules is deployed as shown in Figure 2.11. An optimization was deployed for oxygen molecules to assure that no atom overlapping was caused. Periodical boundary conditions were constructed in all three dimensions. The time step was set as 0.25 fs. Minimization processes were employed on the silicon cluster and oxygen molecules separately and an additional annealing process was conducted for the silicon cluster to produce a more stable configuration. The O_2 gas was added into the simulation box with the Si–O bonding interaction switched off. The whole system was then minimized to remove any bad residual contacts, during which a root-mean-square force threshold,

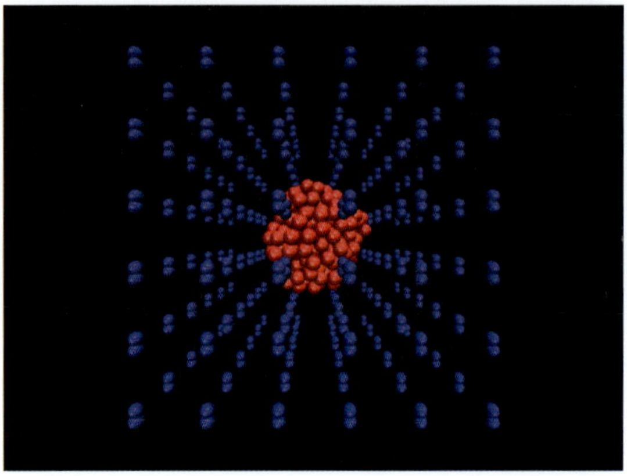

Figure 2.11 Initial configuration of Si–O system (Si atoms in red and O atoms in blue.)

2.0 kJ mol^{-1} Å^{-1}, was set during minimization. Then the full system was ready for equilibration. After being equilibrated for 5000 iterations in NVT, the entire system of O and Si was ready for subsequent simulation. The simulation was initiated by assigning a Boltzmann distribution of velocities at 300 K. The simulation was set at a heating temperature regime of 300–3000 K, and lasted 2 000 000 iterations, leading to nominal heating rates, 1.08×10^{13} K s^{-1}.

2.5.3
Characterization of the Oxidation Process

A snapshot of the oxidation process is shown in Figure 2.12. The nonspherical shape of silicon was caused by the annealing process. Before introduction of

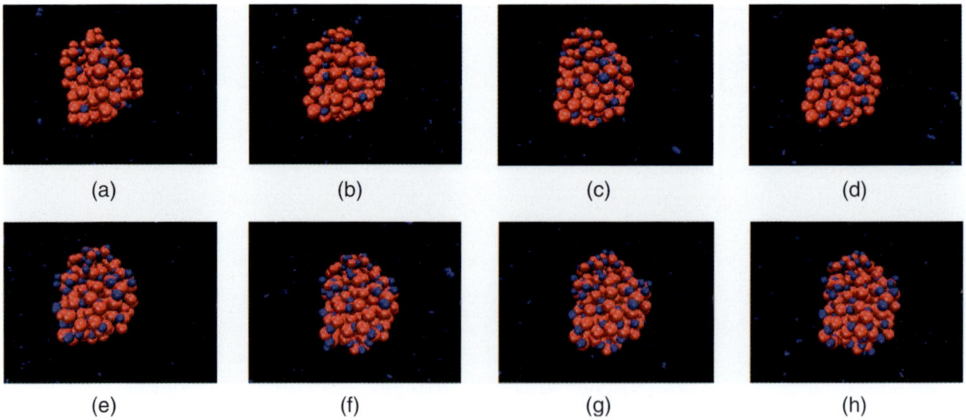

Figure 2.12 Snapshot of the oxidation of silicon cluster in the presence of oxygen. (a) 0.05 ns, (b) 0.10 ns, (c) 0.15 ns, (d) 0.20 ns, (e) 0.25 ns, (f) 0.30 ns, (g) 0.35 ns, and (h) 0.40 ns.

oxygen molecules, the potential energy of the silicon cluster in the cell box was minimized and then equilibrated at 300 K in NVT for 50 000 iterations. After equilibration, the temperature was lifted to 1750 K and kept at that temperature for 5000 iterations with subsequent gradual cooling for 100 000 iterations. The silicon configuration after annealing had been changed from spherical geometry. The reaction was switched on after introducing oxygen gas. Such snapshots show vividly the oxidation and the Si–O bond formation processes. With the reaction timing increase, more and more bonds are formed. Detailed analysis is described subsequently.

An exothermic process is indicated by the declining of the potential energy of the whole system during a heating process illustrated in Figure 2.13. Correspondingly, the oxygen molecules are bonded into silicon clusters shown in the simulation snapshots. Usually the heating process will increase the potential energy as the volume of heated system is increased along the increasing kinetic energy. In an NVT ensemble, the system is connected to a "heat bath" and the kinetic energy is always adjusted to increase to the target temperature. Therefore, the decreased potential has nowhere to go but transfers to other energies (bonding energy, etc.) and the heat released. To calculate the released heat, a separate simulation having the same time span and same heating process was performed, where the only difference was that bonding reactions between Si and O were switched off. The nonreactive heating is an obvious heat-absorbing process where the total energy is increased. Therefore, the total heat released for Si–O oxidation should be calculated from the combination of the heat released at the reaction mode and the heat absorbed at the nonreactive mode, that is, $[\max(E_{\text{reactive}}) - \min(E_{\text{reactive}})] + [\max(E_{\text{non-reactive}}) - \min(E_{\text{non-reactive}})]$,

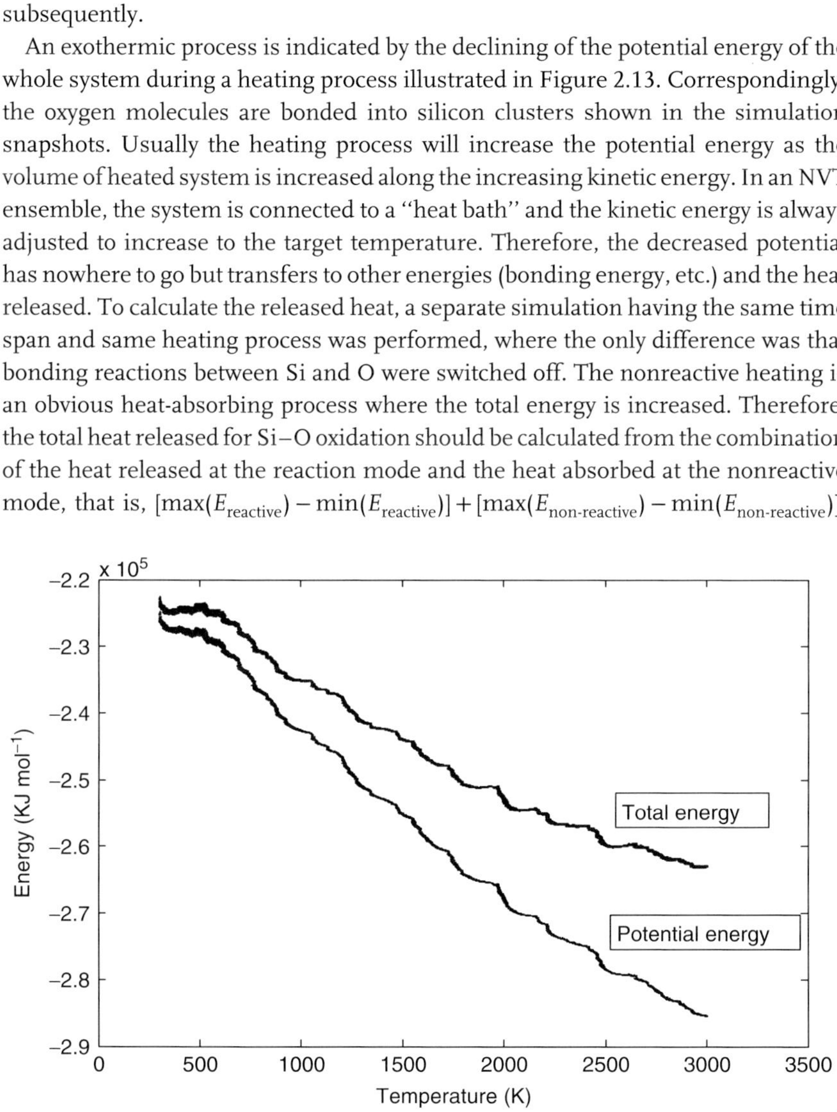

Figure 2.13 Energy variation during the oxidation of silicon with oxygen.

where E represents the total energy. The value is equivalent to 5.95×10^4 kJ mol^{-1}, or 21.1 MJ kg^{-1}, counted as about two-third of the theoretical exothermic value for a complete oxidation of the aluminum nanoparticle, 30.0 MJ kg^{-1}.

MD simulation can provide detailed chemistry when the reaction occurs. For example, the BOs of chemical element are vital indicators of reliability of a reactive MD simulation. In ReaxFF, BO is directly calculated from the instantaneous interatomic distance to simulate the bonding dynamic, that is, the formation and dissociation of bonds. The BO of a particular atom is the sum of partial BOs to its bonded neighbor atoms. The average BO of oxygen and silicon at each chosen time point is calculated to be 2.0243 and 3.9736, respectively, and are in good agreement with the bond valence of oxygen and silicon, 2 and 4. Bond length between Si and O is analyzed once a Si–O is formed, as shown in Figure 2.14. The average of bond length is 1.58 Å that is in good agreement with many theoretically calculated values [37]. Si–O–Si angle analysis shows that most of the Si–O–Si bonds are in the range of 90–110° and 125–145° displayed in Figure 2.15 that is in consistency with others [37]. The respective mean values of bulk SiO_2 are 1.61 Å (Si-O bond length) and 141° (Si–O–Si), respectively [44]. As the Si–O bonds are localized and it is not appropriate to take this reference value of bulk SiO_2 to the MD case where intermediates are likely to be involved. The bond analysis suggests that the ReaxFF creates a credible oxidation scenario.

The bonding process of oxygen into silicon is concluded to involve three mixing mechanisms determined by the simulation of density functional theory (DFT) and kinetic Monte Carlo (KMC) [45]. The three types of mechanisms are as follows: (i) two oxygen atoms from a dissociative oxygen molecule are bonded directly on silicon, that is, Si–O type of bonding; (ii) the incorporation of the oxygen atom in event 1 into silicon to form Si–O–Si bonds; and (iii) the atomic oxygen migration from any previous bonded configurations.

In this study, when one oxygen atom is bonded with only two silicon atoms, it is classified as Si–O–Si. If one oxygen atom from an O_2 molecule is bonded with a silicon atom, that is, an instantaneous oxygen radical, it is denoted as Si–O–O. The

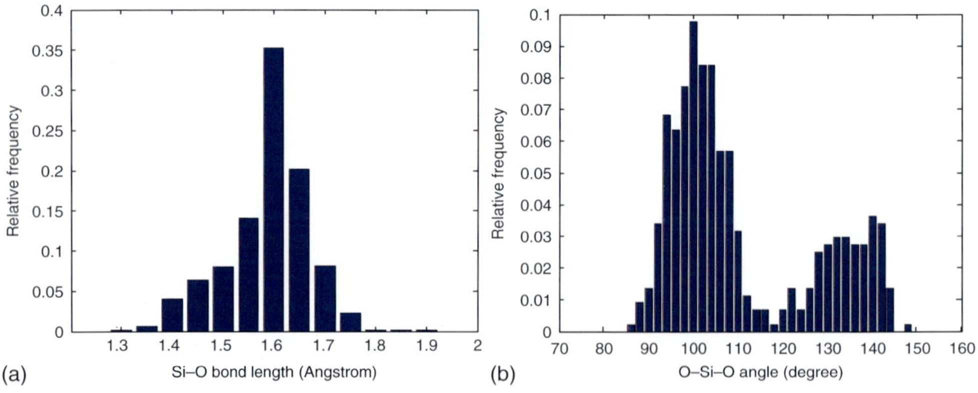

Figure 2.14 Distribution of (a) Si–O bond length and (b) O–Si–O angle.

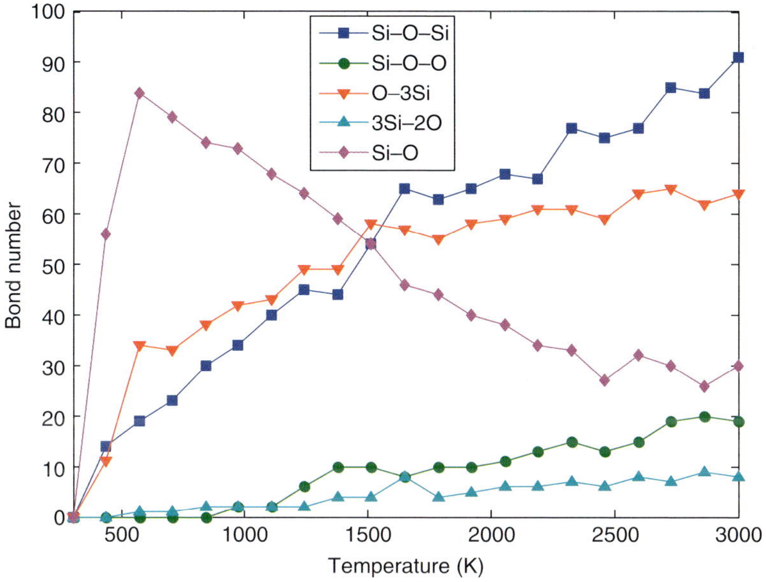

Figure 2.15 Bond configuration evolution during the oxidation of silicon nanoparticle.

existence of the oxygen radical is consistent with the little oversaturated valences of oxygen, 2.0243. The most bonding scenarios of oxygen are observed to be the oxygen radical and single oxygen atom shared by two silicon atoms. Even though the total BO of oxygen is 2, the existence of three-silicon-neighbor oxygen shows that these oxygen atoms are two-coordinated. The configuration is denoted as O–3Si. This bond configuration is believed to be consistent with the nearly 90° Si–O–Si angle found in this study and others [37]. The other configuration of two-coordinate oxygen is defined as a structure where one silicon atom bonds two oxygen and two other silicon atoms, which is labeled as 3Si–2O. Here, 3Si–2O is used rather than Si–2O because it is found that all root silicon atoms in this configuration are bonded to two other peer silicon atoms. The two oxygen atoms can only occupy the rest bonding sites. The trace of two-coordinated oxygen is consistent with van Duin's result obtained by Si–O–Si angle analysis [37]. The two-coordinated oxygen in the form of Si–O is a SiO interstitial and can introduce stoichiometric defects.

The bonding number evolving with temperature is shown in Figure 2.15. One oxygen atom positions itself into bridging in the middle of the dimer bond of Si. This structure forms two Si–O bonds, the length of which is 1.58 Å and the angle of Si–O–Si is 134°. And they also observed that the silicon atom three-coordinated by surrounding oxygen corresponding to the O–3Si bond configuration found in this study and two-coordinated structure. Some oxygen bonded to surface silicon stay unchanged and terminated, similar to the Si–O structure. A distinct structure of a silicon atom inserted into three oxygen atoms on the silicon surface, Si–3O, is found in their simulation but is not observed in this study.

At the early stage of oxidation, the dissociative oxygen atoms absorbed in silicon are predominant in the oxidation. The largest number of Si–O is remarkable at the temperature of 570 K. Then, it can be seen that this type of bonding starts to decline to about 30% of its maximum value at the end of the simulation. Meanwhile, two other configurations, that is, Si–O–Si and O–3Si grow steadily and eventually prevail in the oxidized structure. The explanation of this contrast is that the surface double-bonded oxygen migrates into the silicon matrix and moves into the interstitial space of silicon atoms, that is, the transition from Si–O to Si–O–Si and O–3Si during a diffusive process.

Such detailed chemistry information provides insight into the oxidation at the nanoscale and also can derive useful macroscopic quantities governing the reaction such as the activation energy. As activation energy, ΔE_a, is occurrence frequency dependent, it is usually determined by the MC simulation in which a random series of events occurs. In MD simulation, the activation energy is obtained from a reaction pathway along the energy surface of configurational space associated with reaction coordinates. The reaction coordinates are defined and the energy of atoms under investigation is book-kept. If any energy barrier is observed, the activation energy is subsequently deduced by the energy difference. This task is never easy if the system has many degrees of freedom because the direct mapping from computational simulations of activated processes is profoundly complicated and prohibitively expensive for most molecular systems. The reason is that the identification of the reaction path involves separation of various activated processes at different time scales, such as intramolecular vibrations and intermolecular diffusion. The main difficulty, particularly for MD, is to achieve a sufficient sampling of the relevant range of reaction coordinate values. The umbrella sampling method [46] and constrained MD [47] have been used to achieve efficient sampling of surfaces of configuration space that are otherwise inaccessible in the typical time scale of MD simulations. Any metastable configuration needs to be avoided carefully by looking into local minima associated with it.

However, based on the simplicity that will be discussed subsequently, by skipping all those difficulties, a rough estimation of activation energy can be made for a particular transition. The simplicity used here is that MD is thought to reproduce a genuine physical event of Si oxidation, that is, taking the MD simulation as an "experiment" and all the bonding information obtained as genuine observations. The limitation in theory is obvious: not each configuration of configurational space is covered. The experimental study of bond transformation is operated in a variable-temperature scanning tunneling microscope (STM) by imaging individual molecule and resolving oxygen dissociation and absorption at the atomic scale [48]. The oxygen mechanism is proposed to be site hopping of the intermediate structure of Si and O bonding. The experimentally found structural hopping frequency is about 10^{15}–10^{16} Hz, which is well within the time span of this simulation. Considering that hopping conduction is one basic oxidation mechanism [49], the activation energy of [(Si–O) → (Si–O–Si)] and [(Si–O) → (O–3Si)] can be estimated in the use of the appearing bonding number at each time point. On the basis of Figure 2.15, the maximum lifetime of each bond transformation is taken as

1.25×10^{-11} s. The corresponding average transition rate is R_e. From successive bonding numbers taken at various temperatures, the average transition rate at each temperature, which is equal to the ratio of the total number of bonding to the total elapsed time span in assumption that any bond configuration is only converted from that at previous time step. That is to say, there is no bond transformation longer than one time step, 1.25×10^{-11} s. Similar to the equation of Arrhenius

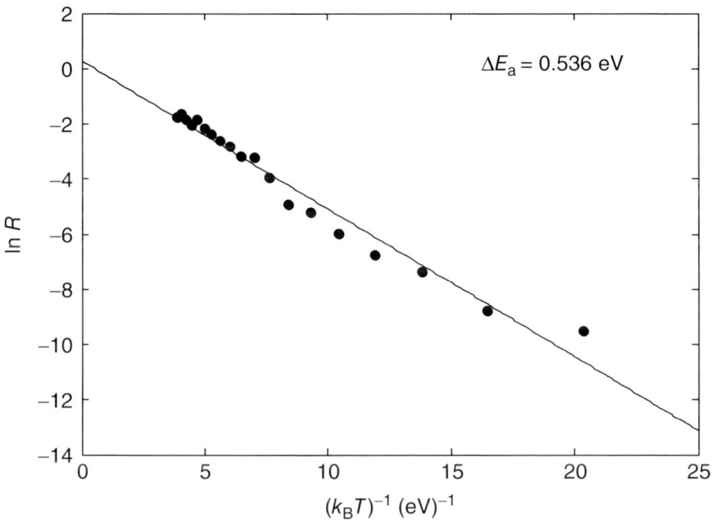

Figure 2.16 Activation energy from Si–O to Si–O–Si.

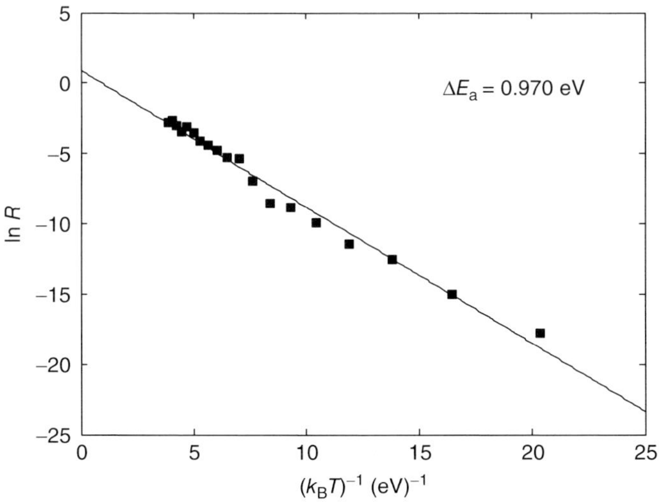

Figure 2.17 Activation energy from Si–O to O–3Si.

theory, the transition rate is expressed as

$$R_e = R_0 \exp\left(-\frac{\Delta E_a}{k_B T}\right) \quad (2.24)$$

The activation energies are deduced from the Arrhenius plots shown as Figure 2.16 and Figure 2.17. The transition from Si–O to Si–O–Si is found to be 0.564 eV and that from strand configuration Si–O to three neighbor configurations O–3Si is 0.970 eV. Therefore, the Si–O–Si bonding is more favorable and costs less energy to achieve thermodynamically.

However, it is worthwhile to point out that the calculation done here does not produce the genuine activation energy of the process but a quantitative estimation based on obtained bonding number, which has a form similar to activation energy [50].

2.6
Heating and Cooling of a Core–Shell Structured Particle

Previous examples show the melting depression phenomenon and high reactivity associated with energetic nanoparticles. For instance, aluminum nanoparticles have pretty low melting, oxidation, and ignition temperature and react easily with surrounding oxidizers. Without any protective passivation layers, their reactivity will be easily lost. Slow low-temperature, oxidation-produced amorphous alumina is the natural passivation layer, which prevents further oxidation of the active aluminum content. In many cases, the passivation layer acts as a barrier and also an influencing parameter to control the oxidation and ignition process. For instance, the ignition temperature of aluminum particles has been found to be significantly dependent on the thickness of its natural forming passivation layer. As we showed earlier that early melting is the catalyst for rapid sintering process, building a passivation layer around reactive metallic nanoparticles will also help prevent forming agglomerates or sintered particles [51].

In this section, MD simulation is used to investigate the heating and cooling behavior of a shell–core-structured Ni–Al functional nanoparticle. The reinforced melting phenomenon, and detailed thermodynamic properties and structure evolution during a continuous heating and under two different solidification rates are investigated.

2.6.1
Simulation Method

Again, the simulations are performed with LAMMPS by classical MD in the use of an EAM FF developed by Angelo *et al.* [10]. Different from the single-element EAM potential (Equation 2.10) for nickel in the previous section, the pairwise interaction potential $\Phi(r_{ij})$ is the same as the Morse potential [52]

$$\Phi(r) = D_M\{1 - \exp[-\alpha_M(r - R_M)]\}^2 - D_M \quad (2.25)$$

Table 2.5 Fitting parameters of alloy EAM potential.

	Ni–Ni	Al–Al	Ni–Al
D_M (eV)	1.5335	3.7760	3.0322
R_M (Å)	2.2053	2.1176	2.0896
α_M (Å^{-1})	1.7728	1.4859	1.6277

where D_M, R_M, and α_M are the depth, distance to the minimum, and a measure of the curvature at the minimum of Morse potential, being listed in Table 2.5.

The density function ρ_i of atom i is defined as

$$\rho(r) = r^6[e^{-\beta r} + 512e^{-2\beta r}] \tag{2.26}$$

where β is an adjustable parameter as 3.6408 and 3.3232 Å^{-1} for Ni and Al, respectively.

The composite nanoparticle is constructed from the perfect crystal measuring 4 nm in diameter with a shell thickness of 0.5 nm. It includes an inner core of aluminum containing 2016 atoms and a shell of nickel with 3620 atoms, separated by a gap of 3.0 Å between the core and shell atoms, shown in Figure 2.18. An energy minimization process is performed upon the whole system before simulation started. The simulation is carried out within Nosé–Hoover thermostat scheme of the NVT canonical ensemble and performed by the EAM FF. The simulation time step is set as 1.0 fs. Nonperiodic boundary condition is imposed to simulate an isolated unsupported bimetallic particle. The temperature range in this study is from 300 to 1600 K. For the heating simulation, the system is kept at 300 K for 40 ps within the NVT ensemble to achieve an equilibrium state. For the solidification simulation, an additional relaxation time of 40 ps at 1600 K

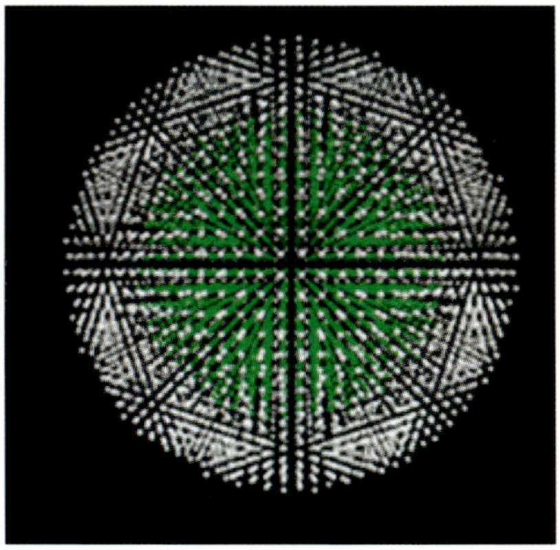

Figure 2.18 Initial configuration of core–shell Al–Ni (Al atoms are colored in green while Ni atoms in gray.)

is conducted before implementing the cooling process. To analyze the geometry-dependent properties, aluminum atoms are classified into five groups with equal distance of $\Delta r = 0.66$ nm each in the radius direction. Two quenching rates for solidification are simulated: 6.5 and 0.13 K ps^{-1}, corresponding to runtimes of 2.0×10^5 and 1.0×10^7 fs. Hereafter, those two quenching rates are termed as the *fast* and *slow quenching rates*, respectively.

A CRH (0.65 K ps^{-1}) is employed in the simulation, that is, the particle is heated gradually for 2×10^6 time steps, which is equivalent to 2000 ps, from 300 to 1600 K. The potential energy, enthalpy, and other parameters of different regimes are calculated every 20 000 steps. The thermodynamic and structure properties of the bimetallic particle are investigated in detail as described here.

2.6.2 Heating Simulation

In the simulation with a CRH of 0.65 K ps^{-1}, the whole system is recentered every five steps and Figure 2.19 displays the two-dimensional cross-sectional view cut through the center of mass. As the volume of the whole system is constant, these snapshots are taken as radial representations of the whole system. With the increase in temperature, the diffusion is clearly visible with aluminum core atoms diffusing outward and shell nickel atoms diffusing inwards. The composite particle becomes slightly deformed. Aluminum atoms reach the outmost surface of the composite particle at ~900 K, and continue spreading over the surface with the increase in temperature. The process speeds up as the aluminum core becomes molten. At ~1200 K, the surface of the composite particle forms an aluminum-rich layer. More molten aluminum atoms migrate to the outer surface as the temperature increases further. The inward diffusion accelerates as the nickel shell melts. A well-mixed liquid aluminum–nickel functional droplet forms at the end of the heating, that is, 1500 K. The outmost region of the liquid particle is Al-rich, which is evidenced by the atom distribution (number of atoms divided by the region volume) of final configuration of the heating mode. However, as the low-energy component, aluminum is not sufficient to seal the surface, which is beneficial for the liquid cluster to nucleate during the cooling process.

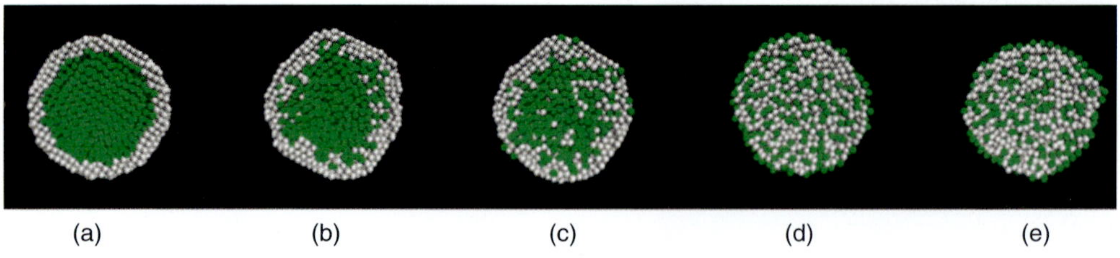

Figure 2.19 Snapshots of the system cross section under heating (temperature are 300, 600, 900, 1200, and 1500 K respectively from (a) to (e).)

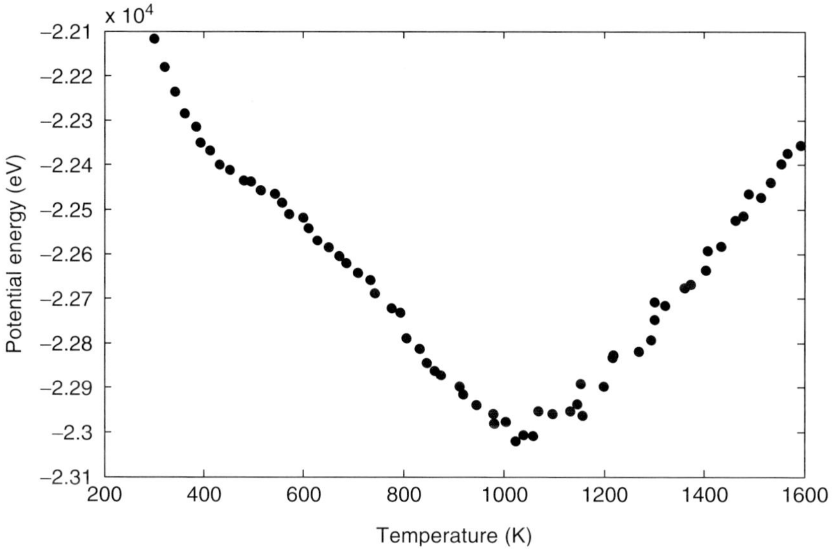

Figure 2.20 Potential energy profile of Ni–Al composite particle under heating.

The profile of potential energy per atom for the composite particle under a CRH 0.65 K ps^{-1}, from 300 to 1600 K, is shown in Figure 2.20. Different from the linear increase followed by an abrupt jump for a pure aluminum particle as shown in Figure 2.4, the average potential energy of the composite nanoparticle exhibits a distinct V-shape with the increase in temperature, that is, the composite particle experiences a nearly linear decline in potential energy at low temperatures, reaching a minimum value at ~1000 K, before a linear increase. A similar behavior is observed for both nickel and aluminum atoms of the composite particle. The minimums for aluminum and nickel atoms are 820 and 1050 K, respectively. The phenomena is believed to be associated with the "hiccup" profile of temperature and pressure for Ni/Al nanolaminate found by Zhao *et al.* [53, 54]. It is due to a synergy between the mixing effort of Ni and Al and a disordering effect under continuous heating. Under the NVT scheme, the system undergoes a thermostat and the temperature of the whole system increases under heating conditions. At relatively low temperatures, the atomic mobility is increased by the external heat source. Consequently, owing to mutual diffusion of two elements and the constraint of a constant whole volume, a two-way movement is caused, that is, nickel atoms move toward the core and aluminum atoms toward the surface. Such a two-way movement introduces the mixing of Ni and Al, which have an exothermic alloying nature. Meanwhile, the external heat source keeps a solid structural disordering. The released energy will either go toward melting that makes atoms disorder or increasing atomic mobility that elevates the temperature. The V-shape of potential energy (Figure 2.20) implies the dominance of melting latent heat. The enthalpy value is calculated directly from its thermodynamic definition of $\Delta H = U + PV$, where U is the total energy, P is pressure, and V is volume. The initial potential energy decrease is mainly due to the substitution of Ni–Ni and Al–Al interaction by Ni–Al interaction, which is confined in the fixed volume. Such a scenario is different from the increase in potential energy for pure element, where the local order of atoms dominates atomic interaction. As the intermixing between Ni and Al is spontaneously driven by a negative ΔH, the enthalpy of this process is a convex function of relative concentration of Ni–Al, which is consistent with other studies [55, 56]. Clearly, the minima in the potential energy are related to the melting phenomena. If one were to take the temperature at the minimum potential energy as a result of global melting phenomenon, the global melting of the nickel-coated aluminum particle is 1020 K, and the aluminum core is ~820 K and nickel shell is 1050 K. As described previously, the pure bare aluminum with the same diameter of the aluminum core, 4 nm, has a melting temperature of 620 K. The core/shell structure thus enhances the thermal stability of the aluminum core.

The minimums for pure aluminum and nickel atoms are ~800 and ~1050 K respectively. The decline in the potential energy and other related thermal properties at low temperature is different from that of single-element particles, which is believed to be associated with the volume reduction due to different diffusion capability of nickel and aluminum atoms. Apart from being temperature sensitive, the potential energy is also a strong function of volume. The initial decline of the average potential energy is believed to be correspondent to the contraction of the whole particle.

During the heating process, the internal pressure builds up with the increase in temperature and the maximum is obtained around the global melting temperature, Figure 2.21. The local stresses are calculated by averaging the per-atom stress tensor in a group over a 2.0 ps interval, which shows a large variation over the temperature, that is, from 0.38 to 0.02 GPa. The per-atom stress tensor for each atom has a general form shown as the following equation [57]

$$
\begin{aligned}
S_{ab} = -\Bigg[& mv_a v_b + \frac{1}{2}\sum_{n=1}^{N_\mathrm{p}} \left(r_{1a}F_{1b} + r_{2a}F_{2b}\right) + \frac{1}{2}\sum_{n=1}^{N_\mathrm{b}} (r_{1a}F_{1b} + r_{2a}F_{2b}) + \\
& \frac{1}{3}\sum_{n=1}^{N_\mathrm{a}} (r_{1a}F_{1b} + r_{2a}F_{2b} + r_{3a}F_{3b}) + \frac{1}{4}\sum_{n=1}^{N_\mathrm{d}} (r_{1a}F_{1b} + r_{2a}F_{2b} + r_{3a}F_{3b} + r_{4a}F_{4b}) + \\
& \frac{1}{4}\sum_{n=1}^{N_\mathrm{i}} \left(r_{1a}F_{1b} + r_{2a}F_{2b} + r_{3a}F_{3b} + r_{4a}F_{4b}\right) + \sum_{n=1}^{N_\mathrm{f}} r_{i_a}F_{i_b} \Bigg]
\end{aligned}
\tag{2.27}
$$

where a, b take on values x, y, z to go through the six components of symmetric tensor: xx, yy, zz, xy, xz, and yz. The first term is a kinetic energy contribution for atom i. The second term is a pairwise energy contribution where n loops over the N_p neighbors of atom i, r_1, and r_2 are the positions of the two atoms in the pairwise interaction, and F_1 and F_2 are the forces on the two atoms resulting from the pairwise interaction. The third term is a bond contribution of similar form for the N_b bonds that atom i is part of. There are similar terms for the N_a angle, N_d dihedral, and N_i improper interactions atom i is part of. Finally, there is a term for

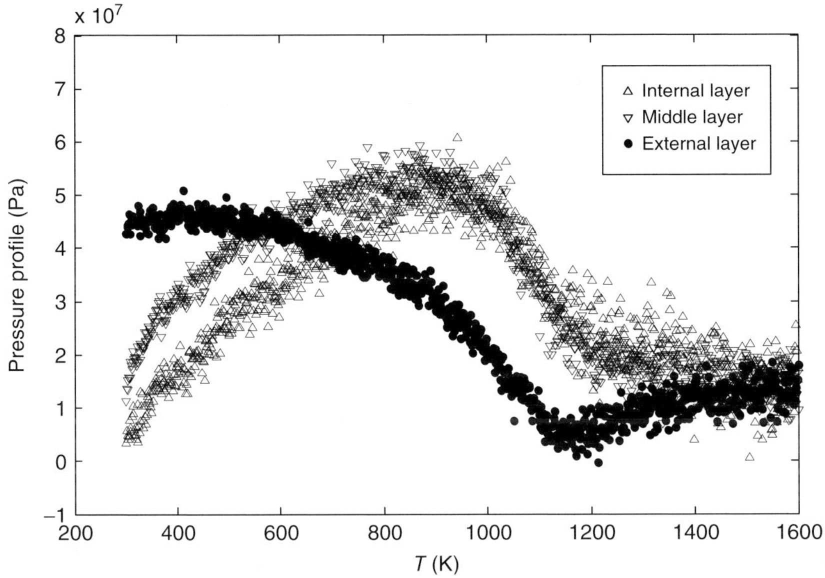

Figure 2.21 Pressure profile for different regimes of Al core.

the N_f fixes that apply internal constraint forces to atom i. For the case of metal system, only the first three terms are included.

The pressure peak of the inner aluminum occurs at 950 K, just before the defined "global melting point" of 1000 K. Consecutively, the pressure declined to reach a plateau of 0.1 GPa. Compared to other studies, that is, 1.0 GPa pressure for a 10 nm radius aluminum core wrapped in alumina [58], and a Ll_2 2315-atom Ni_3Al cluster displayed a pressure up to 5 GPa [18], the MD calculated values are quite small. As shown in Figure 2.20, the migration of molten Al atoms into the solid nickel shell is initiated before global melting of core, and meanwhile the melting of Ni of 0.5 nm in thickness occurs near 1000 K, close to the whole melting point. At temperatures higher than the global melting temperature, both the core and shell are in molten status, which would not produce a high pressure. The competition between alloying reaction and melting, as discussed previously, will ease the pressure increase associated with the phase change. So it is believed that the pressure peak is due to the maximum thermal expansion of solid atoms where the functional particle has shrinking tendency to its minimum volume, just before the global melting, and the decreasing pressure is associated with the global melting process.

2.6.2.1 **Solidification Simulation**

The solidification simulation continues after heating the composite particle to 1600 K. As a long time heating (400 ps) has been performed, a sufficient mixing of aluminum and nickel atoms is obtained. An additional relaxation time of 40 ps at

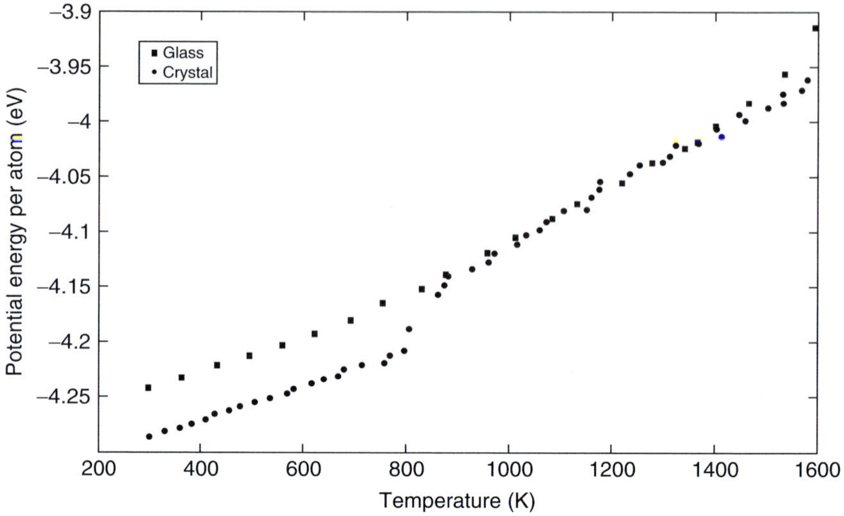

Figure 2.22 Potential energy profile under two cooling rates for Ni–Al particle.

1600 K is conducted before implementing the cooling process. Cooling rates from 6.5 and 0.13 K ps^{-1} were studied.

The average potential energy per atom for two cooling rates is shown in Figure 2.22. With the decrease in temperature, the fast cooling rate results in a smooth decrease in the potential energy, which results in a glassy phase formation. The slow cooling simulation produced an abrupt decrease in the potential energy, beginning at 900 K and finishing at 800 K, a clear indication of the formation of crystal structure. This correlates well with the melting studies as in the heating case.

Radial distribution functions (RDFs) can be employed to monitor particle structure evolutions during the simulation. RDF describes the structural correlation of how the atomic density varies as a function of the distance from one particular atom, which provides a statistical estimation of material local structure. Its general form in spatially homogeneous systems is

$$g(r) = \frac{2V}{N_{\mathrm{m}}^2}\left\langle \sum_{i<j} \delta\left(\vec{r} - \vec{r}_{ij}\right)\right\rangle \tag{2.28}$$

where N_{m} is the numbers of total atom of interest, V is the volume of the atom assembly. The RDF is shown in Figure 2.23a,b for two cooling cases.

For the quenching simulation, the RDF peaks, whose height and positions reflect the local order of structure, increase as the temperature decreases. For the cooling rate of 6.5 K ps^{-1}, the second peaks are not remarkable at high temperatures above ~900 K. Below ~900 K, more peaks besides the primary peak start to emerge and grow. A split of the second peak into two sub-peaks occurs at ~600 K, which is an indication of the glass phase formation [59]. The sub-peaks develop subsequently with the decrease in temperature. The glass formation during fast quenching is

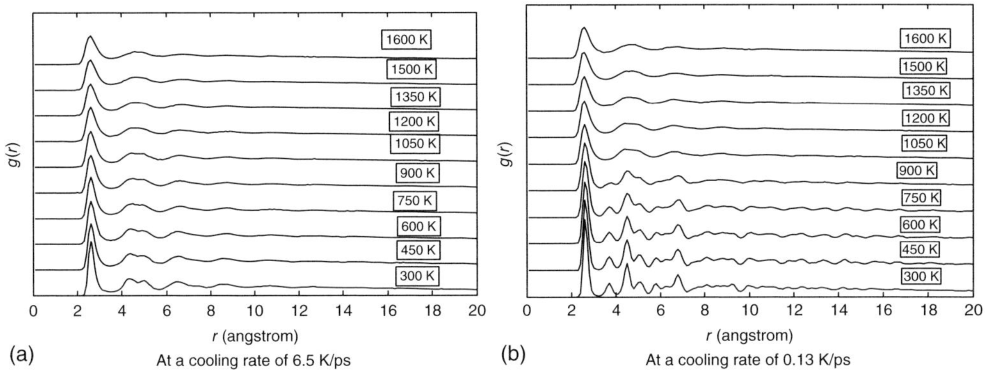

Figure 2.23 RDF of the functional particle under two cooling rates. (a) At a cooling rate of 6.5 K ps^{-1}. (b) At a cooling rate of 0.13 K ps^{-1}.

also reflected by its thermodynamics properties, that is, the decrease in the average potential energy linearly without an abrupt change as shown in Figure 2.22. The glass formation temperature is derived as $T_g = 579$ K.

For the slow solidification of 0.13 K ps^{-1}, RDF displays features of transition from amorphous liquid to ordered crystal. The solidification point is clearly identified at 900 K by the abrupt decrease in potential energy (Figure 2.22). The RDF profile here illustrates the detailed phase evolution during the quench process. The first RDF peak location on the curve corresponds to the nearest interatomic distance, which is estimated as 2.59 Å at 700 K in this study. The location of the second peak, which corresponds to the alloy lattice parameter, is found to be 2.9 Å, which agrees very well with the experimental result of 2.88 Å [60]. This proves that the final structure is Ll_2 martensite, NiAl, an FCC crystal. The RDF profile is also in good agreement with the others for Ni–Al alloy under periodic boundary conditions [61]. On the basis of the phase diagram of Ni–Al, the phase of $Ni_{3620}Al_{2016}$ at 700 K appears to be mainly in the form of Ni_5Al_3. However, different from Kazanc's result [61], there is no autensite phase found during the quench process, which might be caused by insufficient time of nucleation in the relatively short cooling. From the RDF profile, it shows that the martensite nucleated directly from the molten liquid. It was believed that the difference is caused by different atom distribution at the nanoscale.

2.7 Chapter Summary

This chapter aims to provide a fundamental understanding of energetic nanomaterials at the individual particle level based on MD simulations. Related to the ignition and combustion of nanoparticles, this chapter examines the size-dependent properties of individual nanoparticles, the sintering, and agglomeration of metallic nanoparticles, the oxidation of nanoparticle in the presence of oxygen, and the heating and cooling of passivated core–shell nanoparticles. MD is shown to be a powerful tool to study energetic nanomaterials. It repeats well the classical size-dependent property phenomenon and can predict the complicated thermochemical reactions occurring at the nanoscale. In a short summary, the chapter shows the following aspects.

1) The melting temperature of particles at the nanoscale is strongly dependent on its particle size. Using appropriate EAM interactional potentials, MD can predict particle melting depression very well. The tracking of different regimes of nanoparticles reveals the feature of surface melting, which occurs initially at the surface, propagates inwards as heating temperature increases, and influences the subsequent thermochemical properties.
2) The MD investigation of the sintering of two different-sized nanoparticles shows that it occurs in a three-stage scenario. The first stage sintering is independent of temperature, and the late ones are driven mainly by surface diffusion and grain boundary diffusion. The activation energy during the sintering is found to be much smaller than the bulk value but is consistent with other nanoparticle studies.
3) Reactive MD, the ReaxFF, creates a creditable process of oxidation of silicon by reproducing the exothermal reaction, consistent BO, bond length, and bond angle. An atomic picture of Si–O bond configuration evolution is revealed through the MD simulation. The activation energies of bond transformation based on accounting bonding numbers are approximately estimated by the Arrhenius equation. ReaxFF can be used to describe oxidation in an atomic MD level manner with the potentials of simulating mesoscale scenarios for larger particles.
4) Establishing passivation layers around active energetic nanomaterials is a good approach to protect the active content and to engineer their subsequent thermochemical properties. The heating and cooling of a shell–-core-structured Ni–Al composite nanoparticle shows clearly a two-way diffusion phenomenon for both aluminum and nickel atoms during the heating process. A unique V-shape potential energy of the composite particle during the heating process is observed and the global melting point of the nickel-coated aluminum composite particle is identified at 1000 K, much higher than the bare aluminum for a given size. Significant pressure buildup is observed during the heating, but no violate explosion phenomenon is observed. Upon cooling, different structures can be observed depending on the cooling rate. A glass phase is formed for a fast cooling rate of 6.5 K ps^{-1}, with a glass transition temperature of 600 K, while a crystal structure of Ll_2 Ni_3Al is formed beginning at 800 K for a slow cooling rate of 0.13 K ps^{-1}.

Such information will be useful in understanding the mechanism of reactive particles at the nanoscale, being complementary to bulk scale experiments and simulation.

References

1. Verlet, L. (1967) Computer 'experiment' on classical fluids. I. Thermodynamical properties of Lennard-Jones molecules. *Phys. Rev.*, **159**, 98.
2. Andersen, H.C. (1980) Molecular dynamics simulations at constant pressure and/or temperature. *J. Chem. Phys.*, **72**, 2384.
3. Parrinello, M. and Rahman, A. (1980) Crystal structure and pair potentials: a molecular dynamics study. *Phys. Rev. Lett.*, **45**, 1196.
4. Nosé, S. (1984) A molecular dynamics method for simulations in the canonical ensemble. *Mol. Phys.*, **52**, 255.
5. Hoover, W.G. (1985) Canonical dynamics: equilibrium phase-space distributions. *Phys. Rev. A*, **31**, 1695.
6. Berendsen, B.J. (1982) in *Biological Molecules and Membranes, Molecular Dynamics of Rigid System in Cartesian Coordinates: A General Formulation*, Molecular Physics (eds G. Ciccotti and W.G. Hoover), Taylor and Francis Limited, Abingdon.
7. Andersen, O.K. and Jepsen, O. (1984) Explicit, first-principle tight-binding theory. *Phys. Rev. Lett.*, **53**, 2571–2574.
8. Daw, M.S. and Baskes, M.I. (1984) Embedded-atom method: derivation and application to impurities, surfaces, and other defects in metals. *Phys. Rev. B*, **29** (12), 6443–6453.
9. Finnis, M. and Sinclair, E. (1984) A simple empirical *N*-body potential for transition metals. *Phil. Mag. A.*, **5**, 45–55.
10. Angelo, J.E., Moody, N.R., and Baskes, M.I. (1995) Trapping of hydrogen to lattice defects in nickel. *Modell. Simul. Mater. Sci. Eng.*, **3**, 289–307.
11. Jin, Z.H. and Lu, K. (1998) Melting of surface-free bulk single crystals. *Philos. Mag. Lett.*, **78**, 29–35.
12. Mei, Q.S. and Lu, K. (2007) Melting and superheating of crystalline solids: from bulk to nanocrystals. *Prog. Mater. Sci.*, **52**, 1175–1262.
13. Song, P. and Wen, D. (2010) Molecular dynamics simulation of the sintering of metallic nanoparticles. *J. Nanopart. Res.*, **12**, 823–829.
14. Andersen, P.W. (1997) Basic notions of condensed matter physics, Perseus Publishing.
15. Foiles, S.M. (1985) Application of embedded-atom method to liquid transition metals. *Phys. Rev. B*, **32**, 3409–3415.
16. Jose-Yacaman, M., Gutierrez-Wing, C., Miki, M., Yang, D.Q., Piyakis, K.N., and Sacher, E. (2005) Surface diffusion and coalescence of mobile metal nanoparticles. *J. Phys. Chem. B*, **109**, 9703–9711.
17. Alavi, S. and Thompson, D.L. (2006) Molecular dynamics simulations of the melting of aluminum nanoparticles. *J. Phys. Chem. A*, **110**, 1518–1523.
18. Zhurkin, E.E., Hautier, G., and Hou, M. (2006) Stress distribution and mechanical properties of free and assembled Ni_3Al nanoclusters. *Phys. Rev. B*, **73**, 094108.
19. Cao, L.F., Xu, G.F., Xie, D., Guo, M.X., Luo, L., Li, Z., and Wang, M.P. (2006) Thermal stability of Fe, Co, Ni metal nanoparticles. *Phys. Status Solidi B*, **243**, 2745–2755.
20. Weingarten, N.S., Mattson, W.D., and Rice, B.M. (2009) Determination of the pressure dependent melting temperatures of Al and Ni using molecular dynamics. *J. Appl. Phys.*, **106** (6), 063524.
21. Koch, W. and Friedlander, S.K. (1990) The effect of particle coalescence on the surface area of a coagulating aerosol, *J. Colloid Interface Sci.*, **140**, 419–217.
22. Patterson, R.I.A. and Kraft, M. (2007) Models for the aggregate structure of soot particles. *Combust. Flame*, **151**, 160–172.
23. West, R.H., Celnik, M.S., Inderwildi, O.R., Kraft, M., Beran, G.J.O., and Green, W.H. (2007) Toward a comprehensive model of the synthesis of TiO_2 particles from $TiCl_4$. *Ind. Eng. Chem. Res.*, **46** (19), 6147–6156.
24. Zeng, P., Zajac, S., Clapp, P., and Rifkin, J.A. (1998) Nanoparticle sintering simulations. *Mater. Sci. Eng. A*, **252**, 301–306.
25. Arcidiacono, S., Bieri, N.R., Poulikakos, D., and Grigoropoulos, C.P. (2004) On

the coalescence of gold nanoparticles. *Int. J. Multiphase Flow*, **30**, 979–994.

26. Hawa, T. and Zachariah, M.R. (2007) Molecular dynamics simulation and continuum modelling of straight-chain aggregate sintering: Development of a phenomenological scaling law. *Physical Review B*, **76**, 054109.
27. Panigrahi, B.B. (2007) Sintering and grain growth kinetics of ball milled nanocrystalline nickel powder. *Mater. Sci. Eng. A*, **460-461**, 7–13.
28. Koparde, V. and Cummings, P. (2005) Molecular dynamics simulation of titanium dioxide nanoparticle sintering. *J. Phys. Chem.*, **109**, 24280–24287.
29. Koparde, V. and Cummings, P. (2007) Sintering of titanium dioxide nanoparticles a comparison between molecular dynamics and phenomenological modelling. *J. Nanopart. Res.*, **10**, 11069–1182.
30. Sander, M., West, R., Celnik, M., and Kraft, M. (2009) A detailed model for the sintering of polydispersed nanoparticle agglomerates. *Aerosol Sci. Technol.*, **43**, 978–989 (article).
31. Ngoc, H.N., Henning, R., and Wen, J.Z. (2011) Molecular dynamics simulation of iron nanoparticle sintering during flame synthesis. *J. Nanopart. Res.*, **13**, 803–815.
32. Lu, K., Liang, Y.X., and Li, W.L. (2012) Hindered sintering behaviors of titania nanoparticle-based materials. *Mater. Lett.*, **89**, 77–80.
33. Wakai, F. (2006) Modelling and simulation of elementary processes in ideal sintering. *J. Am. Cerm. Soc.*, **89**, 1471–1484.
34. Hawa, T. and Zachariah, M.R. (2006) Coalescence kinetics of unequal sized nanoparticles. *Aerosol Sci.*, **37**, 1–15.
35. Tsyganov, S., Kastner, J., Rellinghaus, B., and Kauffeldt, T. (2007) Analysis of Ni nanoparticle gas phase sintering. *Phys. Rev. B*, **75**, 045421.
36. van Duin, A.C.T., Dasgupta, S., Lorant, F., and Goddard, W.A. III, (2001) ReaxFF: a reactive force field for hydrocarbons. *J. Phys. Chem. A*, **105** (41), 9396–9409.
37. van Duin, A.C.T., Strachan, A., Stewman, S., Zhang, Q., and Goddard, W.A. III, (2003) $ReaxFF_{SiO}$ reactive force field for silicon and silicon oxide systems. *J. Phys. Chem. A*, **107**, 3803–3811.
38. van Duin, A.C.T., Merinov, B., and Goddard, W.A. III, (2008) The ReaxFF reactive force field for solid oxide fuel cell systems with application to oxygen ion transport in yttria-stabilized zirconia. *J. Phys. Chem. A*, **112**, 3133–3140.
39. Beddrov, D., Smith, G., and van Duin, A.C. (2012) Reactions of singly-reduced ethylene carbonate in lithium battery electrolytes: a molecular dynamics simulation study using the reaxFF. *J. Phys. Chem.*, **116**, 2978–2985.
40. Goddard, W.A. III, (2008) Development and application of a ReaxFF reactive force field for oxidative dehydrogenation on vanadium oxide catalysts. *J. Phys. Chem. C*, **112**, 14645–14654.
41. Fogarty, J.C., Aktulga, H.M., Grama, A.Y., van Duin, A.C.T., and Pandit, S.A. (2010) A reactive molecular dynamics simulation of the silica-water interface. *J. Chem. Phys.*, **132**, 174704.
42. Shin, Y.K., Kwak, H., Zou, C.Y., Vasenkov, A.V., and van Duin, A.C.T. (2012) Development and validation of a ReaxFF reactive force field for Fe/Al/Ni alloys: molecular dynamics study of elastic constants, diffusion, and segregation. *J. Phys. Chem.*, **116**, 12163–12174.
43. Kim, S., Yl van Duin, A.C.T., and Kubicki, J.D. (2013) Molecular dynamics simulations of the interactions between TiO_2 nanoparticles and water with Na + and Cl-, methanol, and formic acid using a reactive force field. *J. Mater. Res.*, **28**, 513–520.
44. Schbert, B., Avouris, P., and Hoffmann, R. (1993) A theoretical study of the initial stages of Si(111)-7 × 7 oxidation. I. The molecular precursor. *J. Chem. Phys.*, **98** (9), 7593–7605.
45. Hémeryck, A., Estève, A., Richard, N., Rouhani, M.D., and Chabal, Y.J. (2009) Fundamental steps towards interface amorphization during silicon oxidation: density functional theory calculations. *Phys. Rev. B*, **79**, 035317.
46. Teraishi, K., Endou, A., Gunji, I., Kubo, M., Miyamoto, A., and Kitajima, M. (1999) Use of umbrella sampling in the calculation of the potential of the mean

force for silicon surface oxidation. *Surf. Sci.*, **426**, 290–297.

47. Torre, J.D., Bocquet, J.L., Limoge, Y., Crocombette, J.P., Adam, E., Martin, G., Baron, T., Rivallin, P., and Mur, P. (2002) Study of self-limiting oxidation of silicon nanoclusters by atomistic simulations. *J. Appl. Phys.*, **92** (2), 1084–1094.
48. Hwang, I.S., Lo, R.L., and Tsong, T.T. (1998) Mechanisms and energetics of site hopping and chemical reactions of O_2 molecules at Si(111)-7×7 surfaces. *Surf. Sci.*, **399**, 173–189.
49. Atkinson, A. (1985) Transport processes during the growth of oxide films at elevated temperature. *Rev. Mod. Phys.*, **57**, 437–470.
50. Song, P., Ding, Y., and Wen, D.S. (2013) A reactive molecular dynamics simulation of oxidation of a silicon nanocluster. *J. Nanopart. Res.*, **15** (1), 1–11.
51. Song, P. and Wen, D. (2010) Molecular dynamics simulation of a core-shell structured metallic nanoparticle. *J. Phys. Chem. C*, **114**, 8688–8696.
52. Morse, P.M. (1929) Diatomic molecules according to the wave mechanics. II. Vibrational levels. *Phys. Rev.*, **34**, 57–64.
53. Zhao, S., Germann, T.C., and Strachan, A. (2006) Atomistic simulations of shock-induced alloying reactions in Ni/Al nanolaminates. *J. Chem. Phys.*, **125**, 164707.
54. Zhao, S., Germann, T.C., and Strachan, A. (2007) Melting and alloying of Ni/Al nanolaminates induced by shock loading: a molecular dynamics simulation study. *Phys. Rev. B*, **76**, 104105.
55. Witusiewicz, V.T., Arpshofen, I., Seifert, H.J., Sommer, F., and Aldinger, F. (2000) Thermodynamics of liquid and undercooled liquid Al–Ni–Si alloys. *J. Alloys Compd.*, **305**, 157–171.
56. Ma, E. (2005) Alloys created between immiscible elements. *Prog. Mater. Sci.*, **50**, 413–509.
57. Plimpton, S., Crozier, P. and Thompson, A. (2010) *User Manual of LAMMPS*, Sandia Corporation. *http://lammps.sandia.gov/doc/Manual.pdf* (accessed 16 October 2013).
58. Campbell, T.J., Aral, G., Ogata, S., Kalia, R.K., Nakano, A., and Vashishta, P. (2005) Oxidation of aluminum nanoclusters. *Phys. Rev. B*, **71**, 205413.
59. Tsumuraya, K., Watanabe, M.S., and Ikeda, S.K. (1988) Molecular dynamics study of homogeneous nucleation and growth in a metal liquid. *Trans. Iron Steel Inst. Jpn.*, **28**, 869–875.
60. Miracle, D.B. (1993) Overview No. 104: the physical and mechanical properties of NiAl. *Acta Metall. Mater.*, **41** (3), 649–684.
61. Kazanc, S. and Tatar, C. (2008) Investigation of the effect of pressure on some physical parameters and thermoelastic phase transformation of NiAl alloy. *Int. J. Solids Struct.*, **45**, 3282–3289.
62. Ning, N., Calvo, F., van Duin, A.C.T., Wales, D.J., and Vach, H. (2008) Spontaneous self-assembly of silica nanocages into inorganic framework materials. *J. Phys. Chem. C*, **113** (2), 518–523.

3
Electroexplosive Nanometals

Olga Nazarenko, Alexander Gromov, Alexander Il'in, Julia Pautova, and Dmitry Tikhonov

3.1
Introduction

The methods for metal nanopowders (or nanometals, *nMe*) production can be classified into physical and chemical methods. The physical methods include the high-energy ball-milling, gas condensation method using different vaporization techniques, such as thermal evaporation, laser ablation, and microwave plasma processes [1–7]. The chemical methods include chemical vapor deposition, reduction of salts in aqueous solutions, sol–gel and hydrothermal techniques, and molecular pyrolysis [1–5, 8–10]. Some of the physical methods can include a chemical component, such as mechanochemical processing [11]. The most known among the semi-industrial methods of *nMe* production is plasma-chemical synthesis – the method based on the evaporation of micron-sized powder of metals (μMe) in the stream of high-temperature plasma – and the electrical explosion of wires (EEW) [12–22], as well as the self-propagating high-temperature synthesis (SHS) for some metals [23]. In this chapter, the EEW method is widely discussed as the most productive and propagated method for metal nanopowders, as exists at present. The *nMe* produced by EEW are used as raw materials in the majority of applications [21].

3.2
Electrical Explosion of Wires Technology for Nanometals Production

EEW is a process of explosive destruction of a metal wire under the action of a high-density current ($>10^{10}$ A m^{-2}) [20, 21, 24]. This process is accompanied by the scattering of products, generation of shockwaves, and electromagnetic irradiation. The material of the wire transmutes into the nanoparticles under certain conditions – the density of an entered energy must be high enough to qualitatively disperse the metal wire to nanoparticles. The EEW method, as one for dispersing metals, is characterized by the following physical parameters.

- Current pulse duration is 10^{-5}–10^{-7} s.
- Explosion power is $>10^{13}$ W kg^{-1}.

Metal Nanopowders: Production, Characterization, and Energetic Applications, First Edition.
Edited by Alexander Gromov and Ulrich Teipel.

- Temperature at the explosion is $\sim 10^4$ K.
- Pressure in front of the shock wave is 10^3 MPa.
- Speed of product expansion is 1–5 km s^{-1}.
- Input energy comparable with the energy of metal wire sublimation.

EEW as a method for *nMe* production combines the properties of two approaches to the production of nanopowders: the "top-down" and the "bottom-up" [25]. The "top-down" approach involves the destruction of a bulk material into nanosized structures or particles: the material of a wire is destroyed when electric pulse current passes through the wire. The "bottom-up" approach is based on the particle formation from atoms, molecules, and clusters: during electric explosion, the significant part of the wire material initially transforms into a vapor state and further the vapor condenses into particles.

EEW *nMe* have a high metal content (75–95 wt%) in comparison with *nMe* obtained by other methods [19]. Their cost is comparable to the price of *μMe* in large-scale production [19, 26]. EEW *nMe* exhibit a high chemical reactivity at high temperatures and high stability of properties during their storage. It is possible to regulate the particle size of powders and metal content by changing the parameters of electric explosion [13–17]. Selection of the conditions of *nMe* passivation provides the control of the physical and chemical properties of nanopowders [27, 28].

Intensive works on the study of EEW were started about the middle of the twentieth century. Extensive information on the exploding wire phenomenon, including work on the production of aerosols of metals and their compounds, was compiled by Chase and Moore [29]. The technology of *nMe* production by EEW was developed at Tomsk Polytechnic Institute, USSR (by Ivanov GV, Kotov YuA, Yavorovsky NA) in 1970 and spread all over the world (Russia [13–21], Germany [27, 30], USA [31, 32], Korean Republic [33, 34], Japan [22, 35], India [36, 37], China [38], etc.). Recently, EEW *nMe* were produced with a sufficiently high productivity – up to 1 ton $year^{-1}$ (for *nW*) for a single machine of UDP-4 type [21].

In the first decade of the twenty-first century, EEW *nMe*, nanoalloys, intermetallic and chemical nanocompounds were produced by several companies and institutions in Russia ("Advanced Powder Technology LLC," High-Voltage Institute of Tomsk Polytechnic University, Institute of Electrophysics Ural Branch of Russian Academy of Science, Institute of High Current Electronics Siberian Branch of Russian Academy of Science, etc.), USA ("Argonide Corporation"), Western Europe (Forschungszentrum Karlsruhe), Japan (Nagaoka University of Technology), South Korea (Korea Advanced Institute for Science and Technology – KAIST, Korean Electric Research Institute, University of Ulsan), and China (Jilin University, etc.).

3.2.1
The Physics of the Process of Electrical Explosion of Wires

The EEW method can be realized by using the electrical circuit, including capacitive and inductive components (LC circuit). The electrical LC circuit for *nMe* production by the EEW method is shown in Figure 3.1.

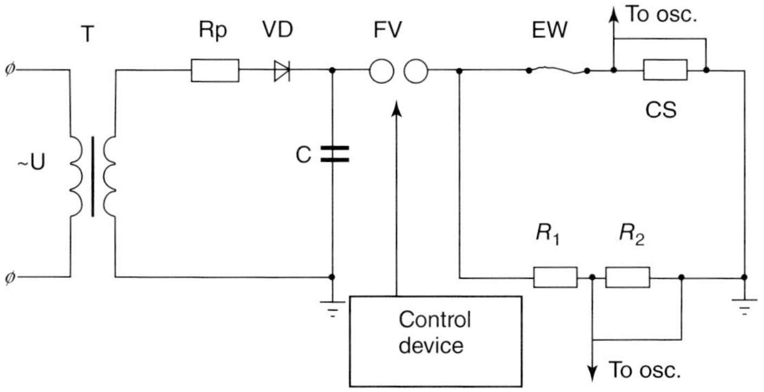

Figure 3.1 Electrical scheme of the EEW LC circuit: T, transformer; R_p, protective resistance; VD, rectifier; FV, switch; EW, exploding wire; C, energy capacitor; CS, current shunt; and R_1, R_2 – voltage divider.

When the switch "FV" closes, the capacitor "C" is discharged on the exploding wire "EW." The diagnostics of electrical parameters are based on the oscillographic recording of the current in the electric circuit and the voltage drop on the exploding wire. Metal wires with a metal content of 99.8 wt% are normally used for *nMe* production.

The energy of the wire explosion process is stored in the capacitor and then converted into the work of heating the wire material in a solid state, a metal melting and heating before the start of vaporization, the expansion of a wire material, the scattering of primary products, and the formation of final particles. After the electric current has passed through a metallic wire, the following primary products are formed: ~30% of ions; ~30% of atoms, molecules, and clusters (gas-like phase); ~40% of drops (superheated liquid-like state) [24, 39]. As a result of the scattering and cooling, the final products have generally a three-modal particle size distribution (Figure 3.2).

The finest fraction (30–50 nm) is formed by the condensation from the gas-like phase, the middle fraction (50–500 nm) – by the liquid-like phase coalescence, and the largest particles are formed because of the decay of the wire ends into large

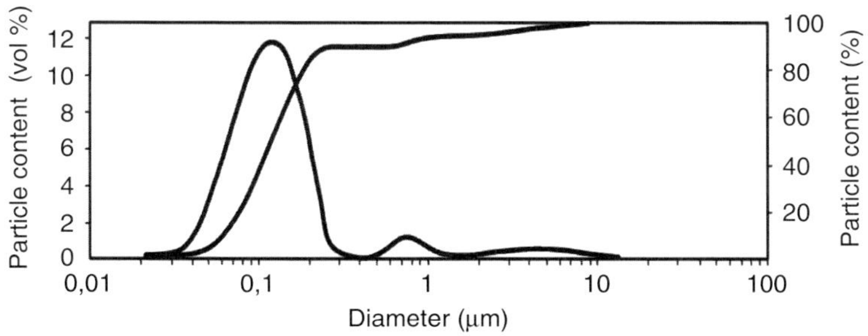

Figure 3.2 Particle size distribution of molybdenum powder produced by EEW, $e/e_s = 0.7$.

droplets. The number of nanosized particles can reach 90% of the total number of particles produced by EEW. But the major part of the powder by weight is the powder consisting of micron-sized particles, which are only a small percentage of the total number of all particles. For example, the mass of the fraction of micron-sized particles (1–3 μm) was ~68 wt% of the total mass for *nAl* sample ($S_{sp.} = 12\,m^2\,g^{-1}$); however, the number of micron-sized particles was only 2% of the total number of the particles [40].

The main parameter determining the particle size of the produced powder is the specific energy consumption or the reached overheat e/e_s [13]. It is the ratio of the volume density of energy e to the energy of sublimation of wire material e_s. e is the ratio of the energy consumed by the wire before explosion to initial volume of the wire. When e/e_s is <0.3–0.7, the dispersion of a liquid wire material takes place mainly. It is due to the development of different inhomogeneities during EEW. When e/e_s is 1.0–1.7, the formation of particulate composition occurs by both mechanisms. When e/e_s is >2, the dispersion of the powder is mainly determined by the mechanism of the condensational formation of particles.

3.2.2
Nonequilibrium State of EEW Products – Nanometals

The formation of *nMe* in the EEW process occurs under strongly nonequilibrium conditions – the large gradients of temperature and speed of cooling (> $10^7\,K\,s^{-1}$), intensive heat and mass transfer, ionization, and influence of electric and magnetic fields. The extreme conditions of the EEW process provide the thermodynamically nonequilibrium structure of *nMe*, which is steady enough at the same time, and the presence of the excess stored energy [20, 41, 42].

During the fast explosion, a metal passes into a nonequilibrium state, which is characterized by phase and temperature nonequilibrium. The thermal nonequilibrium is the result of an electric current passing through a metal wire. The wire destruction leads to the electronic subsystem excitation (up to the temperature of 10^6 K), but the atomic subsystem is colder (10^4 K). The phase nonequilibrium is that the primary products are vapor (clusters), plasma, and overheated liquid drops, which cannot coexist under equilibrium conditions.

Energy is dissipated through all possible channels, including ones that are forbidden under equilibrium conditions. The channels for the energy dissipation are wires heating in a condensed state, melting, evaporation, ionization, and the formation of new surfaces with a chemical bond rupture – metal disintegration on clusters [20, 42].

The formation of a new surface with the rupture of chemical bounds (metal clusterization) is of considerable importance under nonequilibrium conditions. The energy of such a process increases with the decrease in cluster size. The most rapid increase in energy is observed if the cluster diameter is less than 10 nm [41, 42]. The most energy-intensive process for most metals is ionization. For example, the latent heat of melting is $10.8\,kJ\,mol^{-1}$, the heat of sublimation is $327.0\,kJ\,mol^{-1}$, and the first ionization potential is $577.6\,kJ\,mol^{-1}$ for Al. Equilibrium does not

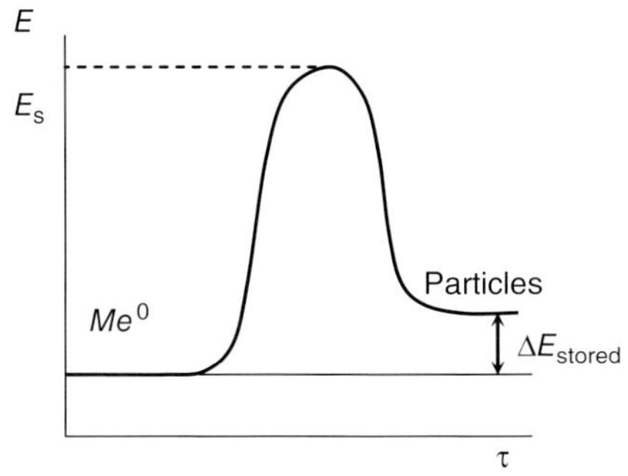

Figure 3.3 Nonequilibrium state of the EEW products: Me^0, initial metal; E_s, energy of metal sublimation; and ΔE_{stored}, stored energy in metal.

attain between the processes under the EEW conditions; it means that the metal exists simultaneously in different energy states [42].

The final products of electrical explosion process in a chemically inert medium are *nMe*. Nonequilibrium processes appear during the preservation of the excess stored energy in the metal during its cooling (Figure 3.3).

nMe are the end products of EEW in a chemically inert medium. Complete relaxation of the obtained powder does not take place in the powder cooling stage. The part of the produced energy is "frozen" in the form of stored energy of a surface, internal defects, and charge states. Under certain conditions, the value of the stored energy can exceed the standard heat of the metal melting [42]. The nonequilibrium of the processes occurring during powder cooling results in the saving of the stored energy in the metal.

Physical parameters of EEW predetermine the *nMe* properties. The characteristics of the EEW *nMe* are totally different from the ones for the *nMe* of the same particle size, but produced by other methods.

In the final stage of the formation of particles during the EEW process, the rapid cooling of the products of wire dispersion leads to substantial and specific defects in the crystal lattices of all types of solid substances – products of EEW. The phases with a low X-ray density (with maximum specific volume, e.g., γ-Fe, β-W, β-Ti, γ-Al_2O_3) stabilize for the metals and chemical compounds having polymorphic modifications [14, 16, 17]. The lattices of metals, the polymorphism of which is not known (Al and Cu), are characterized by static displacement of atoms versus their equilibrium positions [13].

3.2.3 The Equipment Design for *nMe* Production by Electrical Explosion of Wires Method

The functional design of the machine for nanopowder production is shown in Figure 3.4. The machine design is based on LC circuit (Figure 3.1) and has the following action consequences. The capacitor battery 2 is charged from the high-voltage power source 1. The wire driving mechanism 3 is used for automatic feed

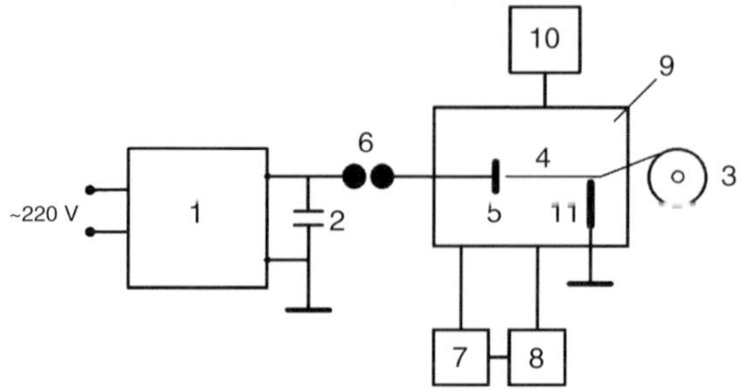

Figure 3.4 Functional scheme of the UDP-4 machine for *nMe* production: 1, high-voltage power source; 2, capacitor battery; 3, wire supply unit; 4, exploding wire; 5, high-voltage electrode; 6, switch; 7, powder collector; 8, ventilator; 9, explosion chamber; 10, gas filling system; and 11, ground electrode.

of the exploding piece of a wire 4 in the gap between the high-voltage electrode 5 and the ground electrode 11. The wire passes through the gap between the electrodes. When the wire reaches the high-voltage electrode 5, the switch 6 closes, the electric discharge of the capacitor battery occurs on this part of the wire, and the wire explodes. The explosion cycle can be repeated with a frequency up to 1.5 Hz. The obtained powder collects in the powder collector 7. The gas cleaned from the powder is given back in the discharge chamber 10 by means of the ventilator 8. The explosion chamber 9 is vacuumed before working, and then the chamber is filled with the working gas by means of the system 10.

The average characteristics of the *nAl* produced by the UDP-4 machine (Figure 3.5) are given in Table 3.1 [14]:

Figure 3.5 UDP-4 machine for *nMe* production: 1, explosion chamber; 2, powder filter; 3, powder collection; and 4, high-voltage power supply.

Table 3.1 Conditions for *nAl* obtained by the EEW method.

Parameter	Value
Wire material (purity) (%)	*Al (99.8)*
Voltage during the explosion (kV)	*18–30*
Input energy e/e_s (au)	*1.2–2.5*
The capacity of capacitor bank (μF)	*2.17*
The electrical resistance of the LC circuit (Ω)	*0.07*
Wire diameter (m)	$(0.3–0.4)\cdot10^{-3}$
Length of the exploding segment of a wire (m)	0.03–0.11
The gas pressure in the explosion chamber (MPa)	*0.2–0.4*
LC circuit inductivity (μH)	0.5
Frequency of explosions (Hz)	0.8–1.5
Productivity ((g *nAl*) h^{-1})	~50

At present, the UDP-4 and its modification machines are equipped with an automatic feed of metal wires of Al, Cu, Zn, Sn, Ti, Zr, In, Fe, Ni, Mo, W, and some alloys.

3.2.4
Comparative Characteristics of the Technology of Electrical Explosion of Wires

The method of EEW has a lot of advantages compared to others for *nMe* production [19, 21]. The important advantage of the electro-explosive technology is the possibility of the control over the properties of EEW products, including particle size, phase, and chemical composition of *nMe* by electrical parameters.

This method is characterized by low energy consumption – less than 20 kW h^{-1} per kilogram of powder, owing to the direct heating of the wire by electric current without additional heating equipments and the high heating rate ($> 10^7$ K s^{-1}) that provides quasiadiabatic conditions of energy transmission to a wire.

The productivity of a single EEW machine is high enough, for example, the yield of *nAl* is 50 g h^{-1} and *nW* is 300 g h^{-1}. EEW in inert gases or hydrogen is used to produce the nanopowders of metals, alloys, and intermetallic compounds [19–21, 43]. EEW in chemically active medium is used to produce nanopowders of chemical compounds [21, 44].

The electro-explosive technology, unlike other technologies, allows using the same installation to produce nanopowders of metals, alloys, intermetallic compounds, and chemical compounds depending on the wire composition and gas medium in the explosion chamber. It is environmentally safe: the process of nanopowder production is carried out in a closed chamber and there is no technological emission.

3.2.5
The Methods for the Regulation of the Properties of Nanometals Produced by Electrical Explosion of Wires

Particles of *nMe* produced by EEW have a spherical shape (Figure 3.6), but they are polydispersed systems at the same time (Figure 3.2) [19–21]. The range of *nMe* particle size is wide – from 5 up to 1000 nm.

Particle size is one of the most important parameters that determines technical characteristics of a powder (apparent density, size distribution, flowability, angle of repose, reaction capacity, and others), and its application. When the size of metal particles becomes less than 100 nm, the reactivity increases, but the metal content in particles decreases simultaneously. Another problem of the nanostate is the agglomeration caused by active particle surfaces. Nanopowders with the particle sizes less than 30 nm are unstable toward oxidation, sintering, and agglomeration. The nanopowders of such a size sinter at room temperature in an inert medium by diffusion mechanism; their interaction is accompanied by an explosion during heating and in chemically reactive mediums. Thus, the problem of *nMe* production with high particle dispersity and narrow size distribution is relevant. On the basis of the investigations carried out, it is possible to define the following methods for the decrease in size and agglomeration degree of the electro-explosive nanopowders:

- increase in the input energy e/e_s;
- decrease in the working gas pressure;
- addition of chemically active gases or reagents to inert gas-medium by explosion;
- decrease in the wire diameter;
- change of the gas medium in the explosion chamber.

The particle size and phase composition of *nMe* depend on all the conditions of explosion – first, on the electrical parameters (the energy consumed by a wire before explosion, the energy of the arc stage, the speed of the input energy [20,

(a) (b)

Figure 3.6 SEM image of *nAl* (a) and *nW* (b) produced by EEW.

21]). The geometry of a metal wire (length and diameter), its microstructure and substructure, and also the environmental properties – pressure, type of a gas, and density (viscosity) – have an influence on the dispersity and other properties of nanopowders. Moreover, the properties of electro-explosive nanopowders are highly dependent on the passivation conditions [45, 46].

The input energy into a wire e/e_s is the most important parameter determining the degree of nanopowder dispersity and the particle size distribution function [13, 20, 21, 32]. The size of the particles produced by EEW decreases with increasing the energy input into a wire, and the particle size distribution becomes narrower. The arc stage of electric discharge provides an additional dispersing effect on the primary products of a wire explosion [20, 43].

Gas in the explosion chamber at high pressure has a compression effect on the explosion products and causes their sintering and coagulation. That is why the particle size increases if the value of the gas pressure rises [15, 20, 47].

The size of the *nMe* produced with the EEW technology decreases with the reduction in wire diameter [15, 16]. However, in this case, the product yield significantly decreases. It should be noted that the particle diameters of 30 nm are critical ones for metal nanoparticles, because smaller particles sinter at room and even lower temperature.

The passivation can be carried out directly in the process of an explosion. The pyrophoric property of powders reduces as well as the time required for passivation, and the technological process is simplified. The addition of chemically active gases (air, O_2, or N_2) to Ar during EEW leads to the formation of passivating films on the particles and to the increase in the dispersion of the obtained powders [21, 48]. The refractory layers on the metal particles' surface prevent their coagulation. In that case, the decrease in the particle size is the result of the reduction in coagulation and sintering at the scattering and cooling of the EEW products.

3.3 Conclusion

The formation of nanoparticles in pulsed fast processes such as EEWs occurs under strongly nonequilibrium conditions that cause some unusual properties of nanopowders. They are steady against oxidation and sintering at room temperature and characterized by high diffusion activity at their heating. The spherical shape of the obtained particles indicates their stabilization through a liquid state. Very fast cooling of molten metal after wire explosion results in significant and specific crystal imperfection.

Acknowledgments

The authors thank all enthusiastic co-contributors of the study of EEW technology for *nMe* production of the past decades for their hard work with impressive results:

Prof. A.P. Il'in, V.G. Domashenko, Dr N.A. Yavorovskii, Dr D.V. Tikhonov, Prof. Yu.A. Kotov, Dr V.S. Sedoi, Dr G.V. Ivanov, and F. Tepper.

References

1. Rempel, A.A. and Gusev, A.I. (2004) *Nanocrystalline Materials*, Cambridge International Science Publishing, Cambridge.
2. Tavakoli, A., Sohrabi, M., and Kargari, A. (2007) A review of methods for synthesis of nanostructured metals with emphasis on iron compounds. *Chem. Pap.*, **61** (3), 151–170.
3. Sergeev, G.B. (2006) *Nanochemistry*, Elsevier, Amsterdam.
4. Gleiter, H. (2000) Nanostructured materials: basic concepts and microstructure. *Acta Mater.*, **48**, 1–29.
5. Andrievski, R.A. (1998) The state-of-the-art of nanostructured high melting point compound-based materials, in *Nanostructured Materials: Science and Technology* (ed. G.M. Chow), Kluwer Academic Publishers, Dordrecht.
6. Suryanarayana, M. (2001) Mechanical alloying and milling. *Prog. Mater Sci.*, **46**, 1–184.
7. Ullmann, M., Friedlander, S.K., and Schmidt-Ott, A. (2002) Nanoparticle formation by laser ablation. *J. Nanopart. Res.*, **4** (6), 499–509.
8. Ryu, T., Sohn, H.Y., Hwang, K.S., and Fang, Z.Z. (2009) Chemical vapor synthesis (CVS) of tungsten nanopowder in a thermal plasma reactor. *Int. J. Refract. Met. Hard Mater*, **27**, 149–154.
9. Uhlmann, D.R., Teowee, G., and Boulton, J. (1997) The future of sol-gel science and technology. *J. Sol-Gel Sci. Technol.*, **8**, 1083–1091.
10. Meisel, D. (1997) Inorganic small colloidal particles. *Curr. Opin. Colloid Interface Sci.*, **2** (2), 188–191.
11. Ricceri, R. and Matteazzi, P. (2003) A study of formation of nanometric W by room temperature mechanosynthesis. *J. Alloys Compd.*, **358** (1), 71–75.
12. Vischnevetsky, II (1974) Research of hydrocarbons decomposition in pulsed electric discharges. PhD thesis, Tomsk Polytechnic University, Tomsk.
13. Yavorovsky, N.A. (1982) Electrical explosion of wires is the method of nanopowder production. PhD thesis. Tomsk Polytechnic University, Tomsk.
14. Davidovich, V.I. (1986) Development of technological process and equipments for the electro-explosive production of metal powders with low conductivity. PhD thesis. Tomsk Polytechnic University, Tomsk.
15. Lerner, M.I. (2007) Electric-explosion-produced nanopowders of inorganic materials: manufacture technology, characteristics, application fields. DSc thesis. Tomsk Polytechnic University, Tomsk.
16. Tikhonov, D.V. (2000) Electro-explosive production of nanopowders of complex composition. PhD thesis. Tomsk Polytechnic University, Tomsk.
17. Nazarenko, O.B. (2006) The processes of the production of nano-sized refractory non-metallic compounds and metals by electric explosion of wires. DSc thesis. Tomsk Polytechnic University, Tomsk.
18. Sedoi, V.S., Valevich, V.V., and Katz, J.D. (1998) Generation of aerosols by the electrical explosion of wires at reduced air pressure. *Aerosols*, **4e** (2), 48–54.
19. Kwon, Y.S., Jung, Y.H., Yavorovsky, N.A. *et al.* (2001) Ultrafine metal powders by wires electric explosion method. *Scr. Mater.*, **44**, 2247–2251.
20. Kwon, Y.-S., Kim, J.-C., Ilyin, A.P., Nazarenko, O.B., and Tikhonov, D.V. (2012) Electroexplosive technology of nanopowders production: current status and future prospects. *J. Korean Powder Metall. Inst.*, **19** (1), 40–48.
21. Nazarenko, O.B. (2005) *Electroexplosive Nanopowders: Production, Properties, Application*, Tomsk University, Tomsk.
22. Jiang, W. and Yatsui, K. (1998) Pulsed wire discharge for nanosize powder synthesis. *IEEE Trans. Plasma Sci.*, **26**, 1498–1501.
23. Nersisyan, H.H., Lee, J.H., and Won, C.W. (2003) SHS for large–scale

synthesis method of transition metal nanopowders. *Int. J. Self Propag. High Temp. Synth.*, **12** (2), 149–158.
24. Burtsev, V.A., Kalinin, N.V., and Luchinski, A.V. (1990) *Electric Explosion of Conductors and its Application in Electro-Physical Installations*, Atomizdat, Moscow.
25. Moriarty, P. (2001) Nanostructured materials. *Rep. Prog. Phys.*, **64** (3), 297–381.
26. V.Y., Ushakov, Ilyin, A.P., Nazarenko, O.B. *et al.* (1997) Ultrafine powders produced with wire electrical explosion (production and properties). Proceedings of the 1st Korea-Russia International Symposium on Science and Technology, Ulsan, Korea, pp. 167–171.
27. Böhme, R, Kempf, B, and Weisenburger, A. (2001) Production of metal submicron- and nano-powders by explosion of wires. Proceedings of the International Congress for Particle Technology PARTEC 2001, Nuremberg, Germany, p. 066 (1–8).
28. Vergara, L.I., Passeggi, M.C.G. Jr., and Ferron, J. (2002) The role of passivation in titanium oxidation: thin film and temperature effects. *Appl. Surf. Sci.*, **187**, 199–206.
29. *Exploding Wires* (1958-1968) Chase WG, Moore HK (ed.) Vol. 1–4, Plenum Press, New York.
30. Hoppé, P, Böhme, R, Frey, W *et al.* (2001) Wire explosion studies performed on DEMON-2 a pulsed power device for the production of nanopowders. Proceedings of the International Conference on Pulsed Power Applications, Gelsenkirchen, Germany, pp. F.01/1–F.01/6.
31. Simpson, R.L., Maienschein, J.L., Swansiger, R.W. *et al.* (1994) Characterization of Ultradispersed Aluminum. Report UCRL-ID-119235, Lawrence Livermore National Laboratory, Livermore, CA.
32. Tepper, F., Lerner, M., and Ginley, D. (2004) Metallic nanopowders, in *Dekker Encyclopedia of Nanoscience and Nanotechnology*, Marcel Dekker, Inc., New York, pp. 1921–1933.
33. Cho, C.H., Park, S.H., Choi, Y.W., and Kim, B.G. (2007) Production of nanopowders by wire explosion in liquid media. *Surf. Coat. Technol.*, **201**, 4847–4849.
34. Bac, L.H., Kwon, Y.S., Kim, J.S., Lee, Y.I., Kim, D.W., and Kim, J.C. (2010) Synthesis and characteristic of FeNi3 intermetallic compound obtained by electrical explosion of wire. *Mater. Res. Bull.*, **45**, 352–354.
35. Hokamoto, K., Wada, N., Tomoshige, R., Kai, S., and Ujimoto, Y. (2009) Synthesis of TiN powders through electrical wire explosion in liquid nitrogen. *J. Alloys Compd.*, **485**, 573–576.
36. Sarathi, R., Sindhu, T.K., and Chakravarthy, S.R. (2007) Generation of nanoaluminium powder through wire explosion process and its characterization. *Mater. Charact.*, **58**, 148–155.
37. Jayaraman, K., Anand, K.V., Bhatt, D.S., Chakravarthy, S.R., and Sarathi, R. (2009) Production, characterization, and combustion of nanoaluminum in composite solid propellants. *J. Propul. Power*, **25**, 471–481.
38. Wang, Q., Yang, H., Shi, J., and Zou, G. (2001) Preparation and characterization of nanocrystalline powders of Cu–Zn alloy by wire electrical explosion method. *Mater. Sci. Eng., A*, **307**, 190–194.
39. Kwon, Y.S., Ilyin, A.P., Tikhonov, D.V., Grigoriev, A.N., and Nazarenko, O.B. (2004) Electrical explosion as a metal treatment by electrical current of high power. Proceedings of the 20th International Conference on Heat Treatment, Czechia, Jihlava, pp. 419–422.
40. Korshunov, A.V. (2012) The features of disperse composition and morphology of the particles of electroexplosive metal powders. *Bull. Tomsk Polytech. Univ.*, **320** (3), 9–16.
41. Kwon, Y.S., Ilyin, A.P., Tikhonov, D.V., Nazarenko, O.B., and Yablunovskii, G.V. (2003) Mechanism of metal destruction and formation of particles in electrical explosion of wires. Proceedings of the 7th Korea-Russian International Symposium on Science and Technology, Ulsan, Korea, Vol. 1, pp. 217–220.
42. Ilyin, A.P., Nazarenko, O.B., Tikhonov, D.V., Ushakov, V.Y., and Yablunovskii, G.V. (2002) Structural and energy

processes in electrically exploded conductors. *Russ. Phys. J.*, **45**, 1176–1180.

43. Ilyin, A.P., Nazarenko, O.B., and Tikhonov, D.V. (2008) Production of tungsten powders by electrical explosion of wires. *J. Optoelectron. Adv. Mater.*, **10**, 645–648.
44. Ilyin, A.P. and Nazarenko, O.B. (2007) Carbide-nanopowders produced by electrical explosion of wires. *J. Optoelectron. Adv. Mater.*, **9**, 1521–1524.
45. Kwon, Y.S., Gromov, A.A., Ilyin, A.P., and Rim, G.H. (2003) Passivation process for superfine aluminum powders obtained by electrical explosion of wires. *Appl. Surf. Sci.*, **211**, 57–67.
46. Kwon, Y.S., Gromov, A.A., Ilyin, A.P., Ditts, A.A., Kim, J.S., Park, S.H., and Hong, M.H. (2004) Features of passivation, oxidation and combustion of tungsten nanopowders by air. *Int. J. Refract. Met. Hard Mater.*, **22** (6), 235–241.
47. Lee, Y.S., Bora, B., Yap, S.L., and Wong, C.S. (2012) Effect of ambient air pressure on synthesis of copper and copper oxide nanoparticles by wire explosion process. *Curr. Appl Phys.*, **12**, 199–205.
48. Kwon, Y.S., Ilyin, A.P., Nazarenko, O.B., and Tikhonov, D.V. (2008) Characteristics of nanopowders produced by electrical explosion of wires in argon with air additives. Proceedings of the 3rd International Forum on Strategic Technology, Novosibirsk, Russia, pp. 214–216.

4
Metal Nanopowders Production

M. Lerner, A. Vorozhtsov, Sh. Guseinov, and P. Storozhenko

4.1
Introduction

Metal powders, including aluminum, boron, magnesium, and some other metals, are widely used in energy-rich condensed systems (ERCS) such as explosives, solid propellants, and pyrotechnic compositions to improve the intended properties of these systems. The studies pursued in the past two decades have demonstrated that ERCS efficiency (in terms of their functionality) could be increased by reducing metal particle size, which results in higher reactivity of metal fuel constituents used in ERCS [1–3]. The advances in state-of-the-art technologies made it possible to produce large-scale submicron-particle powders of aluminum, boron, and other metals and alloys that are now commercially available, a fact that resulted in extensive R&D activities with ERCS-containing metal nanopowders – in Russia, Europe, the United States, and Canada.

The studies demonstrated that aluminum nanopowders, as well as other metal powders, when introduced into explosives and pyrotechnic compositions result in an increased rate of burning [4, 5] and enhanced detonation properties [5, 6].

It is known that extreme conditions of generation of nanoparticles (high temperature and process rate), which are typical for some physical methods, lead to generation of a nonequilibrium structure of dispersed phase. Such conditions are typical for pulse processes of nanopowder production with high rates of change of thermodynamic parameters of the system, particularly the electrical explosion of wires (EEW) method. EEW-based technology is of special interest as a method of production of metal nanopowders (including Al) with high chemical reactivity.

In 1946, Abrams [7] was the one to begin consistent investigation of the possibility of production of superfine powders by means of EEW. This investigation was continued by researchers from different countries, and their works have proved the possibility, in principle, of producing metal powders and a number of chemical compounds of metals (see [8–11]). Some dependencies of properties of nanopowders on EEW parameters were discovered and pilot nanopowder production equipment was created.

Metal Nanopowders: Production, Characterization, and Energetic Applications, First Edition.
Edited by Alexander Gromov and Ulrich Teipel.

The major disadvantage of the EEW technology is the low production capacity of equipment. For example, a modern facility is capable of producing no more than 300 g of Al nanopowder per hour. However, this is compensated for by certain advantages of this technology as well as by properties of the nanopowders produced.

1) High efficiency of energy transmission – energy in EEW technology is transmitted directly to metal volume by pulses and the heat loss is minimal.
2) There is flexible adjustment of process parameters and hence the characteristics of the nanopowders produced.
3) Dispersion of particles is relatively low-size compared to other physical methods.
4) EEW nanopowders show relatively stable properties under normal conditions and high reactivity in various chemical processes.
5) Versatility of the method. The use of conductive material (metal wire) of required diameter is the only limitation of EEW technology. The method makes it possible to produce a wide range of nanomaterials.
6) It has low equipment cost and is marked by its simplicity and small mass and size parameters.

Thus, EEW technology is one of the prospective methods of production of metal nanopowders, alloys, and chemical compositions and can be successfully used in the production of nanopowders for modification of properties of high-energy compounds [12–15].

Another promising nanopowder-producing technique is the plasma-arc recondensation. The process implementation involves (i) closed-loop plant configuration with integrated cooling circuit, (ii) gas-vortex-stabilized plasma-arc zone, (iii) gas-vortex particle separation and product filtration through cloth filter, and (iv) using argon as the plasma-supporting and quenching process gas [16]. One of the known processing plants has been designed by Federal State Unitary Enterprise "State Research Institute for Chemistry and Technology of Organoelement Compounds" (SSC RF GNIIChTEOS, Moscow, Russian Federation) [17]. The plant has been utilized to pursue numerous fundamental research on producing nanodisperse powders (NPs) of various materials using the originally developed chemical counterflow plasma process. Considerable technological advance has been made; in particular, laboratory-scale processes have been created to produce nanopowders with particle size adjustable within 25–100 nm.

The proposed method is distinguished by its versatility, enhanced product performance as well as a potential for obtaining high-purity nanopowders. The method enables control and adjustment of various process-flow-dependent nanopowder parameters including particle size, shape, and so on. Moreover, the method creates the unique opportunity of synthesizing a wide variety of compounds in nanodisperse condition. In particular, the method makes it possible to produce nanopowdered boron and borides of aluminum, the high-energy propellant constituents [18] as well as pyrotechnical elemental silicon [19, 20]. These nanopowders are notoriously challenging products when using EEW.

The expenditure for producing nanopowders is essentially governed by the cost of the starting materials in their compacted state.

4.2
EEW Method of Nanopowder Production

4.2.1
Electrical Explosion of Wires Phenomenon

EEW takes place when a current pulse with a density of 10^6–10^9 A cm^{-2} passes through a metal wire. The metal heats up to melting point, melts and then explodes. The process is accompanied by loss of electrical conductivity, a bright flash of light, metal dispersal, and a shockwave formed in gas medium around the wire. As a rule, an exploding conductor has a cylindrical symmetry (wire); however, some researchers use foils. Depending on the type of gas surrounding the conductor and the metal it is made of, the EEW process makes it possible to produce nanopowders of metals, alloys, chemical compounds as well as composite nanopowders. Usually, an electrical LC circuit is used for production of nanopowders; a circuit diagram is shown in Figure 4.1.

The high-voltage power supply PS charges capacitor bank C to required voltage, monitored by means of a kilovoltmeter kV. By means of discharger D, the capacitor battery C discharges to exploding wire EW, located in the reactor EC. The reactor is previously evacuated and filled up with a working gas (argon, xenon, hydrogen, etc.). The wire explodes, and current and voltage impulses are registered by means of current shunt and voltage divider. Explosion products (i.e., nanopowders) are transferred to the filter by the working gas flow through windows 1 and 2. The required gas atmosphere inside the reactor is created by gas puffing from cylinders, and the pressure is controlled by a pressure gauge.

Characteristic oscillograms of voltage and current are shown in Figure 4.2. Owing to energy absorption, the wire heats up, melts (small step on voltage oscillogram),

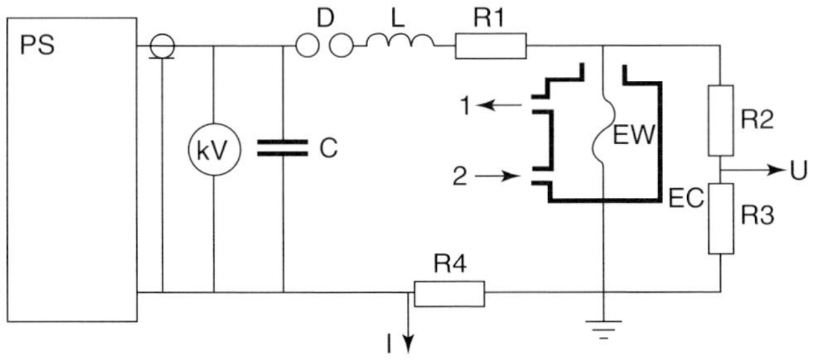

Figure 4.1 Circuit diagram of nanopowder production unit: PS, high voltage power supply; kV, kilovoltmeter; C, capacitor bank; P, discharger; L, inductance of the circuit; R_1, ohmic resistance of the circuit; R_2 and R_3, voltage divider; R_4, current shunt; EC, reactor; EW, exploding wire; 1, reactor output; 2, gas supply.

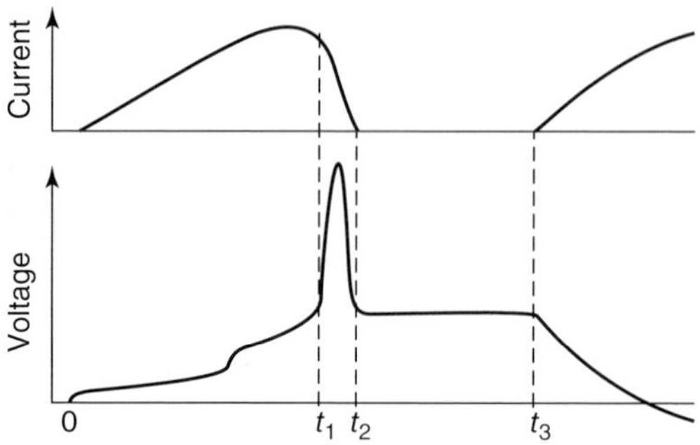

Figure 4.2 Voltage and current characteristic oscillograms for electrical explosion of wires.

and then continues to heat up in liquid state until the moment t_1. After t_1, the wire rapidly expands retaining cylindrical symmetry, loses metallic conductivity, and its resistance increases rapidly while the current rapidly decreases. There is a voltage peak between the moments t_1 and t_2 on the oscillogram. By the moment current t_2 decreases to zero, a "zero-current" pause begins. During this "zero-current" pause, voltage is applied to expanding explosion products (the time between t_2 and t_3). Explosion products while expanding lose their density and cool down. An electrical breakdown of explosion products takes place at t_3 and a secondary discharge (arc stage) develops.

Various scenarios of electrical explosion can take place depending on circuit parameters and properties of the exploding conductor. In case the residual voltage on capacitors is low, the secondary (arc) discharge may not take place and hence there will be no arc phase. In this case, a "zero-current" pause is infinite. In case the situation is quite the opposite, the "zero-current" pause may not take place and an arc discharge may develop before the moment t_2 directly during the current fall. There will be no arc discharge for explosion regimes where all the energy stored in capacitor C (excluding circuit losses) is totally consumed by heating and destruction of the metal. This type of explosion is called a *coherent explosion*. Coherent explosion, which provides higher efficiency of energy transmission to the wire, is frequently used for production of powders of pure metals and alloys. The energy content of EEW (or another term *conductor metal overheating*) depends on the type of metal, which determines the scenario of discharge process in the electrical circuit and the amount of energy stored in the circuit. Metals can be conventionally divided into three groups: metals with high electrical conductivity (Al, Cu, Ag, Sn, etc.); metals with low electrical conductivity (Fe, Ni, etc.); and metals with ionization energy less than bond energy (W, Mo, etc.).

The earlier discussion of the form of oscillograms is relevant to metals of the first group, that is, metals with high electrical conductivity. As for metals of the second group, the scenario of the discharge process and hence the oscillograms will be different. The melting conductors made of these metals significantly increase their resistance and limit current flow. As a result, there is a plateau on the oscillogram instead of a pronounced maximum peak, and the energy content of explosion of these conductors will be lower than the one for metals of the first group. As for the metals of the third group, before the input of binding energy to a wire it is covered by an arc discharge due to intensive electron emission from the surface of the wire, which blocks energy input. It is rather difficult to provide sufficient level of EEW energy content and to produce nanopowders of these metals using the EEW method.

Various approaches to the EEW classification process are described in literature. The most widely used classification is the one offered by Chace and developed by a number of researchers. Depending on the rate of energy input to a conductor (conductor current density), the process of electrical explosion is conventionally divided into the following classes.

1) "*Low-speed explosion*" – the time of energy input necessary for complete destruction of the conductor is greater than the time of development of macroscopic instabilities. A relatively small part of the metal evaporates before destruction of the conductor and the major part is sprayed in the form of drops. "Low-speed explosion" is accompanied by development of constriction-type instabilities (strata). Presumably, there are different causes for the formation of strata: initial structural inhomogeneities, capillary forces, and magnetohydrodynamic (MHD) instability. The characteristic feature of this type of explosion is that development of macroscopic instabilities begins when the conductor is still in a liquid state. Current density typical for this type of explosions does not exceed 10^7 A cm^{-2}.
2) "*High-speed explosion*" – the time of energy input is less than the time of development of macroscopic instabilities. The processes of transition of a metal from a condensed state to a two-phase state develop rapidly. In the process of expansion, explosion products retain a cylindrical symmetry without bends and constrictions and the effect of strata formation is not pronounced. "High-speed explosion," more homogeneous than the "low-speed explosion" discharge current density, exceeds 10^7 A cm^{-2}.
3) "*Ultrahigh-speed explosion*" takes place when the magnetic field does not have enough time to penetrate the inner layers of the conductor, that is, a skin effect is observed with predominant heating of the surface layers of the conductor. The homogeneity of EEW processes can be affected by the skin effect and different thermal expansion conditions on the surface of the conductor and inside its volume. The current increases with evaporation of the skin layers of the conductor and is localized near its evaporating boundary.

At present, many researchers agree that explosive destruction of a wire carrying a current is caused by processes in its volume. A foamy structure consisting of liquid metal and metallic gas bubbles is formed because of pulse heating; this structure rapidly expands, loses integrity, and cools down forming a dispersed structure. Depending on the type of gas in the EEW method, the following products are generated: metal nanoparticles (inert gas) and powders of chemical compounds (active gas).

4.2.2
Nanopowder Production Equipment

The basic design of facilities for production of metal nanopowders (including Al) by means of EEW is shown in Figure 4.3. The facility consists of two parts: high-voltage pulse generator I and nanopowder production device II.

The facility shown in Figure 4.3 operates as follows. High-voltage power supply 1 charges capacitor bank 2 to required voltage, discharger 3 is used for switching high-voltage pulse to the wire. The wire is wound on the reel in block 4. The facility is previously evacuated by means of a roughing pump 7 and filled up with argon from cylinder 6. Vacuum system 7 and gas supply system 6 are isolated from the facility by means of valves V1 and V2. Centrifugal fan 9 is turned on and provides gas circulation along the facility loop "reactor 5 – separator 11 – filter 10 – fan 9 – cooler 8 – reactor 5." After the wire feeding mechanism is turned on, the wire is fed to reactor 5. In the process of its movement inside reactor 5, the wire step-by-step closes the circuit between high-voltage and low-voltage electrodes. Discharger 3 is activated and a high-voltage pulse is applied to the wire; the wire explodes. The obtained aerosol passes with a gas flow from reactor 5 to separator 11,

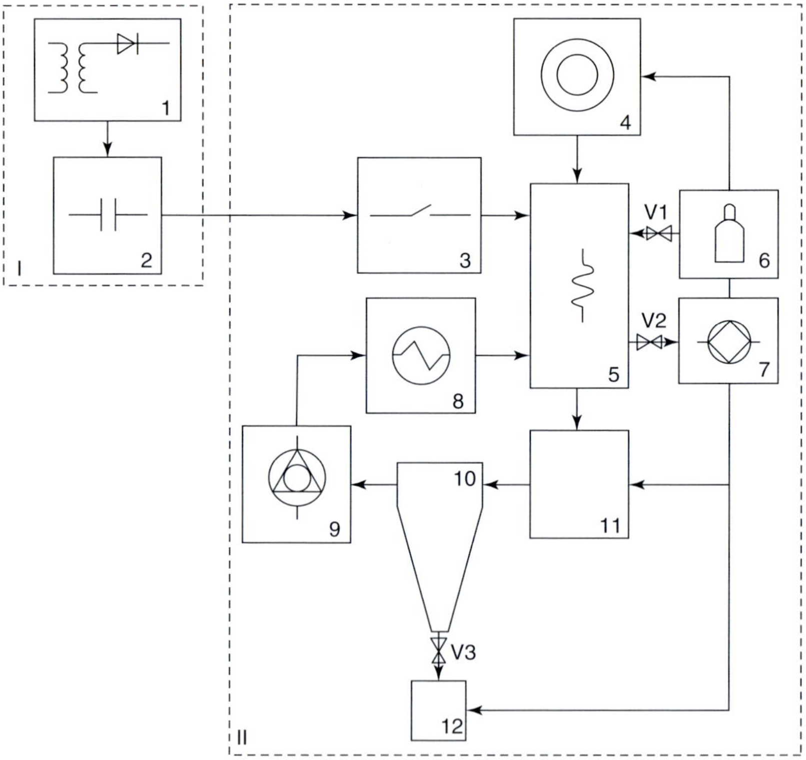

Figure 4.3 Schematic diagram of metal nanopowder production facility.

which captures particles greater than 0.5 μm, and then to filter 10. Filter 10 removes nanoparticles from the gas flow and they gravitate in bowl 12. Purified gas passes through fan 9 to cooler 8, where its temperature decreases to approximately 5 °C and then returns to reactor 5. The process is repeated with a frequency of 1 Hz and higher.

When bowl 12 is filled up with nanopowder, excessive gas pressure is decreased to atmospheric pressure and valve V3 closes. Argon inside the loop is replaced with air, and bowl 12 is removed and packed in a sealed container for further operations.

4.3 Recondensation NP-Producing Methods: Plasma-Based Technology

Materials recondensation, referred to as *distillation*, is one of the most general-purpose NP-producing techniques. The underlying principle is that supersaturated vapor is created upon steaming off the material from the heated melt surface. The vapor so generated is cooled by applying any appropriate technique (i.e., coolant-assisted jet cooling, nozzle flow discharge, etc.), which results in mist formation. Mist drops solidify on aftercooling to form an aerosol system composed of solid material particles, which could be trapped using various techniques.

Until recently, distillation process applicability was restricted by technical availability of high temperatures. At a temperature of 1800 K under atmospheric pressure, the material evaporation rate is so negligible that the process is hard to implement. Currently, such conditions are only adequate to produce Zn and Cd powders. These restrictions are eliminated as far as plasma-based technology advances. Moreover, plasma recondensation adds new specific properties to powders, a fact that makes the approach still more attractive.

Condensation aerosol formation is a complicated process consisting of several consecutive stages. These stages are (i) nucleation, (ii) condensation, cluster formation, (iii) particle growth, (iv) coagulation, and (v) solid particles formation.

Rapid cooling of metal-vapor-doped plasma flows (during quenching) creates conditions to promote metastable nonequilibrium crystal structures. Such powders exhibit higher chemical activity/reactivity of plasma-generated NP because of their metastable crystal structure and stacking fault centers.

The process of producing NP in plasma-generating plants can be illustrated with the following flow chart (Figure 4.4). Obviously, some stages may not be included in the particular process flowsheet depending on target NP properties and associated production objectives. For example, the encapsulation stage is missing if nonpyrophoric powders are to be obtained. The first (preparation) stage may be completed externally as well. The stages 2–10 within plasma-based technology are generally confined to the single process unit (plasma reactor); it is the capacity and performance characteristics of the reactor that govern the overall process flowsheet feasibility. The recondensation process can be implemented in various equipment configurations including, in particular, heating options, vapor transportation, cooling, and disposal from the reaction zone using treatment operations as follows:

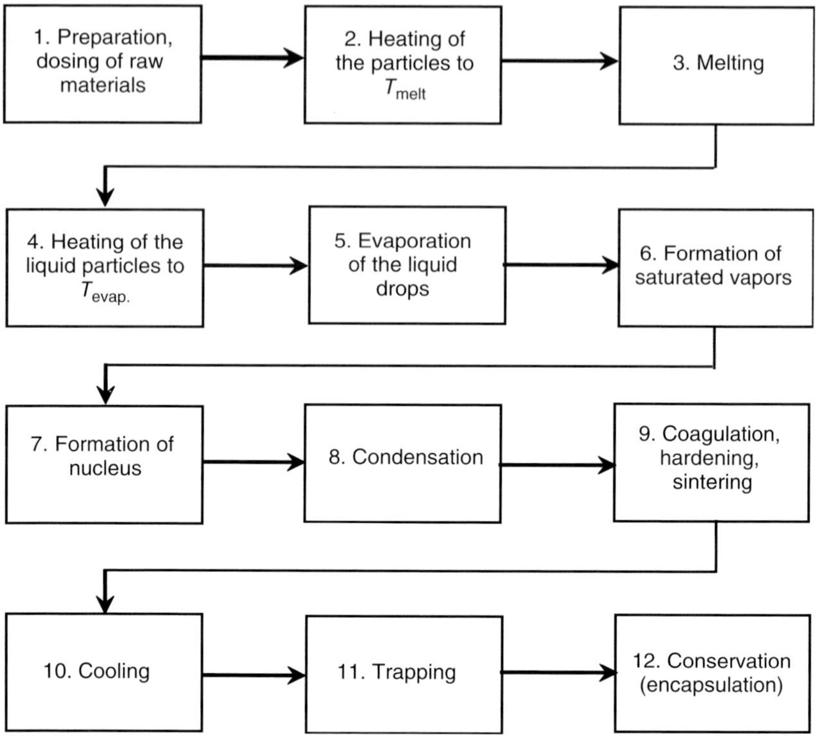

Figure 4.4 Essential stages of plasma-chemical process to produce disperse NP by recondensation.

- metal evaporation from crucibles (containers);
- metal evaporation from automated crucible evaporators;
- plasma-driven evaporation of powders.

The first option normally uses evaporators made of refractory metals such as tungsten, molybdenum, tantalum, and niobium. The evaporant material is heated by the current of appropriate strength passed through the evaporator. In modified processes, the evaporant material is loaded to the crucible and further heated by radiant heat from an external emitter or, alternatively, by an electron beam.

Evaporation is mostly run under reduced pressure because, in this case, lower temperatures are adequate to keep a reasonably high evaporation rate. For example, estimates based on the Langmuir formula [19] predict that the aluminum evaporation rate will be as high as 10^3 kg h^{-1} m^{-2} under 130 Pa pressure at a temperature of 2300 K.

While most researchers tested the evaporators under relatively low weight loads of up to 5×10^{-3} kg, the challenges of evaporating metals from containers are typical for higher output evaporators as well. Many substances form alloys when evaporated from crucibles or react with the evaporator material itself, which accelerates its destruction and results in product contamination.

These difficulties are of particular concern when evaporating nickel, iron, cobalt, and aluminum. Where and insofar the difficulties are avoided or overcome, the material evaporation from crucibles followed by recondensation becomes a commercially viable process. So, the authors of the paper [21] used the induction vacuum furnace to launch pilot production of submicron zinc powder (1 m or smaller particle size). A quantity of 3–7 kg of zinc is loaded to the crucible. Powder is formed for 6–7 h at 720–1000 K and residual pressure of 2.6–4.0 kPa. The optimum specific power consumption is (1–3) kWh kg^{-1}.

The vacuum material recondensation studies suggest that reduced pressure applied in combination with high temperature are factors that dramatically enhance material evaporation.

The commercial NP production process is, however, hard to launch in this case. This is because of technical restrictions – in particular, aerosol trapping and transportation causes some trouble when performed *in vacuo*; the condensation product often deposits on the reactor walls, necessitating a chain of bulky engineering and technical decisions to be accepted when it comes to implementing a continuous process. Nevertheless, the technology has its inherent potential benefits including the opportunity of running evaporation, condensation, and trapping stages in the single integrated unit, the evaporation simplicity, and the high-purity product. These benefits incite researchers to making efforts to eliminate these restrictions.

There are well-known crucible-less metal NP-producing methods including the levitation and conductor electric explosion techniques. In the first case, the evaporant material drop is confined in the volume of gas as a result of the applied electromagnetic field. The plant output is up to $(3–5) \times 10^{-2}$ kg h^{-1} with power consumption of about (150–200) kWh kg^{-1}. In the second case, some section of wire made of the dispersible material explodes inside the reactor induced by the electric resistance heating of the wire under a controlled atmosphere. The disintegration product is ultradisperse powder. The process is implemented in a pilot production facility comprising a battery of several dozens of reactors run simultaneously; it has, however, some restrictions resulting from the raw material preparation method – that is, the powdered conductor wire must be 0.2–0.6 mm in diameter and transmit electricity.

More recently, the methods of producing NP by coarse powder recondensation using high-frequency (HF) or arc reactors are in good progress. Larger heat-transfer areas in this case result in increased plasma gas heat recovery efficiency (up to 20%). More straightforward to implement is the recondensation process in HF reactors, where lower plasma flow velocities provide long-term particle residence in plasma [22, 23]. The resulting metal powders exhibit larger reactivity; the average particle size is about 0.1 m. The specific power consumption is (0.6–1.0) MWh kg^{-1}. The output capacity demonstrated in already mastered processes (SiO_2 and ZrO_2 production) is 0.5 kg h^{-1}.

Analysis of HF processes suggests that despite some success in using HF plasma in materials conversion, the state-of-the-art processes are characterized by high power consumption, a fact that prohibits the dramatic increase in process rate and efficiency. The progress is slowed down by high cost and restricted capacity (up to 160 kW) of the commercially available HF facilities.

A wide variety of the solutions for plasma-arc reactors designed to process disperse materials are covered in several review monographs.

The arc-plasma facilities are essentially advantageous over HF plants in their large unit capacities, a fact that is of great importance when it comes to development of large-scale processes and facilities.

As has been demonstrated by theoretical and experimental studies [24–27], a time of about $(0.5-1) \times 10^{-3}$ s is required to provide plasma evaporation of disperse metal particles (20–40 m size) at 6000 K (for most metals). Taking normal plasma jet velocities into account (i.e., $V = (50-100)$ m s^{-1} in direct-flow systems), the full flashing reaction zone must be extended enough. So, a notorious problem for most facilities is the efficient plasma and disperse phase confinement over the length of the reaction zone. It is the early particle migration toward the walls of the vessel that reduces the material conversion rate and results in occasional process interruptions due to wall encrustation.

From the operational experience of various plasma generators including ascending plasma flow reactors, multiple-arc chambers, batteries of plasma generators connected to the single chambers, and large-diameter chambers with vortex plasma stabilization, it is the multiple-arc reactors and vortex plasma stabilization reactors that seem the most promising solutions. These options have provided the highest output capacity and efficiency in converting almost every solid-phase materials available in coarse-powdered form.

In multiple-arc reactors, however, thermal loads are close to the walls, thus resulting in high specific power consumption and wall encrustation. These restrictions are still far from being properly eliminated.

These difficulties are overcome in plasma-chemical arc reactors equipped with large-diameter chambers where arc and plasma channels are vortex stabilized. In these reactors, the extended plasma spaces are insulated from the walls with a thick cold gas blanket, and so eliminate wall encrustation and prevent heat from migrating to the walls. Arc gas heating and reaction space are integrated in the single vessel. Energy losses in the reactor are mainly due to anode dissipation and radiative transfer. Thermal efficiency (about 65%) is essentially stable against fluctuations in both plasma energy and raw material consumption.

The powders with particle size <40 m are used as raw material. The process gases used are either rare/chemically inactive gases such as argon, helium, and nitrogen or their mixtures with hydrogen, oxygen, and hydrocarbons. The gas supply system is configured as a closed loop, thus providing NP of high purity, reducing process gas consumption, and rendering the technology ecofriendly. Recondensation rate is strongly dependent on the material and particle size of the convertible powder, to vary from 20% to 80%.

The state-of-the-art as deduced from literature therefore suggests that the plasma-based NP production process has a number of advantages.

- It is a single-stage and continuous process.
- It is of lower material demand.
- It provides prolonged service life of the equipment.
- Plasma-generated NPs have many unique characteristics important in chemical and metal industries.

Among a variety of plasma-based processes, it is the use of vortex-stabilized large-volume plasma generators that can become the basis for building high-performance commercial facilities to produce NP from various materials.

The studies performed at SSC RF GNIIChTEOS (on NP production in controlled-shape plasma zone, plasma stabilization, etc.) resulted in formulating engineering philosophy related to plasma generators. In particular, process parameters have been optimized and adopted to produce various NP, and the baseline design standard has been developed for generic plasma reactor including (i) the convertible powder that is fed directly into the active plasma zone (i.e., voltaic arc thermal layer). (ii) Both the arc and "tail-end" plasma flow are gas-vortex-stabilized within the reaction space. Model simulations on the reactor with stabilized arc/plasma zone confined within the arc channel/vortex chamber have demonstrated the inherent efficiency of metal powder vaporization within a high-temperature plasma zone and a potential for long-term operation with no reactor component encrustation. The adjustment options including controlled plasma vapor-phase concentration and vapor-gas mixture quenching rate create a potential for NP specific area control.

The process seems particularly promising compared to the existing processes using current-less plasma flow or unstabilized plasma, in terms of its relatively low energy losses. It therefore could be used as a standard when developing new, more advanced reactors to solve various problems related to plasma-based materials processing.

4.3.1
Fundamentals of Plasma-Chemical NP Production

Plasma process plant development requires that a number of fundamental problems be solved, including efficient plasma-induced raw material vaporization followed by vapor condensation into microparticles of predetermined sizes, creating conditions to trap and classify powder particles under continuous process, and development of the NP encapsulation technology to ensure NP safety during their storage, transportation, and subsequent applications.

A series of calculations and experimental studies made by the authors of the papers [28–30] have demonstrated that full flashing of the material under low-temperature plasma conditions at a characteristic temperature of 5000 K is best achieved with coarse powder fed to the high-temperature area. The evaporation time, however, is about 10^{-3} s even for the starting powder with particles as small as (40–60) m. That is why low-speed extended plasma flows properly heat insulated from the walls are so important. Note that the known plasma-handling equipment – in particular, partitioned-channel plasma generators, multiple-arc vessels, and assemblies with a battery of plasma generators connected to the single chamber – do make it possible, in principle, to obtain the extended low-speed large-diameter plasma jets [31, 32]. Heat loads, however, are close to the walls in the above-mentioned cases, thus resulting in large thermal losses and – if a disperse phase is present in the plasma – wall encrustation becomes unavoidable.

Further studies are needed to search for new vapor-gas flow quenching technique enabling the cooling rate control that is essential to obtain NP featuring target functionalities. Apart from solving these fundamental issues, there are some technical and engineering tasks to tackle when developing the NP production plant. These are (i) NP classification and trapping, (ii) powder transportation between points, and (iii) a series of problems related to fabricating closed-loop gas transportation systems, to supply high-purity process gas, and so on.

4.3.2
Vortex-Stabilized Plasma Reactor

The plasma reactor (Figure 4.5) is designed as a modification of the relatively high aspect ratio vortex chamber ($L/D \approx 3{-}4$, where L is the chamber length and D is its diameter). In this layout, the arc gas heating system and reaction space are integrated within the single vessel.

This design is essentially beneficial in that highly efficient vortex confinement is possible for both arc and plasma flow, under some conditions ($L/D \sim 3{-}4$; $D/dn \geq 3$) [33, 34]. Steady-state vortex flow and stabilized plasma jet criteria were derived from the formula of the equilibrium between pressure gradient and centrifugal forces [34]:

$$\left(\frac{d\rho}{dr}\right) \geq D; \quad \left[\frac{d(\omega\gamma)^2}{dr}\right] \geq 0 \tag{4.1}$$

These stability criteria, however, are in fact useless as (still unknown) vortex/hot core boundary radial gradients are required for both density and velocity.

The following equation has been derived instead:

$$Pe_{0.5}\left(\frac{L}{d}\right) - 1\mathrm{Ri}^* > 7.4 \tag{4.2}$$

where

$$\mathrm{Ri}^* = \frac{V_\mathrm{e}^2}{V_\mathrm{n}^2} \cdot \frac{(\rho x - \rho n)}{\rho n} \tag{4.3}$$

The studies performed earlier at SSC RF GNIIChTEOS have demonstrated that the modified Richardson number Ri* is adequate to describe the stability boundary, the Ri* being expressed through integral characteristics of both stabilizing vortex and plasma flow.

Equation 4.2 enables calculation of the essential hydrodynamic vortex parameters needed to ensure plasma jet stabilization within the near-axial reactor space. Plasma thus gets insulated from the walls with an optimum-thickness cold gas blanket. Two principal problems are tackled in this way, first – we eliminate the convective heat losses into walls and, second – we prevent fused particles from contacting walls (i.e., eliminate wall encrustation).

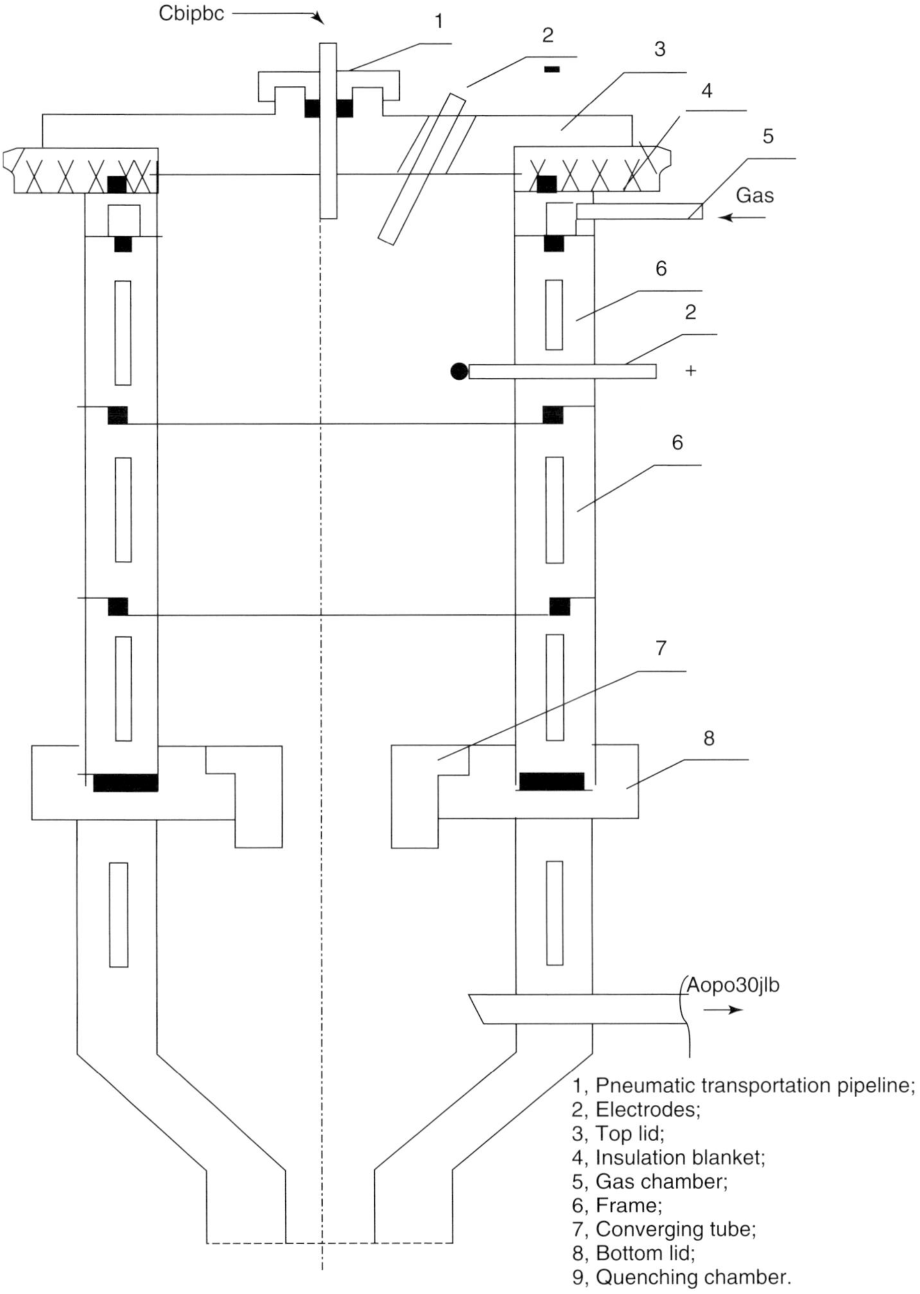

Figure 4.5 Design concept of the plasma-chemical reactor with extended stabilized plasma space. 1, Pneumatic transportation pipeline; 2, electrodes; 3, top lid; 4, insulation blanket; 5, gas chamber; 6, frame; 7, converging tube; 8, bottom lid; and 9, quenching chamber.

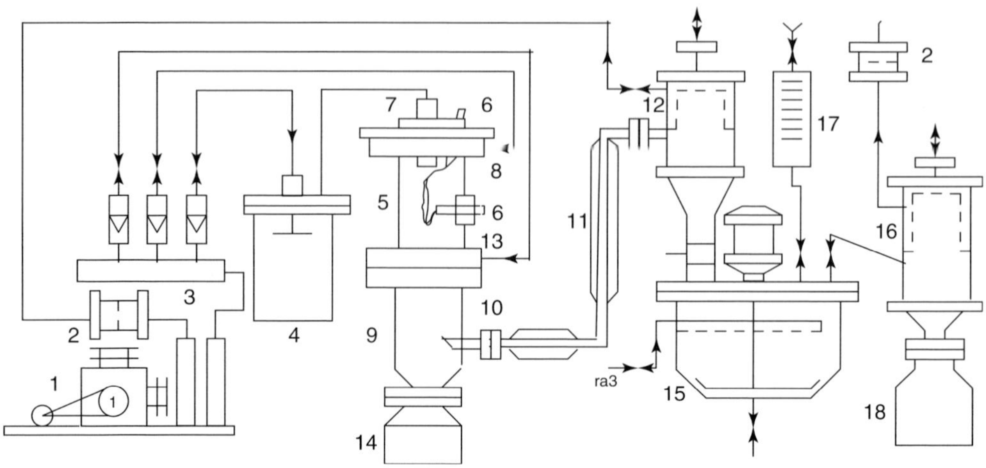

Figure 4.6 Process facility P and I diagram. 1, Compressor; 2, receiver filter; 3, gas distribution ramp; 4, disperse raw material dispenser; 5, 9, reactor body; 6, electrodes; 7, reactor lid; 8, vortex chamber; 10, 11, cooler; 12, filter; 13, quenching chamber; 14, Large grain size receiver and 15–18, encapsulation unit.

This reactor features a negative voltage–current (*VI*) characteristic with slow cooling toward 200 A current values. Energy losses in the compartments 5 and 9 of the body (Figure 4.2) are mainly due to radiative transfer. Energy losses in other points are about 10%. The evaporator efficiency is about 65% when argon is used as process gas.

The erosion of the rod-shaped cathode (which is a tungsten rod) is comparable to the rates of erosion estimated by several workers for voltaic arc-plasma generators (APGs) [35]. The service life of the anode is much shorter (the anode is a tungsten rod press fitted into a copper water-cooled radiator). The anode erosion rate is 5×10^{-10} kg $(\text{A s})^{-1}$ under 150 A current.

Figure 4.6 illustrates the layout of the process cycle designed with the above-mentioned reactor as the core process equipment. Closed-loop gas circuit is used to ensure process gas of the desired purity as well as minimize the gas consumption index. Gas circulation is maintained by a diaphragm-type compressor, which compresses the gas with no contamination. Compressed gas is fed through the receiver 2 to the flow meter ramp 3, which distributes the gas over the units including the dispenser 4 and the reactor where gas vortex is formed (8) and aerosol is then cooled (13). Raw material powder is fed to the reactor by pneumatic transport gas via the top lid 7.

4.3.3
Starting Material Metering Device (Dispenser)

Several meter types have been tested at the process units' development stage, to master the following problems.

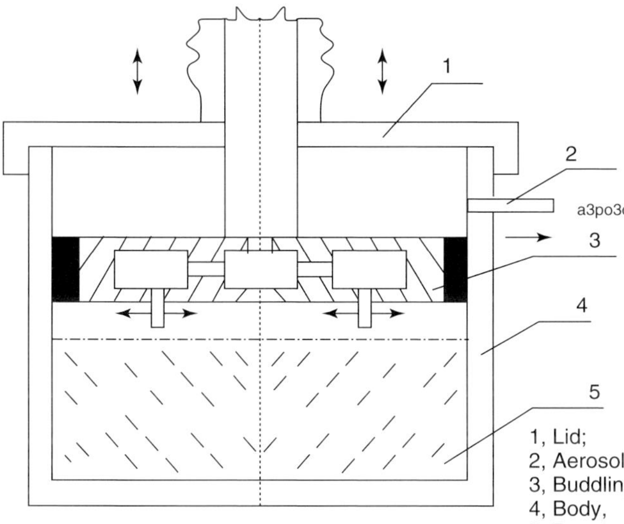

Figure 4.7 Design concept of the disperse raw material dispenser. 1, Lid; 2, aerosol discharge outlet; 3, puddling device; 4, body; and 5, powder raw material.

- Powder consumption shall be mechanically controlled.
- Uniform feed of the starting material shall be guaranteed.

The design shown in Figure 4.7 has proved to be the best choice. In this dispenser design, the cylinder hopper has a fluidifying unit above the powder bed. The unit fluidifies the top layers of the powder bed by spraying multiple jets. The bulk aerosol is fed to the reactor by pneumatic transportation. It is the velocity of the top powder boundary toward the fluidifying unit that governs the powder feed rate. Hopper evacuation is provided to prevent air trapping into the process cycle during the dispenser charging period.

4.3.4
Disperse Material Trapping Devices (Cyclone Collectors and Filters)

The NP inertial-trapping experiments suggest that initial microparticles are aggregated into coarse particles when piping them. Without going into details of microparticle cohesive forces, note that the aggregates are disintegrated relatively easily when small mechanical forces are applied. Mesh-screen analysis suggests that nanodisperse metal powders with particle sizes <0.1 m generated in nitrogen plasma form aggregates sized dozens to hundreds of microns, the largest aggregates being formed from the smaller particles. These fundamental properties of the particles facilitate mastering both classification (with no disaggregation) and trapping problems.

In the above-mentioned design, conventional cyclone collectors are used both to separate ultradispersion powder (UPD) from raw material 14 and to "unload" the trapping filters 12 (Figure 4.6). The cyclone collector, when installed upstream of the filter, expands nearly twice the continuous running period of the filters.

The final NP trapping from the gas flow is achieved because of bag filter 12. Conventional bag filter design was used, in particular: dusty gas is fed into the bag, the gas is filtered through side surface of the bag, and purified gas is removed through the annular space between the bag and the body. Filter shaking was applied under gas load released with no back-purge. Conventional filter cloth was used for filter element.

The above-mentioned filter traps efficiently (upon completion of the "maturity" stage) about 98% of powder particles with particle sizes <0.1 m under specific gas loads of about $100\,m^3\,h^{-1}\,m^{-2}$ of the filter surface area and NP concentrations of up to $0.7 \times 10^{-2}\,kg\,m^{-3}$. Under these conditions, the filter is to be shaken once in 4 h. Differential pressure of filter can increase up to 300 mmHg before shaking.

4.3.5
NP Encapsulation Unit

Reactive metal ultrapowders prepared under nonreactive atmosphere are highly pyrophoric. To prevent them from self-ignition, a nonreactive atmosphere is needed for the storage as well – or, alternatively, the special coating of the particles is required to avoid coming into contact with air. Several particle surface encapsulation techniques are used. One of the simplest methods is partial oxidation resulting in oxide film formation. The oxidation process is run at 20–30 °C under a nonreactive atmosphere with small amounts of oxygen or air, to minimize film thickness and avoid self-ignition.

Metal ultradisperse powders can be also encapsulated by applying a thin film of organic oil or polymer to the surface. This operation is performed by stirring the mixture of the NP and the solvent with the desired quantity of the oil or polymer dissolved. Special-purpose vessels 15–17 have been introduced into the process flowsheet (Figure 4.6), to prevent the resulting NP from coming into contact with air. All encapsulation operations are performed immediately once the NP particles are trapped.

The operation principle of the encapsulation unit using volatile solvent (light ligroin) is as follows: the NP trapped by filter 12 is transported to the vessel 15. The predefined amount of light ligroin with dissolved encapsulation substance is added to the same vessel from the dosage meter 17. The powder is mixed with ligroin solution by an electrically driven stirrer. Then some part of the solvent is drained off through the discharge pipe at the bottom of the vessel. The rest of the solvent is removed from the powder by gas drying under continuous stirring. The dried particles with the encapsulation substance adhered to them are transported by gas to the trapping filter 16 and finally discharged into the hopper 18. The NPs of aluminum, boron, and borides of aluminum, nickel, and iron have been tested under prolonged air storage for 12 months. No changes in observable properties occurred. The storage stability of encapsulated NP had been also confirmed under air storage of the samples at 90 °C for 24 h.

4.4 Characteristics of Al Nanopowders

Characteristic photomicrograph (a) and size distribution histogram (b) for Al particles produced using EEW method are shown in Figure 4.8. The particles have a regular spherical shape.

Experiments [5] have indicated that process parameters that determine the average size of Al nanoparticles are as follows:

1) EEW energy content E/E_c (E, energy delivered to the wire; E_c, energy of metal sublimation);
2) Wire diameter d_0;
3) Gas medium pressure P;
4) Gas medium temperature T.

Dependence of the size of nanoparticles on parameter E/E_c, wire diameter, and gas medium pressure is shown in Figure 4.9. As can be seen in the diagram, with the increase in EEW energy content, average particle size a_s gradually decreases to $E/E_c \approx 1.5{-}2.0$. Subsequently, the change in particle size is insignificant. With the decrease in wire diameter, the value of a_s also decreases and increase in pressure leads to an increase in average particle size.

Electron microscope study of Al powders produced at different gas temperatures in bowl 12 (Figure 4.3) have shown that the average size of nanoparticles depends on the temperature of the gas medium, all other EEW parameters being equal. If specific surface of Al powder at gas temperature $T = 43.8\,°\text{C}$ comprises $S_{sp} = 15\,\text{m}^2\,\text{g}^{-1}$, then at $T = 25.2\,°\text{C}$ it will comprise $S_{sp} = 19.5\,\text{m}^2\,\text{g}^{-1}$.

The majority of nanoparticles in the powder have a pronounced block structure (Figure 4.10). For a particle size of approximately 100 nm, the size of the blocks is between 2 and 10 nm.

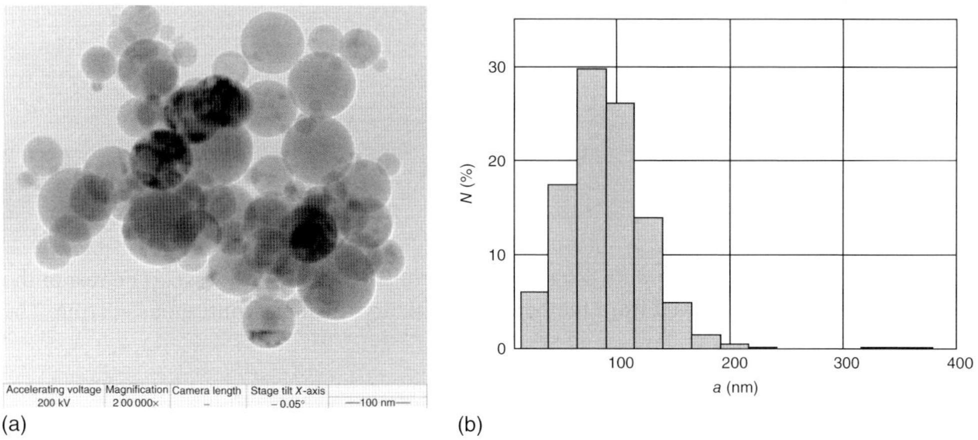

Figure 4.8 (a) Characteristic photomicrograph and (b) size distribution histogram for Al particles: average nanoparticle size – 76.5 nm.

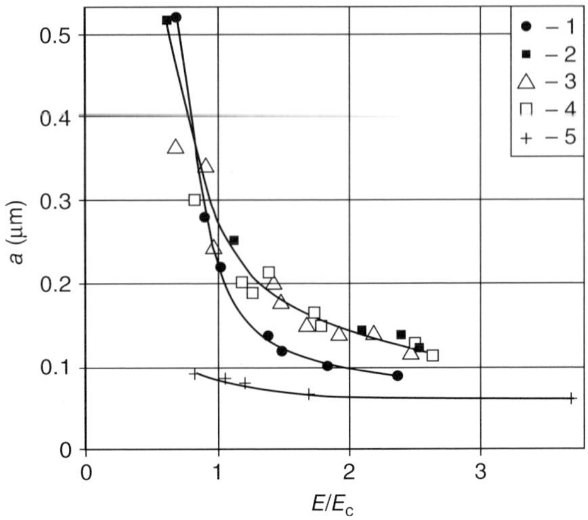

Figure 4.9 Dependence of surface average particle size on a Ar pressure and specific energy content E/E_c: curve 1 – wire diameter $d_0 = 0.43 \cdot 10^{-3}$ m, Ar pressure $P = 10^5$ Pa; curve 2 – $d_0 = 0.43 \cdot 10^{-3}$ m, $P = 5 \cdot 10^5$ Pa; curve 3 – $d_0 = 0.47 \cdot 10^{-3}$ m, $P = 5 \cdot 10^5$ Pa; curve 4 – $d_0 = 0.31 \cdot 10^{-3}$ m, $P = 5 \cdot 10^5$ Pa; and curve 5 – $d_0 = 0.2 \cdot 10^{-3}$ m, $P = 10^5$ Pa.

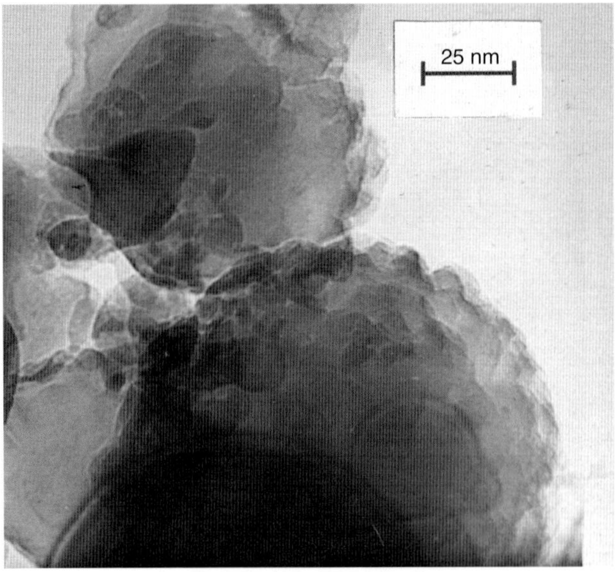

Figure 4.10 Al particles with block structure obtained by means of explosions in cooled gas.

The change in temperatures during the tests is insignificant regarding the temperature of the wire in the EEW process, which comprises, according to a number of authors, thousands of degrees celsius. Therefore, the increase in particle size with the increase in gas medium temperature can be caused only by secondary processes, taking place during scattering of explosion products or in the powder after its particles are formed. For example, coagulation and subsequent coalescence of the already formed but still "hot" particles.

4.5
Nanopowder Chemical Passivation

Metal nanopowders (Al powders in particular) are highly pyrophoric systems. After synthesis, nanopowders exposed to air can ignite as a result of heating caused by metal oxidation by oxygen, which is a component of air. Various passivation methods are used in order to reduce the pyrophoricity of nanopowders. Passivation by oxygen-containing atmospheres (air, in particular) is a widespread method. In this method, an oxide film is formed on the particle surface, which protects the metallic phase from contact with the environment. On the other hand, a major part of existing technologies of metal powder processing are tailored for a certain (usually low) content of oxide phases. Thus, it is necessary to investigate the mechanism of passivation by an oxygen-containing atmosphere and identify dependencies to be used in order to set optimal parameters of the passivation process in order to preserve the chemical activity of nanopowders.

Increase in temperature in the process of passivation can lead to sintering of nanoparticles and changing nanopowder particle size distribution. This was demonstrated by an experimental measurement of nanopowder electrical conductivity as a function of temperature (Figure 4.11). During the tests, nanopowders were heated in argon flow and their electrical conductivity was measured. The point of sharp drop of sample resistance corresponds to the process of sintering of particles with formation of current contacts between them. The change of resistance as a function

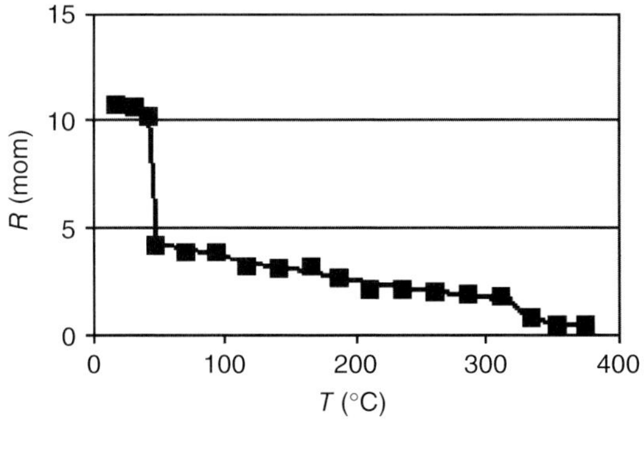

Figure 4.11 The change of resistance of Al nanopowder layer R as a function of heating temperature T.

of the heating temperature of layers of nonpassivated Al nanopowder applied from hexane-based paste to the ceramic base plate is shown in Figure 4.11. As can be seen, the increase in temperature leads to a significant increase of their electrical conductivity approximately at 30 °C, which is probably associated with the sintering of individual nanoparticles. On the basis of obtained data, it may be concluded that sintering of EEW nanopowder is possible at temperatures significantly lower than the melting point of the metal.

Al nanopowder with an average particle size of 100 nm, which was not exposed to air, was used in order to study the process of passivation. The reaction vessel was filled with the powder (volume 25 ml, powder mass 1.75 g) in an inert atmosphere. A gas mixture (O_2 + Ar) was blown through the powder at a specified rate. Powder temperature and active metal content were monitored during the test. Heat exchange with the environment was neglected. Active Al content was analyzed using the volumetric method.

Dependence of maximum temperature T of Al nanopowder samples in the process of oxidation in the flow of O_2 + Ar gas mixture is shown in Figure 4.12. As can be seen, the samples are heated up to temperatures higher than 30 °C even if the gas

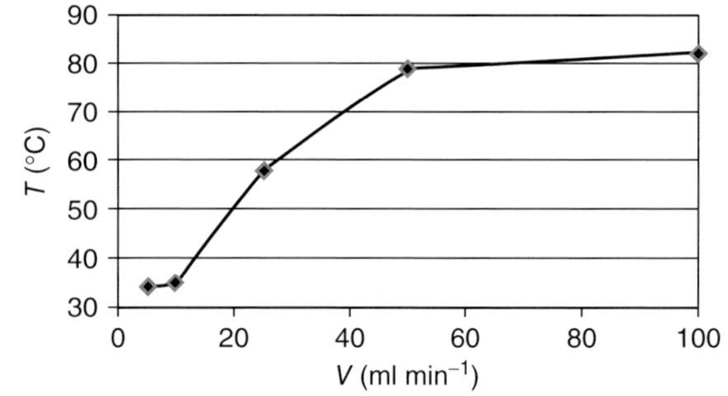

Figure 4.12 Dependence of maximum temperature T of Al nanopowder samples on blowing rate of gas mixture V: [O_2] concentration for all experiments – 5 vol%.

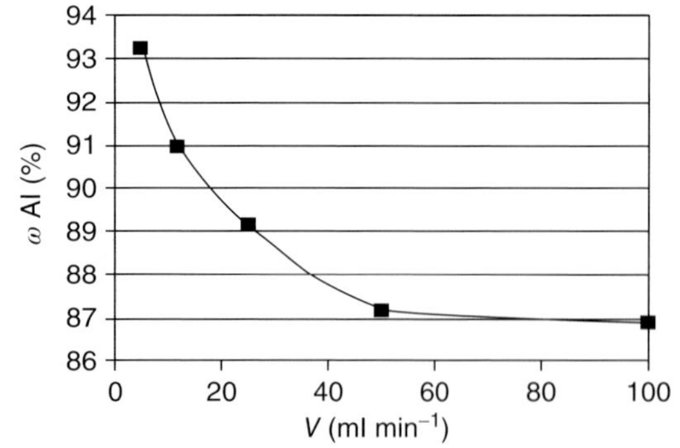

Figure 4.13 Dependence of active Al content ωAl on the flow rate of gas mixture (Ar + O_2) V: [O_2] concentration in gas mixture – 5 vol%.

flow rate and O_2 concentration in the mixture are low. Thus, on the basis of obtained data and the data from Figure 4.7, it may be concluded that there is a high probability of sintering of nanoparticles in the process of air passivation of Al nanopowder.

Active Al content in nanopowders also decreases rapidly with the increase in gas flow rate (Figure 4.13).

Thus, passivation of nanoparticles with an oxygen-containing gas for the purpose of preservation of particle size distribution of nanopowders and maximum content of active metal is not a promising method.

4.6 Microencapsulation of Al Nanoparticles

In order to improve environmental resistance of nanoparticles, the method of nanopowder passivation with organic agents (microencapsulation) – stearic, oleic, and palmitic acids [36, 37] – is used. Adsorption of organic agents creates a protective layer on the surface of nanoparticles, which is more efficient than the oxide layer formed in the process of passivation of nanopowders in an oxygen-containing atmosphere. Adsorption of organic agents also increases affinity of metal nanoparticles to organic components of high-energy materials (HEMs). This has a positive impact on use of nanopowders as HEM components [38]. The kinetic factor plays an important role for HEM in the process of combustion, the use of certain organic agents increases the oxidation rate of Al nanopowders at high temperatures, which leads to an increase in the HEM burning rate [39, 40].

In order to perform microencapsulation tests, the following powders were used: Al nanopowder ALEX passivated in the atmosphere with controlled partial pressure of O_2 and nonpassivated Al nanopowders produced by EEWs in an argon medium. Average particle size for ALEX nanopowder comprised 100 nm, active Al content – 92% by mass, and specific surface of Al nanopowder – $12\,m^2\,g^{-1}$.

The study of the morphology of initial and microencapsulated Al nanoparticles was performed by means of transmission electron microscopy (TEM) using JEM-100 CXII (JEOL) electron microscope. The study of the morphology of Al nanopowder hydrolysis products was performed by means of scanning electron microscopy (SEM) using LEO EVO 50 scanning electron microscope. Phase composition of conversion products was investigated using DRON-3M X-ray diffractometer (CuKα-radiation). Specific area of Al nanopowders and conversion products was determined by thermal desorption of nitrogen using specific area analyzer Sorbtometer-M (Katakon) and calculated using the Brunauer–Emmett–Teller (BET) method.

Protective coatings were represented by low-molecular-weight organic substances capable of forming a hydrophobic protective layer on the surface of metal nanoparticles – stearic acid, 8-hydroxyquinoline, glyoxal, and polymers: polyethylene glycol adipate (PEGA), vinylidene difluoride copolymer, and hexafluoropropylene (Viton) in combination with a monoether (ME) produced by means of interaction of maleic anhydride with 1,1,11-trihydroeicosa-fluoroundecyl alcohol [41].

Organic compound was applied to Al nanoparticles using its solution. A quantity of 20.0 g of Al nanopowder was suspended in 200 ml of ethyl acetate in the process of intensive mixing by H6-15D submersible dispenser (5000 rpm, 3 min) with subsequent addition of ethyl acetate solution of organic compound containing an estimated amount of organic agent to obtain a suspension followed by 30 min of mixing.

As for Viton coating, Al nanopowder suspension was premixed with a solution of 1 g of ME in 10 ml of ethyl acetate with subsequent 30 min mixing and addition of the solution of 2 g of Viton in 50 ml of ethyl acetate followed by additional 30 min of mixing.

Then the solvent was removed by low-pressure vaporization using an IKA RV-10 rotary evaporator. Microencapsulated nanopowders were dried at low pressure 1 mmHg for 16 h. Characteristics and TEM images of microencapsulated powders are shown in Table 4.1 and Figure 4.14.

Table 4.1 Characteristics of microencapsulated Al nanopowders.

No.	Coating	Quantity of organic matter (%)	Specific surface ($m^2\,g^{-1}$)
1	W/o coating	—	11.8
2	Stearic acid	3	13.9
3	8-Hydroxyquinoline	0.5	—
4	Glyoxal	1	—
5	Stearic acid[a]	3	11.3
6	PEGA	1	—
7	Viton	15	11.0

[a]The coating was applied to nonpassivated Al nanopowder in an inert atmosphere of petroleum-ether solution.

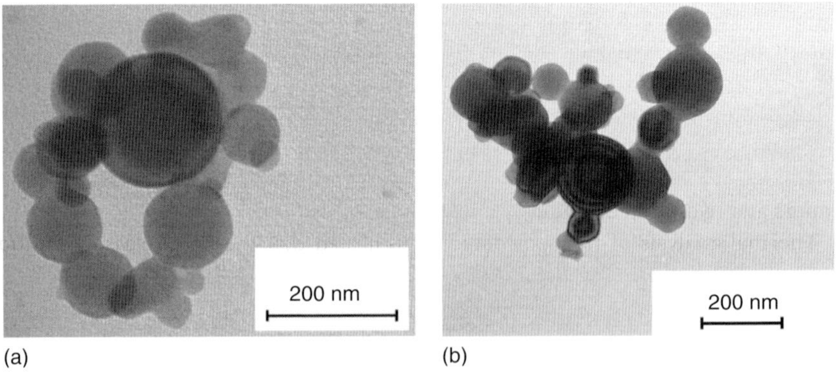

Figure 4.14 Characteristic TEM images of microencapsulated nanopowders coated with (a) stearic acid and (b) Viton.

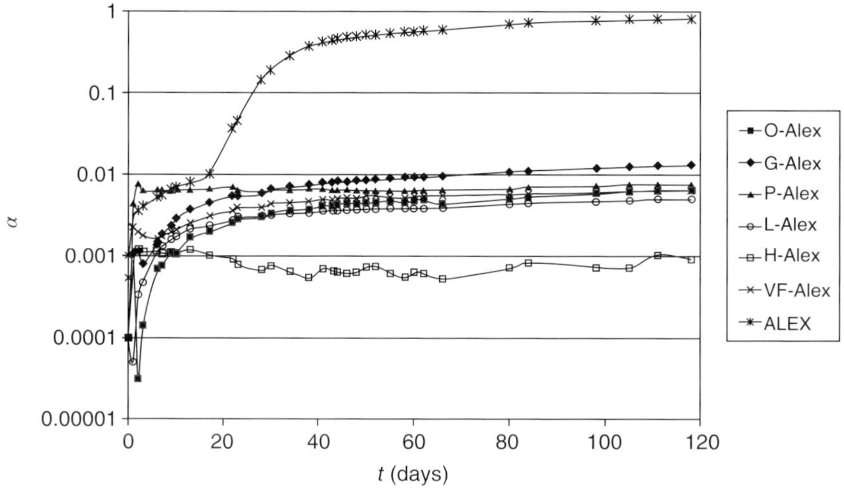

Figure 4.15 Kinetic curves of hydrolysis of microencapsulated nanopowders in humid atmosphere: $\alpha = \alpha(t)$.

In order to test barrier properties of protective coating, the samples of microencapsulated nanopowders were exposed to relative humidity of 90% and temperature of 25 °C for 180 days (Figure 4.15). Conversion of nanopowders $\alpha(t)$ was determined on the basis of the change in sample mass and chemical equation of the chemical reaction that takes place:

$$Al + 3H_2O = Al(OH)_3 + (3/2)H_2$$

Thermal analysis of microencapsulated nanopowders in air medium was performed using NETZSCH STA 449F3 while heating samples up to °C at the rate of 10 °C min^{-1}.

Interaction between microencapsulated Al nanopowders and water vapors is significantly slower compared with Al nanopowder (ALEX). Conversion of microencapsulated Al nanoparticles for the same period of interaction (150 days) is 50–400 times lower than conversion of Al nanopowder. The value of $\alpha(t)$ after 150 days was $<10^{-4}$, the reaction of microencapsulated Al nanopowders with water vapors during the period of study has not proceeded beyond the induction period. The change in mass of nanopowder samples can be explained by adsorption of water vapors and hydration of oxide layer of nanoparticles.

Thermal analysis of nanopowders has indicated that the heat release rate and hence oxidation rate increases in the following sequence: ALEX, L-Alex, VF-Alex (Figure 4.16).

Thus, coating Al nanoparticles with organic compounds at the rate of 0.5–3% creates on the surface of Al nanoparticles a protective layer, which entirely blocks interaction between Al nanoparticles and water vapors and increases the oxidation rate of nanoparticles in the process of heating.

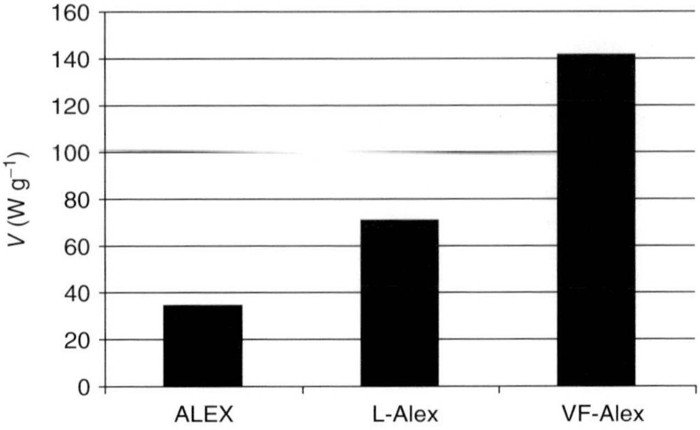

Figure 4.16 Heat release rate of nanopowders according to thermal analysis data.

4.7 The Process of Producing Nanopowders of Aluminum by Plasma-Based Technology

4.7.1 Production of Aluminum Nanopowder

A part of the R&D activities to develop reactive metal NP production process was the study of the influence that the vortex-stabilized reactor design and gas conditions could have on the disperse properties of each type of nanopowder as well as reactor output capacity. A potential for higher reactor capacity when running the powder recondensation process is directly driven by the uniform raw material distribution over plasma space. In the conventional plasma generators used to convert predispersed materials, special-purpose input points are designed above the particular plasma jet diameter. The main objective of the input points' modification is to distribute the raw material evenly over the cross section of the jet and thereby increase the reactor's thermal efficiency.

Conventionally, the problem is solved by feeding the raw material through several peripheral channels into the plasma jet at a characteristic angle. For the reactors used in this study, however, the reagent injection to the reaction zone from the periphery is hindered by the fact that the plasma boundary is the area where maximum rotation speed is observed. Owing to the interaction of the vortex with the pneumatic transportation jet, some part of the raw material cannot reach the hot internal plasma area or is rejected to the reactor walls. The top lid of the reactor has been modified to promote more efficient plasma jet–raw material interaction. Raw material is now fed through two pneumatic transportation nozzles from two independently operated dispensers or, alternatively, through the single axial nozzle through the single dispenser.

Two flows are injected into the stabilized plasma space, to eliminate the effect of the rotational cool gas motion component on the flow path; in addition, the channel centerline are inclined toward the reactor axis at an angle of 15–20°, that is, to counterbalance centrifugal force effect on the particles suspended in the plasma jet.

The experiments were performed with plasma diameters of 0.04–0.07 m. The reactor efficiency criteria are conversion rate and specific power consumption. The results suggest that specific power consumption is a few kilowatt-hour per kilogram higher in the case of dissipated material injection than in the case of axial material injection. Increased specific power consumption is mainly due to reduced processing time. Plasma jet diameter has no effect on these regularities within 0.04–0.07 m.

The following optimal electric and gas parameters of the plasma facility used for producing aluminum NP were established in the results of the researches conducted:

- Electric current, A = 80–100;
- Arc voltage, V = 120–180;
- Reaction chamber length, m = 0.15;
- Plasma zone diameter, m = 0.04;
- Flow rate of stabilization gas, $nm^3\,h^{-1}$ – 25;
- Flow rate of gas for pneumatic transport, $nm^3\,h^{-1}$ – 3.5–4;
- Flow rate of quenching gas, $nm^3\,h^{-1}$ – 30.

These parameters allow obtaining stable burning of electric arc; the plasma flow is stabilized in the reactor near-axis zone, and occlusion of the reactor units with the disperse phase is eliminated. By changing the flow rate of the stock fed to the reactor, one can change the average size of the particles of aluminum NP produced. In the course of capsulation of particles with oxygen, a film of aluminum oxide is generated on them; the thickness of this film, in certain investigators' opinion, is 0.5–4 nm, predominantly 2–3 nm [42]. Judging by computations for particles of the ideal spherical shape (Table 4.1) at their average size 100 nm, the content of active aluminum is 84.3%, which reduces energetic efficiency of nanoaluminum. Further, diminishing of particle size at the unchanged oxide film thickness will lead to still lesser content of active aluminum, which results in problematic use of such nanoaluminum in ERCS.

Thus, as we think, optimal aluminum nanopowders for use in ERCS are those with particle sizes over 150 nm (specific surface: 6–14 $m^2\,g^{-1}$).

4.7.2
Some Properties of Produced Nanopowders of Aluminum, Boron, Aluminum Boride, and Silicon

Nanopowders were passivated with S2-80/120 petroleum solvent adsorbed on their surface. It is worth noting that the adsorbed hydrocarbon film protects the nanopowder quite reliably from air oxidation. The produced powders, even being exposed to pressure 3 MPa and temperature 20 °C in pure oxygen atmosphere for 24 h, did not demonstrate any increase in their weight.

Figure 4.17 SEM photomicrograph of nanopowders of aluminum. Magnification: 40 000×.

Depending on parameters of the mode, metal powders produced by this process diagram shall have specific surface (BET method) 6–15 $m^2\,g^{-1}$ for aluminum at particle size 50–500 nm at their spherical form. Figure 4.17 shows photomicrographs of aluminum produced by the SEM method at 40 000× magnification.

It can be clearly seen on the photographs that particles of all powders have essentially a spherical shape, their size ranges from 80 to 500 nm, most part of them are agglomerated, and the size of the agglomerates is as big as several micrometers.

Thus, the method of producing NPs by plasmatic recondensation developed in RF State Scientific Centre – FSUE "GNIIHTEOS" and the facility constructed allow producing a wide range of NPs, in particular, powders of aluminum, boron, aluminum borides, and silicon. NPs of iron, titanium, iron and aluminum oxides, as well as oxides of other elements and metal hydrides, which all are out of the scope of this survey, were also produced at this facility.

A distinctive feature of the method created is its universality, which allows, depending on the process mode, adjustment of not just dispersity but also particle shape and other features of the nanopowders produced.

References

1. Dreizin, E.L. (2009) Metal-based reactive nanomaterials. *Prog. Energy Combust. Sci*, **35**, 141–167.
2. Sakovich, G., Arkhipov, V., Vorozhtsov, A. *et al.* (2010) *Nanotechnol. Russ.*, **5** (1–2), 91–107.
3. Fedorov, S., Guseinov, S., and Storozhenko, P. (2010) *Nanotechnol. Russ.*, **5** (7-8), 1–13.
4. Mench, M.M., Yeh, C.L., and Kuo, K.K. (1998) Propellant burning rate enhancement and thermal behavior of ultra-fine aluminum powders (Alex). 29th International Annual Conference of ICT, Karlsruhe, Germany, June 30 – July 3, 1998, pp. 30-1–30-15.
5. Lefrancois, A. and LeGallic, C. (2001) Expertise of nanometric aluminum

powder on the detonation efficiency of explosives. 32nd International Annual Conference of ICT, Karlsruhe, Germany, July 3–6, 2001, p. 36-1.

6. Brousseu, P. and Cliff, M.D. (2001) The effect of ultrafine aluminum powder on the detonation properties of various explosives. 32nd International Annual Conference of ICT, Karlsruhe, Germany, July 3-6, 2001, p. 37-1.
7. Abrams, R. (1946) *Production and Analysis of Radioactive Aerosols*, CH3629, University of Chicago, Chicago, IL.
8. Karioris, F.G., Fish, B.R., and Royster, G.W. (1962) Aerosols from exploding wires, in *Exploding Wires*, vol. **II**, Plenum Press, New York.
9. Jonson, R. and Siegel, B. (1970) Chemical reactor utilising successive multiple electrical explosions of metal wires. *Rev. Sci. Instrum.*, **42** (6), 854–859.
10. Karioris, F. and Fish, B. (1962) An exploding wire aerosol generator. *J. Colloid Sci.*, **17**, 155–161.
11. Lerner, M.I. (2011) *Electrical Explosion Nanopowders of Inorganic Materials: Production Technology, Characteristics and Some Applications*, LAP LAMBERT Academic Publishing, p. 332ISBN: 978-3-8433-2075-7 (in Russian).
12. Ivanov, V. and Gavrilyuk, O. (1999) The regularities in oxidation and self-ignition of electroexploded ultradispersed metal powders in air. *Fiz. Goreniya Vzryva*, **35** (6), 53–60(in Russian).
13. Gusev, A. (1998) *Nanocrystallized Materials – Production Methods and Properties*, UrO RAN, Ekaterinburg (in Russian).
14. Gromov, A., Iliyn, A., and Teipel, U. (2006) On influence of passivated coating kind, sizes of particles and time of storage for oxidation and nitriding of aluminum particles. *Fiz. Goreniya Vzryva*, **42** (2), 61–69 (in Russian).
15. Kwon, Y.S., Jung, Y.H., Yavorovsky, N.A. *et al.* (2001) Ultrafine metal powders by wires electric explosion method. *Scr. Mater.*, **44**, 2247–2251.
16. Storozhenko, P., Guseinov, S., and Malashin, S. (2009) *Nanotechnol. Russ.*, **4** (1-2), 27–39.
17. Pranda, P., Prandova, K., and Hlavacek, V. (2000) Particle size and reactivity of aluminum powders. *Combust. Sci. Technol.*, **156**, 81–96.
18. Yagodnikov, D., Voronetskiy, A., Devyatukha, D. *et al.* (2009) The peculiarities of use of nano- and ultradispersed powders of aluminum diboride in pyrotechnical compounds. Vestnik of MGTU named Bauman. *Sect. Mashinostroenie.*, **2**, 115–121 (in Russian).
19. Thiruvengadathan, R., Belarde, G.M., Bezmelnitsyn, A. *et al.* (2012) Combustion characteristics of silicon-based nanoenergetic formulations with reduced electrostatic discharge sensitivity. *Propellants Explos. Pyrotech.*, **37**, 359–372.
20. Koch, E.-C. and Clement, D. (2007) Special materials in pyrotechnics: VI. Silicon – an old fuel with new perspectives. *Propellants Explos. Pyrotech.*, **32** (3), 205–212.
21. Gribovskiy, S.V. Frishberg I.V., and Kvater L.I. (1978) *Gas-phase method for production of powders*, M. Nauka, pp. 224 (in Russian).
22. Kuelleron, J. and Croushat, B. (1974) The production of high pure boron, reduction of trichloride of boron by hydrogen in induction argon plasma, in *Boron, Production, Structure, Properties*, Nauka, Moscow(in Russian).
23. Miller, T. and Grabis, Y. (1980) *Methods of Production, Properties and Area of Application of Nitriles*, Riga, p. 5 (in Russian).
24. Chizhikov, D., Tsvetkov, Y., and Tagirov, I. (1970) The peculiarities of reduction processes in conditions of plasma temperatures, in *Mechanism i Kinetika Vosstanovleniya Metallov*, Nauka, Moscow, pp. 3–18(in Russian).
25. Thevenot, F. (1990) in *The Physics and Chemistry of Carbides, Nitrides and Borides* (ed R. Freer), Springer, Dordrecht, p. 87.
26. Koroleva, E., Lokhov, Y., and Pegruchev, V. (1976) On calculation of heating and evaporation of dispersed particles in plasma. *Phys. Chem. Mater. Treat.*, **5**, 25–27 (in Russian).
27. Luzhnova, M. and Reichbaum, Y. (1970) *Inzh-Phys. Z.*, **18** (1), 77–80 (in Russian).

28. Mosse, A. and Burov, I. (1980) *The Treatment of Dispersed Materials in Plasma Reactors*, Nauka i Tekhnika, Minsk (in Russian).
29. Kulagin, I.D. and Sorokin, L.M. (1969) *Proced of All-Union Scientific and Technical Conference in Generators of Low-Temperature Plasma*, Energiya, Moscow (in Russian).
30. Krapivina, A. (1981) *Plasmachemical Technological Processes*, Khimiya, Leningrad (in Russian).
31. Mosse, A. and Pechkovsky, V. (1973) *The Use of Low-Temperature Plasma in Technology of Inorganic Substances*, Nauka i Tekhnika, Minsk (in Russian).
32. Sheer, C., Korman, S., and Angier, D. (1976) Method and apparatus for energizing materials in an electric Arc. US Patent 3931542 – Published in Izobreteniya za rubezhom, Vol. 33. No. 8.
33. Chigier, B. and Laminori, D. (1971) Zation of turbulent flames in rotating environments. *Combust. Flame*, **16** (1), 39–45.
34. Jonson, S. (1972) The stability of rotating stratified liquids. *Raketnaya Tekh. Cosmonautika*, **10**, 131–132 (in Russian).
35. Zhukov, M., Smolyakov, V., and Uryukov, B. (1973) *Electroarc Heaters of Gas (plasmotrons)*, Nauka, Moscow (in Russian).
36. Sossi, A., Duranti, E., Manzoni, M., Paravan, C., DeLuca, L.T., Vorozhtsov, A.B., Lerner, M.I., Rodkevich, N.G., Gromov, A.A., and Savin, N. (2013) Combustion of HTPB-based solid fuels loaded with coated nanoaluminum. *Combust. Sci. Technol.*, **185** (1), 17–36.
37. Kwon, Y.-S., Gromov, A.A., and Strokova, J.I. (2007) Passivation of the surface of aluminum nanopowders by protective coatings of the different chemical origin. *Appl. Surf. Sci.*, **253**, 5558–5564.
38. Vorozhtsov, A., Gromov, A., Lerner, M., and Rodkevich, N. (2010) Characterization and analysis of Al nanoparticles passivated with organic layers for energetic applications. Energetic Materials for High Performance, Insensitive Munitions and Zero Pollution: 41 International Annual Conference of ICT – Karlsruhe, Germany: Fraunhofer ICT, June 29 – July 02, 2010, pp. c.1-10 (90474673).
39. Manelis, G.B., Nazin, G.M., Rubtsov, Y.I., and Strupin, V.A. (1996) *Thermal Decomposition and Combustion of Explosives and Gunpowders*, Nauka, Moscow, p. 223 (in Russian).
40. Kubota, N. (2007) *Propellants and Explosives: Thermochemical Aspects of Combustion*, Wiley-VCH Verlag GmbH, Weinheim, p. 509.
41. Toroptseva, A.M., Belogorodskaya, K.V., and Bondarenko, V.M. (1972) *Laboratornyj Praktikum po Khimii i Technologii Vysokomolekulyarnyh Soedinenij(Laboratory Course of Chemistry and Technology of Macromolecular Compounds*), Khimia, Leningrad, p. 416 (in Russian).
42. Trunov, M.A., Schoenitz, M., Zhu, X., and Dreizin, E.L. (2005) Effect of polymorphic phase transformations in Al_2O_3 film on oxidation kinetics of aluminum powders. *Combust. Flame*, **140** (4), 310–318.

5
Characterization of Metallic Nanoparticle Agglomerates

Alfred P. Weber

5.1
Introduction

Owing to their extraordinary optical, electronic, and catalytic properties, individual metal nanoparticles have been in the focus of intensive research over the past few decades. The great potential of nanoparticles is based on quantum size effects such as surface plasmon resonance [1], large specific surface area in catalysis [2], or the mere geometrical extension for membrane transfer [3], for instance. Applications enclose the fields of functional materials and coatings, pigments, biomedical sensors, heterogeneous catalysis, waste water and exhaust air cleaning, and many more. While the properties have mostly been studied on idealized single particles in elaborate laboratory experiments, nanoparticles in technical processes are usually present as agglomerates. This is on one hand the consequence of the high number concentration required in technical relevant loading and, on the other hand, the result of the increasing importance of adhesion forces with decreasing particle size compared to separating forces such as inertia. As an example, an agglomerate of platinum nanoparticles is shown in Figure 5.1a where the single constituents of the agglomerate are called *primary particles*. Being far from classical geometries, the characterization of the ramified structure of these agglomerates became only feasible on the basis of fractal geometry introduced by Mandelbrot [4]. As found in experiments, for example, by Forrest and Witten [5], and numerical simulations [6], these mass fractal structures are generated in aggregation processes where the growth mechanism is reflected in the agglomerate structure. Therefore, as long as the growth regime does not change, the agglomerates exhibit the same structural features independent of their stage of growth, that is, independent of their size. This property of fractal objects, which is called *self-similarity*, is indicated in Figure 5.1 by magnifying parts of a platinum agglomerate, which may correspond to earlier stages of agglomerate formation.

The agglomerate structure will influence optical [7], electrical [8], and chemical behavior [9] of the particles. Obviously, physical and chemical properties of nanoparticle agglomerates will differ significantly from the ones of spherical nanoparticles

Metal Nanopowders: Production, Characterization, and Energetic Applications, First Edition.
Edited by Alexander Gromov and Ulrich Teipel.

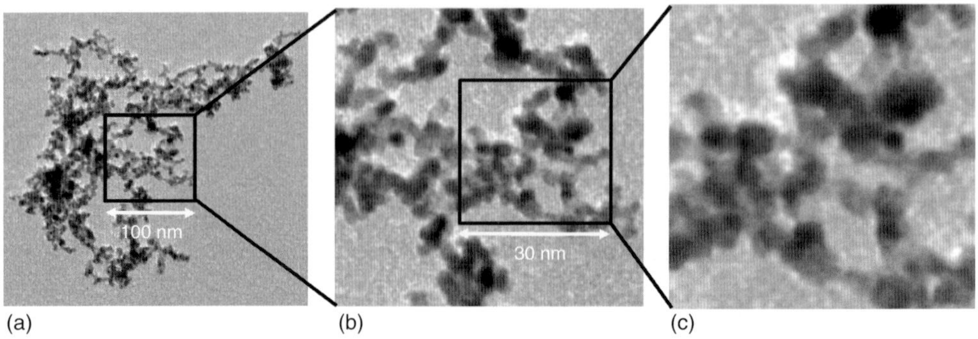

Figure 5.1 TEM micrograph of a platinum agglomerate generated by spark erosion shown at different magnifications ((a) 1 : 150 000, (b) 1 : 500 000, and (c) 1 : 1 000 000).

with the same volume. Therefore, first, appropriate parameters for the characterization of the agglomerate structure are needed and, second, the particle properties have to be related to these structural parameters (structure–property relationship). In addition to the relevance of the agglomerate structure for the material properties, processing and handling of nanopowders also depend critically on the particle structure, for instance, in filtration [10], mixing [11], redispersion [12], and packaging [13], to name only a few (structure–processing relationship). Therefore, this chapter first gives a concise review on the standard techniques to characterize and measure the agglomerate structure. Then, the influence of the agglomerate structure on the particle properties is presented with a few examples including the mechanical and thermal stability and gas-phase reactions on nanoparticle agglomerates.

5.2 Description of the Structure of Nanoparticle Agglomerates

Agglomerates of metal nanoparticles such as shown in Figure 5.1 are typical for particles formed in high-temperature processes such as welding [14] or exhaust gases from smelters [15]. While the dendritic structure of the agglomerates have been observed for a long time, the description of the highly nonsymmetric structures became possible only with the introduction of the concept of fractals in the mid-1970s [4]. Before that the physical behavior of agglomerates was described by shape factors that related to a certain particle property. For instance, the dynamic shape factor is the ratio of the actual drag on an agglomerate to the drag on a sphere of equivalent volume moving at the same velocity [16]. This approach relies on the determination of empirical parameters, which are valuable over a limited size range and apply for a special particle material and form only. For a more general consideration of the agglomerate structure with respect to material and handling properties, the concepts of fractals not only provides a tool to characterize the agglomerate morphology with a few parameters but it was also shown that these parameters are related to the way of agglomerate formation as outlined here.

As a consequence of the self-similarity of their ramified structure, the agglomerates exhibit scaling laws between agglomerate mass and diameter over a certain size range as found in experiments [17] and numerical simulations [18]. This means that the agglomerate morphology appears to remain the same – in a statistical sense – independent of the applied magnification (Figure 5.1). There are two approaches to characterize the fractal nature of the particles: (i) by dividing the particle ensemble in size classes and determining the average mass for particles of each size class or (ii) by establishing the autocorrelation function for individual particles, that is, considering the density of the agglomerate within an imaginary counting sphere [19]. The lower limit for the applicability of the scaling law is given by a few primary particles only [17]. For the upper limit, the statistical termination of the agglomerate is taken into account by a cut-off function [20, 21]. In between these limits, the agglomerate mass M is related to the agglomerate radius R and the fractal dimension D_f by

$$M \propto R^{D_f} \tag{5.1}$$

It was shown that not only for monodisperse but also for polydisperse primary particles the mass fractal dimension D_f can be translated into a number fractal dimension with the same value [22]. If we assume that the monodisperse primary particles are all spherical with radius r_0, Equation 5.1 can be rewritten as radial distribution of the number N of primary particles within a counting sphere of radius R:

$$\frac{M}{m_0} = N = k \cdot \left(\frac{R}{r_0}\right)^{D_f} \tag{5.2}$$

where $m_0 = (4\pi/3) \cdot r_0^3 \cdot \rho_p$ is the primary particle mass, ρ_p is the particle density, and k is called *structure prefactor* [23]. For the characterization of the agglomerate size the radius of gyration R_g is often used, which, for monodisperse primary particles, is given by

$$R_g^2 = \frac{1}{N} \cdot \sum_{i=1}^{N} r_i^2 \tag{5.3}$$

where r_i is the distance of the ith primary particle from the center of mass.

According to Equation 5.3, the radius of gyration is easily accessible in numerical simulations of the agglomerate behavior. Moreover, the radius of gyration represents the relevant size measured in scattering experiments performed in the Rayleigh–Debye–Gans (RDG) regime as outlined subsequently. As a simple measure for the outer agglomerate radius, the minimal radius R_c of the sphere, which completely encloses the agglomerate, can be employed. This radius is relevant when describing the growth of the agglomerates by collision with other particle clusters. If the particle motion occurs in the Stokes regime ($Re < 0.25$), then it is assumed for a mass fractal agglomerate that radius of gyration and collision radius are proportional to each other, where the proportionality factor χ is a function of the fractal dimension only [24]:

$$\chi = \frac{R_c}{R_g} = \left(\frac{D_f + 2}{D_f}\right)^{1/2} \tag{5.4}$$

As every member of the ensemble has a fractal structure, the correlation $g(r)$ between all the points of an individual agglomerate can, alternatively to the consideration of the agglomerate ensemble, be used to evaluate the fractal dimension. The correlation function $g(r)$, which uses the density function $\rho(\vec{r}_i)$ within the agglomerate, gives the probability of finding a particle at a distance r from any other particle belonging to the agglomerate:

$$g(r) = \frac{1}{4\pi} \cdot \sum_{i=1}^{N} \sum_{j=1}^{N} \frac{\rho(\vec{r}_i) \cdot \rho(\vec{r}_j)}{|\vec{r}_i - \vec{r}_j|} \propto r^{D_f - 3} \tag{5.5}$$

where $r = |\vec{r}_i - \vec{r}_j|$ is the distance between the particle i and the particle j with the position vectors $\vec{r}_i$ and $\vec{r}_j$, respectively.

For a sufficiently large number of primary particles in the agglomerates and without rearrangement during growth, the fractal dimensions obtained from the ensemble analysis (Equation 5.2) and from the correlation analysis (Equation 5.5) are identical. However, in both cases the proportionality constants need to be known for quantitative results. In addition, the increasingly stochastic character of the agglomerate mass distribution toward the outer end (see also Figure 5.1) has to be taken into account. There have been various cut-off functions proposed to describe the decay of $g(r)$ for distances r approaching the outer shell of the agglomerate. While mere exponential decay functions [25] seem to perform rather poorly, the overlapping spheres form of the cut-off function represents a good approximation [21]. However, in most practical cases, the cut-off effect will be neglected either by restricting the evaluation of fractal agglomerates to the range where it does not come into play or by averaging over a large number of large agglomerates.

Although the fractal dimension could, in principle, vary between 1 (stringlike agglomerate) and 3 (densely packed agglomerate) there are some values that occur more frequently than others. This is a consequence of the agglomerate formation consisting of collision and sticking processes. In aerosols and colloids during the growth process, clusters meet other clusters via random motion. For a sticking probability of one, this growth regime is called *diffusion-limited cluster aggregation* (*DLCA*). This type of agglomerate formation leads to fractal dimensions of about 1.8, which was found in many numerical simulations [26] and experimental investigations [27]. Also, the agglomerate shown in Figure 5.1 grew by a DLCA mechanism leading to a fractal dimension of 1.81. For long mean free path of the fluid molecules compared to the size of the attaching clusters, as it is possible in aerosols, the clusters may approach each other rather on straight lines, that is, in a ballistic particle motion, resulting in a somewhat higher fractal dimension of 1.9 [28]. Finally, for reduced sticking probabilities, as they are encountered in colloids by repulsive double layers, the growth may shift into the so-called reaction-limited cluster aggregation (RLCA) regime with values of about 2.1 for the fractal dimension [29]. Much higher fractal dimensions are only encountered in special growth regimes such as the ballistic addition of single primary nanoparticles to a growing cluster ($D_f = 2.97$) [29] or by restructuring of agglomerates with initially low fractal dimensions, for instance, induced by heat treatment [20]. Especially, the possible rearrangement of the primary particles after the agglomerate formation prevents a universal bijective correlation between aggregation process and the resulting fractal dimension. Jullien and Meakin [30] showed that DLCA with a simple restructuring algorithm produces structures with the same fractal dimension as the RLCA.

Another important parameter for the description of the agglomerate properties such as mechanical strength and electronic conductivity is the *coordination number*, which can be defined as the average number of contacts of the primary particles in an agglomerate. Compared to the extensive literature on the fractal dimension, the knowledge of the coordination number is scarce [19]. Therefore, Monte Carlo simulations of DLCA and ballistic particle–cluster aggregation were performed [31]. It was found that for agglomerates consisting of <50 primary particles ($N < 50$), the coordination number increased with increasing N. However, for larger values of N, the coordination number approached an asymptotic value depending on the growth mechanism (Figure 5.2). Surprisingly not even for the ballistic particle–cluster aggregation, which leads to high fractal dimensions ($D_f = 2.97$), a high value for the coordination number was observed. For comparison, the coordination number of a random packing of the primary particles is shown with coordination numbers between 5 and 6.

In general, the restructuring as described by Jullien and Meakin [30] is accompanied by an increase of the coordination number, resulting in higher agglomerate strength. Intuitively, one would expect that in restructured agglomerates the fractal

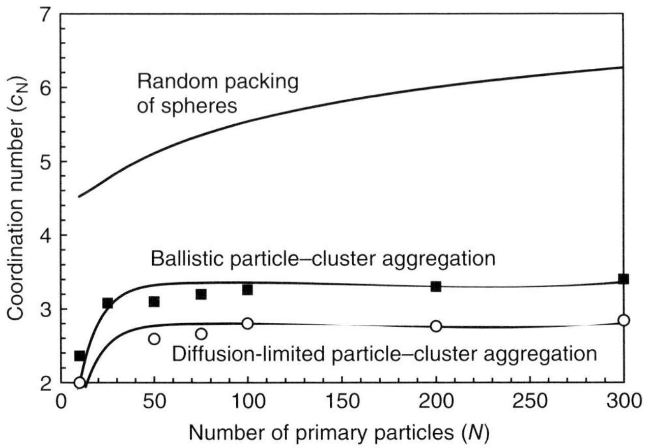

Figure 5.2 Coordination number for ballistic and diffusion-limited aggregation based on random walking models and coordination number calculated for random packing• [31].

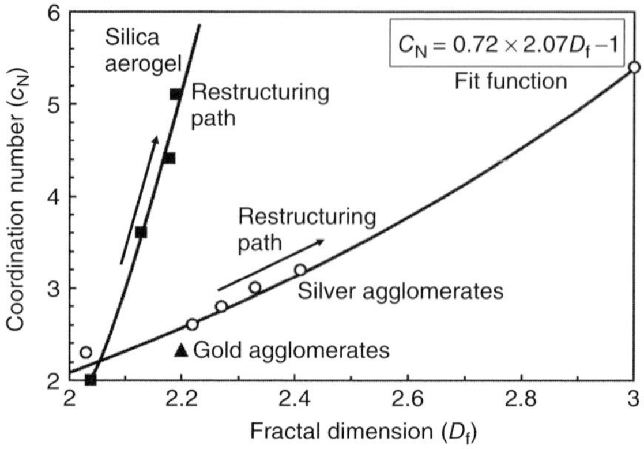

Figure 5.3 Coordination number as a function of the fractal dimension for various particle systems: (■) restructuring of silica aerogels (from [32]), (▲) colloidal gold aggregates measured with small-angle X-ray scattering (SAXS) (from [33]), and (○) restructuring of silver aggregates measured with TEM analysis (adapted from [31]).

dimension correlates with the coordination number. Although there are some experimental and simulation results that indicate such an interdependence (cf. Figure 5.3), there is currently no relationship established on theoretical grounds.

Besides the number of bonds within an agglomerate, the strength of the individual bonds will also influence the stability of aggregated particle systems. While many molecular dynamics (MD) simulations have been performed to determine the contact forces between nanoparticles [34, 35], there are only a few direct measurements of the adhesion forces of nanoparticles available [36]. In particular, most of the results obtained by using atomic force microscopy (AFM) relate to the force between a nanoparticle and an extended surface. There is only one AFM experiment where the interparticle bond strength was measured by pulling titania nanoparticle chains apart until breakage occurred [37]. However, because the breaking position within the chain could not be identified unambiguously, the particle size corresponding to the measured force of 2.5 nN was not exactly known. When assuming that the smallest particles detach first, then the particle diameter was in the range of about 4 nm. Interestingly, before final chain breakage some displacements at an applied force of only 0.5 nN were observed, which were attributed to internal rearrangement in the particle chain possibly due to rolling and shifting of primary particles [37]. This conclusion agrees with the observation that thermally induced restructuring of the agglomerates requires energies that are significantly below the expected bond energies as discussed later in more detail [38].

5.3
Experimental Techniques to Characterize the Agglomerate Structure

Among the structural parameters of nanoparticle agglomerates the fractal dimension is of general interest as it appears in the scaling laws given earlier. Other structural parameters may be significant for special applications such as the agglomerate density in separation techniques. In this chapter, standard imaging and scattering techniques are outlined first, followed by the various methods to determine the agglomerate density.

5.3.1
TEM and 3-D TEM Tomography

Image analysis has been employed for particle characterization for more than a century. In powder technology, the analysis of photographical images of particles larger than a few microns with visible light has reached online capabilities with 500 evaluated and analyzed frames per second based on the enormous speed of processing units [39]. For nanoparticles, the recording of micrographs requires techniques with higher size resolution such as AFM [40], scanning tunneling microscopy (STM) [41], scanning near field optical microscopy (SNOM) [42], or transmission electron microscopy (TEM) [43], on which the focus will be laid here. While the possibilities of TEM analysis to obtain morphology, size distribution, and crystal habit of the primary particles have been addressed, here the ways to characterize the agglomerate structure are of interest. However, the methods of image analysis outlined here are universal and can be applied to micrographs recorded with any of the available imaging techniques.

There are two main strategies to obtain the fractal dimension from micrographs as the one shown in Figure 5.4, namely, box counting methods and nested circles methods. In the box counting technique, the agglomerate is covered with objects (named boxes), for example, small circles or squares, so that every piece of the agglomerate is part of at least one box [21]. Then the minimum number N_b of boxes needed to include all parts of the agglomerate is plotted against the characteristic

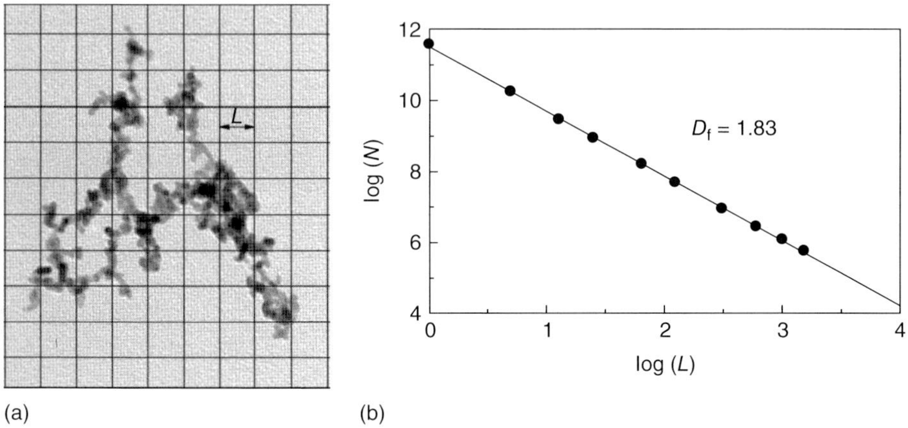

Figure 5.4 (a) Application of the method of "box counting" to a TEM micrograph of a platinum nanoparticle agglomerate and (b) the logarithm number of primary particles within a counting box versus logarithm of a characteristic length.

length L of the counting box. The fractal dimension is obtained from the negative slope in the log–log plot of N_b versus L (cf. Figure 5.4b):

$$D_f = \lim_{L \to 0} \left(-\frac{\log(N_b(L))}{\log(L)} \right) \tag{5.6}$$

In the nested circle technique, a center of mass is chosen within the agglomerate from which circles of radius R are drawn and the number of primary particles N within this circle is counted [22]. The fractal dimension is directly obtained as the slope in the log–log plot of N versus R:

$$D_f = \frac{\log(N(R))}{\log(R)} \tag{5.7}$$

The determination of mass fractal dimension of agglomerates with image analysis is limited to values of D_f of two or less (geometrical transparency condition). Agglomerates with mass fractal dimensions greater than two are geometrical opaque. Therefore, the fractal dimension measured with image analysis D_{im} will be 2, because an electron micrograph provides a 2D projection of a 3D object [21].

$$\begin{aligned} D_{im} &= D_f \qquad D_f \leq 2 \\ D_{im} &= 2 \qquad D_f > 2 \end{aligned} \tag{5.8}$$

However, in reality even for fractal dimensions above two reasonable values can be deduced from projection analysis as long as $D_f \leq 2.4$ [22]. In order to see if the orientation of the agglomerate has a significant influence on the fractal dimension obtained from the projection analysis, a platinum agglomerate was oriented under different angles with respect to the electron beam as illustrated in Figure 5.5.

In Figure 5.6, three TEM micrographs of the same platinum agglomerate under different tilting angles are shown. For these three images, the fractal dimension was measured with the box counting method. The fractal dimensions are nearly the same for the three images, indicating that for geometrical transparent agglomerates no bias due to the orientation is found.

The tilting of the agglomerate alignment is also the basis for the characterization of geometrical opaque agglomerates with transmission electron tomography (3D-TEM). This method is suitable for the structural characterization of inorganic

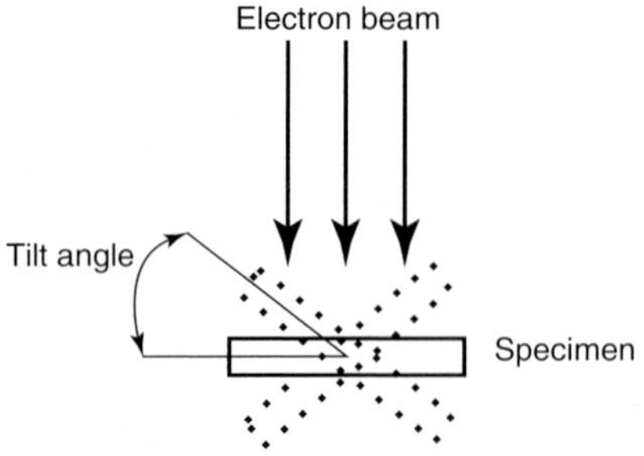

Figure 5.5 Schematic illustration of TEM image acquisition under different angles.

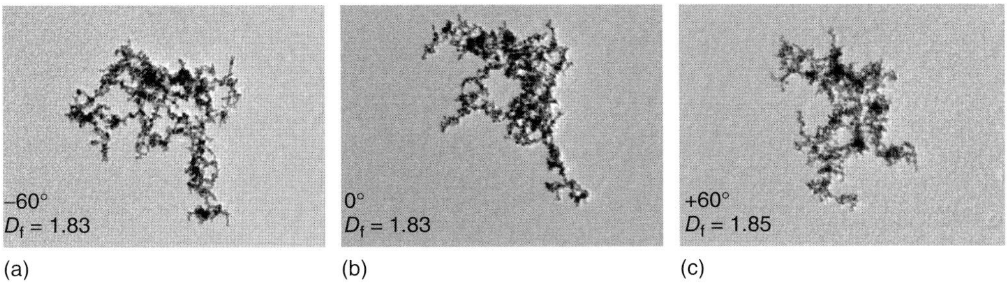

Figure 5.6 TEM images of a platinum agglomerate with different tilt angles; (a) −60°, $D_f = 1.83$; (b) 0°, $D_f = 1.85$; and (c) 60°, $D_f = 1.85$.

materials in the nanometer range [44]. In electron tomography, images are recorded under different tilting so that a 3D reconstruction of the examined agglomerate can be computed. Therefore, it is necessary to take a series of images under different angles. The specimen is tilted in the range of ±60° to ±80° with a small increment. Two different tilting schemes are possible, linear and the Saxton Scheme. Owing to mechanical imperfections, not every image is centered. Thus, an additional step, the alignment, is necessary to recenter the images. This step is the most important for a successful reconstruction [45]. The image series can be aligned with fiducial markers, a cross-correlation function or with a combination of both. For the reconstruction, a backprojection of the 2D projections into 3D is done. Another method of reconstructing is the simultaneous iterative reconstruction technique. Here, the reconstructed object is projected in 2D again. The reprojected image is compared to the real image, although deviations will be corrected.

The reconstruction enables a structural characterization of opaque agglomerates. Furthermore, reconstructed agglomerates allow a more accurate determination of their properties [46]. They compare the surface area and volume of soot agglomerates calculated from 2D projections and 3D reconstruction. The surface areas derived from the 2D projections are 5–16 times larger than the values calculated with the reconstruction method. Calculations of the agglomerate volumes show also an overestimation when using the 2D projections.

5.3.2 Scattering Techniques

The scattering of electromagnetic radiation on particles was treated, for instance, by van de Hulst [47]. Using the size parameter $\alpha = \pi d_p/\lambda$, which is the ratio of particle circumference and wave length of the radiation, the scattering of visible light and X-ray can be described with the same theory. However, for a fixed scattering angle, the scattering on nanoparticles will be performed in different regimes for light and X-rays owing to the large difference in wave length. Therefore, light is more suitable for probing particles in the micrometer range, whereas X-rays are appropriate for much smaller length scales. The scattering on agglomerates is marked by the superposition of scattering events on single particles (primary particles) and by the interference of the scattered radiation due to the spatial arrangement of the primary particles. The simplest theory to describe such a situation is the Rayleigh–Gans–Debye (RGD) theory, where it is assumed that the agglomerate consists of individual Rayleigh scatterers having no interaction with each other. The Rayleigh scattering regime is characterized by the condition $a \ll 1$. The assumption of the RGD theory that the individual scatterers are independent implies that second- and higher order scattering is negligible. On the basis of numerical simulations on fractal aggregates, Farias *et al.* [48] concluded that the scattered intensity predicted by the RGD approximation is accurate within 10% when the following conditions are fulfilled:

$$|m-1| \leq 1 \quad \text{and} \quad \frac{2\pi \cdot n_0}{\lambda_0} \cdot L \cdot |m-1| \leq 0.6 \tag{5.9}$$

where m is the relative refractive index of the scatterers, n_0 the refractive index of the dispersion medium, λ_0 the *in vacuo* wavelength of the radiation, and L the length of the scattering body. Owing to their relatively weak interaction with matter X-rays and neutrons, which are also used in scattering experiments, they comply always with the requirements for the application of RGD theory. In order to include also the scattering angle θ in the description, the *scattering wave vector* $\vec{q}$ is defined as the difference of incident wave vector $\vec{k}_{in}$ and scattered wave vector $\vec{k}_{sca}$:

$$q = |\vec{q}| = |\vec{k}_{in} - \vec{k}_{sca}| = \frac{4\pi \cdot n_0}{\lambda_0} \cdot \sin\left(\frac{\theta}{2}\right) \tag{5.10}$$

When considering two scatterers i and j, which are separated by the vector $\vec{r}_{ij}$, the difference in path length of radiation scattered at $\vec{q}$ from both particles is given by $\vec{r}_{ij} \cdot \vec{q}$. Using this product, the scattering character can be divided into coherent and incoherent scattering. For $\vec{r}_{ij} \cdot \vec{q} < 1$: in phase scattering leads to constructive interference of the wave amplitudes so that the intensity is proportional to the square of the sum of the scattering particle masses. On the other hand, for $\vec{r}_{ij} \cdot \vec{q} > 1$: the intensity is proportional to the sum of the squares of the scattering particle masses. Translated to fractal aggregates, this means that scattering from agglomerates is a linear superposition of contributions from volumes in which the primary particles scatter coherently ($r_{ij} < 1/q$). Therefore, the scattering wave vector q can be considered as the probing scale for structures of size $1/q$. This picture serves as the pattern for the factorization of the scattering intensity into a form factor $P(q)$ describing the scattered intensity from a single primary particle and a structure factor $S(q)$ taking into account the interference of the single scattering events:

$$I(q) \propto P(q) \cdot S(q) \tag{5.11}$$

Using the scaling law (Equation 5.2), the number of primary particles within a coherently scattering volume is given by

$$N_{coh} \propto \left(\frac{1/q}{r_0}\right)^{D_f} \tag{5.12}$$

and the number of such volumes within the agglomerate is given by

$$N_{\mathrm{incoh}} \propto \left(\frac{R_{\mathrm{c}}}{1/q}\right)^{D_{\mathrm{f}}} \tag{5.13}$$

with the outer radius R_{c} of the agglomerate.

For small values of q the form factor $P(q)$ is constant and the scattering intensity I is influenced mainly by structural effects. When taken into account that coherent scattering contributes quadratically and incoherent scattering linearly to the structure factor, then for small q the following relation is obtained:

$$I(q) \propto S(q) \propto N_{\mathrm{incoh}} \cdot N_{\mathrm{coh}}^{2} \propto q^{-D_{\mathrm{f}}} \tag{5.14}$$

From a double logarithmic plot of the scattering intensity versus q, the mass fractal dimension can be obtained from the negative slope.

5.3.3 Direct Determination of Agglomerate Mass and Size

As an alternative to scattering techniques that may be involved in the mathematical extraction of the agglomerate structure, a straightforward approach to obtain the fractal dimension is the determination of the size-dependent agglomerate density. On the basis of Equation 5.1, the mass density is given by

$$\rho(R) = \frac{M(R)}{V(R)} = \rho_0 \cdot \left(\frac{R}{r_0}\right)^{D_{\mathrm{f}}-3} \tag{5.15}$$

where M is the mass of an agglomerate with radius R, $V(R) = (4\pi/3)R^3$ its volume, ρ_0 and r_0 density and radius of the primary particles, respectively. In Figure 5.7, different possibilities to determine the fractal dimension based on the measurement of particle density or mass are summarized. They all depend on the fractionation of a monodisperse size class from the original aerosol obtained with a differential mobility analyzer (DMA), which consists of two concentric cylindrical electrodes (Figure 5.7). Between the inner and outer cylinder, a voltage difference is applied leading to an electric field perpendicular to the direction of the gas flow. While the inner gas stream between the two cylinders contains particle-free sheath gas, the outer stream consists of the aerosol. In order to classify the particles in the electrical field according to their electrical mobility, the particles are charged in a radioactive source (bipolar charging) or with a corona charger (unipolar charging). Then, the particles experience an electrostatic force toward the inner electrode, which induces a friction force with the carrier gas leading to a constant migration velocity perpendicular to the convective transport of the particles in the gas flow.

The number concentration of the classified particles can be detected by a condensation particle counter (CPC) or a Faraday cup electrometer (FCE). In the CPC, the particles act as condensation nuclei for an alcohol vapor and grow to a final size of about 10 μm. These large particles are then counted one by one using single-particle light scattering. In the FCE, the charged particles leaving the DMA are collected in an isolated particle filter, which is enclosed in a Faraday cup. The

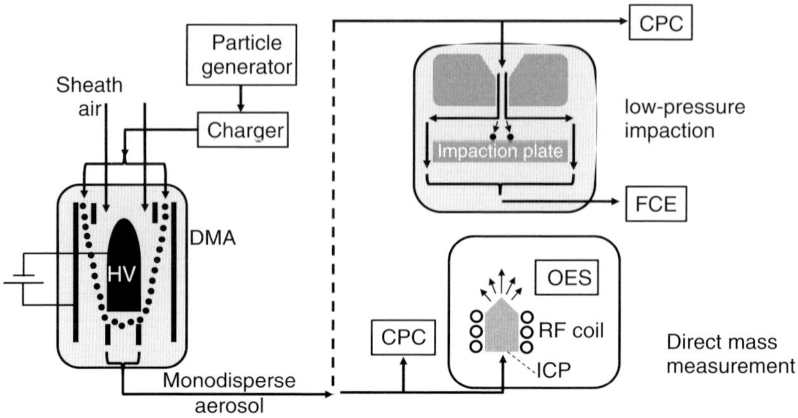

Figure 5.7 Principle of the particle classification with a differential mobility analyzer (DMA) (left), which is used for (i) the direct measurements of the agglomerate mass with inductively coupled plasma optical emission spectroscopy (ICP-OES) (right – bottom) and (ii) for the density determination with low pressure impaction (right – top). As particle counters, condensation particle counters (CPC) and Faraday cup electrometer (FCE) are employed.

influenced current to the Faraday cup is amplified and measured with an operation amplifier.

Direct measurements of the mass of silver agglomerates were performed with inductively coupled plasma optical emission spectroscopy (ICP-OES) [49]. In this study, silver nanoparticle agglomerates were produced with a spark discharge generator [50], size selected with a DMA, and introduced into the ICP-OES where the particles evaporated in argon plasma at temperatures of about 6000 K. The resulting silver atoms and ions were excited in the high-temperature zone and, during subsequent cooling and returning to the ground state, emitted light characteristic of silver, which was analyzed and quantified with an optical spectrometer. To relate the ICP-OES signal to an absolute silver mass, particle samples were taken in parallel on membrane filters, which were quantified by neutron activation analysis (NAA) [51]. Owing to the high sensitivity of NAA to silver, minute amounts could be quantified, which would have not been possible with microbalances. Even with the less sensitive optical spectrometry (compared to mass spectrometry), a detection limit for Ag nanoparticles of less than a femtogram (10^{-15} g) per cubic centimeter of carrier gas was achieved corresponding, for example, to about one 74 nm silver agglomerate per cubic centimeter of gas. The data in Figure 5.8 represent the behavior of the density with agglomerate size for silver particles. Interestingly, the density stays constant up to the mean free path of the carrier gas and decays only for larger agglomerates. Taking into account the modeling of the agglomerate growth discussed earlier, this result indicates that up to the mean free path agglomeration was dominated by ballistic particle–cluster aggregation ($D_f = 2.97$) and only for larger agglomerates changed to diffusional cluster–cluster aggregation ($D_f = 1.80$).

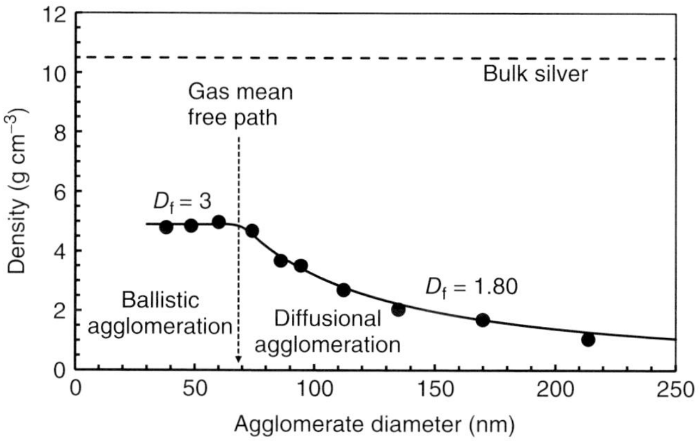

Figure 5.8 Density of silver agglomerates exhibiting two growth regimes separated by the gas mean free path.

Such a transition of the agglomerate growth regime was observed for the first time in these experiments.

The small silver agglomerates with $D_f = 3$ have an effective density of $4.9\,\mathrm{g\,cm^{-3}}$, which is only 47% of the silver bulk density of $10.5\,\mathrm{g\,cm^{-3}}$. For monodisperse spherical primary particles, the densest packing corresponds to a hexagonally closed packing with a volume filling degree of 76% (i.e., $8\,\mathrm{g\,cm^{-3}}$), which may be somewhat higher for polydisperse primary particle ensembles. When comparing this maximum value to the observed density, it is obvious that the ballistic particle–cluster agglomeration process leads to high fractal dimensions, but not to a closed packing.

Low-pressure impaction is another technique that delivers the particle density. Here, the particles classified with the DMA pass first a critical orifice and are then accelerated in a converging nozzle toward an impaction plate (Figure 5.7, right – top). In the 90° bending of the gas flow in front of the impaction plate, particles with high inertia cannot follow the deviation and are separated from the gas stream. The main parameter determining the degree of particle deposition is the Stokes number, which relates the particle inertia to its drag force with the carrier gas. By varying the pressure in the impaction chamber, the Stokes number (Stk) can be scanned and the fraction of deposited particles can be determined from the FCE signal behind the impactor by counting the charge of the nonseparated particles. For this type of impactor, the following expression applies [52]:

$$\mathrm{Stk}_{50} = 0.178 \cdot \frac{d_p \cdot \rho_{\mathrm{eff}}}{p_{50}^2} = \mathrm{constant} \tag{5.16}$$

where Stk_{50} is the Stokes number, at which 50% of the particles with diameter d_p and density ρ_{eff} are separated, and p_{50} is the corresponding pressure in the impaction chamber. Typical separation curves as a function of the applied pressure is shown in Figure 5.9a. By varying the agglomerate size the particle density can be

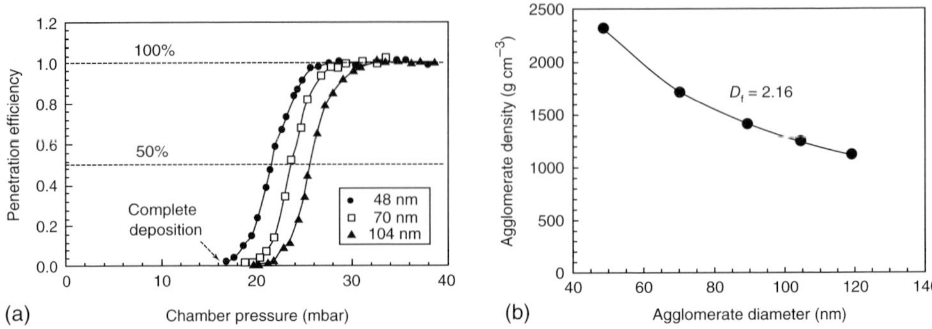

Figure 5.9 Separation curves for silver agglomerates in a low-pressure impactor (a) and agglomerate density as a function of particle diameter (b).

evaluated as a function of particle diameter, from which the fractal dimension can be obtained from Equation 5.15 (Figure 5.9b).

The low-pressure impaction was also employed for the determination of the thickness of coatings of nanoparticles [53]. From the separation curve the effective density of the core-shell-nanoparticles is determined according to Equation 5.16. From the particle size known from the DMA classification and the effective density, the layer thickness can be calculated. However, as it is assumed that the overall density of the composite particles lies between the densities of core and shell material, this technique works only for core and shell materials with different densities.

5.4 Mechanical Stability

In sampling and processing, particles may experience significant stressing by acceleration, shear stress, and wall collisions. Stahlmecke *et al.* [54] observed fragmentation of industrial nanopowders when passing a nozzle where the degree of fragmentation (DoF) depended on the particle material. However, they could not clearly relate the disintegration to a main type of stressing. Also, in tube flow agglomerate fragmentation was observed [55]. While Peng and Williams [56] found fragmentation along the total tube, others concluded stress-induced fragmentation to occur preferentially on the flow contraction at the tube entrance [57, 58]. Zumaeta *et al.* [55] explained this discrepancy with the different flow regimes, implying that for laminar conditions the acceleration dominates, while for turbulent flow the agglomerate fragmentation is mainly due to shear and impaction on the tube wall. In order to quantify the individual contributions, Wengeler and Nirschl [59] included different stressing mechanisms in a fluid-dynamic model and compared the simulation results with experiments. Nevertheless, owing to the lack of a reliable model for impact fragmentation, wall collisions were not considered. They found experimentally that the detachment of single primary particles or clusters of them dominated for silica agglomerates, which was called *erosion*. However, in all these studies, various stress mechanisms were at work simultaneously and, in addition, the drag force on the complicated agglomerate structures could only approximately be taken into account, which inhibited the correct description of the particle motion and of the experienced shear stress. There are several indications that impaction is the most effective stressing mechanism for fragmentation, however, mostly investigated for larger particles [60–63]. Therefore, to deepen the understanding of the relationship between applied stress and structural changes of the agglomerates, impact fragmentation at a known velocity and known particle trajectory was employed here for nanoparticle agglomerates. The few simulations of the fragmentation behavior of nanoparticle clusters predict a logarithmic increase in the DoF, defined as the fraction of broken bonds, with the impact energy [64]. Interestingly, simulation results for impact fragmentation [65] as well as for shear-induced fragmentation [66] predict some induction time, that is, some delay between the time of maximum stressing and the resulting fragmentation. While these induction times are too short to be observed in experiments, the logarithmic dependence of the DoF was indeed corroborated by the experimental results presented in the following.

Agglomerates of metallic nanoparticles were produced by spark discharges and subjected to wall collisions in a low-pressure impactor shown in Figure 5.7 (right –top). The impaction velocity was adjusted by the pressure and flow rate of the carrier gas. On the basis of self-similarity of the flow pattern, dimensionless equations were established, which allow the direct prediction of the impact velocity [67]. The correctness of the approach was verified by comparison with elaborate computational fluid dynamics (CFD) calculations and confirmed by the experimental results such as separation curves and spot size of the deposited particles [67]. To evaluate the fragmentation patterns of the impacted agglomerates, a TEM grid coated with a carbon film was placed in the center of the impaction plate. As the agglomerates were size selected in a DMA (cf. Figure 5.7), the projection area of the particles before impaction was known. For a given mobility class, the initial number of bonds per agglomerate was determined by image analysis of the TEM micrographs for particles sampled by diffusional deposition ($v \rightarrow 0$) where fragmentation could be neglected. Then, for agglomerates impacted at a known velocity, the average total number of remaining bonds was determined. In addition, the size distribution of the fragments was evaluated. To obtain reliable data, the coverage on the TEM grid was so low that all agglomerates were clearly separated and a statistically significant number of agglomerates was evaluated (>200). The DoF is defined as

$$\text{DoF} = \frac{\#\,\text{broken bonds}}{\#\,\text{initial bonds}} \tag{5.17}$$

The DoF was determined as a function of the impact velocity for silver agglomerates of primary particles with different average sizes (6.5, 7.7, and 8.3 nm) and for platinum agglomerates ($d_0 = 4.8$ nm) both produced by spark discharge. The results are shown in Figure 5.10. It becomes obvious that for smaller primary particles higher impact velocities are required to obtain a given DoF. This is a consequence of the ratio of binding forces (usually proportional to the diameter)

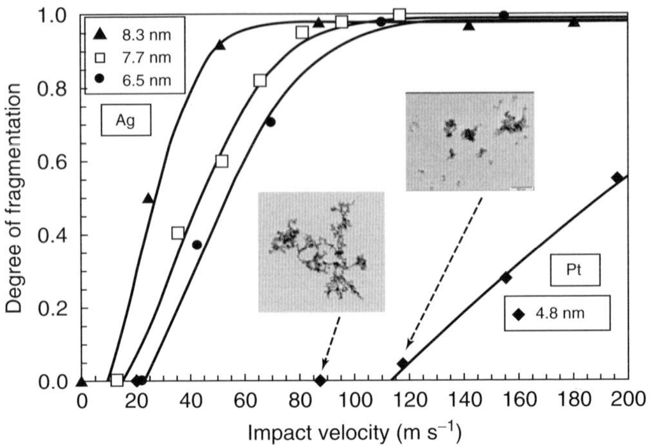

Figure 5.10 Degree of fragmentation as a function of impact velocity for various primary particle sizes of silver agglomerates and for platinum agglomerates.

to separating forces (proportional to the diameter cubed), which increases with decreasing particle size. There is also a clear influence of the particle material observed. Although the Pt primary particles are smaller than the Ag primary particles, the inset of fragmentation occurs at significantly higher impact velocities and the increase in the DoF is more moderate for Pt than for Ag. The reason for the higher mechanical stability of the platinum agglomerates is likely due to its surface state. Pt is only marginally covered with oxygen, while experiments performed with aerosol photoemission spectroscopy [68] revealed that the surface of silver easily acquires oxygen even under high purity conditions. In fact, it was found that even minute surface contaminations of metal nanoparticles retard very much their sintering kinetics [69], known as *pinning effect*. The contaminants on the silver surface may increase the separation distance of the primary particles, reducing the bond strength. In contrast, the high purity of the Pt surface allows close contact, which may even result in partial sintering in the contact points even at strongly reduced temperatures, a phenomenon observed in MD simulations for other metallic nanoparticles [70]. The conclusion that in the Pt agglomerates the bond strengths are much higher than in the Ag agglomerates is also corroborated when the minimal kinetic energy for fragmentation is compared with the van der Waals bond energy. Assuming the usual separation distance of 0.4 nm, the kinetic energy of the Pt particles exceeds the van der Waals energy by two orders of magnitude ($\approx$80) before fragmentation occurs. In the case of silver, there is only a factor of about 4.

As shown in Figure 5.10, the predicted logarithmic increase in the DoF with impact energy is also observed. Sator and Hietala [64] argue that with increasing kinetic impact energy, an increasing fraction of the energy is transferred to the fragments as rebound energy, leading to more inefficient energy utilization for bond breaking. To show that this behavior applies to all agglomerates independent of the primary particle size and the material, the so-called Weibull statistics [71] was applied, which is based on the idea that for uniform stressing a chain breaks at its weakest link. Developed originally for the failure prediction of brittle materials [72], Weichert [73] introduced the Weibull statistics to the field of particle comminution and Seipenbusch *et al.* [74] extended it to the fragmentation of nanoparticle agglomerates. They proposed the following form for the DoF:

$$\mathrm{DoF} = \mathrm{DoF}_{\max} \cdot (1 - \exp(-f(d_0, N, \text{material}) \cdot (v^2 - v_0^2)) \tag{5.18}$$

where $\mathrm{DoF}_{\max}$ is the maximum achievable fragmentation degree; f is a function depending on particle material, primary particle diameter d_0, and number N of primary particles in the agglomerate; while v corresponds to the agglomerate impact velocity and v_0 to the speed at which fragmentation starts. The maximum DoF may be limited by the fraction of hard contacts, which exhibits very strong interparticle bonds originating from local sintering, for instance [75]. For commercial silica nanoparticles, the fraction of "unbreakable" bonds may be as high as 65% [76].

From fitting of Equation 5.18 to the fragmentation curves presented in Figure 5.10 the values for f and v_0 were obtained, leading to a quite uniform fragmentation behavior for silver and platinum agglomerates (Figure 5.11). The good fit justifies the assumption that the stressing over the agglomerate is quite homogeneous. This might be surprising when considering the dendritic random structure of the agglomerates (Figures 5.1 and 5.6). However, as mentioned earlier, the interparticle bonds show some flexibility such as sliding or rotating before breaking, so that possibly the agglomerate is strongly flattened during the first stage of collision. In addition, the predicted induction time between maximum stressing and fracturing events favors an equalization of stress before bond breaking occurs. In contrast to

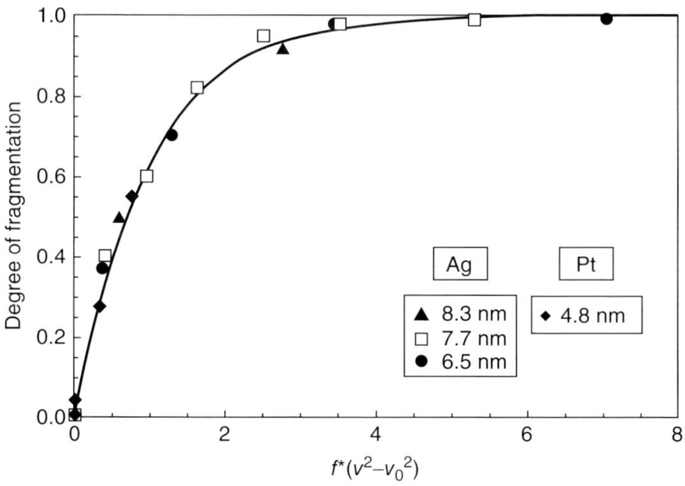

Figure 5.11 Degree of fragmentation of agglomerates of different materials and with different primary particle diameters as a function of the Weibull parameter $f * (v^2 - v_0^2)$ (cf. Equation 5.18).

the results of [74], the Weibull statistics describes all the agglomerates investigated here rather well. While in [74] some agglomerates consisted of a very limited number of primary particles, here all agglomerates were made up of at least several hundreds of primary particles. This indicates that the function f becomes independent of the number of primary particles for sufficiently large values of N. While the material dependence of f has been discussed earlier, its size dependence cannot be deduced from the limited diameter variations available so far.

5.5 Thermal Stability

Depending on the interparticle forces, nanoparticle agglomerates can behave differently when heated *in situ* (in the carrier gas stream). For strong bonds, the agglomerate size decreases, while the thickness of the branches increases because of local sintering of primary particles and the fractal dimension stays constant [77]. The driving force is the minimization of the surface free energy by coalescence [78]. For weak bonds, the primary particles undergo restructuring, resulting in more compact arrangements [20]. This latter type of restructuring is accompanied by an increase in agglomerate density and coordination number. The driving force here is the maximization of the number of bonds, resulting in a lowering of the agglomerate Gibbs free energy. The kinetics of these two mechanisms are described by the following equations:

$$\text{Coalescence of primary particles:}\quad \frac{\mathrm{d}(a - a_{\mathrm{sph}})}{\mathrm{d}t} = -\frac{1}{\tau_{\mathrm{c}}} \cdot (a - a_{\mathrm{sph}}) \tag{5.19}$$

a = surface of two initial particles
a_{sph} = surface of completely coalesced particles (single sphere)
τ_{c} = characteristic coalescence time

$$\text{Rearrangement of agglomerate:}\quad \frac{\mathrm{d}(c_{\mathrm{N}} - c_{\mathrm{N}\infty})}{\mathrm{d}t} = -\frac{1}{L} \cdot (c_{\mathrm{N}} - c_{\mathrm{N}\infty}) \tag{5.20}$$

c_{N} = initial coordination number
$c_{\mathrm{N}\infty}$ = coordination number in closed packed structure
$L = A * \exp(-E_{\mathrm{a}}/RT)$ = coefficient for activated restructuring
A = preexponential factor
E_{a} = activation energy

As in real metal nanoparticle agglomerates, the bond strengths are distributed, as reflected in the resistance against mechanical stress (cf. Figure 5.10), a superposition of both mechanisms may be at work as will be shown subsequently.

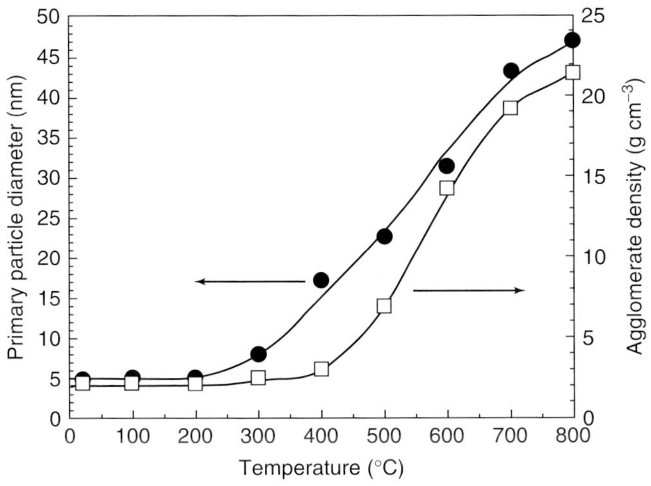

Figure 5.12 Thermal restructuring of Pt nanoparticle agglomerates: simultaneous coalescence of primary particles and compactization of the agglomerate indicated by the agglomerate density.

For gold nanoparticle agglomerates, Nakaso *et al.* [79] found a transition between the two mechanisms. They concluded that for small primary particles coalescence dominates, while for larger ones rearrangement is more pronounced. The size dependence of coalescence is well known and also reflected in Equation 5.19. Rearrangement seems to be less influenced by the particle size. However, both processes are activated processes. Therefore, the rate of coalescence and restructuring is expected to increase with temperature and, for different activation energies, their relative importance should change. In Figure 5.12, the increase in the primary particles of Pt agglomerates and the agglomerate density with heating temperature is shown.

Using the relationship given in Figure 5.3, Weber and Friedlander [38] deduced corresponding coordination numbers from experimentally determined fractal dimensions as a function of the heating temperature. From fitting Equation 5.20 to experimental results, they obtained the activation energy for restructuring. The obtained values were about one order of magnitude lower than the van der Waals bond energies, indicating that restructuring does not proceed over bond breaking but rather particle dislocation by sliding and rolling. In particular, from TEM micrographs of the different restructuring degrees, they concluded that whole branches, likely to be attached in a late stage in the agglomerate formation when temperatures had dropped and contact sintering ceased, may rotate in a hinging motion, leading to a more compact arrangement. However, minute amounts of surface contaminants are able to suppress effectively the motion of the primary particles against each other due to sliding, rolling, and rotating. This was, for instance, observed for platinum nanoparticle agglomerates when they were used as catalyst for the dehydrogenation of propane, which led to the deposition of carbon atoms on the Pt surface (Figure 5.13).

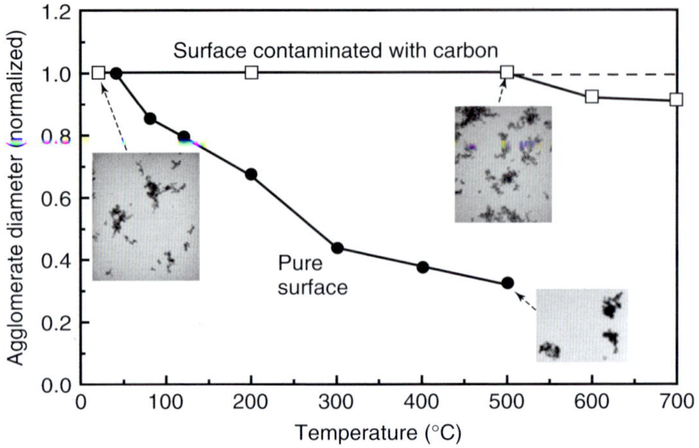

Figure 5.13 Thermal restructuring of Pt nanoparticle agglomerates in nitrogen as carrier gas (indicated by the agglomerate diameter): for pure Pt surfaces, the restructuring starts for temperatures below 100 °C, whereas for carbon contaminated surfaces the heating temperatures need to be above 500 °C to induce rearrangement.

The contamination of the contact zone by impurity deposition may be an effective way of suppressing sintering and restructuring of nanoparticle agglomerates. Recent approaches to extend the application range of noble metal catalysts follow up on this route [80] together with the immobilization effect of highly textured support materials [81].

5.6
Rate-Limiting Steps: Gas Transport versus Reaction Velocity

As mentioned in the Introduction, the structure of nanoparticle agglomerates can have a substantial influence on their catalytic activity. Especially for fast gas-phase reactions, transport limitations may come into play and only parts of the total surface will participate on the reaction. In the following, the active surface fraction will be determined for the hydrogen oxidation on gas-borne platinum nanoparticle agglomerates. The particles were produced in nitrogen by spark discharge and mixed with the educt gases in two different ways: (i) hydrogen and oxygen were added at the same time and (ii) oxygen was added first to saturate the Pt surface and hydrogen was added with a delay of some seconds. The educt gas concentrations were only a few percent and the reaction was performed at room temperature to avoid excessive release of heat (avoiding sintering) and preventing the risk of explosion. The reaction time was varied and the amount of reaction product was determined by measuring the water concentration in the exhaust gas with Fourier transformed infrared (FTIR) spectroscopy. From the cumulative water concentration, the time-dependent reactivity was deduced, given as turnover rate (TOR), which is the number of product molecules formed per time and per active

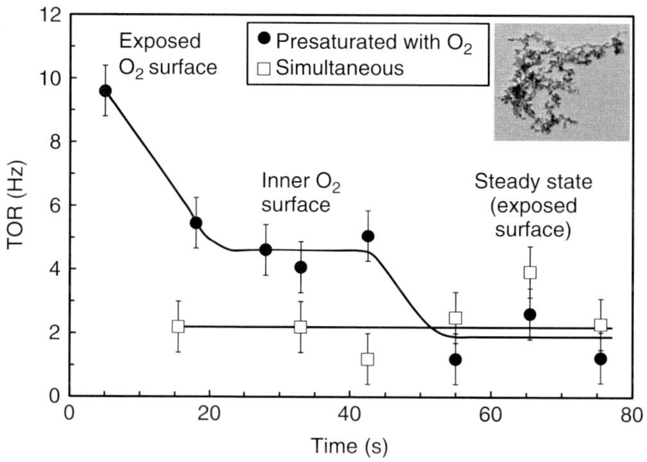

Figure 5.14 Reactivity of platinum nanoparticle agglomerates: (i) for simultaneous addition of hydrogen and oxygen and (ii) for presaturation of the platinum surface with oxygen before adding the hydrogen. The inset shows the typical ramified structure of the Pt agglomerates.

surface site. The results are presented in Figure 5.14. For simultaneous addition of the educt gases, the TOR stays constant over the investigated reaction time. This observation is explained by the fact that hydrogen and oxygen molecules are competing for the available surface sites [82]. However, the hydrogen will cover very quickly nearly the whole available platinum surface as hydrogen diffusivity in the gas phase and on the surface is much faster than the ones of oxygen. In addition, a hydrogen molecule needs only one free surface site for adsorption, whereas an oxygen molecule requires two free sites. This leads to the situation that more than 95% of the surface is covered with hydrogen and the rate-limiting step is the adsorption of oxygen at two free neighboring sites.

In the case of oxygen presaturation of the platinum surface, the hydrogen starts to propagate from the exposed surface of the agglomerate and spreads toward the inner surface, always releasing water molecules. During this spreading, the exposed surface is already in the steady state of the simultaneous addition. Finally, when all preadsorbed oxygen is consumed, the TOR drops to the level of the steady state where no more oxygen reaches and adsorbs onto the inner surface. By adding all the contributions from inner and exposed surface reactions, it can be concluded that in the final steady state only one-third of the total agglomerate surface participates in the fast reaction of hydrogen oxidation [83].

5.7 Conclusions

Agglomerates of metal nanoparticles exhibit dendritic structures, which can be described by the concept of fractals. Characterization methods for the fractal dimension enclose image analysis, scattering techniques, and density scaling measurements. When exposed to mechanical stress the agglomerates may disintegrate into fragments. A method to determine the threshold for impact stressing was presented here. Using Weibull statistics, it is possible to give a general description of the fragmentation behavior as a function of applied stress, taking into account interparticle bond strength and primary particle size. At elevated temperatures, the primary particles of an agglomerate may undergo coalescence and rearrangement, leading to a more compact structure. It was observed that for metal nanoparticle agglomerates, sintering and restructuring processes are superimposed because of the broad distribution of interparticle bond strength. Such a loss of accessible surface is undesired in applications such as heterogeneous catalysis. Therefore, methods are investigated to extend the thermal stability of catalyst nanoparticles to higher temperatures. A part of these stabilization mechanisms is based on the interface passivation by introducing contaminants into the contact zone as shown here for C-stabilized platinum nanoparticles. Finally, it was shown that the agglomerate structure can also influence the reactivity of catalysts particles when the transport of educt molecules is slow compared to the reaction speed. Using the oxidation of hydrogen on platinum nanoparticle agglomerates, it was found that only one-third of the total surface participates in the reaction.

Acknowledgments

The author thanks his coworkers, Dr.-Ing. K. Legenhausen, Dipl.-Ing. S. Rennecke, M.Sc. M. Gensch, Dipl.-Wi-Ing. L. Zeng, Dipl.-Ing. D. Koch, for their contributions and the German Science Foundation (DFG) for financial support.

References

1. Yeshchenko, O.A., Dmitruk, I.M., Alexeenko, A.A., Kotko, A.V., Verdal, J., and Pinchuk, A.O. (2012) *Plasmonics*, **7**, 685–694.
2. Köhler, J.M., Abahmane, L., Wagner, J., Albert, J., and Mayer, G. (2008) *Chem. Eng. Sci.*, **63** (20), 5048–5055.
3. Urban, A.S., Pfeiffer, T., Fedoruk, M., Lutich, A.A., and Feldmann, J. (2011) *ACS Nano*, **5** (5), 3585–3590.
4. Mandelbrot, B.B. (1977) *Fractals, Form, Chance, and Dimension*, Freeman, San Francisco, CA.
5. Forrest, S.R. and Witten, T.A. (1979) *J. Phys. A*, **12**, L109.
6. Meakin, P. (1983) *Phys. Rev. A*, **27**, 1495.
7. Sattler, K. (2002) in *Handbook of Thin Film Materials* (ed H.S. Nalwa), Academic Press.
8. Riefler, N. and Mädler, L. (2010) *J. Nanopart. Res.*, **12**, 853–863.
9. Maillard, F., Schreier, S., Hanzlik, M., Savinova, E.R., Weinkauf, S., and Stimming, U. (2005) *Phys. Chem. Chem. Phys.*, **7**, 385–393.
10. Kim, S.C., Wang, J., Emery, M.S., Shin, W.G., Mulholland, G.W., and Pui, D.Y.H. (2009) *Aerosol Sci. Technol.*, **43**, 344–355.
11. Hogg, R. (2009) *KONA Powder Particle J.*, **27**, 3–17.
12. Müller, F., Peukert, W., Polke, R., and Stenger, F. (2004) *Int. J. Miner. Process.*, **74S**, S31–S41.

13. Prescott, J.K. and Barnum, R.A. (2000) *Pharm. Technol.*, **24**, 60–85.
14. Brand, P., Lenz, K., Reisgen, U., and Kraus, T. (2012) *Ann. Occup. Hyg.*, **56** (8), 948–58. doi: 10.1093
15. Hoflich, B.L.W., Weinbruch, S., Theissmann, R., Gorzawski, H., Ebert, M., Ortner, H.M., Skogstad, A., Ellingsen, D.G., Drablos, P.A., and Thomassen, Y. (2005) *J. Environ. Monit.*, **7** (5), 419–424.
16. Fuchs, N.A. (1964) *The Mechanics of Aerosols*, Pergamon Press, New York.
17. Schmidt-Ott, A. (1988) *J. Aerosol Sci.*, **19**, 553–563.
18. Weitz, D.A., Huang, J.S., Lin, M.Y., and Sung, J. (1985) *Phys. Rev. Lett.*, **54** (13), 1416–1419.
19. Friedlander, S.K. (2000) *Smoke, Dust and Haze*, Oxford University Press, New York.
20. Sorensen, C.M. (1997) in *Handbook of Surface and Colloid Chemistry* (ed K.S. Birdi), CRC Press, New York.
21. Bushell, G.C., Yan, Y.D., Woodfield, D., Raper, J., and Amal, R. (2002) On techniques for the measurement of the mass fractal dimension of aggregates. *Adv. Colloid Interface Sci.*, **95** (1–50).
22. Tence, M., Chevalier, J.P., and Jullien, R. (1986) *J. Phys.*, **47**, 1989.
23. Wu, M.K. and Friedlander, S.K. (1993) *J. Colloid Interface Sci.*, **159** (1), 246–248.
24. Rogak, S.N., Flagen, R.C., and Nguyen, H.V. (1993) *Aerosol Sci. Technol.*, **18**, 25.
25. Tang, S. (1999) *Colloids Surf., A*, **157**, 185.
26. Meakin, P. (1988) *Adv. Colloid Interface Sci.*, **28**, 249–331.
27. Schmidt-Ott, A. (1988) *Appl. Phys. Lett.*, **52**, 954.
28. Samson, G.W.S.J., Mountain, R.D., and Ernst, M.H. (1988) *Energy Fuels*, **2**, 481–486.
29. Schaefer, D.W. (1988) *MRS Bull.*, **13** (2), 22.
30. Jullien, R. and Meakin, P. (1989) *J. Colloid Interface Sci.*, **127**, 265.
31. Weber, A.P. and Friedlander, S.K. (1997) *J. Aerosol Sci.*, **28** (Suppl. 1), S765.
32. Hasmy, A., Foret, M., Pelous, J., and Jullien, R. (1993) *Phys. Rev. B*, **48** (13), 9345–9354.
33. Dimon, P., Sinha, S.K., Weitz, D.A., Safinya, C.R., Smith, G.S., Varady, W.A., and Lindsay, H.M. (1986) *Phys. Rev. Lett.*, **57**, 595–598.
34. Miesbauer, O., Götzinger, M., and Peukert, W. (2003) *Nanotechnology*, **14**, 371–376.
35. Alimohammadi, M. and Fichthorn, K.A. (2009) *Nano Lett.*, **9**, 4198–4203.
36. Vakarelski, I.U. and Higashitani, K. (2006) *Langmuir*, **22**, 2931–2934.
37. Salameh, S., Schneider, J., Laube, J., Alessandrini, A., Facci, P., Seo, J.W., ColombiCiacchi, L., and Mädler, L. (2012) *Langmuir*, **28**, 11457–11464.
38. Weber, A.P. and Friedlander, S.K. (1997) *J. Aerosol Sci.*, **28**, 179.
39. List, J., Köhler, U., and Witt, W. (2011) Particulate Systems Analysis, Edinburgh UK, pp. 1–5.
40. Viguié, J.-R., Sukmanowski, J., Nölting, B., and Royer, F.-X. (2007) *Colloids Surf. A*, **302**, 269–275.
41. Gloaguen, F., Léger, J.M., Lamy, C., Marmann, A., Stimming, U., and Vogel, R. (1999) *Electrochim. Acta*, **44**, 1805–1816.
42. Betzig, E., Trautman, J.K., Harris, T.D., Weiner, J.S., and Kostelak, R.L. (1991) *Science*, **251**, 1468–1470.
43. Park, K., Kittelson, D.B., Zachariah, M.R., and McMurry, P.H. (2004) *J. Nanopart. Res.*, **6** (2), 267–272.
44. Koster, A.J., Ziese, U., Verkleij, A.J., Janssen, A.H., and de Jong, K.P. (2000) Three-dimensional transmission electron microscopy: a novel imaging and characterization technique with nanometer scale resolution for material science. *J. Phys. Chem. B*, **104**, 9368–9370.
45. Weyland, M., Janssen, P.A.H., and de Jong, K.P. (2007) in *Electron Tomography in Nanocharacterisation* (eds J. Hutchison and A. Kirkland), Chapter 6, The Royal Society of Chemistry.
46. van Poppel, L.H., Friedrich, H., Spinsby, J., Chung, S.H., Seinfeld, J.H., and Buseck, P.R. (2005) Electron tomography of nanoparticle clusters: Implications for atmospheric lifetimes and radiative forcing of soot. *Geophys. Res. Lett.*, **32**, 24.

47. van de Hulst, H.C. (1957) *Light Scattering by Small Particles*, John Wiley & Sons, Inc., New York.
48. Farias, T.L., Köylü, Ü.Ö., and Carvalho, M.G. (1996) *Appl. Opt.*, **35**, 6560.
49. Weber, A.P., Baltensperger, U., Gäggeler, H.W., and Schmidt-Ott, A. (1996) *J. Aerosol Sci.*, **27**, 915–929.
50. Schwyn, S., Garwin, E., and Schmidt-Ott, A. (1988) *J. Aerosol Sci.*, **19**, 639–642.
51. Weber, A.P., Keil, R., Tobler, L., and Baltensperger, U. (1992) *Anal. Chem.*, **64**, 672–677.
52. Kütz, S. and Schmidt-Ott, A. (1990) *J. Aerosol Sci.*, **21** (Suppl. 1), 47–50.
53. Seipenbusch, M., Heel, A., Weber, A.P., and Kasper, G. (2002) *Chem. Eng. Technol.*, **25**, 77–82.
54. Stahlmecke, B., Wagener, S., Asbach, C., Kaminski, H., Fissan, H., and Kuhlbusch, T.A.J. (2009) Investigation of airborne nanopowder agglomerate stability in an orifice under various differential pressure conditions. *J. Nanopart. Res.*, **11**, 1625.
55. Zumaeta, N., Byrne, E.P., and Fitzpatrick, J.J. (2007) Predicting precipitate breakage during turbulent flow through different geometries. *Colloid Surf. A*, **292**, 251.
56. Peng, S. and Williams, R. (1994) Direct measurement of floc breakage in flowing suspension. *J. Colloid Interface Sci.*, **166**, 321.
57. Higashitani, K., Inada, N., and Ochi, T. (1991) Floc breakup along center line of contractile flow to orifice. *Colloid Surf.*, **56**, 13.
58. Blaser, S. (2000) Break up of large aggregates in contracting and swirling flows. *Colloid Surf. A*, **166**, 215.
59. Wengeler, R. and Nirschl, H. (2007) Turbulent hydrodynamic stress induced dispersion and fragmentation of nanoscale agglomerates. *J. Colloid Interface Sci.*, **306**, 262.
60. Kousaka, Y., Okyama, K., Shimizu, A., and Yoshida, T. (1979) Dispersion mechanisms of aggregate particles in air. *J. Chem. Eng. Jpn.*, **12**, 152–158.
61. Zahradnicek, A. (1976) Untersuchung zur Dispergierung von Quarz- und Kalksteinfraktionen im Korngrößenbereich 0.5–10 µm in strömenden Gasen. Dissertation University Karlsruhe.
62. Kousaka, Y., Endo, Y., Horiuchi, T., and Niida, T. (1993) Dispersion of aggregate particles by acceleration in air stream. *KONA*, **11**, 199.
63. Niedballa, S. (1999) Dispergierung von feinen Partikeln in Gasströmungen –Einfluss von Dispergierbeanspruchung und oberflächenmodifizierenden Substanzen. Dissertation TU Freiberg.
64. Sator, N. and Hietala, H. (2010) Damage in impact fragmentation. *Int. J. France*, **163**, 101.
65. Antony, S.J., Moreno-Atnasio, R., Musadaidzwa, J., and Williams, R.A. (2008) Impact fracture of composite and homogeneous nanoagglomerates. *J. Nanomater.*, **2008**, 7. doi: 10.1155/2008/125386(2008)
66. Eggersdorfer, M.L., Kadau, D., Herrmann, H.J., and Pratsinis, S.E. (2010) Fragmentation and restructuring of soft agglomerates under shear. *J. Colloid Interface Sci.*, **342**, 261.
67. Rennecke, S. and Weber, A.P. (2013) *J. Aerosol Sci.*, **55**, 89–103.
68. Weber, A.P., Seipenbusch, M., and Kasper, G. (2001) *Chem. Eng. Technol.*, **24**, 702–705.
69. Seipenbusch, M., Weber, A.P., Schiel, S., and Kasper, G. (2003) *J. Aerosol Sci.*, **34**, 1699–1709.
70. Dalis, A. (2005) *Nanotechnology*, **16**, 626–631.
71. Weibull, W. (1951) *J. Appl. Mech.*, **9**, 293–297.
72. Almeida, J.B. and Mater, J. (1999) *Proc. Technol.*, **92–93**, 257–263.
73. Weichert, R. (1992) *Zement-Kalk-Gips*, **45**, 1–8.
74. Seipenbusch, M., Toneva, P., Peukert, W., and Weber, A.P. (2007) *Part. Part. Syst. Charact.*, **24**, 193–200.
75. Tsantilis, S. and Pratsinis, S.E. (2004) *Langmuir*, **20**, 5933–5939.
76. Seipenbusch, M., Rothenbacher, S., Kirchhoff, M., Schmid, H.J., Kasper, G., and Weber, A.P. (2010) *J. Nanopart. Res.*, **12** (6), 2037–2045.
77. Sempéré, R., Bourret, D., Woignier, T., Phalippou, J., and Jullien, R. (1993) *Phys. Rev. Lett.*, **71**, 3307.

78. Koch, W. and Friedlander, S.K. (1990) *J. Colloid Interface Sci.*, **140**, 419.
79. Nakaso, K., Shimada, M., Okuyama, K., and Deppert, K. (2002) *J. Aerosol Sci.*, **33**, 1061–1974.
80. Binder, A. and Seipenbusch, M. (2011) *Appl. Catal.*, **396**, 1–8.
81. Morrow, B.H. and Striolo, A. (2007) *J. Phys. Chem. C*, **111** (48), 17905–17913.
82. Seipenbusch, M., Binnig, J., Heim, M., Weber, A.P., and Kasper, G. (2001) *Helv. Chim. Acta*, **84**, 3686–3701.
83. Weber, A.P., Seipenbusch, M., Binnig, J., and Kasper, G. (2002) *Part. Part. Syst. Charact.*, **19**, 300–305.

6
Passivation of Metal Nanopowders

Alexander Gromov, Alexander Il'in, Ulrich Teipel, and Julia Pautova

6.1
Introduction

After production by any of the existing methods (condensation from plasma, electrical explosion of wires (EEW), mechanochemistry, etc.) all metallic nanopowders (Al, Cu, Fe, etc.) are pyrophoric, and surface passivation procedure for particles is required. The passivation procedure for metallic nanoparticles has these features compared to that for bulk metals and micron-sized powders: high amount of heat released in the process of the interaction of the surface of metal nanoparticles with passivating reagents as a result of the limitation of heat dissipation inside the particle, short time, fire and explosion hazard, and so on. Immediate stabilization is necessary for aluminum nanopowder (*nAl*), for example: storage of *nAl* in an inert gas is inexpedient as it is sintered spontaneously (Figure 6.1), and it is self-ignited on contact with air/water. In this chapter, mainly the problems of *nAl* passivation are discussed: *nAl*, among other active nanometals, was selected as the most propagated in the world market of energetic nanopowders since the 1970s [1].

The mechanism of *nAl* passivation and stabilization in different gas and liquid media is widely discussed in periodicals [2–7]. The thickness, composition, and structure of the passivating layer on particles are the main differences between *nAl* and micron-sized powders regarding their oxidation in heterogenic medium, because they define the diffusion rate of a gaseous oxidizer at low temperatures and, therefore, an ignition temperature and the delay period of the particles' ignition [8].

The structure of passivation coating for *nAl* defines its chemical stability and activity in the further oxidation processes. The traditional method of passivation of micron-sized powders is coating with hydrocarbons. For example, industrial aluminum powder with micron-sized flake particles contains 2–3 wt% of paraffin [9].

The particles of *nAl* as well as other metal nanoparticles are passivated, as a rule, with oxide layers having an amorphous or crystal structure [10–12]. The oxide layers are formed on the *nAl* particles at slow oxidation by air during

Metal Nanopowders: Production, Characterization, and Energetic Applications, First Edition.
Edited by Alexander Gromov and Ulrich Teipel.

6.2
Theoretical and Experimental Background

6.2.1
Chemical and Physical Processes in Aluminum Nanoparticles during Their Passivation by Slow Oxidation under Atmosphere (Ar + Air)

Directly after the wire explosion, *nAl* coagulated and precipitated in the collector normally or are exposed to the cyclone-type unit for powder fractions of different size separation [15]. Self-sintering of nanoparticles is not observed in comparison with the methods of aluminum nanoparticles obtained under quasi-equilibrium conditions [16]. When the electric explosion is realized at slightly higher than the atmospheric pressure, the nanoparticles have an adsorbed argon gas on their surface that does not allow the particles to contact and prevents them from sintering in the first minutes after synthesis [10]. Aluminum nanoparticles are exposed to the oxidation of their surface by slow reaction with air for 72 h during the passivation process. The *nAl* placed in a hermetic box filled with argon (air concentration ~0.1 vol%) are periodically mixed at room temperature. The air concentration is kept constant in the passivation box. The temperature control of the passivation process is realized with alumel/chromel thermocouple placed into the powder.

The temperature change in the *nAl* volume versus passivation time is shown in Figure 6.4. The argon gas remains on the particle surface at the primary moment after the powders are obtained. Argon desorption is an endothermic process (Figure 6.4). Then, argon is gradually replaced by atmospheric components. First, these components adsorb on the particle surface and then chemically interact, leading to the protective oxide-hydroxide layer formation that accompanied the increase in temperature [17].

The mass-spectrometric analysis result of the gases evaluated from *nAl* after prior incomplete passivation by air portions is presented in Table 6.2. Gaseous N_2, CO_2, and water are the main gases evolved from the mentioned fractions under heating. Hydrogen starts evolving at 200 °C and then becomes more intensive from 600 °C. The gas with the ratio $m/z = 17$ appears in the thermal desorption spectra at 770 °C; the gas peak intensity exceeded that of water. If the intensity ratios of the water peak ($m/z = 18$) and OH ion peak ($m/z = 17$) must be equal to 3/1, then the exceeding species, obviously, can be concerned with the NH_3 appearance in the mixture of the thermoreleased gases. The ammonia appearance was possible

Table 6.1 Aluminum nanopowder production conditions.

Wire material (purity, %)	Al (99.8)
Voltage by explosion (kV)	26
Entered energy (a.u.)	1.4
Wire diameter (mm)	0.35
Length of exploded bit (mm)	70
Gas pressure (MPa)	0.15

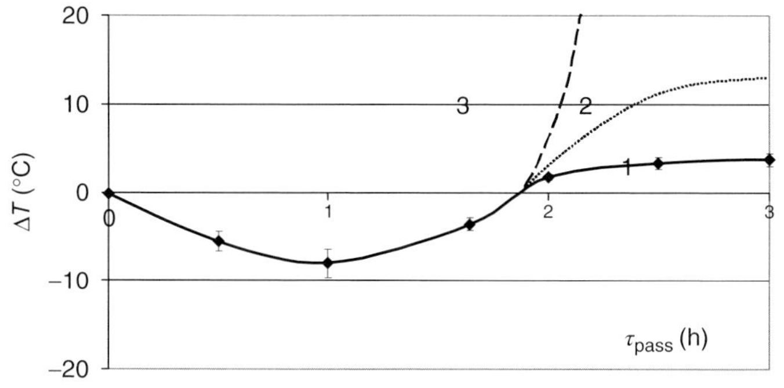

Figure 6.4 Passivation period duration for *nAl* versus temperature change in gas mix (Ar + 0.1 vol% air): 1 – optimal regime; 2 – self-heating regime; and 3 – self-ignition.

Table 6.2 Mass-spectrometric analysis of gases evolved from aluminum nanopowder, obtained in H_2.

T (°C)	*T* of gas evolving (°C)			
	$m^a/z^b = 28$ (CO^+, N_2^+)	$m/z = 44$ (CO_2^+)	$m/z = 18$ (H_2O^+)	$m/z = 17$ (OH^+, NH_3^+)
240–460	460	400	240	400

[a] *m* – ion weight in atomic units.
[b] *z* – ion charge.

because of AlN hydrolysis, which took place as a result of the *nAl* reaction with air nitrogen during the passivation. The catalytic reaction of nitrogen and hydrogen on the aluminum or aluminum oxide surface could be another source of ammonia formation. The hydrogen presence in the composition of thermodesorbed gases was probably caused by the hydrogen accumulation at the aluminum nanoparticle oxidizing by water vapor during the passivation process (proton mechanism of oxidation). The sharp increase in the hydrogen emission rate was observed at temperatures higher than 200 °C. It could be because of the hydride decomposition or hydrogen desorption. Thus, the oxide and hydroxide layers were formed on the aluminum nanoparticle surface during the passivation process. These layers could adsorb atmospheric gases in large quantities.

The electron diffraction spectroscopy (EDS) data of the *nAl* samples after their passivation showed the presence of the aluminum compounds with oxygen deficiency in the surface and near-surface layers of the particles. The atomic concentration ratio Al/O changed from 0.42/1 up to 0.61/1 for partially passivated *nAl* (Table 6.3). The oxygen content on the particles' surface increased in the last stage of passivation. The singlet signal was observed in the spectrum of *nAl*, according to electronic paramagnetic resonance data [18]. This signal intensity was dependent on oxygen partial pressure over the particles' surface. In accordance

Table 6.3 XPS data for passivated aluminum nanopowder.

No	Powder	Binding energy (eV)		Storage time (min)	Atomic ratio Al/O (related formula)
		O 1s	Al 2p		
1	*n*Al	531.7	74.3	800	0.44/1 ($Al_2O_{4.5}$)
2	*n*Al	531.8	74.3	700	0.61/1 ($Al_2O_{3.3}$)
3	*n*Al	531.6	74.2	400	0.71/1 ($Al_2O_{2.8}$)
4	*n*Al	531.9	72.7	300	0.95/1 ($Al_2O_{2.1}$)
5	*n*Al	531.1	73.0	200	0.94/1 ($Al_2O_{2.1}$)
6	*n*Al	530.9	72.2	48	0.86/1 ($Al_2O_{2.3}$)
7	micon-sized Al (etalon)	531.6	74.3	15 000	0.53/1 ($Al_2O_{3.7}$)

with the infrared (IR) spectroscopy data (Figure 6.5), the absorption bands with maximum in the range 950–1440 cm^{-1} could be related to chemisorbed oxygen molecule. Semi-empirical calculation with the data from IR spectroscopy and electronic paramagnetic resonance were the reason to suppose that the oxygen molecule joined with the particle surface by the one linkage (Al–O–O$^\bullet$) and remained a paramagnetic radical with the corresponding singlet signal.

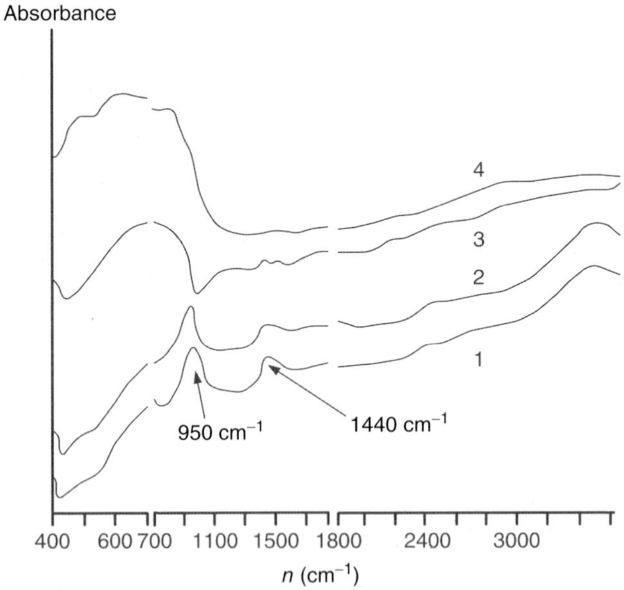

Figure 6.5 IR spectrum of aluminum nanopowder, produced by electrical explosion of wires, heated in air: 1 – $T = 30\,°C$; 2 – $T = 400\,°C$; 3 – $T = 550\,°C$; and 4 – IR spectrum of Al_2O_3.

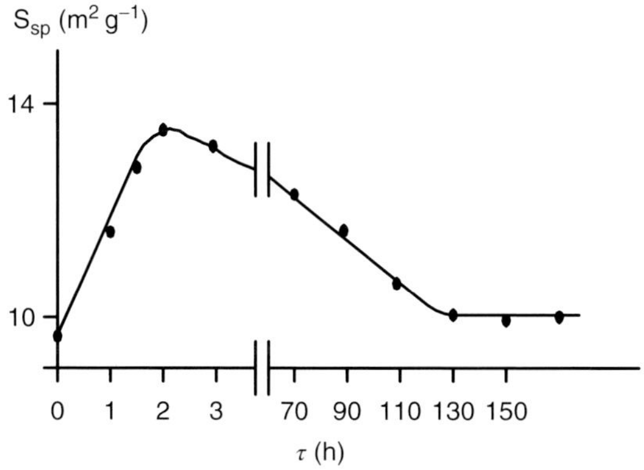

Figure 6.6 S_{sp} of aluminum nanopowder, produced by electrical explosion of wires, versus passivation duration in air.

The processes in the oxide-hydroxide layer were studied for *nAl* samples subjected to incomplete passivation (sample No 6, Table 6.3). The measurement of the specific surface area for this nanopowder during the passivation process showed that it increased relatively fast during the initial passivation stage – maximum of 30–40% for 2 h (Figure 6.6). After that, the specific surface area was found to slowly decrease at 20–30% for a few days. Probably, the increase in S_{sp} resulted from the formation of the oxide-hydroxide layer with the friable structure on the metal surface. Later, the layer densified and crystallized and the specific surface area decreased.

The growth of the protective oxide-hydroxide layers on the metal nanoparticles was satisfactorily corresponding to the model of very thin layer formation [19]. The nanoparticles had an amorphous oxide layer according to XRD (X-ray diffraction) data and transmission electron microscopy data (Figure 6.2b) – the only metallic aluminum reflexes were on the XRD pattern. The oxide layer thickness was calculated on the basis of oxide specific surface values and the masses of oxide and metal (Table 6.4) by Equation 6.1.

$$\varphi = R - r = \sqrt[3]{\left(\frac{m_{Al} + m_{Al_2O_3}}{(\rho_{Al} \cdot x_{Al} + \rho_{Al_2O_3} \cdot x_{Al_2O_3})(4/3 \cdot \pi)}\right.} - \sqrt[3]{\frac{m_{Al}}{4/3 \cdot \pi \cdot \rho_{Al}}}, \qquad (6.1)$$

where φ – oxide layer thickness, R – particle radius (outer), r – metal nucleus radius (inner), m_{Al} and $m_{Al_2O_3}$ – metal and oxide weights in the particle, ρ_{Al} and $\rho_{Al_2O_3}$ – metal and oxide density in the particle, and x_{Al} and $x_{Al_2O_3}$ – metal and oxide weight parts in the particle.

The calculated thickness of the oxide layer (φ) showed a minimum value for the particles with ~100 nm in diameter (Figure 6.3). These particles had optimal combination of the most important physical properties such as high stability and high metal content. The smaller aluminum particles had a lower metal content and, consequently, a thicker oxide layer. The oxide layer thickness increased up to 80 nm for large particles with 1000 μm diameter. The calculated oxide layer thickness was 0.8–1.7 nm for the oxide content 1.1–3.7 wt% for *nAl* (a_s = 140–290 nm) obtained in

Table 6.4 Aluminum nanopowder characteristics (exploded in $Ar + N_2$).

No	Voltage (kV)	Entered energy (e/e_c)	C_{Al} (%)	S_{sp} (BET) (m^2/g)	a_s (nm)	Oxide content (wt%)	Oxide layer thickness (φ) (nm)
1	30	1.82	88.0 ± 1.4	10.8 ± 0.3	210	9.0	1.84
2	28	1.71	87.9 ± 0.3	9.9 ± 0.3	220	9.1	1.96
3	26	1.62	88.1 ± 1.1	9.9 ± 0.9	220	8.9	1.91
4	24	1.45	88.5 ± 0.9	9.3 ± 0.3	240	8.4	1.95
5	22	1.30	90.9 ± 0.8	8.8 ± 0.3	250	6.1	1.43
6	20	1.13	90.0 ± 0.6	6.7 ± 0.2	330	7.0	2.20
7	18	0.92	91.0 ± 0.7	7.7 ± 0.3	290	6.0	1.63

($Ar + H_2$) gaseous mixture. The calculated oxide layer thickness was 1.5–2.1 nm and the aluminum oxide content – 8.4–8.8 wt% for *nAl* ($a_s = 120$–160 nm) obtained in gaseous Ar. The calculated oxide layer thickness was 1.6–2.2 nm and the aluminum oxide content – 6.0–9.1 wt% for *nAl* ($a_s = 210$–290 nm) obtained in the ($Ar + N_2$) mixture (Table 6.4).

Thus, the EEW aluminum nanoparticles showed the oxide layer of thickness to be about 500% thinner than on the particles of industrial spherical aluminum powders (Table 6.5). This fact was confirmed by electron microscopy data (Figure 6.2b). The real thickness of the oxide layer on the *nAl* particles was twice as that calculated. This was caused by the friability and relief of the passivating layers.

6.2.2 Chemical Mechanism of Aluminum Nanopowder Passivation by Slow Air Oxidation

If one were to take the oxide layer (thickness 20–40 nm) covering the particle of micron-sized powders ($a_s = 10\,\mu m$) and calculate the residual metal content for spherical nanoparticles with equal thickness of oxide layer, the metal content will be more than 40–60 wt%. The real metal content in *nAl* is 95 wt%. This indicates the mechanism of oxidation process inhibition in *nAl*, which is different from the diffusion one. In the thin layer case [19], the oxidation process is limited by the electric charge of double electric layer (DEL) [20]. Thus, the oxidizer diffusion to the metal surface is controlled by DEL charge. Being different from inert substances, metals in the colloid condition form the oxide-hydroxide layers on their surface [21]. The DEL structure will have some features depending on the chemical processes proceeding on the metal particles [21]. Metals are reducing agents and they are charged positively in red-ox processes. The electric current running during the electrochemical reaction – metal oxidation by oxygen or water (Faraday current) – gives the extra potential [22], larger "capacity," which is named "pseudocapacity" in contrast to ordinary DEL capacity. Thus, the total polarization capacity of the interfacial boundary "metal-gas" could be calculated using Equation 6.2

Table 6.5 Calculated oxide layer thickness for aluminum particles.

Particle diameter (nm)	Powder	S_{sp} (BET) ($m^2\,g^{-1}$)
10	Aluminum nanopowder	222.222
20		111.111
30		74.074
40		55.556
50		44.444
60		37.037
70		31.746
80		27.778
90		24.691
100		22.222
200		11.111
500	Aluminum nanopowder (coarse fraction)	4.444
1 000	Aluminum micron-sized powder (fine fraction)	2.222
10 000	Aluminum micron-sized powder	0.222
100 000	Aluminum micron-sized powder	0.022
1 000 000	Aluminum micron-sized powder	0.002

$$C_{total} = C_{DEL} + C_{pdc}, \tag{6.2}$$

where C_{DEL} – DEL capacity and C_{pdc} – pseudocapacity are caused by chemical reaction proceeding.

In contrast to bulk aluminum (electrically neutral), aluminum nanoparticles produced by the EEW method were divided by the charge in the electric filter. During the EEW, the particle charge could be formed as a result of thermoemission and ionization by electric field, and the charge structure stabilization was the result of a high rate product cooling and an electromagnetic field presence [21]. In comparison with neutral particles, charged ones held the polarized Ar molecules stronger because of the stronger "dipole–induced dipole" interaction (Figure 6.7). The uncharged particles will interact with argon as a result of the weaker dispersion attraction mechanism. The electron transfer (penetration through oxide) from metal to oxidizer or the oxidizer (protons) drift to interface "metal-oxide" occurred during the passivation by slow oxidation in air. The total charge of the positively charged particles increased and reached the necessary value of an electric static potential for the inhibition of the oxidation process faster than that of the uncharged particles. In the case of negatively charged particles, the oxidation process proceeded till complete compensation of this charge. Then, the additional metal quantities were subjected to oxidation to achieve a positive potential, which is necessary for stabilization. So, the nature of a potential barrier providing the stability of aluminum nanoparticles is connected to the presence of

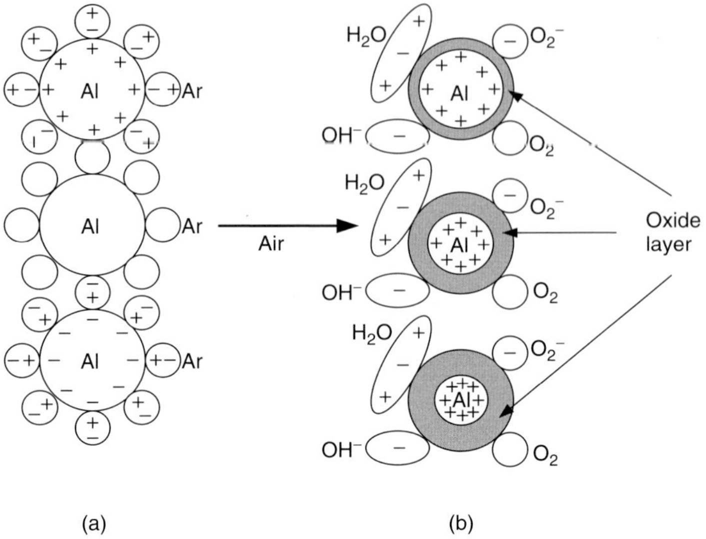

Figure 6.7 Chemical diagram of *nAl* passivation: (a) particle structure (not in a real scale) before passivation and (b) particle structure after passivation.

an electric static field on the surface and the above-surface particle layers having different charges.

The main reason for *nAl* deposition on electric filter electrodes was the particles' static charge arising directly at the time of wire explosion. Both the *nAl* fractions from positively and negatively charged electrodes and the non-settled fraction on the electrodes were studied using bomb calorimetry (Table 6.6). The data showed the *nAl* fractions non-settled on the electric filter electrodes and having the highest specific surface value and the lowest ΔH^{298} (Al_2O_3) value during combustion in the bomb. The least oxide formation enthalpy (less than standard) was typical for aluminum in this kind of fraction. The calorimetric data of aluminum fractions obtained by charge division in the electric filter confirmed the suggested scheme of the electric explosive particle structure (Figure 6.7). The energy content was higher for negatively charged particles (Table 6.6). The positively charged nanoparticles had a higher stability to oxidation in comparison with the particles having not enough charge. The possibility of the formation of complicated (flaky) charging structures, dependent only on the conductor matter, was initiated by the dispersion mechanism of the conductors during the EEW process (wire disintegration to cluster and gas-like phases) [21]. On the basis of the EEW mechanism, the gas-like phase had a negative charge on the primary stage of dispersion, and being the lightest one it retired from the conductor axle faster and was broken faster, too. In the next step, the positively charged liquid-like phase overtook and permeated the layers of the gas products. Taking into account the EEW, the products showed a marked temperature decrease during the period of their relatively separate existence; there was an increased possibility of the stabilization of charging structures.

Table 6.6 Enthalpy for aluminum nanopowder combustion.

Sample (gas media)	S_{sp} (BET) (m^2/g)	ΔH^{298} (Al_2O_3) (kJ mol^{-1})
1. *nAl* (H_2) average:	7.4	1728.4 ± 6.0
from the positively charged electrode	5.4	1804.1 ± 6.0
from the negatively charged electrode	6.1	1737.5 ± 6.0
neutral fraction	9.0	1595.2 ± 5.5
2. micon-sized Al (etalon)	0.8	1667.7 ± 5.0

The stability of EEW nanoparticles during the storage was significantly determined by their obtaining conditions. Optimal electrical explosion regime and passivation conditions are necessary to get powders with a high metal content. The increase of the extra metal content in nanopowders was possible owing to the passivation with the layer-formative liquids. The choice of the passivation method of metal nanopowders is determined by their application field [22, 23]. When the *nAl* are used in high-temperature processes, it is necessary to apply the particles covered with protective layers to prevent an interaction with the environment during storage.

At present, the majority of EEW *nAl* ($a_s = 30-300$ nm) on the world market are passivated with oxide layers formed during the slow *nAl* oxidation by air. In fact, this method of passivation can be considered as a process of "passivation until self-saturation" including the reactions (6.3–6.4).

$$2\,Al(s) + 3/2\,O_2(g) \rightarrow Al_2O_3\,(s) \tag{6.3}$$

$$2\,Al(s) + 3\,H_2O\,(g) \rightarrow Al_2O_3\,(s) + 3\,H_2\,(solved) \tag{6.4}$$

It is obvious that the water vapor always present in air is more reactive to passivating agents than oxygen. Therefore, aluminum nanoparticles passivated by air contain up to 2 wt% of hydrogen (hydrogen analyzer RHEN 602). The main idea of having the nonoxide passivating layers applied on the electroexplosive *nAl* particles is the possibility of increasing the metal content in them.

6.3 Characteristics of Passivated Particles

The EEW *nAl*[1)] after obtaining and before contact with air were passivated by the following substances: (i) liquids – nitrocellulose (NC), oleic acid (Ol Ac, $C_{17}H_{33}COOH$), and stearic acid (St Ac, $C_{17}H_{35}COOH$) dissolved in kerosene and ethanol; fluorine-polymer; (ii) solids – amorphous boron, nickel; and (iii)

1) *nAl* samples were obtained in the High Voltage Institute of Tomsk Polytechnic University and the Institute of High-Current Electronic (Siberian Branch, Russian Academy of Science).

Table 6.7 Aluminum nanopowder characteristics.

No	Sample	Wire composition	Gas media	Passivation condition	S_{sp} (BET) ($m^2 g^{-1}$)	C_{Al} (wt%)	C_{Al}, wt% (1 year age)
1.	ALEX	Al	Ar	Air	11.3	86	85
2.	Al (Ni)	Al (Ni)	Ar	Air	40.7	53	53
3.	Al (B)	Al (B)	Ar	Air	12.0	84	82
4.	Al ($Ar + H_2$)	Al	Ar + 10 vol% H_2	Air	9.4	92	92
5.	Al (N_2/CO_2)[a]	Al	N_2	CO_2	19.5	84	n/a
6.	Al ($N_2 + CO_2/N_2$)[a]	Al	$N_2 + CO_2$	N_2	30.1	67	n/a
7.	Al (St Ac) ethanol	Al	Ar	Stearic acid in ethanol	12.1	74	70
8.	Al (St Ac) kerosene	Al	Ar	Stearic acid in kerosene	7.3	79	59
9.	Al (Ol Ac) ethanol	Al	Ar	Oleic acid in ethanol	14.3	45	43
10.	Al (F)	Al	Ar	Fluoropolymer	11.6	81	81
11.	Al (NC)	Al	Ar	Nitrocellulose in ethanol	12.6	68	58

[a] Reference [26].

gases – N_2, CO_2, and air (for comparison). The *nAl* passivation method was chosen after reasoning, from the prior experiment results – if the nanoparticles contacted with air after obtaining, the formed oxide layers (~10 wt% Al_2O_3 for particles with the diameter ~100 nm) could not be removed by any subsequent treatment.

characteristics of studied *nAl* samples are presented in Table 6.7. The powders' specific surface was determined by BET (Brunauer–Emmett–Teller) method, metallic aluminum content – by volumetric analysis [24].

The powder ALuminum EXploded (ALEX) (sample No 1, Table 6.7) with known characteristics [25] was chosen as a comparison sample. Samples No 2 and 3 were obtained from the composite wires Al–Ni and Al–B, respectively. During the EEW process, H_2 was added to Ar to obtain sample No 4, N_2 – for sample No 5 and ($N_2 + CO_2$) mixture – for sample No 6. The *nAl* passivation by gases was carried out for 72 h at $T = 30 \pm 2$ °C and $P = 0.11$ MPa (samples No 2–6). Samples No 7–11 were passivated before the contact with air by the following organic substances:

- 7 – 0.1 wt% stearic acid in ethanol;
- 8 – 0.1 wt% stearic acid in kerosene;
- 9 – 0.1 wt% oleic acid in ethanol;
- 10 – 0.1 wt% fluorine-polymer solution;
- 11 – 0.1 wt% NC solution in ethanol.

Ethanol and kerosene were chosen as solvents because they do not react with *nAl*. The passivating reagent solutions were added to the just obtained powders and were mixed for ~2 h. The passivation temperature was kept at 30 ± 5 °C to prevent the powder self-heating. The solvent residue was removed out of *nAl* by vacuum drying at room temperature. After passivation, all powders were kept in open air for two months to reproduce the industrial production conditions. After the first stage of analysis, the samples (Table 6.7) were kept in air at room temperature and a relative

Table 6.8 Elemental and phase composition (Table 6.7).

No	Sample	Element content (wt%) (XPS data)[a]		Phase composition (XRD data)[b]
		O	Al	
1.	ALEX	10 ± 2	90 ± 5	Al (04-0787)
2.	Al (Ni)	22 ± 3	$74 \pm 2\% + 4\%$ Ni	Al
3.	Al (B)	11 ± 2	89 ± 4	Al
4.	Al ($Ar + H_2$)	9 ± 2	91 ± 5	Al
5.	Al (N_2/CO_2)	18 ± 3	82 ± 3	Al, γ-Al_2O_3 (47-1308)
6.	Al ($N_2 + CO_2/N_2$)	33 ± 4	67 ± 2	Al, γ-Al_2O_3
7.	Al (St Ac) ethanol	15 ± 2	85 ± 3	Al, traces of Al_4C_3 (35-0799)
8.	Al (St Ac) kerosene	15 ± 2	85 ± 3	Al
9.	Al (Ol Ac) ethanol	18 ± 3	82 ± 3	Al, traces of Al_4C_3
10.	Al (F)	12 ± 2	88 ± 4	Al
11.	Al (NC)	n/a		Al, γ-Al_2O_3

[a] Powder surface for analysis ~20 × 20 μm. XRD pattern for sample No 4 and corresponding analyzed surface is on Figure 6.10.
[b] Number of JCPDS card (PCPDFWIN v.1.30).

humidity ~70% for two months. The particles' composition and morphology (Table 6.8) were studied by XPS (X-ray photoelectron spectroscopy) (PHILIPS CM 200 FEG) and XRD (Rigaku MAX-B, CuK_α-radiation). Differential scanning calorimetry and thermogravimetry (thermal analyzer 2.4 F TA instruments) were used to test the *nAl* for their nonisothermal oxidation (Table 6.9). According to XRD data, the oxidation products of all *nAl* samples consisted of α-Al_2O_3 after heating (10 °C min^{-1}) in air up to 1500 °C. Isothermal *nAl* oxidation in air was studied by microcalorimetry (Figure 6.8a,b).

The best suitable etalon for *nAl* with nonoxide protective layers is the powder passivated by Al_2O_3. Thermal properties of such powders are well known [26, 27]. That is why all EEW *nAl* properties were compared with the industrial aluminum powder ALEX (Tables 6.7–6.9). All *nAl* samples excepting Nos 4 and 8 had larger specific surface value than ALEX powder. "Nonoxide" passivation had a stronger influence on the metal content in *nAl*. Metallic aluminum remained in quantity of 45–81 wt% in the samples passivated by organic substances. The reduction of metallic aluminum content was maximal (up to 45 wt%) for *nAl* passivated by oleic acid. Therefore, the passivation by liquid organic reagents resulted in the considerable decrease of metal content in powder (Table 6.7). Besides, the material of the passivating layer permanently reacted with *nAl*, for example, metallic aluminum content reduction after 12 months of storage for samples No 7–11 was higher for the samples passivated by gaseous and solid substances.

The XPS and XRD results of the *nAl* samples are summarized in Table 6.8. All samples contained more than 10 wt% of oxygen (in the form of γ-Al_2O_3) on the particles' surface. The aluminum carbide traces were seen in samples No 7

and residual kerosene introduced the additional heat contribution to the oxidation process.

The conversion degree (α) for the *nAl* samples was relatively high for all samples up to $T = 1400\,^\circ C$ (Table 6.9), except for the fully oxidized sample No 6 and for sample No 11 in which Al interacted with NC (Table 6.8).

The isothermal calorimetry results of the chosen powders are shown in Figure 6.8. The samples were kept in open air for 60 days at the temperature $T = 100\,^\circ C$ (Figure 6.8a) and 150 °C (Figure 6.8b), respectively. Uncommon behavior of the samples No 7 and 9 (they showed mass decrease) could be explained by residual solvent evaporation. The residual solvent (kerosene) presence was relatively less for sample No 8 in the isothermal heating condition, that is, kerosene adhered on the nanoparticles surface stronger than ethanol.

6.3.1
Characteristics of Aluminum Nanopowders, Passivated by Gaseous and Solid Reagents (Samples No 1–6, Table 6.7)

The highest metallic aluminum content was found in the samples ALEX, Al ($Ar + H_2$), and Al (N_2/CO_2) (Table 6.7). The specific surface of the sample No 4 (Al ($Ar + H_2$)) was two times higher than that for the Al (N_2/CO_2), that is, for the powder formation with larger particles (Tables 6.7 and 6.8), the EEW in the ($Ar + H_2$) medium resulted in less oxidation after passivation by air. The parameters C_{Al} and S_{sp} had the inverse correlation – the higher S_{sp} the less C_{Al} (especially for samples No 2 and 6). The *nAl* s passivated by N_2- and CO_2-contained Al_2O_3 (Table 6.8). Hence, the metastable compounds "Al–N" and "Al–C" formed on the particle surface at the production and storage in N_2 and CO_2 mediums oxidized at the following powder contact with air. Therefore, the temperature at the beginning of an intensive oxidation was minimal for samples No 5 and 6 (Table 6.9). Both very fine and coarse fractions were contained in the sample Al (Ni). According to XPS, nickel (4 wt%) did not accumulate on the *nAl* particle surface but settled in the form of a fine NiO powder. Thus, the "Al–Ni" wire explosion resulted in the formation of finer particles than aluminum wire without cover and the particle size was less than the *nAl* stability boundary (~30 nm). The residual nonoxidized aluminum particles completely reacted with air at the temperature $T < 1000\,^\circ C$ and Ni used in this case did not increase the fine aluminum fraction stability against oxidation in air.

All powders contained 2–4 wt% of adsorbed gases (Table 6.9). But the sample Al (Ni) contained 9 wt% of gases that was the result of its high specific surface value. Besides, the dissolved hydrogen evolving from *nAl* particles could be the reason of a low oxidation beginning temperature (~190 °C) for the sample Al ($Ar + H_2$) in comparison with the ALEX one. *nAl* stabilized by boron [14] are a promising fuel component because the particles coated with boron are able to increase the powder burning enthalpy. The C_{Al} and S_{sp} values for the sample (covered with boron) were comparable to the values for the sample No 1 (ALEX, Table 6.7) and with its average value of fractional conversion up to 1400 °C (Table 6.9). The powder covered with boron had the highest value of ΔH_{ox} among "dry" samples No 1–6 (Table 6.9).

The increase in mass for the sample Al (Ni) during the first day of *nAl* isothermal oxidation in air at $T = 100$ and 150 °C can probably be attributed to the metallic nickel oxidation (Figure 6.8a,b). For the other "dry" samples, the weight increase during the first day of the low-temperature isothermal oxidation was 1–3 wt%, which corresponded to the slow aluminum oxidation process by absorbed water.

6.3.2 Characteristics of Aluminum Nanopowders, Passivated by Gaseous and Solid Reagents (Samples No 7–11, Table 6.7)

Stearic and oleic acids as passivating reagents chemically reacted with the aluminum of *nAl*. The deeper chemical interaction was in the case of oleic acid, the aluminum content decreased to 45 wt% after passivation (sample No 8) and kept on decreasing during storage (Table 6.7). The aluminum interaction with stearic and oleic acids resulted in particle surface carbonization (Table 6.8). NC was permanently oxidizing the *nAl* particles – the powders contained the metallic aluminum on 10 wt% less after one year of storage than just-obtained ones. The samples Al (F) and Al (NC) are the most prospective as a composition component for use in fuels. Nevertheless, they had a sufficiently low metal content. Adsorbed organic layers on the powder particles did not protect them from the oxidation that followed for *nAl* passivated in organic liquids; intrinsic oxide layers fixed under the organic layer (Figure 6.9). The oxygen content in the form of Al_2O_3 was higher for the sample No 9 than for the other ones passivated by liquids (Table 6.8). Smooth

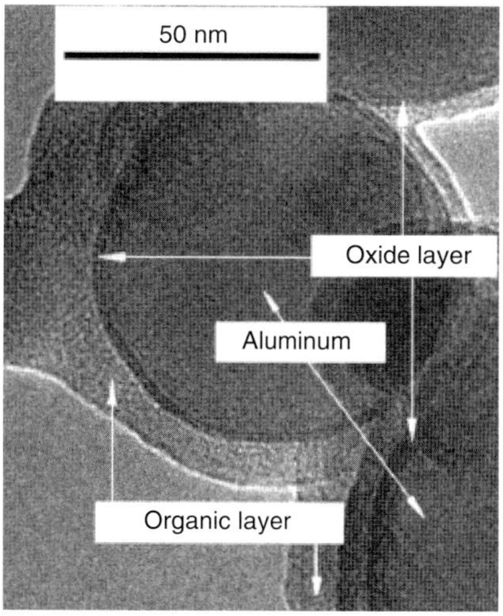

Figure 6.9 Transmission electron microscopy (TEM) image of aluminum nanoparticle, passivated with stearic acid (×500 000).

7
Safety Aspects of Metal Nanopowders

M. Lerner, A. Vorozhtsov, and N. Eisenreich

7.1
Introduction

Various types of nanoparticles are used in many industries with strongly growing volumes, especially many enterprises that contribute to the technology transfer from science to industrial applications [1–4]. While applications in energetic materials are still in a development stage, in civil industry commercialization has already taken place. However, the additional risks concerning reactivity and explosion induced by the nanosize and nanostructure of the particles in reactive atmospheres are yet to be either identified or investigated. Therefore, these risks are not taken into account in the design of processes of nanoparticle applications. Basic research results, however, gave strong indications that there are unknown effects, which mainly and substantially enhance reactivity [5–13]. These effects cannot be understood by a straightforward extrapolation of bigger to smaller size alone. For example, slow diffusion-controlled oxidation of metallic particles is substituted by a fast-reaction kinetically dominated oxidation when reaching nanosize. Some effects not observed for particles of bigger size are [5–13]:

- self-ignition and thermal explosions together with air, oxygen, or other reactive gases;
- substantially decreased ignition temperatures and energies even with other normally inert solid materials (oxides);
- sharply rising combustion velocities; and
- possibility of deflagration detonation transfer (DDT).

As often observed, nanoparticles react easily with many fluids – atmospheric oxygen is just a dominant one – or induce – catalyze – new, unexpected, and powerful reactions. As a consequence, interactions of nanoparticles with those fluids that are expected to be mixed together in the manufacturing, storing, handling, and application of specific nanomaterials or nanoparticles partially exposed to reactive environments (i.e., atmosphere) are challenging researchers to study potential hazards and reaction mechanisms that might lead to combustion and explosions.

Metal Nanopowders: Production, Characterization, and Energetic Applications, First Edition.
Edited by Alexander Gromov and Ulrich Teipel.

This chapter presents some results of risk criteria of nanoparticles and their classification with respect to handling and transportation. Basic understanding of the mechanisms concerned some basic findings and the classification of the nanoparticles with respect to reactions at small scale, classification of combustibility, ignition when initiated by low-energy ignition sources, rates of flame propagation, intensity of hydrogen generation on interaction with water, and inflammability initiated by electrostatic discharge (ESD).

Results included nanopowders of Al(Ar), Al(N_2), Cu(Ar), Cu(CO_2), Ni(Ar), Zn(Ar), Fe(Ar). Given in brackets is the gas in which the electric explosion was realized. All nanopowders were initially passivated by slow exposure to air.

7.2
Some Basic Phenomena of Oxidation of Nanometal Particles in Air

Nanoparticles show a different behavior on reaction with surrounding atmospheres than bigger particles. One effect concerns the oxidation at thin layers on the surface. In contrast to a normally assumed picture of smooth surfaces at the nanoscale, nanostructures are generated by the oxide on oxidation. These form only a barrier for oxygen when a certain thickness is surpassed. For Al, this thickness is about 8–10 nm. The reaction till this thickness is achieved occurs by chemical kinetics not hindered by diffusion. The effect was investigated in detail earlier [6, 7] (Figure 7.1).

A similar situation is found for Cu/5/. Bulk Cu as plate oxidizes is controlled by diffusion. The results of *in situ* X-ray diffraction are shown in Figure 7.2. The kinetics follows the Janda equation, which is typical for diffusion. The oxidation of nanopowders of Cu (Figure 7.3) occurs at 150 K at lower temperature. The reaction occurs in two steps, forming $Cu_2=$ and then CuO. The reaction can be fitted by chemical first-order kinetics without an influence of diffusion.

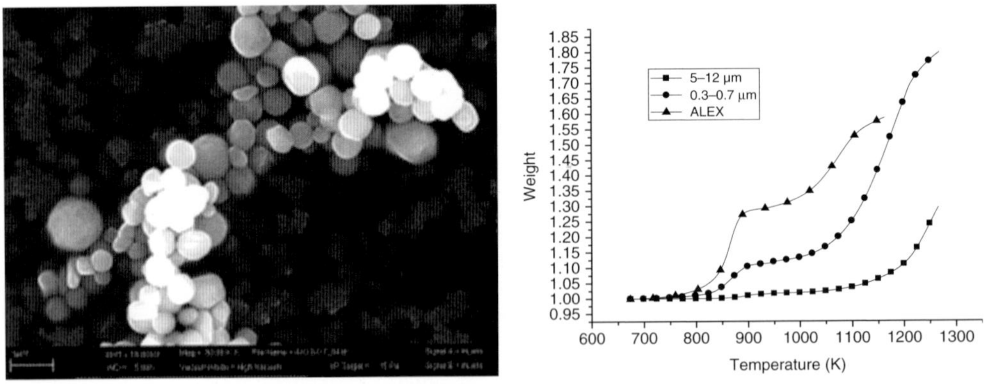

(a) (b)

Figure 7.1 (a) SEM (scanning electron micrography) picture of Al of a mean size between 0.3 and 0.7 nm and (b) TG curves on oxidation, which reveal a step increasing with decreasing particle size.

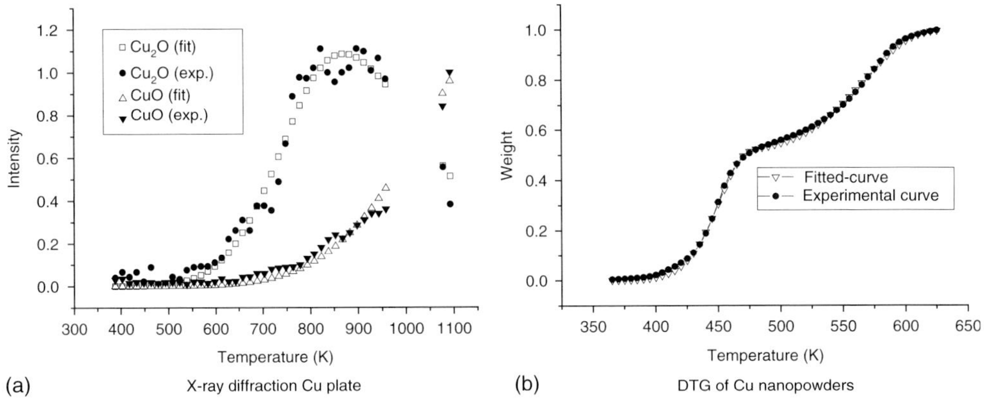

Figure 7.2 (a) Comparison of the surface oxidation of bulk copper of a plate (X-ray diffraction) with (b) the oxidation of copper nanopowder (thermogravimetry/differential scanning calorimetry, TG/DSC) and fit of the curves by using the Janda equation and first-order kinetics.

Figure 7.3 Copper nanoparticles with size of about 50–100 nm.

7.3 Determination of Fire Hazards of Nanopowders

For the classification of spontaneous inflammability by short-term application of ignition by a low-energy source the particles Al(Ar), Al(N_2), Cu(Ar), Cu(CO_2), Ni(Ar), Zn(Ar), and Fe(Ar) were investigated. This test was realized by placing samples of the metal powder into a furnace of temperature 500 °C under air or oxygen.

Al(Ar) formed a layer-to-layer (smoldering) combustion in a bulked layer. On oxygen exposure, the combustion became more active, and the brightness of the burning zone increased. The powder ignited with air, establishing a flame front with bright light emission.

$Al(N_2)$ (content of AlN in powder is around 40%) ignited during 1 s pronounced layer-to-layer combustion in bulked mode. After a fully developed combustion the flame covered the entire powder surface with bright light emission of burning zone. A white material was generated (alumina). The burning was more active compared with passivated Al(Ar). An air–powder mixture (stored in air) ignited, with fast generation of a full flame front and bright light emission.

$Cu(CO_2)$: no burning occurred in the bulked powder. It sintered, keeping the crucible form. An air–powder mixture was ignited easily with the generation of a full flame front.

Cu(Ar) did not ignite. At the tests, the samples sintered keeping the crucible form. An air–powder mixture was ignited with intensive sparking and development of a flame front.

Ni(Ar) formed a layer-to-layer combustion (smoldering) in the agglomerated powder with the powder totally consumed. The tests samples were partially sintered keeping the crucible form. An air–powder mixture was ignited with intensive sparking and generation of a flame front.

Zn(Ar) reacted in a manner similar to that of Ni(Ar). No visible flame occurred; however, the powder was completely consumed.

Fe(Ar) also developed a layer-to-layer combustion (smoldering) in the agglomerated powder, covering the total volume. The powder was totally consumed with a continued bright glow. An air–powder mixture ignited with intensive arcing and generation of a flame front.

Table 7.1 summarizes the results of the qualitative analysis and Table 7.2 lists the burning rates for the various powders investigated. The results of the inflammability by a low-energy source of ignition are presented in Table 7.3.

It is assumed and confirmed by preliminary experiments that nanopowders such as Al(Ar), ($Al(N_2)$) generate hydrogen on interaction with water. Besides, on interaction of $Al(N_2)$ with water, ammonia is evolved.

Table 7.1 Classification of nanopowder combustibility when subjected to furnace temperature of 500 °C.

Nanopowder	Ignition time delay in the furnace (s)	Burning time in the furnace (s)	Burning time outside (s)	Burning of air–powder mixture	Classification with respect to combustibility
Fe(Ar)	1.5	3	35 (light emission up to 600)	Burns	Combustible
Zn(Ar)	5.5	2	45	Burns	Combustible
Ni(Ar)	7	5	12	Burns	Combustible
$Al(N_2)$	7	1	2	Burns	Combustible
Al(Ar)	11	8	180	Burns	Combustible
$Cu(CO_2)$	No burning	No burning	No burning	Burns	Hardly combustible
Cu(Ar)	No burning	No burning	No burning	Burns	Hardly combustible

Table 7.2 Determination of flame spreading rate.

Nanopowder	Size of front (mm)	Linear rate (mm s^{-1})
Zn(Ar)	24–25	4.42
$Cu(CO_2)$	6.5–7	0.43
Al(Ar)	25–26	4.39
$Al(N_2)$	10–10.5	12.76
Cu(Ar)	7.5–8	0.48
Ni(Ar)	22–23	0.79
Fe(Ar)	18–18.5	2.0

Table 7.3 Inflammability at short-term influence of low-energy source of ignition.

Nanopowder	Ignition delay time from low-energy source	
	By flame (Bunsen burner) (s)	By spiral wire heating (s)
$Al(N_2)$	Ignite; 1.5	Ignite; 1.5
Al(Ar)	Ignite; 2.0	Ignite; 0.5
Zn(Ar)	Ignite; 1.0	Ignite; 1.0
$Cu(CO_2)$	Ignite; 6.5	Ignite; 3.0
Cu(Ar)	Ignite; 2.5	Ignite; 2.0
Ni(Ar)	Ignite; 5.0	Ignite; 0.5
Fe(Ar)	Ignite; 0.5	Ignite; 0.5

The trials reported were repeated three times, whereas the exposure of passivated nanopowder Al(Ar) to water occurred at $T = 20\,°C$. The exposure to air was $t = 72$ h. The procedure is described here.

- *Stage 1.* We prepared 15 ml of distilled water in a cup and put 0.1 g of Al nanopowder into it. We found no gassing and spontaneous ignition. The same trial was repeated for samples 0.5 and 1.0 g.
- *Stage 2.* We dispersed 2.0 g of Al nanopowder on filter paper with diameter 80 mm and put it in a cup with distilled water. No spontaneous ignition occurred.
- *Stage 3.* We put 10.0 g of Al nanopowder into a cup and filled the center of the sample with 3.0 ml of distilled water. No spontaneous ignition occurred.
- *Stage 4.* We put 2.5 g of Al nanopowder into a retort and added 50 ml of distilled water. Gassing (hydrogen) started after 5.65 h, the intensity of gassing was $I = V/t{\cdot}m = 16.9\ dm^3\ kg^{-1}\ h$. It follows that the risk is low (Stage 3).
  - The same procedures of stages 1–3 were repeated for $Al(N_2)$. No spontaneous ignition occurred for this nanopowder.
- *Stage 5.* We put 2.5 g of $Al(N_2)$ into a retort and added 50 ml of distilled water. Gassing (hydrogen and ammonia) started after 2 h, the intensity of gassing

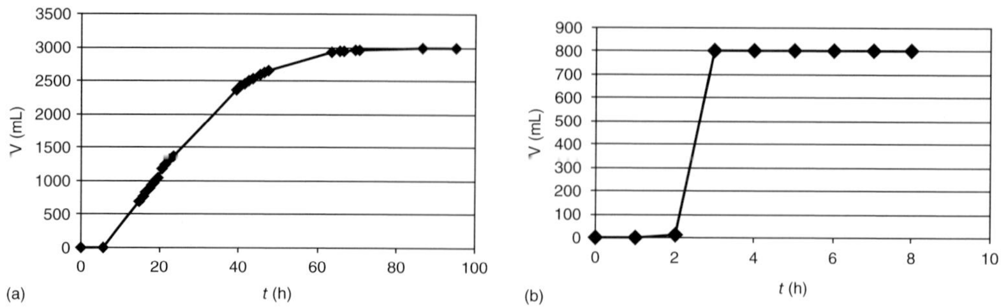

Figure 7.4 Gas volume evolved at reaction of (a) Al nanopowder and (b) aluminonitride composition with water ($T = 20\,°C$).

being $I = V/t{\cdot}m = 106.6\,dm^3/kg{\cdot}h$. The derived risk is medium (Stage 2). Thus, nanopowders of aluminonitride composition (Al (N_2)) exhibit a higher hazard and lower induction times compared to Al nanopowder (Figure 7.4).

7.4 Sensitivity against Electrostatic Discharge

Static electricity storage and formation of ESD accompany the utilization of many devices. The body of men is also the source of electric discharges. Therefore, it is necessary to determine the behavior of nanopowders by the influence of electrostatic discharge.

We tested (Al(Ar)), and nanopowder Al(Ar), coated by palmitic acid (L-ALEX) after storage at different times in air. The results are listed in Table 7.4, where t is time of storage in air, N the number of tests, H the distance from electrode to powder surface, S_s the value of capacitor, U the voltage of powder ignition, and W is the energy of powder ignition.

The average energy that a man's body can keep is around $78{\times}10^{-3}$ J. From the data in Table 7.4 it is evident that an ESD from a man's body can ignite nanopowders. The energy of ignition depends on the amount of active metal in the nanopowder. The content of active metal decreases from 92% mass (5 days (and nights)) up to 86% mass (30 days (and nights)) at storage in air. For nanopowder L-ALEX, the content of active metal is 96% mass. We found also during the trials that at the

Table 7.4 Nanopowders ignited by electric discharge.

Nanopowder	t (d)	N	H (mm)	S_s (pF)	U (kV)	W (J)
Al(Ar)	5	12	1	340	4.4	3.276×10^{-3}
Al(Ar)	5	10	0	340	4.53	3.488×10^{-3}
Al(Ar)	30	10	0	1100	4.83	12.83×10^{-3}
L-ALEX	20	21	0	227	3.13	1.11×10^{-3}

formation of aerosol above surface of nanopowder (e.g., at a first discharge when ignition does not happen), its ignition takes place at energies of $<10^{-3}$ J with a fast combustion.

7.5
Ranking of Nanopowders According to Hazard Classification

These results enable to classify nanopowders according to hazard classes. The classification was conducted in accordance with Russian GOST 12.1.044-84, GOST R 50587-93, GOST R 1052-97, and GOST R 19433-88, international requests ICAO (International Civil Aviation Organization), and IATA (International Air Transport Association) (Dangerous Goods Regulations). The international requirements were preferably applied.

For example, Cu(Ar) was classified according to GOST 19433-88 and requests of ICAO and IATA.

Nanopowders ignited by a low-energy ignition source with ignition time <30 s are:

Class 4 – highly inflammable solid: subclass 4.1 with one type of risk. The rate of flame spread is low: the grade of danger is low: 3. In accordance with GOST 19433-88 it gets category 411, classification number 4113, and the number of the label is 4a.

In accordance with instructions of ICAO and IATA, Cu nanopowder cannot be classified independently. Therefore, it is possible to consider it a metal powder,

Table 7.5 Hazard classification of nanopowders.

Name powder	UN #	Class or division	Subsidiary risk	Hazard label	PG	Packing instruction	Maximum net[a]
Ni(Ar) metal powder	3089	4.1	No	RFS flammable solid	III	419	25 kg
Al(Ar) (aluminum powder, uncoated)	1396	4.3	No	RFW dangerous when wet	II	415	15 kg
Su(Ar) (metal powder, flammable, n.o.s.)	3089	4.1	No	RFS flammable solid	III	419	25 kg
Zn(Ar) (Zinc powder)	1436	4.3	4.2	RFW dangerous when wet; RSC spontaneous combustion	III	419	25 kg
Ti wetted by hexanes	1208	3	No	RFL flammable liquid	II	Y305	5 l
Fe(Ar) pyrophoric metal, n.o.s.	1383	4.2	Prohibited for transportation				

[a] For air transport.

flammable, n.o.s., UN – 3089; class or category 4.1; group for packing – III; the instruction for packing is 419; max. net for one package at passenger air transportation is 25 kg. Mark label – RFS, black flame against a background of vertical lines and inscription FLAMMABLE SOLID.

In Table 7.5, the classification of different nanopowders is presented. Powders from electric wire explosion can be sorted according to the classification groups:

- Nanopowders that are dangerous because they are highly inflammable substances (all nanopowders of metals).
- Nanopowders that can ignite on contact with air (all nonpassivated powders in critical environment).
- Nanopowders that react with water with hydrogen generation.
- Pyrophoric nanopowders.

One powder of the above-mentioned groups is presented in Table 7.5. Some of them belong to two groups, for example, Zn(Ar). It is necessary to note that this classification does not take into account the ignition on ESD. Therefore, (in particular, Al powder) measures to avoid ESD are necessary during transportation, handling, and application.

7.6 Demands for Packing

Based on our research and according to demand for minimization of interaction products of metal reaction with water vapours and with oxygen (air) we can suggest that more best packing for nanosized unpassivated metal powders is hermetically sealed capacity- glass or metal, f.e. ampules (Figure 7.5a).

Polymer capacities are not perfect in the point of view prevention of air penetration and mechanical strength. When copper nanopowder kept in the polymer

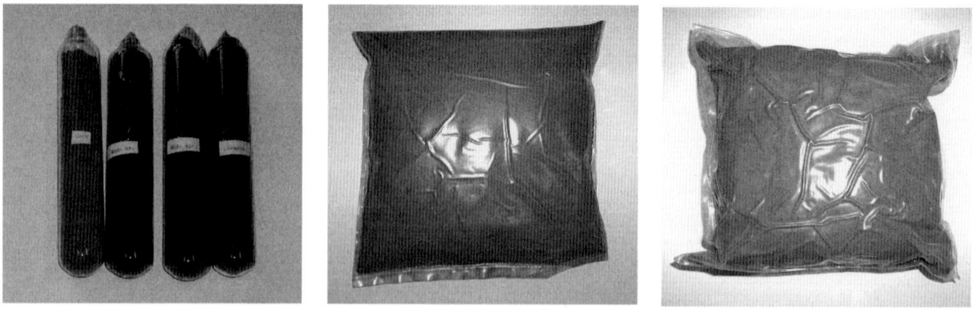

(a) (b) (c)

Figure 7.5 Examples of packing: (a) ampoules with powder packed in inert medium, (b) polyethylene (PE) package with passivated Al powder right after package was sealed, and (c) the same package after five days (and nights).

hermetically sealed capacities after short time it was found compression of capacities and cracks on the case.

We guess that nanopowder oxidation process leads to depression inside capacity and to destruction under air pressure. Therefore for storage of passivated metals bilayer PE film is used. When internal pressure decay, film outlines powder and keep safe (Figure 7.5b,c). Individual packages of Al powder it is better to store within sacks of metallized polymer for prevention of accumulation of static electricity.

References

1. Ivanov, G.V. and Tepper, F. (1996) Activated aluminium as a stored energy source for propellants. Fourth International Symposium on Special Topics in Chemical Propulsion, Challenges in Propellants and Combustion 100 Years after Nobel, Stockholm, Sweden, May 27–28, 1996.
2. Ivanov, G.V. (1995) Uniformity of burning and detonation of pyrotechnic mixtures based on activated aluminum. JANNAF Propulsion Meeting, CPIA Publication 630, 1995, Vol. 1, pp. 85–95.
3. Biryukov, Y, Voroztshov, A., and Bogdanov, L. (1998) Obtaining submicron abrasive powders by pneumatic processing of electric corundum and products of solid rocket propellant combustion. 29th International Annual Conference of ICT, Karlsruhe, Germany, June 25–28, 1998, p. 162.
4. Ivanov, Y.F., Osmonoliev, M.N., Sedoi, V.S., Arkhipov, V., Bondarchuk, S., Vorozhtsov, A., Korotkikh, A., and Kuznetsov, V. (2003) Productions of ultra-fine powders and their use in high energetic compositions. *Propellants Explos. Pyrotech.*, **28**, 319.
5. Pritchard D.K. (2004) Literature Review – Explosion Hazards Associated with Nanopowders. HSL/2004/12.
6. Eisenreich, N., Fietzek, H., Juez-Lorenzo, M., Kolarik, V., Weiser, V., and Koleczko, A. (2005) Influence of nano-particle size on the oxidation behaviour of Al, Fe, and Cu. *Mater. High Temp.*, **22** (3–4), 329–333 (Microscopy of Oxidation 6, Birmingham, 4–6. April 2005).
7. Eisenreich, N., Fietzek, H., Juez-Lorenzo, M., Kolarik, V., Koleczko, A., and Weiser, V. (2004) On the mechanism of low temperature oxidation for aluminum particles down to the nano-scale. *Propellants Explos. Pyrotech.*, **29**, 137.
8. Weiser, V., Kelzenberg, S., and Eisenreich, N. (2001) Influence of the metal particle size on the ignition of energetic materials. *Propellants Explos. Pyrotech.*, **26**, 284–289.
9. Trunov, M.A., Schoenitz, M., and Dreizin, E.L. (2006) Effect of polymorphic phase transformations in alumina layer on ignition of aluminium particles. *Combust. Theor. Model.*, **10**, 603.
10. (a) Bockmon, B.S., Pantoya, M.L., Son, S.F., and Asay, BW (2003) Burn rate measurements of nanocomposite thermites. 41st Aerospace Sclences Meeting and Exhibit, Reno, NV, AIAA Paper 003-241; (b) Danen, W.C., Jorgensen, B.S., Busse, J.R., Ferris, M.J., Smith, B.L. (2001) Los alamos nanoenergetic metastable intermolecular composite (super thermite) program. 221st ACS National Meeting, San Diego, CA, April 1–5, 2001; (c) Wilson, D.E. and Kim, K.A. Simplified model for the combustion of Al/MoO_3 nanocomposite thermites. 39th AIAA/ASME/SAE/ASEE Joint Propulsion Conference, Huntsville, Alabama, July 20–23, 2003, AIAA-2003-4536.
11. Tulis, A.J., Sumida, W.K., Dillon, J., Comeyne, W., and Heberlein, D.C. (1998) Submicron aluminum particle

size influence on detonation of dispersed fuel-oxidizer powders. *Arch. Comb.*, **18** (1–4), 157–164. Zhang, F., Murray, S.B., and Gerrard, K.B. (2005) Aluminum particles–air detonation at elevated pressures. *Shock Waves*, **15**, 313.

12. Mench, M.M., Kuo, K.K., Yeh, C.L., and Lu, Y.C. (1998) Comparison of thermal behavior of regular and ultra-fine aluminum powders (Alex) made from plasma explosion process. *Combust. Sci. Technol.*, **135**, 269–292.

13. Jones, D.E.G., Turcotte, R., Fouchard, R.C., Kwok, Q.S.M., Turcotte, A.-M., and Abdel-Qader, Z. (2003) Hazard characterization of aluminum nanopowder compositions. *Propellants Explos. Pyrotech.*, **28**, 120–131.

8
Reaction of Aluminum Powders with Liquid Water and Steam

Larichev Mikhail Nikolaevich

8.1
Introduction

The study of the reaction of aluminum oxidation with liquid water and steam is of interest as much from the standpoint of fundamental academic science as it is from the standpoint of many possible practical applications of this reaction. Indeed, for academic science, this reaction is a very interesting example of a complex multistadia and multifactorial high enthalpy heterogeneous process in which film of oxidation products covering Al surface inhibits the start of the oxidation process and newly formed solid products control the subsequent oxidation process. The understanding of this mechanism gives the possibility of effectively using this process for energy purposes in practice.

The energy accumulated in metal aluminum can be used in a wide spectrum of applications including metallized solid [1], liquid [2], and metallized gelled propellants and fuels for power sources such as standby power sources and portable devices, and fuels for transport engines [3–5]. The liberation of energy stored in metal aluminum can be carried out by oxidation reaction using different oxidants (O_2, N_2, CO_2, and H_2O) and at different process temperatures. It is assumed that aluminum production will be organized in places containing renewable energy sources, while energy utilization of aluminum will be organized in the most energy-loaded regions (megalopolises), in decentralized regions, for example, without electricity. Even with the existing technology of aluminum production, its use as an energy source can improve the local ecological situation in the cities, because the oxidation of aluminum with water is an environmentally friendly process.

Metal Al can successfully play at least two roles in future power engineering:

- As a power accumulator especially for renewable energy sources;
- As a high-effective intermediate energy carrier for the needs of hydrogen power engineering considering it as an alternative to traditional hydrocarbon power engineering at present [6].

Metal Nanopowders: Production, Characterization, and Energetic Applications, First Edition.
Edited by Alexander Gromov and Ulrich Teipel.

High-purity metal Al oxidation in pure liquid water can be used for production of not only high-purity hydrogen but also for production of ultrapurity aluminum oxides.

The possibility of Al being used as a high-effective energy carrier is justified by its

- wide natural occurrence (third element in the Earth's crust);
- high specific heat value (31.1 MJ kg^{-1} in the oxidation reaction by oxygen that corresponds to specific heat value of high-rank coals);
- high chemical activity;
- low cost of Al metal production;
- easy storage and transportation (unlike hydrogen – another perspective but very explosive energy carrier);
- products of oxidation being "green" compounds.

Aluminum oxidation by water can be used to generate heat and hydrogen, simultaneously producing solid oxidation products (aluminum oxides and hydroxides) with high marketable properties. This process can be realized at a wide range of temperatures (from room temperature up to 3500 °C). Of special interest is the use of the aluminum–water oxidation process to produce high-temperature hydrogen (up to 3500 °C). The last represents a perspective "working body" for a new generation of magnetohydrodynamic generators (MHDs). Another perspective variant of this process is the low-temperature Al oxidation (up to 100 °C) with liquid water. It is the simplest, safest, and cheapest variant of this process but it requires the use of activation methods to a great extent. The advantage of the practical use of aluminum oxidation process is the possibility of providing reasonable values of the chemical reaction rate of aluminum oxidation, that is, to provide rates of hydrogen generation correlating with gas consumption of energy plants. In practice, it is reached by using dispersed aluminum (aluminum powders) with simultaneous use of activation methods for the oxidation process.

The use of predispersed powders with different specific surface areas can provide the necessary contact area of reactants. Predispersed aluminum powders with different particle sizes, including nanoparticles, can be chosen depending on the required rate of the oxidation process [7]. However, the metal can be dispersed to the required particle size during the oxidation process. For realization of this possibility, the thermal energy evolved at the oxidation of aluminum can be used. The amount of heat evolved at oxidation of 1 g of Al exceeds 31 kJ g^{-1}. The amount of heat needed for fusion of 1 g aluminum is equal to 0.4 kJ g^{-1} only and heat of evaporation of 1 g of Al is equal to 11.2 kJ g^{-1}. It is possible to assume that the metal oxidation heat can be effectively used to provide the dispersion of solid metal. The amount of energy is enough because the specific thermal effect of reaction is greater than the specific heat of Al melting. Perhaps this process can be realized at the increased pressure and temperatures.

The surface of Al in contact with atmospheric air is usually covered by a thin and tight passivating oxide coating (POC). This is a product of "soft" oxidation (long-time slow oxidation) of aluminum surface layer in reaction with atmospheric oxygen and water vapor. Usually, this coating may consist of the mixture of aluminum oxides and hydroxides. The concrete structure of POC depends on conditions of a "soft" oxidation and has, as a rule, a "crazy quilt" structure.

The presence of POC on the Al surface requires the special efforts necessary to initiate the reaction and at the same time it is necessary to make special efforts to overcome the passivating effect of solid reaction products forming continuously on the surface during the metal oxidation. The last allows maintaining a stable rate of H_2 formation and assuring 100% oxidation of metal aluminum.

Activation methods are used to ensure the effective contact of water molecules with nonoxidized metal surface by means of the following:

- Destruction of POC usually covering the Al surface to accelerate the start of oxidation process
- Minimization of effect of metal Al surface passivation as a result of deposition of solid oxidation products produced continuously during the oxidation of Al.

Currently, activation of low-temperature aluminum oxidation is carried out by a number of methods.

- Use of aqueous alkali solutions as oxidizing agents [8, 9]. It provides the dissolution of solid oxidation products.
- Use of aluminum pretreated by alloys on the basis of liquid gallium and other metals [10, 11]. It provides the weak fixation of solid oxidation products on the metal surface and facilitates the access of water molecules to the surface of the oxidizing metal.
- Mechanochemical activation of aluminum [12, 13], which provides the creation of structure defects on the surface and in the bulk of metal, that increases the metal reactivity.
- Use of water-soluble organic and inorganic substances, which allow managing the condensation process of the solid oxidation products. It leads to formation of porous structures, which are permeable to water molecules [14].

Each of above-mentioned techniques has both advantages and disadvantages. Although the use of aqueous alkali solutions accelerates the oxidation of aluminum particles and can provide, in some cases, complete aluminum oxidation even at room temperature, it requires special precaution. Application of aluminum nanoparticles is limited by the high cost of nanoparticles. The aluminum activated by alloys on the basis of gallium creates problems in environmental aspects. The techniques of activation using special additives can in some cases create problems of utilization of solid oxidation products containing activators that were used.

The choice of activation method for low-temperature oxidation of aluminum in water should be determined by a specific device, which is serviced by the oxidation process and tasks of production of hydrogen, heat, and solid oxidation products.

The limited natural resources necessitate their integrated use. In this regard, the use of metallic Al as an energy carrier, in particular, for hydrogen production raises the questions about the rational use of not only evolved heat energy but also solid oxidation products. The last can be easily used as the raw material for the reproduction of the metal Al. But the solid oxidation products can be used successfully as initial components in the creation of new composite and ceramic materials with unique properties as well as in medicine and biology. Fabrication of high-purity hydroxides with high consumer cost is of particular interest. Such an oxide (boehmite) can be fabricated by means of hydrothermal aluminum oxidation [15]. But certainly, the method of low-temperature aluminum oxidation (up to 100 °C) is simpler, safer, and cheaper than the process of hydrothermal aluminum oxidation. In the case of low-temperature oxidation of aluminum micron powders, the solid product represents aluminum hydroxide. To provide the production of nonpolluted aluminum hydroxide at low-temperature conditions, ultrasonic (US) activation was proposed to be used, as the activating effect of ultrasound is well known for a number of chemical reactions [16].

For several years, the V.L. Talrose Institute has been performing investigation of energy properties of metal aluminum. Some interesting results of this work are presented in this chapter.

8.2 Experimental Technique for Studying Reaction Al Powders with Liquid and Gaseous Water

Powder samples were investigated by means of thermogravimetry (TG) and differential thermal analysis (DTA) analyses combined with measurements of rate of hydrogen evolution for linear heating up to 900 °C at atmospheric pressure in flows of inert gas Ar, air, and water vapor.

The scheme of apparatus designed for investigation of the reaction of aluminum particles with water vapor at temperatures up to 1000 °C is shown in Figure 8.1. This apparatus permits studying the reaction of Al powder with liquid water at temperatures up to 130 °C at atmospheric pressure in an "open" system. The last

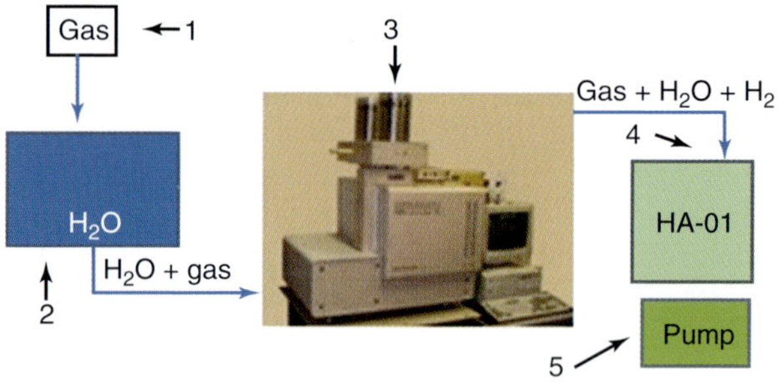

Figure 8.1 The scheme of apparatus for investigation of reactivity of Al powders with liquid water at temperatures up to 130 °C and water vapor at temperatures up to 1000 °C at atmospheric pressure. 1, gas carrier flow system (Ar); 2, saturator; 3, STA; 4, analyzer of hydrogen (HA-01); and 5, pump for gas flow formation.

means that evaporation of water from composition of water suspension changes the composition of the tested suspension.

This apparatus is a combination of two devices: apparatus for simultaneous thermal analysis (STA) (Derivatograph C) and the hydrogen analyzer AH-01 [5]. The linear heating of the sample was made in a stream of pure inert gas Ar or the inert gas carrier containing water vapor. In the last case, the carrier gas was analyzed for the presence of hydrogen by means of AH-01 after passing through the reaction zone.

The apparatus provides simultaneous measurement of thermal analytic characteristics of the aluminum samples and the rate of formation of gaseous hydrogen. The quantitative calibration of the sensitivity of AH-01 was made by using aluminum hydride (AlH_3) thermal decomposition. Thermal decay of a known mass of AlH_3 permits creation of a known amount of hydrogen. The possibility of TG measurement permits verification of this procedure.

The behavior of Al powders in a mixture with liquid water and gelled water were studied also at temperatures down and up to the temperature of water boiling at atmospheric pressure. The polyacrylamide at a concentration of 3 wt% was used as gelled agent. Reaction of Al powders with liquid H_2O was studied by using a special apparatus constructed for this purpose (Figure 8.2).

The Al/H_2O slurry or dispersion samples were exposed to linear heating or exposure at a certain temperature in an atmosphere of saturated water vapor. The temperature range was between room temperature and boiling temperature of Al powder containing slurry or dispersion. Measurements were performed at atmospheric pressure. The use of saturated water vapor atmosphere allows preventing

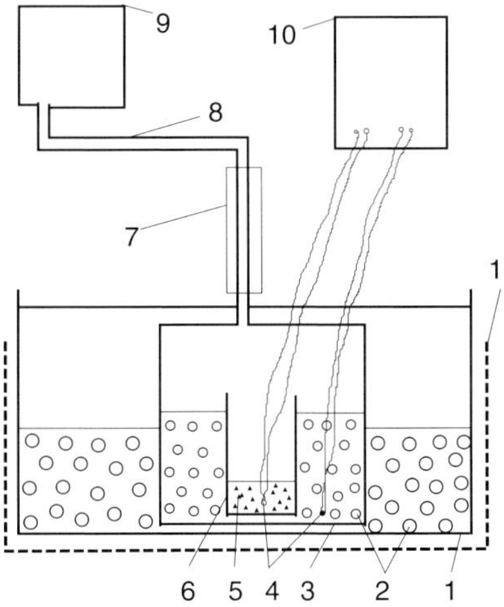

Figure 8.2 The scheme of apparatus for investigation of reactivity of Al powders with liquid water at linear heating in saturated water vapor up to temperature of water boiling point. 1, external liquid thermostat; 2, distilled water; 3, internal liquid thermostat; 4, thermocouples; 5, sample (Al powder + liquid water); 6, reactor; 7, cooler; 8, connective tube; 9, volume meter (flow meter); and 10, differential thermometer.

the change of water content in samples as the result of water evaporation during the sample heating (exposition). The construction of the apparatus gave the opportunity of simultaneous measurement of both temperature of investigated sample and temperature of water heated in this sample. It permits to realize the measurement of temperature at the start of the reaction between the metal aluminum and the liquid water in sample compositions from analysis of the difference between temperature of reaction media and temperature of heating water. This difference is negative at the initial step of sample heating (exposition) and changes to positive after the start of oxidation reaction and increases significantly as the reaction progresses. The possibility of measurement of the rate of evolution or the volume of gas-phase reaction products (hydrogen and methane) allows determining both the start temperature of the reaction of Al powder with the liquid in slurry composition and fractional conversion of metal aluminum as result of the reaction.

8.2.1
Oxidation of Aluminum Powder with Distilled Water

The heterogeneous reaction of metal Al with water can be used as a source of gaseous hydrogen being produced along with thermal energy and solid oxidation products (oxides and hydroxides). Three channels with different thermal effects can be realized depending on the conditions of reaction. The composition of solid reaction products is different, but the amount of gaseous hydrogen formed of the mass unit of metal Al is the same (more than 11% from the total mass of Al) for these three channels.

$$2Al + 3H_2O = Al_2O_3(\text{oxide}) + 3H_2 + Q1 \quad (817\,\text{kJ}) \tag{8.1}$$

$$2Al + 4H_2O = 2AlOOH\,(\text{boehmite}) + 3H_2 + Q2 \quad (846\,\text{kJ}) \tag{8.2}$$

$$2Al + 6H_2O = 2Al(OH)_3(\text{bayerit}) + 3H_2 + Q3 \quad (871\,\text{kJ}) \tag{8.3}$$

This value is a completely good parameter in comparison with other solid substances: reactants or adsorbents proposed now for production and storage of hydrogen.

The oxidation process that starts at close to room temperature can be considered as the sequence of several overlapping stages. They are listed subsequently (Figure 8.3) for micron Al powder with particle size about 4 μm:

- The stage of initial oxidation (1) is not intensive and quickly damped. It results in the generation of a small amount of hydrogen during the first minutes of contact of aluminum powder with water. The initial oxidation can lead to oxidation up to a small percentage of metal and to a brief temperature increase (up to several degrees) of the reaction medium.
- The induction stage (2) lasts from several minutes to several days depending on the composition and structure of POC that covers the Al particles, the values of both starting reaction temperature and pH of reaction medium. The process of degradation (hydrolysis) of POC takes place during the induction

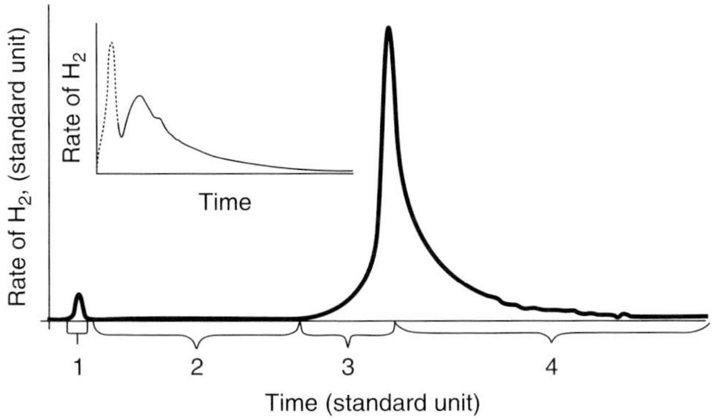

Figure 8.3 Kinetic curve of the rate of hydrogen evolution for low-temperatures oxidation of micron-sized Al powder starting at near-room temperature. 1, stage of initial oxidation; 2, induction stage; 3, stage of intensive oxidation; and 4, stage of oxidation decay.

stage. The hydrolysis promotes the disruption of POC continuity and increases its permeability for water molecules. The considerable hydrogen generation and increase in temperature of the reaction mixture is observed at the end of the induction stage. It indicates the beginning of the stage of intensive oxidation. The border between the induction stage and the stage of intensive oxidation is conditional and depends on the sensitivity of a method used for measurement of the rate of hydrogen release.

- During the stage of intensive oxidation (3), the fast increase in the rate of hydrogen generation is observed. It is connected with the increase of the reaction mixture temperature and with the intense destruction of the POC. Increase in both rate of hydrogen generation (V_{H_2}) and temperature (T) of the reaction medium is observed simultaneously (Figure 8.4) and is accompanied by the accumulation of solid products of aluminum oxidation.
- The stage of oxidation decay (4) is caused by the fact that as a result of the oxidation process, the surface of the aluminum particles decreases. In addition, oxidation products fix on the surface of the metal aluminum and complicate the access of water to the reaction zone. Owing to the combination of these factors, the oxidation rate begins to decrease.

Sometimes (depending on oxidation process parameters), the kinetic curve may have not one but two (Figure 8.3, the inset) and even three peaks, indicating the complex, multistage nature of aluminum oxidation.

The size and the surface topography of aluminum particles as well as the method of forming aluminum powder have a very important influence on the observed specific rate of hydrogen generation. The kinetic curves of hydrogen evolution for Al powders with the different sizes and shapes of particles are shown in Figure 8.5. Several methods were used for forming these powders: the Gen–Miller method [17] of metal vapor condensation in inert gas carrier flow (for nanosized (NS) Al (100 nm)); the method for spraying molten metal for micron-sized Al (3.8 μm); and the method for mechanical-chemical treatment for Al with a flake-like particle

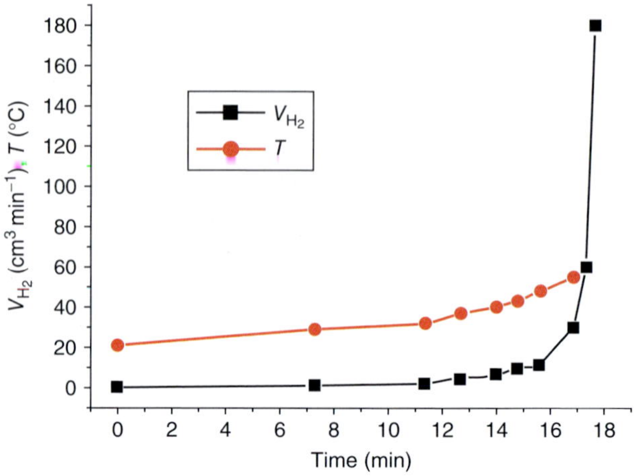

Figure 8.4 Kinetic curves of the volume of generated hydrogen and of reaction medium temperature observed at the Al nanopowder oxidation in distilled water.

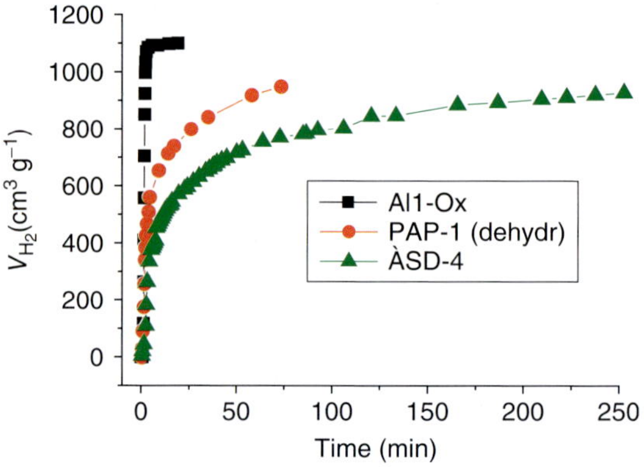

Figure 8.5 Kinetic curves of H_2 evolution for Al powders with different sizes and shapes of particles (the temperature of distilled water −95 °C): nano-sized Al (100 nm); micron-sized Al (3.8 μm); and Al flakes (80 μm × 40 μm with dehydrophobizated surface of particles).

(80 μm × 40 μm). The temperature of distilled water was 95 °C. It is clear that NS Al particles react most rapidly with water. The next conclusion that reactivity of flakes with water is higher than that of spherical particles is not so evident. The scanning electron microscopy (SEM) microphotographs of Al powder of different particle sizes and forms are shown in Figure 8.6.

Figure 8.6 Microphotographs of Al powder of different particle sizes and forms: (a) ASD-4 micron Al powder image and particle size distribution (D = 3.8 μm). (b) Al1-Ox nanosized Al powder image and particle size distribution (D = 100 nm). (c) PAP-1 flake-like particles of Al images of average sizes of 80 μm × 40 μm and with the submicron thickness (<1 μm).

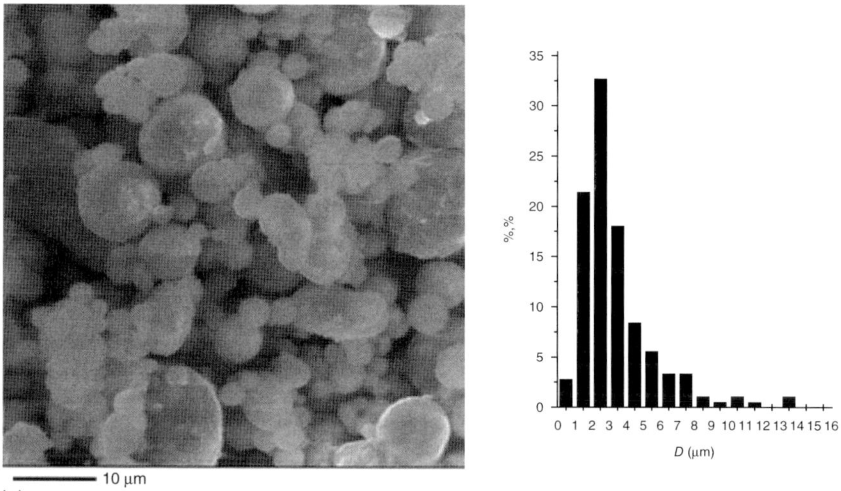
10 μm
%,%
0 5 10 15 20 25 30 35
0 1 2 3 4 5 6 7 8 9 10 11 12 13 14 15 16
D (μm)

(a)

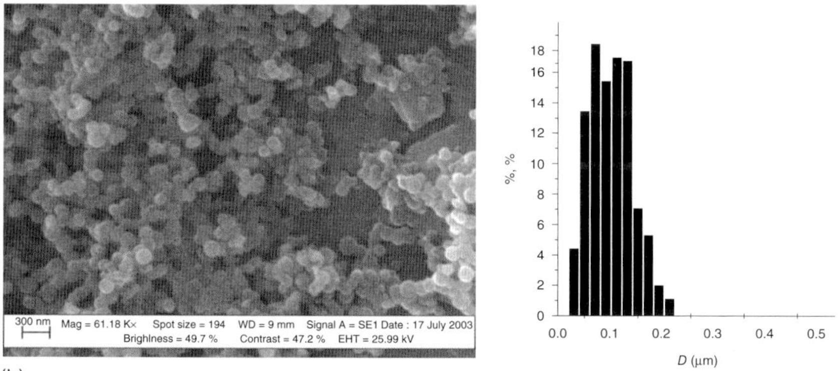
300 nm Mag = 61.18 K× Spot size = 194 WD = 9 mm Signal A = SE1 Date : 17 July 2003
Brighlness = 49.7 % Contrast = 47.2 % EHT = 25.99 kV
%, %
0 2 4 6 8 10 12 14 16 18
0.0 0.1 0.2 0.3 0.4 0.5
D (μm)

(b)

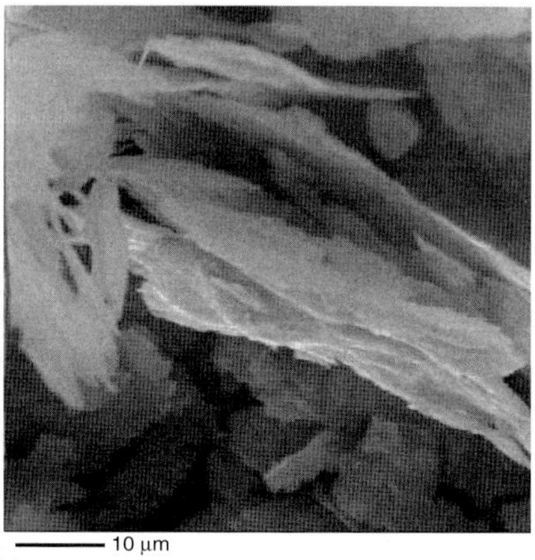
10 μm

(c)

The high reactivity of the flake Al particles is the result of the submicron thickness (much less than 1 μm) of particles. Comparison of shapes of kinetic curves shown in Figure 8.3 for these powders reveals that even one submicron dimension parameter can essentially increase the rate of Al oxidation. It is an important conclusion to practice because this type of aluminum powder is the most inexpensive among those produced by the Russian industry. In addition, Figure 8.6c shows that the basic mechanism of mechanical activation of metal Al is the shear destruction of metal.

The next important parameter that determines the rate of oxidation of the aluminum particles is the state of its surface. The fracture resistance of water for POC, usually covering the surface of aluminum, depends on its structure and phase composition. Hydrolysis is mainly responsible for the destruction of the POC contacting with liquid water. For example, POC containing more hydroxyl groups (chemically adsorbed water) in its structure is faster subjected to hydrolysis. For this POC, the duration of its induction stadia may be significantly (several times) lower.

In some cases, the surface of Al particles after POC formation can be subjected to a special chemical modification that changes the surface properties of the particles (hydrophobization, hydrophilization, etc.) [18]. This treatment provides to Al oxides properties that are essential for effective utilization of the powder. For example, it may be anti-agglomeration properties for Al particles and, simultaneously, the properties that provide the necessary adhesion between Al particles and the given liquid or solid matrix.

Figure 8.7a shows the effect of a monolayered hydrophobic coating on the surface of the nanopowder with a particle size of about 150 nm (the particle surface treatment by hexamethyldisilazane (HMDS) vapors). Curve 1 presents the kinetic dependence of hydrogen evolution for a nanopowder covered with trimethylsilyl

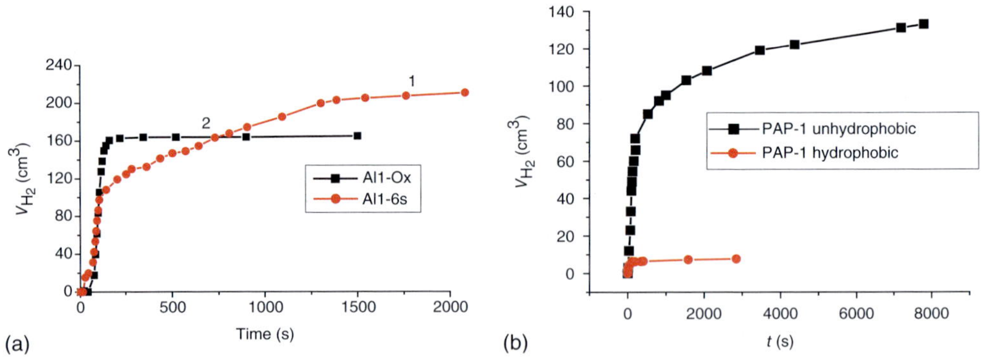

(a) (b)

Figure 8.7 Kinetic curves of H_2 evolution for Al powders oxidized in distilled water at a temperature of 95 °C: (a) Al nanopowder with surface covered by monolayer trimethylsilyl hydrophobic coating (curve 1) and nanopowder covered by nonhydrophobic POC (curve 2) and (b) Al powder (flake particles) covered by hydrophobic multilayer coating on the base of fatty acids and same powder covered by nonhydrophobic POC (curve 2).

monolayer hydrophobic coating and curve 2 shows the same kinetic dependence for powder with the surface covered with POC. Both powders were manufactured by the Gen–Miller method. The oxidation was performed in distilled water at 95 °C and atmospheric pressure. For hydrophobic powder the chemically bound trimethylsilyl umbrellas screen the surface to protect it from contact with water.

However, owing to the steric factor, a part of the surface may contact and react with water molecules penetrating under the umbrellas. At the beginning, the oxidation of both samples is similar but later the hydrophobic surface oxidizes slower.

Figure 8.7b demonstrates the hydrophobic effect of the multilayer coating on the base of fatty acids. Aluminum powder PAP-1 (flakes) with a hydrophobic surface is practically nonoxidable, at least during the 50 min contact with hot water. The same powder with a nonhydrophobic surface (POC) reacts quickly at the start of the process but takes several hours to reach full metal oxidation.

The acidity of the suspension of aluminum powder in distilled water remarkably depends on the time after preparation of this suspension. At the beginning, its value is near the value of acidity of the distilled water used. With time, this value changes. It is the result of the interaction of the surface layer of particles with distilled water. Of course, the growth rate and the magnitude of this change depend on the ratio of the volume of distilled water and the surface area of the aluminum powder wetted with this water. They depend also on the POC composition. For example, the increased amount of chemically bound water in POC provides a higher pH of reaction medium. The kinetic curves of change of the reaction medium acidity (pH increase) for a freshly prepared suspension of micron Al powder in distillated water with different mass ratio of reactants are shown in Figure 8.8. Figure 8.8 indicates both the value of pH and the equilibrium time are higher for higher value of the mass ratio of Al to H_2O. The experiments were performed at room temperature.

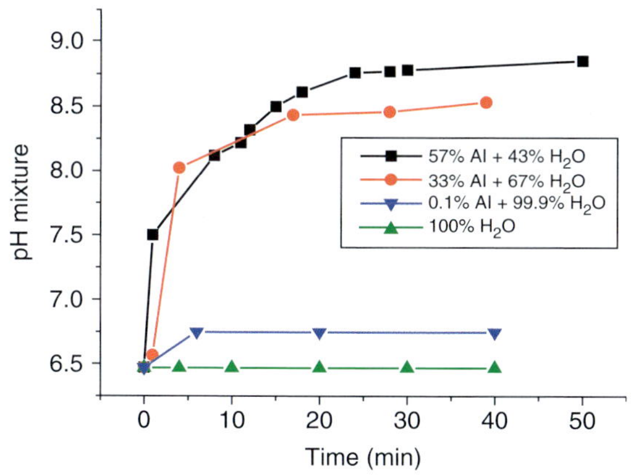

Figure 8.8 The kinetic curves of pH change for the reaction medium for fresh prepared suspension of micron-sized Al powder in distilled water with different mass ratio of reactants (under the room temperature).

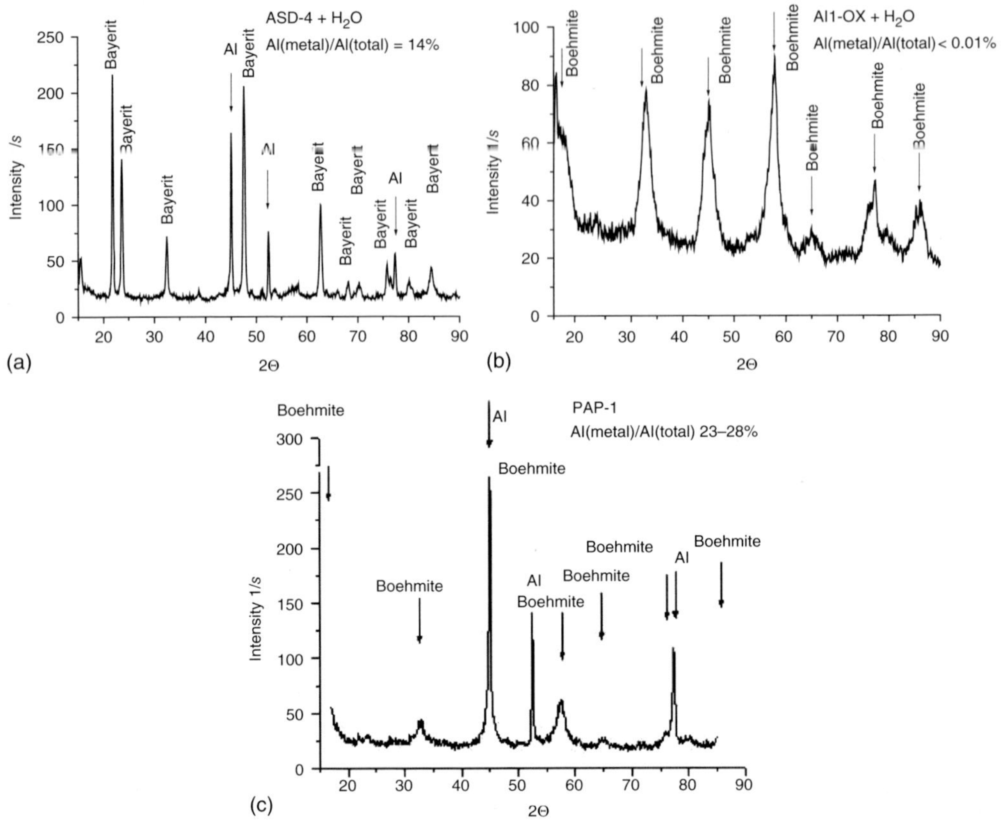

Figure 8.9 Results of X-ray analysis of solid reaction products of liquid water with (a) micron-sized Al powder (15 h of oxidation) – bayerit ($Al(OH)_3$); (b) nanosized Al powder (10 min of oxidation) – boehmite (AlOOH); and (c) flake-like Al particles with submicron thickness (2 h of oxidation) – boehmite (AlOOH).

The results of X-ray analysis (Figure 8.9) demonstrate that the content of solid oxidation products depends on the size of the aluminum particles. Particles having at least one nanoscale parameter are oxidized to form boehmite (AlOOH) in excess of the oxidizer. Micron powders are oxidized to form bayerit ($Al(OH)_3$) under the same conditions. The decrease in the amount of reacting water to stoichiometric or less for micron powder results in the formation of boehmite (AlOOH) admixture. This fact indicates the possibility of local overheating (more 300 °C) of the surface of the aluminum particle during its oxidation in liquid water.

8.3 Oxidation of Aluminum Powder in Water Vapor Flow

The behavior of Al powder heating linear in water vapor at temperatures lower 800–900 °C is mostly the same as the behavior of Al powder heating linear in air

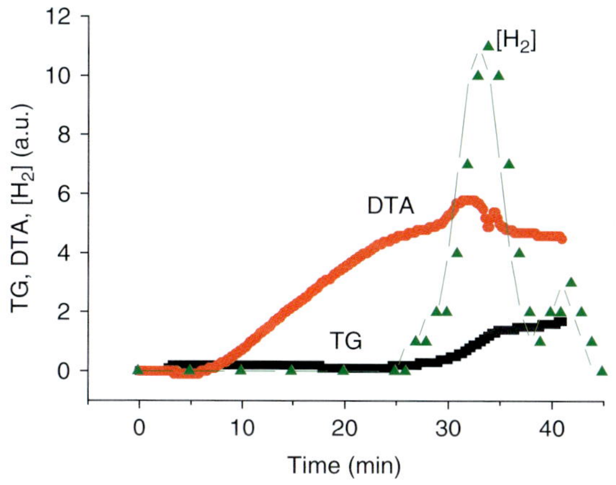

Figure 8.10 TG, DTA curves, and rate of hydrogen formation for Al powder (flake-like particles) linearly heated (rate 20 grad min^{-1}) in atmosphere of water vapor at atmospheric pressure up to 800 °C.

under the same conditions [19]. The interesting feature of the oxidation process in water vapor is the inequality of both temperatures achieved: maximum of heat evolution and maximum of rate of hydrogen evolution. The value of the first temperature is lower than that of the second. The observed difference cannot be explained by the feature of the registration system of the apparatus used.

TG, DTA curves, and kinetic curve of rate of hydrogen formation for flake-like Al particles linearly heating (rate 20 °C min^{-1}) in an atmosphere of water vapor at atmospheric pressure up to 800 °C are shown in Figure 8.10. The temperature of the maximal hydrogen evolution (collapse of POC) is about 640 °C, and the difference between temperature maximum of heat evolution and temperature maximum of hydrogen evolution is about 30 °C.

8.4 Nanopowders Passivated with Coatings on the Base of Aluminum Carbide

As noted earlier, aluminum is an extremely active chemical element. In contact with air, the aluminum surface is usually covered with POC – a thin (several nanometers) oxide film. This film prevents the process of further oxidation of a metal aluminum in air atmosphere at near-room temperature and provides to Al particles a sufficient unreactiveness, which is necessary for the organization process of their energy application. Aluminum particles with their surface coated with POC are very convenient for practical use. They are relatively safe. Storage and work with them do not require creating a special atmosphere. The simplicity and low cost of POC synthesis are its main advantage.

At the energy use of aluminum powder, POC plays a dual role: first, it facilitates work with Al powder and the possibility of convenient long-term storage and, second, it delays the start of oxidation and increases the duration of the induction period. In addition, POC is an inert ballast in the reaction medium.

As the average thickness of POC is about 3–4 nm, it means that for aluminum particles with diameter of 100 nm, the mass of POC is about 15% that of the mass of the particle. Because the decrease in diameter of a particle increases the mass of chemically inert POC inversely proportional than for NS Al powders, the mass of chemically inert passivating oxide layer will constitute a significant part of the mass of the particle.

The use of fine (F), ultrafine (UF), and NS aluminum powders as a source of thermal or chemical energy is of special interest. The common distinguishing characteristic of these powders is a high specific surface area, which determines extensively the behavior of Al powders in energy systems. A high percentage of inert oxides may call into question the feasibility of energy uses of aluminum nanopowders coated with POC.

This situation can be overcome in several ways. For example, storage of Al particles with pure metal surface and manipulation by them can be carried out in an inert medium. But an inert medium can prevent oxidation of the particles and cannot restrict their coalescence.

Another possibility is to cover the metal Al surface with a passivating coating (PC), ensuring safety in an atmosphere of storage and operation of the Al powder, and reacting with the evolution of thermal or chemical energy during a process of utilization of Al powder in a given reaction medium. A particular case is the coating, which permits both storage and operation with Al powder at atmospheric conditions and is chemically active at a condition of Al powder oxidation. It is possible if PC is oxidized with energy release or a catalyst for a process in which particles are used. An example of such a coating is the nickel coating on the surface of the Al particle.

Unfortunately, it is practically impossible to design a universally "reactive" PC for different energy applications of Al powders. Such PCs should be developed directly under each specific energy use. "Reactive" PCs may be organic or inorganic films deposited to the surface of particles. But it may be a film formed on the surface with participation of Al surface atoms [19]. It can include Al atoms in its structure. Representatives of the latter type are PC coatings on the base of carbides, nitrides, oxycarbides, salts, and other compounds of aluminum.

Coatings on the base of aluminum carbide and aluminum nitride are of interest for energy processes in which the water is an oxidizing agent. These coatings can react with water (liquid or vapor). They can participate actively in oxidation (burning) processes with evolution of heat and combustible gas.

A method of carbothermy (reaction of the carbon or carbon-containing compounds with the metal aluminum or aluminum oxides at high temperatures) was proposed for synthesis of PC coatings on the base of carbides and oxycarbides on the nanopowder surface in V.L. Talrose Institute for Energy Problems of Chemical Physics RAS [20]. These reactive coatings on the base of aluminum carbide were designed on the basis of aluminum carbide (Al_4C_3) for Al powders produced by condensation Gen–Miller method and by the electrical explosion of wires [21].

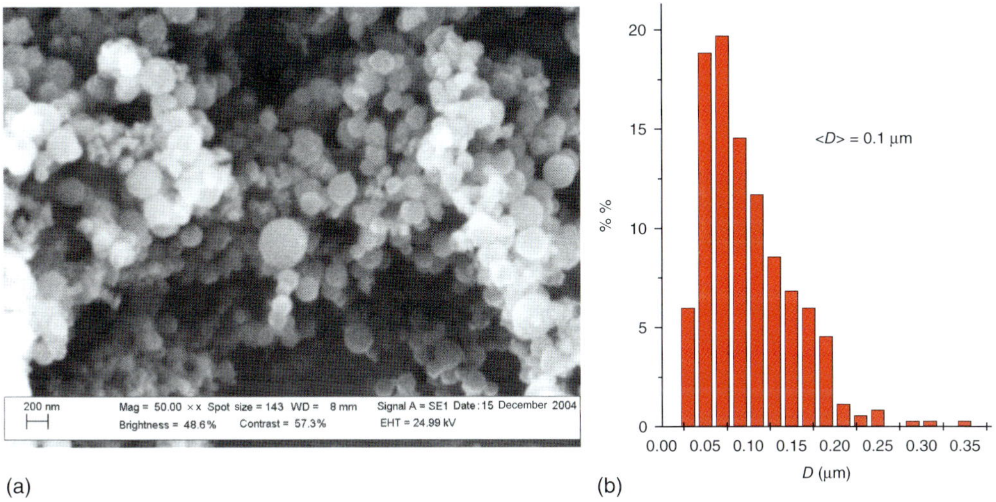

Figure 8.11 Al powder produced by the condensation method of a Gen–Miller and passivated with aluminum carbide: (a) microphotographs of particles and (b) particle size distribution ($\langle D \rangle = 0.1\ \mu m$).

Al_4C_3 is inert with respect to dry air under normal atmospheric conditions but reacts with the water, water vapor, and oxygen at high temperatures. Reaction of Al_4C_3 with water leads to evolution of heat and fuel gas methane. Aluminum powders passivated with these reactive coatings can be used in composition with water or gelled water as slurries or dispersions designed for energetic application, in particular, as "green" propellants.

SEM images and particle size distribution for NS Al powder manufactured by condensation Gen–Miller method ($D = 0.1\ \mu m$) are shown in Figure 8.11. SEM images and particle size distribution for NS Al powder manufactured by the method of electrical explosion of wires ($D = 0.16\ \mu m$). It is clear that the use of the carbothermy method permits producing of submicron powders.

The carbide of aluminum was detected in the composition of surface coating by means of gas chromatography analysis of the gas-phase product of reaction of Al powders with hot liquid water. It was established that the evolving gas was a mixture of two gases: hydrogen and methane. Methane is a result of the chemical reaction of aluminum carbide with water:

$$Al_4C_3 + 12H_2O = 4Al(OH)_3 + 3CH_4(gas) + Q1 \tag{8.4}$$

$$Al_4C_3 + 8H_2O = 4AlO(OH) + 3CH_4(gas) + Q2 \tag{8.5}$$

$$Al_4C_3 + 6H_2O = 2Al_2O_3 + 3CH_4(gas) + Q3 \tag{8.6}$$

It is generated along with thermal energy and solid oxidation products (aluminum oxides and hydroxides). Three channels with different thermal effects can be realized depending on the conditions of reaction. The composition of solid reaction

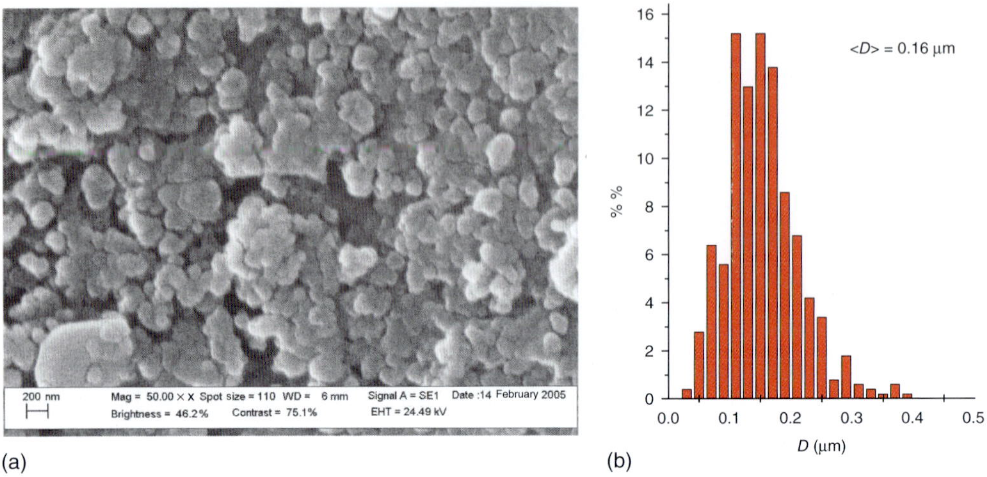

Figure 8.12 Al powder produced by the method of electrical explosion of wire and passivated with aluminum carbide: (a) micro-photographs of particles and (b) particle size distribution ($\langle D \rangle = 0.16\,\mu m$).

products is different, but the amount of gaseous methane formed of the mass unit of metal Al is the same (more 33% from the mass of Al_4C_3 and more 44% from the mass of Al) for these three channels. Each molecule of Al_4C_3 can produce three molecules of CH_4.

The presence of methane in the composition of gaseous reaction products shows the presence aluminum carbide in the composition of Al particles. Crystal aluminum carbide was detected by means of X-ray structural analysis only in composition of Al–CO_2–CO_2 sample. Figure 8.12 shows micro-photograph of particles (a) and particle size distribution (b) for this sample. SEM image of Al particles indicates that the particles have an angular shape unlike the shape of particles of Al-CO_2-1 sample having regular spherical shape (Figure 8.11). It is possible that the observed faceting of particles is the result of the presence of crystal Al carbide in its content. The content of PC in this sample was extremely high (about 80 wt%). This sample was manufactured by the exploding wire method. For other samples, the weight percent of PC was between 15% and 25%. This value is comparable to a value of percentage of oxide PC measured for UF powders with the same size of particles. It is evident that Al_4C_3 has an amorphous structure for these samples.

The experimental results demonstrate that the PCs designed on the basis of aluminum carbide provide effective protection for the metal core of aluminum nanoparticles from oxidation on contact with atmospheric air at room temperature and natural humidity. The long-time storage of aluminum powders (manufactured by both methods) for several months in contact, from time to time, with the atmospheric air in a laboratory room did not result in any noticeable decrease in the amount of metal aluminum in the powders.

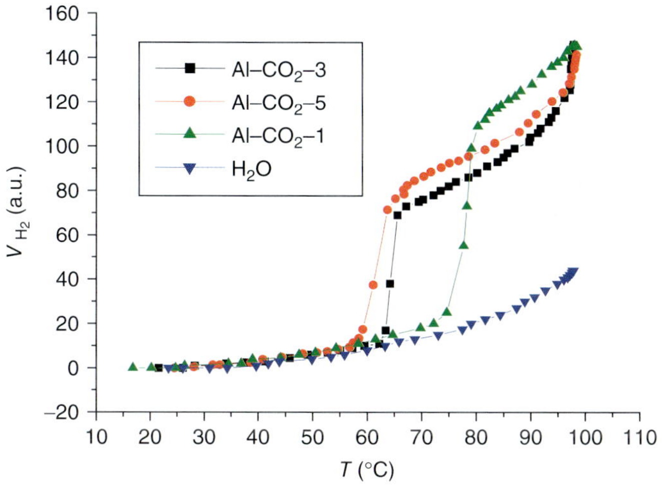

Figure 8.13 Kinetic curves of H_2 evolution for Al nanopowders with passivated with aluminum carbide reacted with liquid water. The rate of linear heating of samples −5 grad min^{-1}.

The investigation of the reactivity of Al nanopowders passivated by aluminum carbide with respect to different gases, water vapor, and liquid water (in "open" system) was performed by using the apparatus shown in Figure 8.1. The study of water suspensions or slurries heated in an atmosphere of saturated water vapor was performed using the apparatus shown in Figure 8.2.

The shapes of the kinetic curves of the temperature of the reaction mixture testify that the reaction between Al particles and liquid water passes at least through two stages. The start of the first stage of reaction was observed in the temperature region 57–73 °C. Kinetic curves of H_2 evolution for several samples of NS aluminum powders with PC on the base of aluminum carbide are shown in Figure 8.13. The rate of linear heating of samples was 5 grad min^{-1}. Samples were heated at the atmosphere of saturated water vapor.

The value of the start temperature depends on the composition of the PC. The admixture of aluminum oxide increases this value. For nanopowders covered with aluminum oxide coatings (POC), the first stage of reaction with liquid water starts at temperatures 74.5–76.0 °C. The second stage of reaction of Al oxidation, which is responsible for hydrogen evolution, appears at the temperature near 100 °C for carbide aluminum PC and for POC.

The experimental results demonstrate that contact of aluminum powder with liquid water can destroy the PCs designed on the basis of aluminum carbide even at room temperature. The evolution parameters of the oxidation process of aluminum powder Al–CO_2-1 as a result of the preliminary contact with water at room temperature is given in Table 8.1.

Sample Al–CO_2-1 with metal content 77.2% was investigated immediately after water slurry preparation (line 1) and after 4 and 5 h of preliminary exposition of

Table 8.1 Influence of preliminary contact time of aluminum powder Al–CO_2-1 with distilled water on the properties exhibited by the linear heating of suspension.

N	Al powder Al–CO_2-1	T_{oini} (°C)	T_{ofin} (°C)	$Al_{met}1$ (%)
(1)	(2)	(3)	(4)	(5)
1	Fresh	72.3	83.3	~100%
2	4 h	72.5	80.9	59.6%
3	5 h	72.3	78.2	55%

T_{oini} (°C), start temperature of the first stage of oxidation process; T_{ofin} (°C), finish temperature of the first stage of oxidation process; $Al_{met}1$ (%), percentage of metal aluminum oxidized in reaction with water; Column (2), time between suspension preparation and testing.

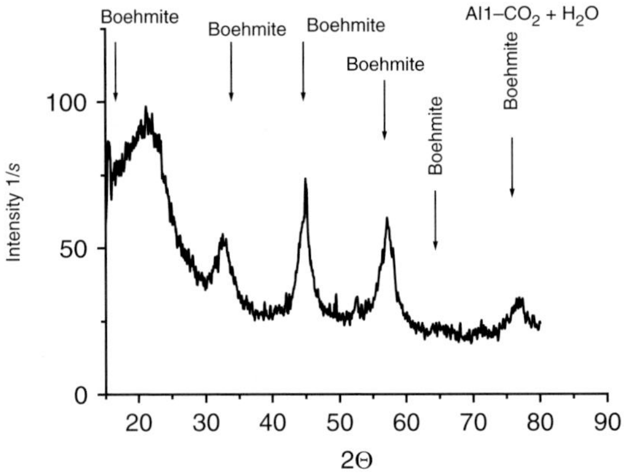

Figure 8.14 XRD pattern of oxidation solid products obtained in result of oxidation of submicron Al powder Al–CO_2-1 with distilled water at temperature 95 °C. The main product is boehmite (AlOOH).

aluminum powder in contact with water at room temperature (lines 2 and 3). Test samples were heated linear with the rate 5 grad min^{-1}. The difference in the values of $Al_{met}1$ (percentage of metal Al reacting at the first stage of reaction (column 5)) for line 1 and for lines 2 and 3 shows that the reaction between liquid water and metal Al in content of particles covered with aluminum carbide is remarkable even at room temperature. It means that additional efforts are necessary to increase the time of life of Al particles in contact with liquid water at near-room temperatures. For example, surface hydrophobization of aluminum carbide coating may be used.

The results of X-ray analysis show that the origin of PC has no effect on the mechanism of solid oxidation product formation (Figure 8.14). The main crystal product is boehmite (AlOOH), as in the case of POC.

Study of Al powders passivated with coatings containing the aluminum carbide by TG and DTA analyzes by means of the apparatus in Figure 8.2 shows

1) Linear heating of Al powders with carbide coating in inert flow of argon up to 900 °C gives no way of the identification of composition of PCs. The only difference observed in properties of Al powders manufactured by the condensation Gen–Miller method and by the exploding wire method is the amount of gases solved in volume of powders. This amount is larger for powders manufactured by the exploding wire method.
2) Linear heating of Al powder samples in the flow of atmospheric air up to 800 °C demonstrates that collapse of coatings starts at temperatures up to 370 °C. The observed collapse temperature can be different for different samples, depending on the coating structure and composition and is generally lower than the collapse temperature of oxide coating. The value of the start temperature decreases with the increase of Al carbide in the coating composition. DTA, TG, and derivative thermogravimetric analysis (DTG) curves for powder Al–CO_2-1 heated linearly in flow of air are shown in Figure 8.15. The collapse of coatings starts at temperatures lower than 500 °C. Maximum of heat evolution appears at 600 °C.
3) Linear heating of Al powder samples in the water vapor flow up to 900 °C shows that the reaction of metal Al with water vapor is remarkable already at temperatures higher than 100 °C. The rate of hydrogen evolution is MAX at temperature near the melting point of Al. It is higher than the temperature of MAX of heat evolution.

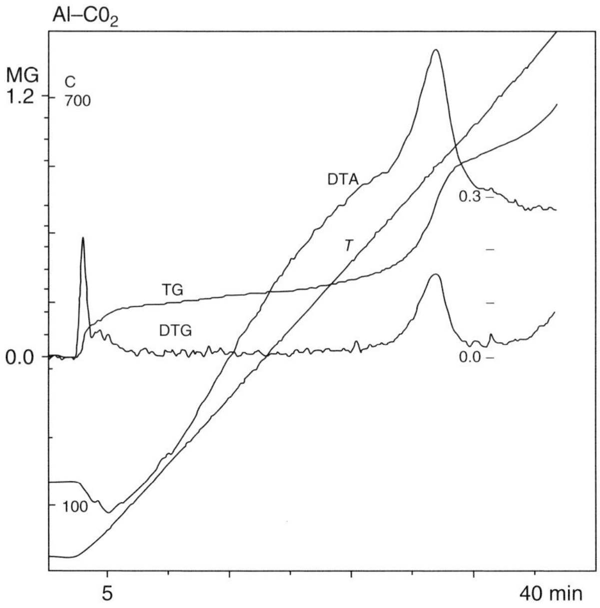

Figure 8.15 TG, DTA, DTG, and *T* curves for Al powder sample passivated with aluminum carbide coatings (Al–CO_2-1) linear heating in a flow of atmospheric air up to 800 °C.

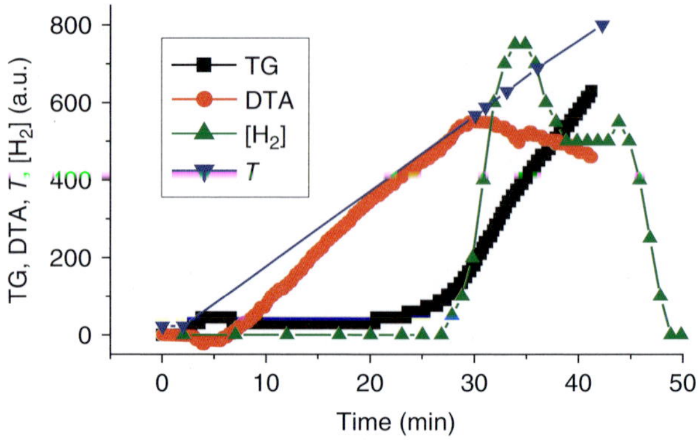

Figure 8.16 TG, DTA curves, and rate of hydrogen formation for Al powder Al–CO_2-1 linearly heated in a flow of water vapor at atmospheric pressure up to 800 °C.

TG, DTA curves, and rate of hydrogen formation for Al powder Al–CO_2-1 linearly heated in flow of water vapor at atmospheric pressure up to 800 °C are shown in Figure 8.16. MAX of heat evolution was observed at temperature about 580 °C and MAX rate of hydrogen evolution at temperature about 660 °C.

The thermoanalytic curves TG, DTA, and the kinetic curve of rate of H_2 evolution (Figure 8.16) lead to the conclusion that the increase in the sample mass caused by Al reaction with the water vapor becomes appreciable already at temperatures above 200 °C. The sample was linear heated in a flow of water vapor at the rate of 20 K min^{-1}. Two competitive processes were observed: the first is the destruction of the POC as a result of Al particle heating and as a consequence the acceleration of metal oxidation and the second is the formation of solid reaction products protecting the metal Al from oxidation and as a consequence the reconstruction of PC and deceleration of metal oxidation. As a result, both the DTA curve and curve of H_2 evolution have the maximum located at the similar temperature. The collapse of coating is reached at temperatures around 620 °C and is accompanied by the maximum rate of hydrogen evolution.

Similar kinetic curves were observed for all investigated samples of powders and flakes irrespective of the average size of particles and the structure of surface PCs. Distinctions were observed in the values of temperatures of the beginning of PC destruction (induction period), in the values of temperatures of maxima of heat and H_2 evolutions, in percentage of relative increase in sample mass as a result of oxidation of metal. The values of characteristic temperatures of oxidation process depend on both the average size of Al particles and the origin of PCs. The maxima temperatures are located between 450 and 620 °C. The rise in temperature of the sample above the melting point (T_M) of Al leads to transformation of backformed Al hydroxide coating in Al oxide coating, to it redestruction, and to oxidation of already liquid metal Al.

The existence of low-temperature collapses of PCs designed on the base of Al carbide permits to realize the low-temperature oxidation (burning) of Al nanopowders in air or in water vapor atmosphere. Moreover, it allows to expect that at an average size of particles less critical full oxidation (burning) of metal Al contained in the particle is possible at temperatures below T_M of Al, which was earlier observed in an atmosphere of air and carbon dioxide [20, 22, 23].

8.5 Study of Al Powder/H_2O Slurry Samples Heated Linear in "Open System" by STA

Samples of Al powder/H_2O suspensions were studied by derivatography in an "open system." They were placed in a crucible covered with a cover limiting the rate of water evaporation process but not excluding the process of evaporation itself. The crucible was heated linearly in an inert argon gas to a temperature of 700 °C.

Samples of slurries were investigated immediately after preparation (mixture) and after a given time of slurry exposure at room temperature in an atmosphere of saturated water vapor.

TG, DTA, and T curves for freshly prepared Al powder/H_2O slurry heated linearly in inert Ar flow with the rate 10 °C min^{-1} are shown in Figure 8.17. Powder Al–CO_2-1 was used in this experiment. The shape of the DTA curve indicates that PC undergoes collapse at temperatures 97 and 107 °C (maxima of heat evolution, Figure 8.17a). Despite the fact that hydrogen evolution starts at 88.5 °C, it has the only maximum at 108 °C (Figure 8.17b). This maximum of hydrogen evolution accompanies to MAX at 107 °C for heat evolution. This fact allows to propose that the first peak of heat evolution (at 97 °C) is responsible for the reaction of aluminum carbide with liquid water.

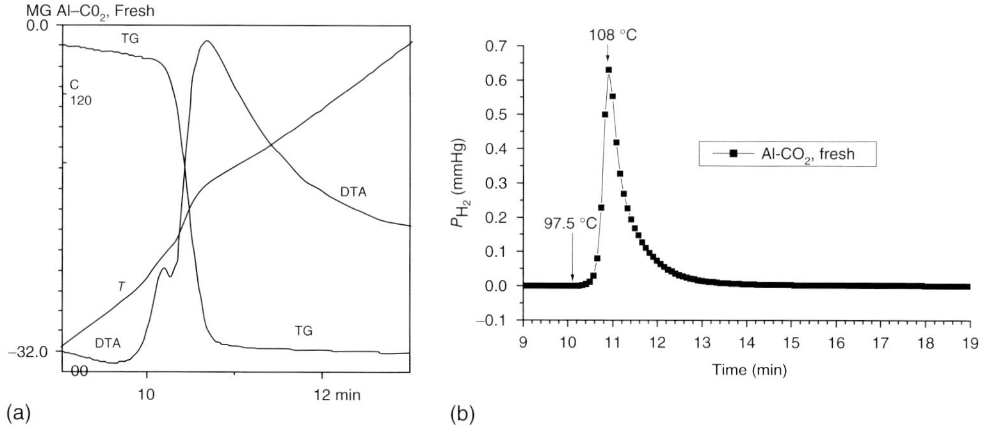

Figure 8.17 (a) TG, DTA, T curves for Al/H_2O slurry linear heating in an inert argon flow up to 190 °C in "open" system. Rate of heating –10 °C min^{-1}. Sample Al–CO_2-1. (b) Kinetic curve of rate of hydrogen evolution for Al/H_2O slurry linear heating in inert argon flow up to 190 °C in "open" system. Rate of heating –10 °C min^{-1}. Sample Al–CO_2-1.

As was mentioned (Figure 8.13), the experiments carried out with different samples of nanopowders covered with an aluminum carbide coating shows that reaction with hot water (hydrogen evolution) starts at 57–73 °C and depends on the amount of Al carbide in the coating content. Experiments were conducted under samples heated linearly in an atmosphere of saturated water vapor. Schema of the apparatus used are shown in Figure 8.2.

Long-time exposure of Al powder/H_2O slurries at room temperatures not only decrease the metal Al content of slurry but also results in change of slurry properties such as viscosity and temperature of PC collapse. A similar behavior was observed for Al powders passivated with oxide coating. Storage temperature for prepared mixtures of Al powder with water should be reduced to minimize this effect. In particular, Al/H_2O slurries may be frozen and stored and then used in solid state.

The comparative study of reactivity of Al powders passivated with different types of PCs covering the surface of Al particles shows that its stability to contact with liquid water can be placed in the following sequence: $(\mathrm{Al}_{\mathrm{carbides}}) < (\mathrm{Al}_{\mathrm{nitrides}}) = (\mathrm{Al}_{\mathrm{oxides}}) \ll (\mathrm{Al}_{\mathrm{oxides-Hydrophobic}})$.

8.6
Ultrasound (US) and Chemical Activation of Metal Aluminum Oxidation in Liquid Water

High-purity metal Al oxidation in pure liquid water can be used for production not only of high-purity hydrogen but also for production of ultrapurity aluminum oxides.

The process of Al oxidation can be realized at a wide range of temperatures (from room temperature up to 3500 K). But, certainly, the method of low-temperature aluminum oxidation (up to 100 °C) is the simplest, safest, and most inexpensive variant of the process of metal aluminum oxidation with liquid water. Aluminum hydroxides are the solid product of low-temperature oxidation of aluminum micron powders in this case. The activation methods should be used to provide a necessary value of specific rate of hydrogen generation to produce the necessary amount of hydrogen or nonpolluted aluminum hydroxides.

The US activation is proposed to be used for these purposes, as the activating effect of ultrasound is well known for a number of chemical reactions [16]. Preliminary studies [14, 24] showed the perspective of ultrasound application in activation of oxidation reaction of metal aluminum in water.

Mechanisms and special features of sonochemical reactions have been studied intensively and for long periods [16, 25]. It is established now that the main reason for physical and chemical processes occurring in the liquid under the US field action is cavitation. It is realized through the transformation of low-density energy of the ultrasound field into high-density energy near and inside the gas bubbles. Hydrogen formed during US irradiation of the mixture of aluminum powder and water creates favorable conditions for the formation of cavitational bubbles. The collapse of these bubbles leads to the shock wave appearance, which propagates into the reaction media causing microscale turbulence. Symmetric cavitation seems to be responsible for the mass transfer intensification. Asymmetrical cavitation leads to the emergence of microjets directed to the solid surface. It leads to solid surface destruction. Cavitation is the reason for a number of phenomena such as heating of reaction medium, intensification of mass transfer, mechanical destruction of solid, and decomposition of molecules into radicals.

The activating factors of US field application for reaction of oxidation of aluminum powder in liquid water are

- cavitation effect;
- thermal effect;
- effect of mass transfer;
- mechanical destruction of solid surface.

Factors of US fields slowing the reaction of oxidation of aluminum powder in liquid water are

- stimulation of polycondensation processes of solid oxidation products ($\text{sol} \rightarrow \text{gel} + nH_2O \rightarrow \text{solid body} + mH_2O$), which was observed at some temperature regions;
- surface Al particle consolidation.

What kind of US factors will have a crucial effect depends on the combination of many parameters including configuration of US field; its intensity, nature, and purity of reactants; value of ratio of Al powder and liquid water in the reaction mixture; size of reacting particles; and so on.

The effect of US activation was investigated for oxidation of pure micron-sized Al powder – distilled water suspension and for this suspension along with chemical activation. A small admixture of CaO (several percent from the mass of oxidized aluminum) was used as chemical activator. CaO purity was 99 wt%. The content of aluminum metal in the powder is 99.1%. Main impurities are Fe (0.1 wt%), Si (0.08 wt%), and Mg (0.008 wt%). The average particle size is about 4 μm. The micrographs of the surface of aluminum powder particles in delivery state are shown in Figure 8.18.

The effect of US activation was investigated at various configurations of US fields and the different powers. Two types of experimental equipments were designed and used for these purposes. In the first type, US power was supplied directly in chemical reactor. For this purpose were used apparatuses Muza-0.1/22-M and ASU-0.1 (Russia). Muza-0.1/22-M has working frequency 22 kHz and feeding electrical power about 100 W. ASU-0.1 has three working frequencies – 48 kHz, 22 kHz, and 100 Hz and total feeding electrical power about 15 W. The scheme of the experimental device is shown in Figure 8.19.

In the second type of apparatus, the chemical reactor was immersed in water filling the volume of the US bath. US bath with a feeding electrical power about 50 W was used. Thermoregulation system was allowed to maintain the desired temperature of the water filling the bath. The scheme of the experimental devices is shown in Figure 8.20.

The process of aluminum powder oxidation was investigated at near-atmospheric pressure. Various modes of oxidation (isothermal, adiabatic, a mode of partial heat exchange with surrounding medium) were used in these experiments.

The starting temperature was between room temperature and boiling point of water (100 °C). The used experimental installation provides the control of the temperature of a reactant mixture in the reactor as a function of time, to measure formed hydrogen volume and rate of hydrogen formation as a result of aluminum oxidation. Primary experimental data were kinetic curves of rate of hydrogen evolution or volume of produced hydrogen, and also time dependencies of the temperatures characterizing specificity of oxidation process.

After the end of the experiments, the solid products of oxidation (aluminum hydroxides) were analyzed by X-ray diffraction technique, SEM, measurement of specific surface Brunauer–Emmett–Teller (BET), the original dynamic desorption porometry (DDP) technique [26], and volumetric technique for residual metal measurement.

The experimental equipment gives the possibility of sampling the reagent mixture at different stages of oxidation and studying the process of formation of oxidation products. In a number of experiments, the sampling of reaction media was carried out at different oxidation times to study the Al particle transformations and the formation of oxidation products during the oxidation process. After quick drying, these samples were studied by means of the above-mentioned methods. The results clarify the impact of different parameters on the process of oxidation of Al particles. The same experimental equipment was applied to study the oxidation processes without US activation.

As mentioned earlier, the process of aluminum powder oxidation by the distilled liquid water [5] passes through the several consecutive stages if the reagent ratio is near *stoichiometric* and the initial temperature of the mixture is near room temperature. The same stages were observed during oxidation flowing in the presence of chemical activation and US activation. Moreover, both the

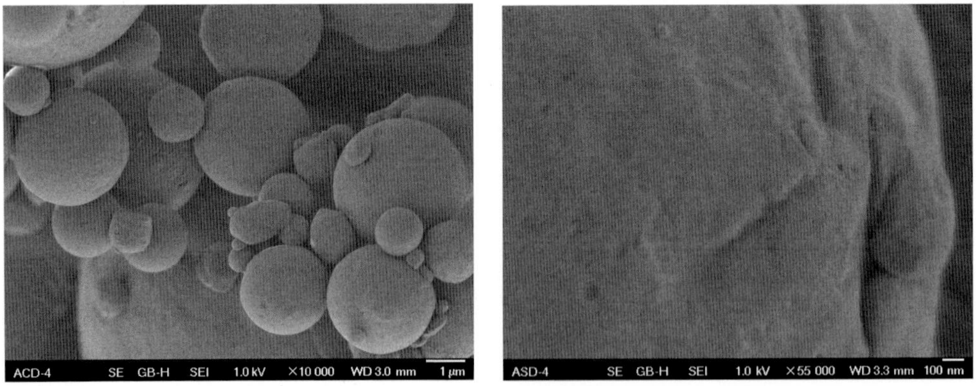

(a) (b)

Figure 8.18 The microphotographs of particles of micron Al powder at delivery condition.

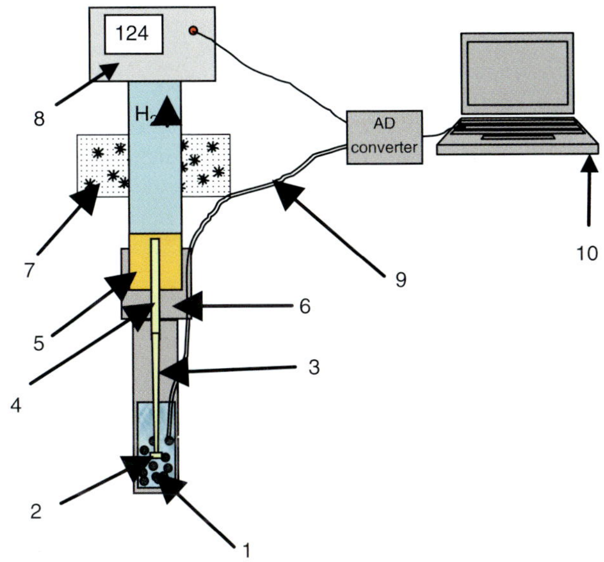

Figure 8.19 The scheme of chemical reactor with US working body inside reactor: 1, reactor with reagents; 2, working body of US generator; 3, US vibrations concentrator; 4, US vibrations transport system; 5, US vibrations converter; 6, sluice for loading of reagents in the reactor; 7, refrigerator for clearing of hydrogen; 8, flow meter; 9, thermocouple; and 10, PC system for control and registration of oxidation process parameters.

chemical activation and US activation increase the intensity the stage of initial oxidation.

An attempt was made to accelerate by means of US activation the initial oxidation stage and convert it directly into the stage of intense oxidation, avoiding the stage of induction. US activation with increased power was used for this purpose. It reduces significantly duration of induction period but increases simultaneously the temperature of reaction medium.

The relative amount of the oxidized metal at the stage of initial oxidation for oxidation in distilled water was measured for micron powder. It is equal to about 0.5% for pure distilled water and increases up to 2–3% when using chemical activation. The SEM image of Al particles sampled after initial oxidation in distilled water is shown in Figure 8.21. The mass of the oxidized metal does not exceed 0.5 wt%. The comparison of surface microrelief of this particle with surface microrelief of particle at delivery condition (Figure 8.18) shows that contact with the distilled water results in the small erosion of the external surface oxide POC and in formation of separate nano knolls on the surface. Al oxidation products formed these knolls on the surface of particles at location areas through defects in the oxide coating.

The microphotographs of micron Al particle sampled at the beginning of the period of intense oxidation in distilled water is given in Figure 8.22. It clearly

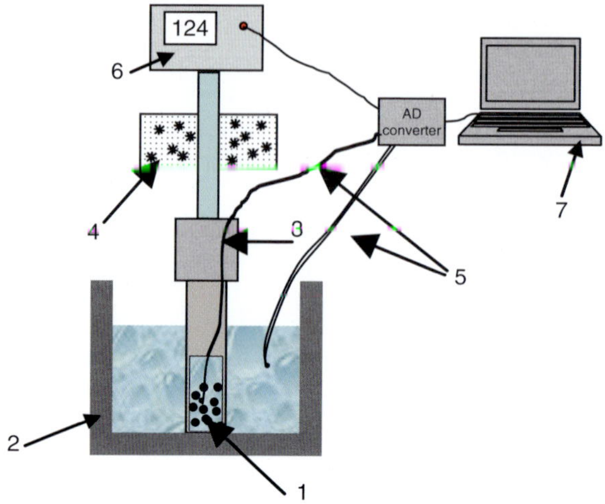

Figure 8.20 The scheme of chemical reactor with US activation of Al powder oxidation through the reactor walls: 1, reactor with reagents; 2, US bath with controlling temperature; 3, sluice for loading of reagents in the reactor; 4, refrigerator for clearing of hydrogen; 5, system for temperature control; 6, flow meter; and 7, PC system for control and registration of oxidation process.

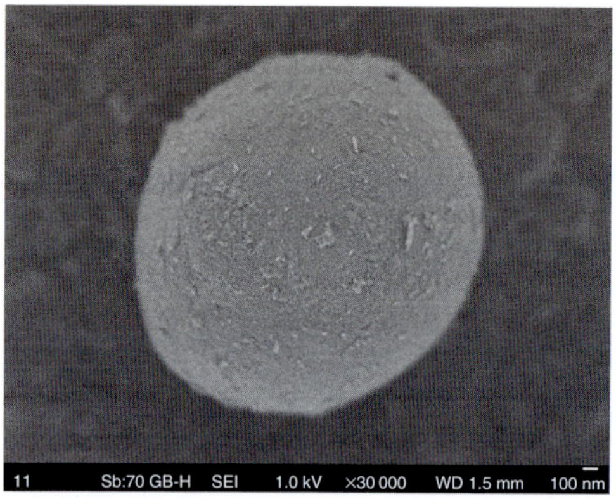

Figure 8.21 Microphotographs of micron Al particle sampled after stage of initial oxidation in distilled water.

demonstrates the growth of size of separate nano knolls. Electronic microphotographs of micron Al particle sampled at the end of intense oxidation in distilled water (Figure 8.23) shows agglomerates of particles with localization of oxidation products on the particle surface. For these particles, metal Al conversion reaches up to about 30% after intense oxidation or thermal explosion in distilled water. Usually, aluminum hydroxides are formed on the surface of Al particle structures, which are often called *nanowalls*.

Figure 8.22 Microphotographs of micron Al particle sampled at the beginning stage of intense oxidation in distilled water.

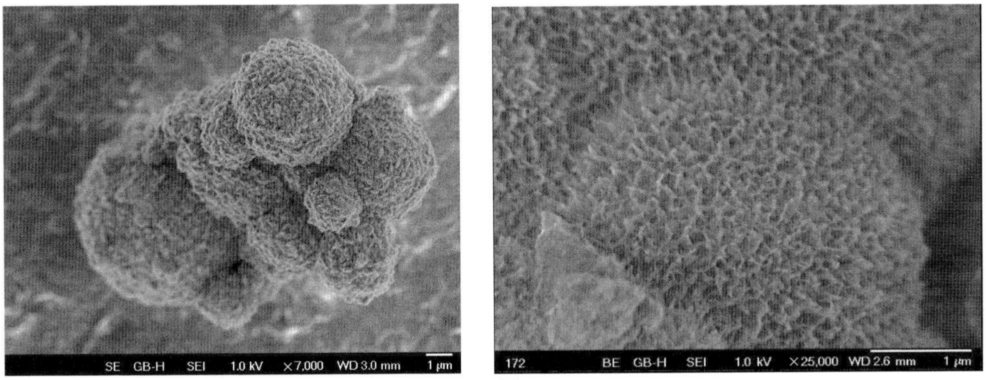

(a) (b)

Figure 8.23 Microphotographs of micron Al particle sampled at the end of intense oxidation in distilled water.

The use of chemical activators (CaO) to intensify the oxidation process provides much fuller oxidation (more than 90%) of metal aluminum during period of intense oxidation or thermal explosion. X-ray analysis of solid reaction products of micron particles oxidizing with water in the presence of CaO shows that calcium aluminates (products of reaction: CaO + Al) were detected distinctly only at the initial time of the intense oxidation period. The observed increase in efficiency of metal aluminum oxidation can be connected with the change in structure of formed solid products (Figure 8.24). As can be seen from the photographs, the increase in degree of oxidation is the result of recrystallization of the formed products and the rodlike structure formation. Apparently, growth of these crystals occurs from the liquid phase.

The surface of solid products (Figure 8.24) is much more friable than of solid products formed by oxidation of aluminum in the absence of the activator

Figure 8.24 Microphotographs of solid products result in the oxidation of micron Al particle in the presence of CaO activator.

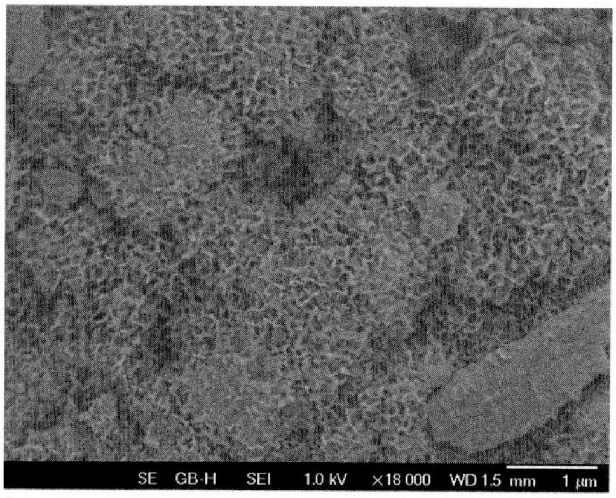

(a)

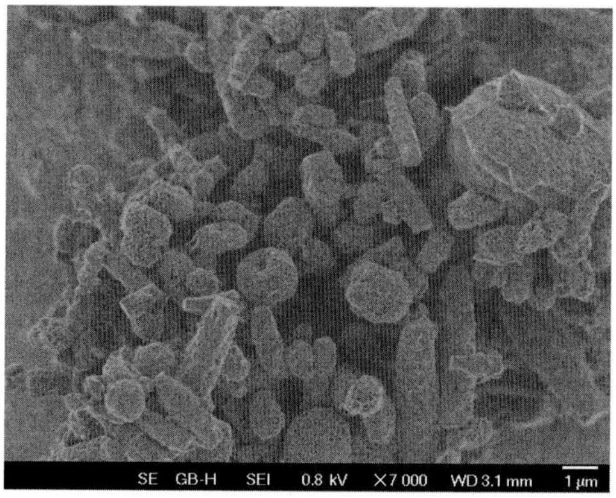

(b)

Figure 8.25 Microphotographs of micron Al particle sampled at the end of intense oxidation in distilled water with US activation.

(compare with Figure 8.23). A comparative study of the porosity of these solid oxidation products performed by the method DDP [26] confirms this conclusion.

The microphotographs of micron Al particle sampled at the end of intense oxidation in distilled water with US activation are shown in Figure 8.25. They indicate clearly that US activation intensifies the processes proceeding during the oxidation in distilled water. The "nanowalls" are observed at oxidation of Al particles without activators dissolving and forming crystals. As a result, the conversion of metal Al in hydroxide Al is about 98%.

The result of SEM analysis for the solid products of Al particle oxidation at the simultaneous impact of chemical activator (CaO) and physical (US field) activator

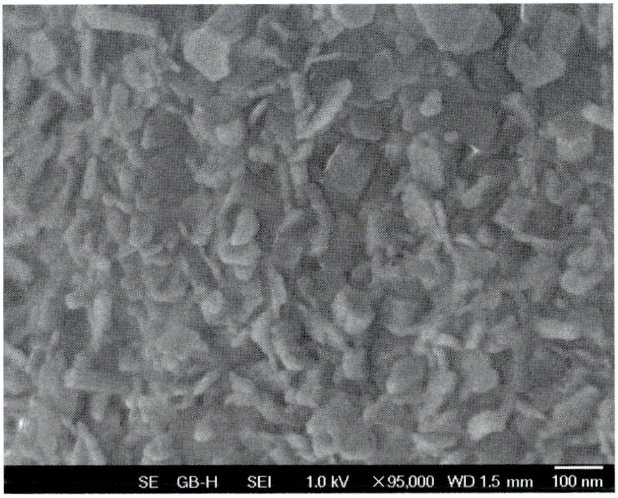

Figure 8.26 Microphotographs of solid products result in the oxidation of micron Al particle in the presence of CaO activator and US activator.

at $T = 20\,°C$ is shown in Figure 8.26. The aluminum oxidation degree is 100%. Application of US activation in combination with a CaO activator allows changing essentially the crystallization process (intensification of formation of crystallization nuclei). It leads to formation of nanoparticles with rodlike structures.

Determination of kinetic parameters of the oxidation processes was carried out in the experiments under isothermal conditions. The series of kinetic curves of the rate of hydrogen evolution obtained under isothermal conditions for the oxidation of aluminum powder with distilled water without activation are shown in Figure 8.27 [27]. The lowest tested temperature in this series of experiments was 55 °C. It is the result of very strong temperature dependence of the rate of hydrogen evolution.

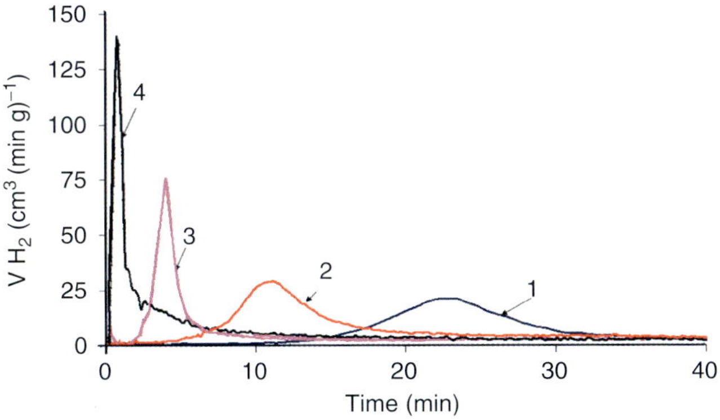

Figure 8.27 Kinetic curves of specific hydrogen evolution rate as a function of time at different temperatures in isothermal experiments in the absence of activation at the temperature: 1, 55 °C; 2, 60 °C; 3, 70 °C; and 4, 100 °C.

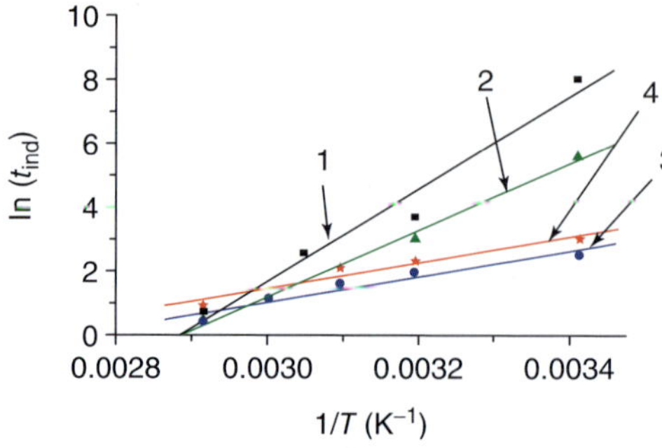

Figure 8.28 Dependence of ln(t_{ind}) versus reciprocal value of oxidation temperature for distilled water (curve 1), US activation (curve 2), CaO activator (curve 3), and US + CaO activator (curve 4).

When the temperature decreases from 100 °C to 20 °C, the rate changes four orders of magnitude. Simultaneously, the maximum of the rate of hydrogen evolution shifts to the right with the temperature decrease. The treatment of these data by using the Arrhenius equation allows estimating the values of effective activation energies for the different stages of the Al oxidation process. Similar results were obtained for the cases of the use of CaO and US activators and simultaneous use of both activators.

The dependence of logarithm of t_{ind} (time of induction period) versus reciprocal value of oxidation process temperature for distilled water (curve 1), US activation (curve 2), CaO activator (curve 3), and US + CaO activator (curve 4) is shown in Figure 8.28.

An average rate of hydrogen evolution was considered as a parameter that characterizes the stage of intensive oxidation. The effective activation energies of this parameter were evaluated in the same manner (Table 8.2).

The mechanisms for US activation and CaO activations are different (Figure 8.29). You can see the difference in durations of induction periods, in values of the maximum rate of hydrogen evolution, and also in the shapes of kinetic curves. All these indicate clearly that the mechanism of activation for two considered activators is not the same. Chemical activator CaO provides the destruction of POC, fast start of the oxidation process, and does not prevent the deposition of $Al(OH)_3$ film on the

Table 8.2 Effective activation energies characterizing the process of oxidation of micron Al powder during the induction stage and at the stage of intensive oxidation.

Parameter	Pure H_2O (kJ mol^{-1})	US activation (kJ mol^{-1})	CaO activation (kJ mol^{-1})	CaO + US activation (kJ mol^{-1})
Induction time	121 ± 21	88 ± 13	33 ± 7	33 ± 7
Average rate of H_2 evolution	126 ± 21	33 ± 7	21 ± 6	8.5 ± 3

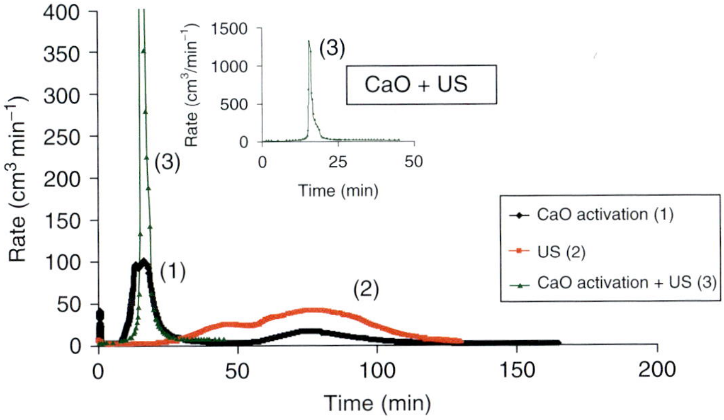

Figure 8.29 Kinetic curves of rate of H_2 evolution for oxidation process of Al particle ASD-4 in liquid water in the presence of CaO activator (curve 1), US activation (curve 2), and CaO + US activation (curve 3). Al : H_2O – 2 : 22 (mass).

metal surface. For this reason, curve (1) has two maximums of hydrogen evolution and long time passes between these two maxima. In contrast, US activation results in the destruction of POC being not as effective, but prevention of surface deposition of solid oxidation products is more significant. For this reason, curve 2 has two unresolved maxima. But the increase in rate of hydrogen evolution is several orders for both activation modes.

Simultaneous use of these two modes of activation increases the rate of H_2 evolution profoundly (Figure 8.29, curve 3). This effect cannot be explained as the simple sum of effects of two types of activations. It is a completely synergetic effect. The rate of PC decay increases dramatically in the case of combined action of US field and chemical activation.

The results given in Figure 8.30 allow comparing the shapes of kinetic curves of the rate of H_2 evolution shown in Figure 8.29. Analysis of shapes of kinetic curves confirms previous conclusions.

The rate of H_2 evolution versus degree of metal Al conversion for using CaO activator, US activation, and simultaneous US + CaO activator are shown in Figure 8.31. It is clear that use of US activation and US activation in combination with CaO activator is convenient for technical application in case generation of stable hydrogen flow is necessary for a long time.

Both provide near 100% conversion of metal aluminum and relatively stable rate of hydrogen evolution during most of the oxidation time. The simultaneous US + CaO activator use is interesting if you need huge emissions of hydrogen in a short time.

It is necessary to note that the use of US activation is effective in the temperature range of 20–70 °C. It leads to the reduction in the induction period and to the increase in the average rate and degree of aluminum conversion in comparison with the values observed in the absence of US activation. Hydrogen yield is close to

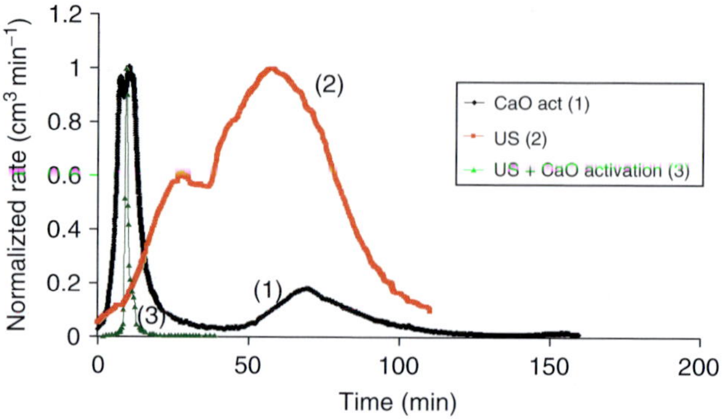

Figure 8.30 Comparison of shapes of kinetic curves of rate of H_2 evolution for CaO activator (1); US activation (2); and US + CaO activator (3).

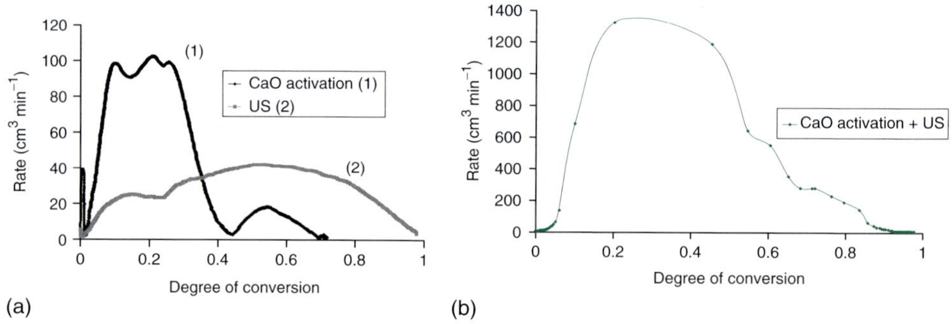

Figure 8.31 The rate of H_2 evolution versus degree of metal Al conversion for (a) CaO activator; US activation and (b) US + CaO activator.

100%. When the temperature of reaction medium exceeds 70 °C, the efficiency of US activation dramatically drops. This is because of the fact that when temperature approaches the boiling point of water, the gas cavitation passes into the steam cavitation and its efficiency decreases [25]. The increasing of US power introduced into the reactor leads to acceleration of oxidation process; however, at high specific power, this effect can be explained, at least partially, as the result of thermal activation.

8.7 Conclusion

The oxidation ability of Al powders was investigated with respect to water vapor and to distillated water in dispersion and slurry composition. The results show that this reaction can be used as the source of gaseous hydrogen.

The full oxidation of metal Al can be achieved (during reasonable time) for the micron powders and for the flakes with thickness <1 μm by using the activation modes.

If reaction of Al particles passes in excess of distillated boiling water, the composition of solid reaction products depends on the average size of Al particles.

- For the average size larger than 1 μm, the main oxidation product is byerit ($Al(OH)_3$).
- For UF (submicron, nano) powders, the main oxidation product is boehmite (Al(OOH)).
- For two-dimensional micron particles (flakes), the determinative parameter is thickness. For particles with thickness below 1 μm, the main solid reaction product is bemit (Al(OOH)).

The oxidation time of particles depends on the average size and the state of surface of particles. It is in the range of 1–60 min. For micron powder, the decrease in the amount of reacting water to stoichiometric results in mainly boehmite (Al(OOH)) formation.

The surface hydrophobization of Al particles increases its stability in contact with water at temperatures lower than 100 °C.

The use of activation: chemical activation (admixtures of CaO), physical activation (US field), and the combination of these two results in both dramatically increasing the rate of hydrogen evolution and increasing the metal aluminum oxidation degree up to 100%. Application of US activation in combination with CaO activator allows changing essentially the crystallization process of oxidation products (intensification of formation of crystallization nuclei). It provides formation nanoparticles with rodlike structures.

Acknowledgments

The author thanks his coworkers Laricheva Olga O. and Shaytura Natalia S. for their contributions and the RFBR (Russian Foundation for Basic Research) for financial support.

References

1. Ivanov, V.G., Leonov, S.N., Savinov, G.L., Gavrilyuk, O.V., and Glazkov, O.V. (1994) Combustion of mixtures of ultradisperse aluminum and gel-like water. *Combust. Explos. Shock Waves*, **30**, 569–570.
2. Choudhury, P.R. (1992) Slurry fuels. *Prog. Energy Combust. Sci.*, **18**, 409–427.
3. Zhuk, A.Z., Sheindlin, A.E., and Kleymenov, B.V. (2006) Use of low-cost aluminum in electric energy production. *J. Power Sources*, **157**, 921–927.
4. Wang, H.Z., Leung, D.Y.C., Leung, M.K.H., and Ni, M. (2009) A review on hydrogen production using aluminum and aluminum alloys. *Renewable Sustainable Energy Rev.*, **13**, 845–853.
5. Larichev, M., Laricheva, O., Leipunsky, I., Pshechenkov, P., and Shkolnikov, E.

(2007) Metal aluminum as the high-performance energy accumulator for alternative power engineering/high energetic materials: demilitarization and civil application (HEMs 2007). Proceedings of the International Conference – Arcachon, France, October 1–3, 2007.

6. Varshavskii, I.L. (1980) *Energoakkumulirujushie veshestva i ich ispolzovanie*, Naukova Dumka, Kiev (in Russian).
7. Bunker, C.E. and Smith, M.J. (2011) Nanoparticles for hydrogen generation. *J. Mater. Chem.*, **21**, 12173–12180.
8. Soler, L., Macanas, J., Mumoz, M., and Casado, J. (2007) Aluminum and aluminum alloys as sources of hydrogen for fuel cell applications. *J. Power Sources*, **169**, 144–149.
9. Soler, L., Candela, A.M., Macanas, J., Munoz, M., and Casado, J. (2009) In situ generation of hydrogen from water by aluminum corrosion in solutions of sodium aluminate. *J. Power Sources*, **192**, 21–26.
10. Ilyukhina, A.V., Kravchenko, O.V., Bulychev, B.M., and Shkolnikov, E.I. (2010) Mechanochemical activation of aluminum with gallams for hydrogen evolution from water. *Int. J. Hydrogen Energy*, **35**, 1905–1915.
11. Ziebarth, J.T., Woodall, J.M., Kramer, R.A., and Choi, G. (2011) Liquid phase-enabled reaction of Al–Ga and Al–Ga–In–Sn alloys with water. *Int. J. Hydrogen Energy*, **36**, 5271–5279.
12. Dupiano, P., Stamatis, D., and Dreizin, E.L. (2011) Hydrogen production by reacting water with mechanically milled composite aluminum-metal oxide powders. *Int. J. Hydrogen Energy*, **36**, 4781–4791.
13. Zhao, Z., Chen, X., and Hao, M. (2011) Hydrogen generation by splitting water with Al-Ca alloy. *Energy*, **36**, 2782–2787.
14. Shaytura, N.S., Laritchev, M.N., Laritcheva, O.O., and Shkolnikov, E.I. (2010) Study of texture of hydroxides formed by aluminum oxidation with liquid water at various activation techniques. *Curr. Appl. Phys.*, **10** (Suppl. 2), S66–S68.
15. Vlaskin, M.S., Shkolnikov, E.I., Bersh, A.V., Zhuk, A.Z., Lisicyn, A.V., Sorokovikov, A.I., and Pankina, Y.V. (2011) An experimental aluminum-fueled power plant. *J. Power Sources*, **196**, 8828–8835.
16. Suslick, K.S. and Price, G.I. (1999) Applications of ultrasound to materials chemistry. *Annu. Rev. Mater. Sci.*, **29**, 295–326.
17. Gen, M.Ya. and Miller, A.V. (1981) A technique for generation of aerosols. Patent USSR N814432, Inventions Bulletin 11, Mar. 23 1981, p. 25.
18. Jigatch, A.N., Leipunsky, I.O., Kuskov, M.L., Pshechenkov, P.A. *et al.* (2002) *Khim. Fiz. (Russia)*, **21** (4), 72–78.
19. Gromov, A., Förter-Barth, U., Teipel, U., and Ilyin, A. (2005) DTA-TG study of aluminum nanopowders (ANP) passivated by reactive coatings. Pros. of 36 ICT Conference, Energetic Materials. Performance and Safety, Karlsruhe, Germany, 28 June-01 July, 2005.
20. Laritchev, M.N., Jigatch, A.N., Leipunsky, I.O., Kuskov, M.L., *et al.* (2003) Air oxidation of ALUMINIUM nano particles. Influence of size and surface coating of particles on the mechanism of aluminium oxidation process. Proceedings of the 9 International Workshop on Combustion and Propulsion " Novel Energetic Materials and Applications", La Spezia, Italy, September 14-18, 2003, pp. 14-1–14-14.
21. (a) Laritchev, M.N., Leipunsky, I.O., Laricheva, O.O., Pshechenkov, P.A. *et al* (2006) New reactive surface coatings for Al particles reacting with water. *J. Chem. Rhys. (Russia)*, **25** (10), 72–79. (b) Laritchev, M.N., Leipunsky, I.O., Laricheva, O.O., Pshechenkov, P.A., *et al.* (2005) New reactive surface coatings for Al metal nanoparticles. Proceedings of the 36 ICT Conference, Energetic Materials – Performance and Safety, Karlsruhe, Germany, 28 June-01 July, 2005.
22. Laritchev M.N., Jigatch, A.N., Leipunsky, I.O. *et al.* (2002) Aluminum nanoparticles as the energy source for multi-sample mars return mission. Proceedings of International Conference – NanoTech 2002 – At the Edge of Revolution, AIAA Paper 2002-5734.

23. Laritchev, M.N., Jigatch, A.N., Leipunsky, I.O., Kuskov, M.L., *et al.* (2003) Aluminum nanoparticles as a basis for fuel for mars conditions. Proceedings of the 10 International Workshop on Combustion and Propulsion, La Spezia, Italy, September 21–27, 2003, pp. 29-1–29-13.
24. Larichev, M.N., Shaitura, N.S., and Laricheva, O.O. (2008) The influence of ultrasonic field on the oxidation of Al powders with water. *Russ. J. Phys. Chem. B.*, **2**, 757–758.
25. Margulis, M.A. (1985) Sonoluminescence and sonochemical reactions in cavitation fields. A review. *Ultrasonics*, **23**, 157–169.
26. Shkol'nikov, E.I., Elkina, I.B., and Volkov, V.V.; (Method of analysis of the porous structure); RF Patent 2141642; 17.04.1998
27. Larichev, M.N., Shaytura, N.S., Kolokol'nikov, V.N., Laricheva, O.O., and Shkolnikov, E.I. (2010) Oxidation of ASD-4 aluminum powder with water. Possibilities of the chemical and physical activation of the process. Obtaining nanodimensional oxidation products. *Izv. Russ. Akad. Nauk Energetika.*, **2**, 85–104 (in Russian).

9
Nanosized Cobalt Catalysts for Hydrogen Storage Systems Based on Ammonia Borane and Sodium Borohydride

Valentina I. Simagina, Oksana V. Komova, and Olga V. Netskina

9.1
Introduction

Research into the nanosized catalytic systems is a rapidly developing field of catalysis. Transition to nanosized metal particles allows a growth in catalytic activity owing to their high content of surface active centers [1, 2]. A substantial influence of nanosize effects takes place in structure-sensitive catalytic reactions because of a change of the crystalline structure and electronic properties of the active component [3, 4]. The use of nanoparticles of metals in catalysis is therefore promising and the development of methods for the synthesis of nanosized catalysts with a controlled functional activity is a great challenge.

Among the approaches to the synthesis of catalysts, the low-temperature processes allowing obtainment of a highly dispersed state of the active component are of the most interest. Reduction of salts of metals in aqueous and nonaqueous media of hydrides has been the most common procedure to prepare massive and supported catalysts [5–7]. Sodium borohydride ($NaBH_4$) [8] and, more recently, ammonia borane (NH_3BH_3) containing up to 20 wt% hydrogen [9] were generally used as reducing agents. After the formation of nanoparticles of Group VIII in aqueous solutions of these boron-containing hydrides on the particle surface, there takes place interaction of the hydride with the water to release a large volume of hydrogen [10]. Such an interaction may proceed at ambient temperatures so that the hydrides may act not only as reducing agents but also as sources of hydrogen for the low-temperature fuel cells with a proton-exchange membrane [11, 12]. Realization of the catalytic hydrolysis of boron-containing hydrides makes possible an autonomous controlled generation of hydrogen and eliminates the problems of hydrogen gas storage and transportation.

Several types of hydrogen generators based on $NaBH_4$ hydrolysis have been developed. However, they require the use of expensive catalysts containing platinum group metals [13]. It is therefore important to develop inexpensive catalytic materials whose active phase forms *in situ* in the boron hydride medium. This simplifies the preparation of catalysts, making unnecessary the high-temperature reduction

Metal Nanopowders: Production, Characterization, and Energetic Applications, First Edition.
Edited by Alexander Gromov and Ulrich Teipel.

of catalysts by hydrogen. Among the catalysts under study currently, the low-temperature amorphous borides of cobalt (Co_xB) are the most promising ones [14, 15]. Their unique electronic, magnetic and catalytic properties, and the possibility of their synthesis directly in the hydrolysis of the boron-containing hydrides (*in situ*), allow them to be used instead of the Ru-, Pt-, and Rh-containing catalysts in the hydrolysis of both $NaBH_4$ and NH_3BH_3. But this requires an insight into the composition, structure, and properties of the nanosized particles of cobalt boride forming *in situ* and the ways of controlling their catalytic activity and stability.

9.1.1
Experimental

The salts of cobalt tested in the hydrolysis of sodium borohydride and ammonia borane included commercial $CoCl_2{\cdot}6H_2O$, $Co(NO_3)_2{\cdot}6H_2O$, $Co(CH_3COO)_2{\cdot}4H_2O$, Co_3O_4, and CoO, synthesized CoOOH [16] (PDF No261107), $Co(OH)_2$ [17] (PDF No480083), cobalt borate [18, 19], and nanosized borides of cobalt reduced by sodium borohydride and ammonia borane (Table 9.1).

The hydrolysis of $NaBH_4$ or NH_3BH_3, or of a mixture of them, was carried out at 40 °C in a glass temperature-controlled reactor with an internal magnetic stirrer at a stirring rate of 800 rpm. The volume of the releasing hydrogen was measured volumetrically with a 100 cm^3 gas burette connected to the reactor. A freshly prepared solution (1.2 mmol in 10 ml) of the hydride was charged into the reactor and then 0.0117 g of the catalyst was added. The system was kept tight and the volume of the releasing hydrogen was determined as a function of time. The volume of the released hydrogen was corrected to normal temperature and pressure (NTP):

$$V^0_{H_2} = \frac{V \cdot (P - P_{H_2O}) \cdot 273}{(273 + T) \cdot 760} \tag{9.1}$$

where $V^0_{H_2}$ is the volume reduced to the normal conditions (cm^3), V (cm^3) is the volume of the gas released at temperature (T) and atmospheric pressure (P), P_{atm} is the atmospheric pressure (mm Hg), P_{H_2O} is the saturated vapor pressure of water (mm Hg), and T is the temperature of the experiment (°C).

Table 9.1 List of catalyst samples.

Sample	Preparation
$Co_xB{-}SBH$	Reduction of $CoCl_2$ by $NaBH_4$
$Co_xB{-}AB$	Reduction of $CoCl_2$ by NH_3BH_3
$Co_xB{-}H_2O$	Treatment of $Co_xB{-}SBH$ by water
$Co_xB{-}NaBO_2$	Treatment of $Co_xB{-}SBH$ by a solution of sodium borate (0.12 M)
$Co_xB{-}NaOH$	Treatment of $Co_xB{-}SBH$ by a solution of NaOH (5 wt%)
Co–B–O	Amorphous cobalt borate
Co–B–O–hydr	Treatment of amorphous cobalt borate by water (5 h)

The cyclic stability tests of the catalysts in $NaBH_4$ hydrolysis was performed at 40 °C. After the processing of the first batch of the hydride, the catalyst was separated from the reaction medium and the next $NaBH_4$ batch was added. In some specified cases, the catalyst was prewashed with distilled water and stored in the wet state in air for 1.5 h or dried in a drying box at 110–130 °C for an hour.

The reaction products were collected, dried, and analyzed. According to IR (infrared) spectroscopy, the products of the $NaBH_4$ and NH_3BH_3 hydrolysis in the presence of $CoCl_2{\cdot}6H_2O$ were sodium tetraborate $Na_2B_4O_7{\cdot}nH_2O$ and $NH_4B_5O_8{\cdot}4H_2O$, respectively.

The cobalt borides forming in the medium of boron-containing hydrides were studied by physicochemical methods.

The bulk concentration of Co and B in the samples of the catalysts was determined using inductively coupled plasma atomic emission spectrometry on an Optima 4300 DV instrument.

The FTIR (Fourier transform infrared) spectra were recorded in air at room temperature on a Bomem MB-102 Fourier-spectrometer. For the analysis, 2 mg of the samples were pressed to pellets with CsI.

High-resolution transmission electron microscopy (HRTEM) images were obtained on a JEM-2010 microscope with an accelerating voltage of 200 kV and a resolution of 0.14 nm. The images were analyzed and filtered using the DigitalMicrograph program (GATAN). The elemental analysis of the samples (with a locality of 10 nm) was carried out using energy-dispersive X-ray microanalysis (EDX) on an EDAX Phoenix spectrometer with a Si(Li) detector and an energy resolution of 130 eV or higher. The samples to be analyzed were supported onto a holey carbon film on a standard copper grid.

The thermal analysis was performed on an STA 449 C Jupiter® apparatus in a flow of helium (10 ml min^{-1}) at temperatures from 25 to 800 °C, with the temperature raised at a rate of 2.5, 5, 10, and 15 °C min^{-1}. The conversions taking place upon heating were judged by analyzing the differential thermal analysis (DTA) curves. The activation energy of crystallization was determined by the Kissinger method using the equation:

$$\ln\left(\frac{T_{\mathrm{max}}^2}{\beta}\right) = \left(\frac{E_{\mathrm{a}}}{RT}\right) + C \tag{9.2}$$

where β is the rate of heating, T_{max} is the temperature of the exothermic maximum, E_{a} is the activation energy of crystallization; R is the universal gas constant; and C is a constant. The $\ln(T_{\mathrm{max}}^2/\beta)$ dependence on $1/T_{\mathrm{max}}$ has the form of a straight line, from the slope of which E_{a} was determined.

The high-temperature diffraction studies in air and under a vacuum were performed on a D8 (Bruker Germany) diffractometer with a Cu K_α radiation in a high-temperature Anton Paar (Austria) chamber. The diffractometer was equipped with a Gebel (Bruker, Germany) mirror forming a parallel X-ray beam. The studies under a vacuum were performed at a residual pressure of $5{\cdot}10^{-6}$ bar. The diffraction patterns were recorded in the form of angles from 15 to 70° range with 0.05° increments and a counting time of 5–20 s per point. The heating rate was

$10\,°C\,min^{-1}$. The reflections due to metallic Pt in the diffraction patterns appear because of the use of a Pt support. The particle (crystallite) size was calculated by the Scherrer formula from the (111), (103), and (311) reflections for Co, Co_3B, and Co_3O_4. In these calculations, the physical line-broadening characteristic of the Bruker diffractometer (0.07°) was subtracted from the observed line width. The calculation error was 10%. The quantities the crystalline phases in the samples were determined using the PCW program.

The magnetic measurements were carried out using a VSM-5T (Cryogenic Ltd.) magnetometer. The ambient temperature measurements were performed in fields (H) in fields up to 2 T. To prevent sample interaction with the atmosphere, the powder under study was wrapped in a Parafilm film whose weight and magnetic susceptibility were determined before the measurement.

9.1.2
Study of the Activity of Nanosized Cobalt Boride Catalysts Forming in the Reaction Medium of Sodium Borohydride and Ammonia Borane

Reduction of the $CoCl_2{\cdot}6H_2O$, $Co(NO_3)_2{\cdot}6H_2O$, and $Co(CH_3COO)_2{\cdot}4H_2O$ salts at 40 °C in aqueous solutions of borohydride and ammonia borane proceeds quickly with the formation of a black precipitate of cobalt boride (Co_xB). Hydrogen released during the formation of the catalytically active phase makes a contribution to the initial rate of hydrogen generation (Figure 9.1), after which the rate begins to decrease as the reaction goes on probably as a result of aggregation of the forming boride particles. According to the results shown in Figure 9.1, the catalytic activity of cobalt borides forming *in situ* in the medium of $NaBH_4$ or NH_3BH_3 hydrolysis depends on the nature of the salt anion and decreases in the series $Co(CH_3COO)_2 > CoCl_2 > Co(NO_3)_2$.

It should be noted that ammonia borane is a milder reducing agent than sodium borohydride. Thus, at pH = 14 the following half-reaction potentials were observed [20, 21]:

$$NH_3BH_3 + 6OH^- \rightarrow BO_2^- + NH_4^+ + 4H_2O + 6e \quad E_0 = -1.216\,V \text{ vs. SHE} \quad (9.3)$$

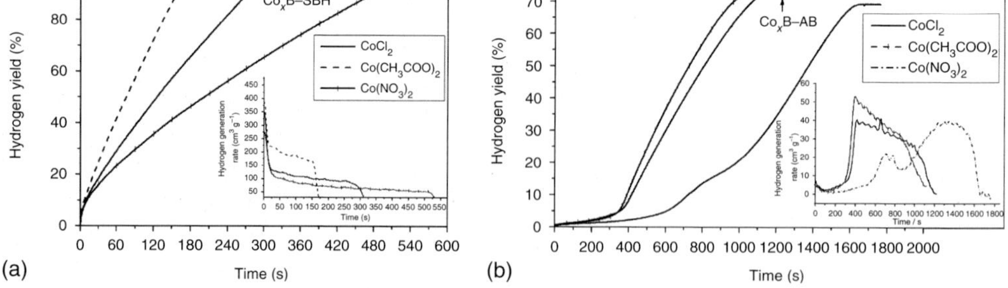

Figure 9.1 The catalytic activity of cobalt salts in the hydrolysis of $NaBH_4$ (a) and NH_3BH_3 (b). $T = 40\,°C$.

$$BH_4^- + 8OH^- \rightarrow B(OH)_4^- + 4H_2O + 8e \quad E_0 = -1.24\,V \text{ vs. SHE} \tag{9.4}$$

Therefore, the rate of reduction in the solution of NH_3BH_3 is smaller than in $NaBH_4$. This shows itself in the kinetic hydrogen yield versus time curves as an activation (or induction) period during which time there takes place the *in situ* formation of the active component and the release of hydrogen is inconsiderable (Figure 9.1b). After the end of this period, the generation of hydrogen proceeds at a high rate until all hydride is converted. The length of this period and the rate of hydrogen generation from aqueous solutions of NH_3BH_3 depend on the nature of the salt (Figure 9.1). In the case of cobalt nitrate, there was an inflection in the curve of the hydrogen generation rate after approximately 900 s, which may be associated with the formation of cobalt hydroxide in the reaction medium as was described in the literature for the system $Co(NO_3)_2 + NaBH_4$ [22, 23]. Indeed, after the first 600–900 s, the reaction medium was observed to have a blue color characteristic of unstable α-$Co(OH)_2$.

The washing and drying conditions of the cobalt borides are known to have an appreciable influence on their composition, structure, and catalytic properties [24, 25]. Thus, a prolonged washing by water of the cobalt borides prepared from cobalt chloride leads to a decrease in their activity (Figure 9.2) because of the oxidation of cobalt and a leaching of boron. Thermal treatment, too, has a negative influence on the activity of cobalt borides (Figure 9.2). Thermal treatment under argon at moderate temperatures (150 and 300 °C), the conditions at which no crystallization takes place, does not lead to any noticeable changes in activity but at temperatures of as high as 500 °C there is a substantial deactivation of the initial amorphous Co_xB because of the crystallization processes which lead, in this particular sample, to the formation of crystalline phases Co_3B (30 wt%, PDF No. 120443) and Co^0

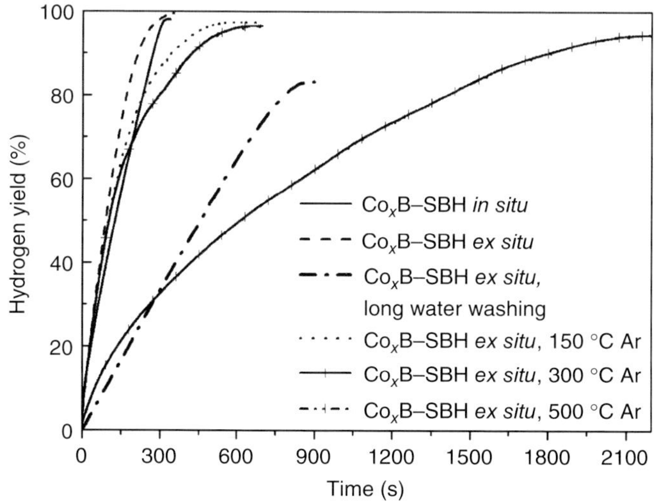

Figure 9.2 Effect of Co_xB synthesis conditions on the catalytic activity in $NaBH_4$ hydrolysis. $T = 40\,°C$.

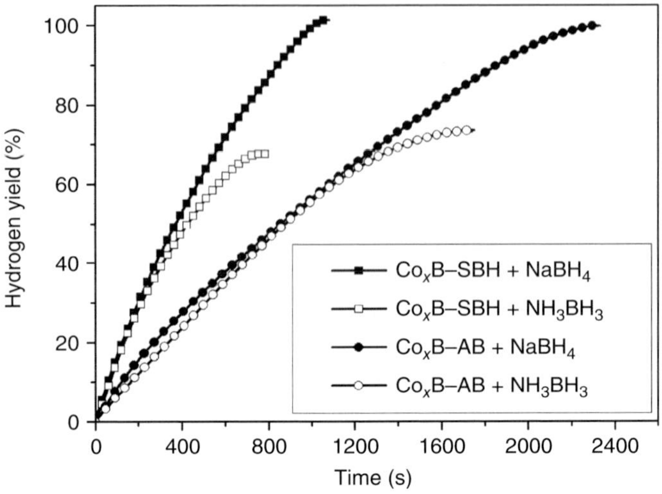

Figure 9.3 Catalytic activity of Co_xB–SBH and Co_xB–AB in $NaBH_4$ and NH_3BH_3 hydrolysis. $T = 40\,°C$.

(70 wt%, PDF No. 150806) under the annealing conditions used. Thus, with the right choice of the treatment conditions, it was possible to synthesize *ex situ* a sample of cobalt boride with an activity not inferior to and even considerably higher than that of the cobalt boride formed *in situ* in the reaction medium (Figure 9.2).

It should be noted that in hydrogen generation in the presence of cobalt borides, the catalysts do not need to be pre-dried and Co_xB prepared *in situ* in the catalytic reactor can be used after a short washing by water to remove the reaction products. For a physicochemical characterization, cobalt borides were quickly separated from the reaction medium and after several washings by acetone were dried in a vacuum box at 70 °C for 2 h and stored under argon. It was shown that these procedures do not lead to catalyst deactivation (sample Co_xB *ex situ* in Figure 9.2) or a change in the physicochemical properties.

The influence of the nature of the reducing agent on the catalytic properties of Co_xB–SBH and Co_xB–AB prepared from $CoCl_2$ in solutions of $NaBH_4$ and NH_3BH_3, respectively, has been studied. As can be seen from Figure 9.3, the rate of hydrogen generation is practically independent of the hydride nature and is determined by the properties of the catalyst. Thus Co_xB–SBH is a more active catalyst than Co_xB–AB. To understand the reasons for such a behavior, physicochemical properties of Co_xB–SBH have been studied and compared with those of the less active Co_xB–AB.

9.2
A Study of Nanosized Cobalt Borides by Physicochemical Methods

The prepared samples, Co_xB–SBH and Co_xB–AB, are X-ray amorphous ferromagnetic powders (a broad diffraction line at $2\Theta = 40–53°$) of a black and gray color,

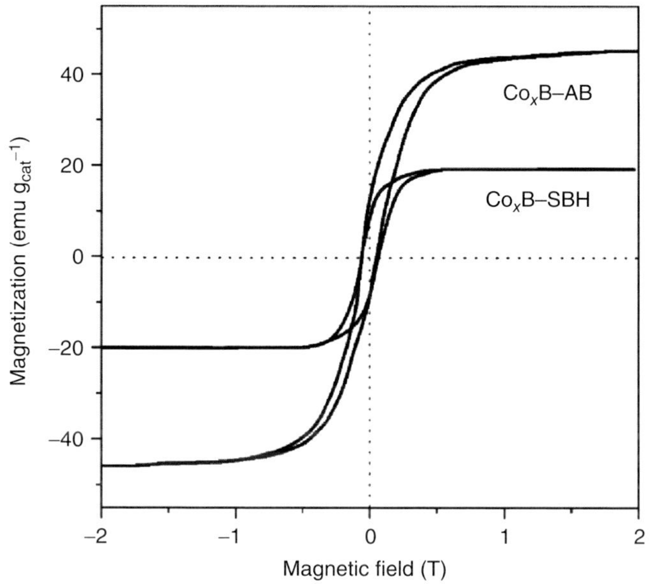

Figure 9.4 Magnetic hysteresis loop of Co_xB–SBH and Co_xB–AB samples.

respectively, Co_xB–AB showing stronger magnetic properties (Figure 9.4). The chemical analysis indicated Co_xB–SBH to contain 81.7 wt% Co and 7.4 wt% B. It can be seen that the content of Co and B is <100%. The remaining 10.9 wt% appears to be the sample's content of oxygen. The presence of oxygen in cobalt borides prepared in an analogous way has also been discussed in [26, 27]. The calculated molar ratio in the synthesized Co_xB–SBH sample is Co : B : O = 2.0 : 1.0 : 1.0. It should be noted that the formation of cobalt boride with Co : B = 2 : 1 was most typical when $CoCl_2$ and $NaBH_4$ were used as the initial reagents [6, 18, 26, 27]. According to the chemical analysis, Co_xB–AB contains 94.3 wt% Co and 5.7 wt% B. The Co : B : O molar ratio is 3.0 : 1.0 : 0.

The IR spectroscopy also reveals differences between the samples. The IR spectra of both Co_xB–SBH and Co_xB–AB (Figure 9.5) show a structureless absorption at 4000 to 200 cm^{-1}. The absorption bands at 3475 and 1635 cm^{-1} correspond to the valence vibrations of the O–H group and deformational vibrations of H–O–H, indicating the presence in the sample of a small amount of water. In addition to this, the IR spectrum of Co_xB–SBH shows absorption bands at 1392, 1066, 857, and 704 cm^{-1} characteristic of B–O bond vibrations [28]. These absorption bands point to the presence in Co_xB–SBH of B–O containing compounds not found in Co_xB–AB. These oxygen-containing compounds of boron have traditionally been described as surface borate anions and oxides of boron [24, 29, 30]. However, the high intensity of the absorption bands of the borate anions characteristic of Co_xB–SBH and their similarity to the absorption bands of the amorphous cobalt borate prepared by us under conditions most closely approaching those of the catalytic $NaBH_4$ hydrolysis in the presence of $CoCl_2$ may also point to the presence

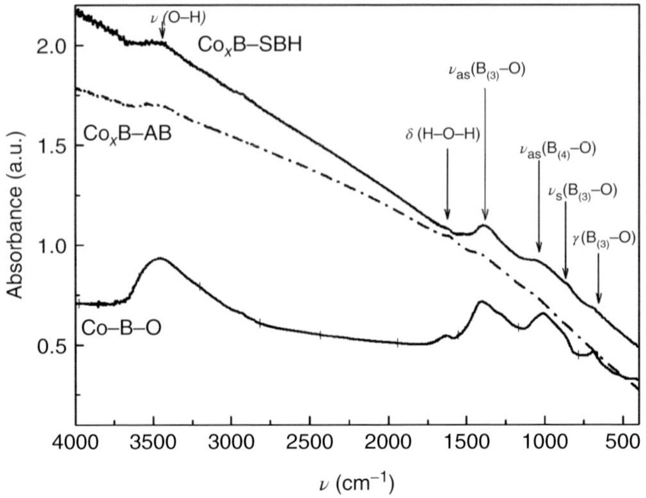

Figure 9.5 FTIR spectra of Co_xB–SBH, Co_xB–AB, and amorphous cobalt borate (Co–B–O).

in the cobalt boride sample of an amorphous phase of cobalt borate. Indeed, formation under some experimental conditions of cobalt borate as a by-product has also been described in the literature [18, 26, 31].

An HRTEM study of Co_xB–SBH and Co_xB–AB (Figure 9.6) has shown that the particle size of Co_xB depends on the nature of the reducing hydride used in the synthesis of the samples. The Co_xB–AB sample consists of aggregated particles of a globular shape with a diameter from 150 to 500 nm and the prevailing particle size of about 250 nm (Figure 9.6a). In Co_xB–SBH, the prevailing particles have a size of about 30 nm with a size scatter from 20 to 60 nm (Figure 9.6b). Characteristic of the globular particles in Co_xB–SBH is the presence of block intergrowth boundaries, which are seen in PEM as short light strips (Figure 9.6e).

The EDX spectra (Figure 9.6c) from the near-surface regions of cobalt boride particles in Co_xB–SBH and Co_xB–AB show that, along with the cobalt, they also contain boron and oxygen. The content of boron is difficult to determine quantitatively because of the limited sensitivity of the instrument to this element. The content of oxygen is higher in Co_xB–SBH (Figure 9.6c), which is in agreement with the chemical analysis and IR spectroscopy data.

It is seen from Figure 9.6d–f that cobalt boride particles are enclosed by a shell with a thickness of 2–3 nm. In the high-resolution TEM (transmission electron microscopy) images (Figure 9.6d,f), one can see fine crystalline particles of a size of ~2 nm included into this shell. The values of the interplanar spaces determined for these crystallites are assignable both to metallic Co and Co_3O_4. In the Fourier image shown in the inset to Figure 9.6d, the reflections correspond to $d_{100} = 0.22$ nm of the hexagonal Co, whereas for some of the particles in Figure 9.6f one can see the lattice of either CoO or Co_3O_4 ($d = 0.24$ and $d = 0.47$ nm). The crystallites are stabilized on the surface of cobalt borides by carbon, which may be left after washing in acetone.

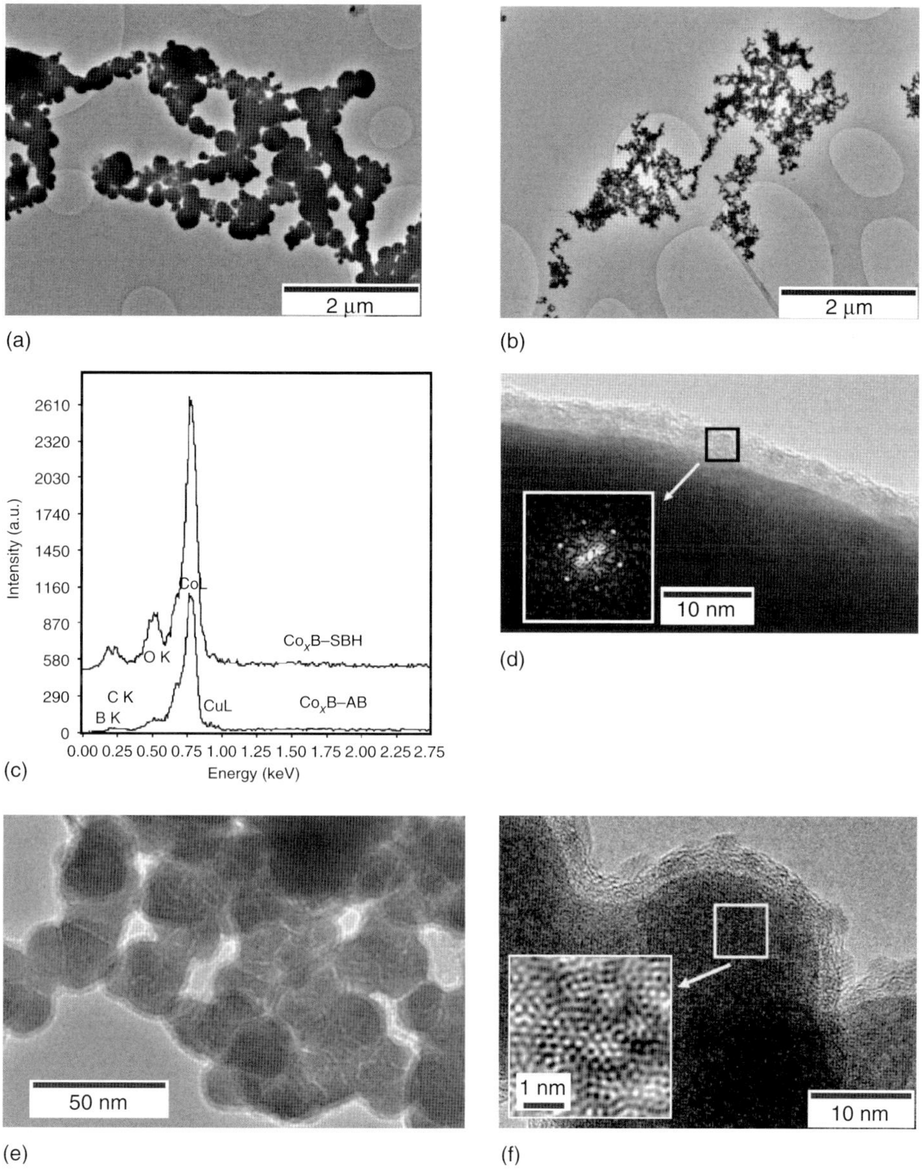

Figure 9.6 TEM images and EDX spectra of the two cobalt boride samples: (a,b) morphology of Co_xB–AB and Co_xB–SBH particles, respectively; (c) EDX spectra of Co_xB–AB and Co_xB–SBH; (d) HRTEM image of the near-surface zone of a Co_xB–AB particle coated with a shell [inset: FFT (fast Fourier transform) from the shell, which can be assigned to the hexagonal cobalt with $d_{100} = 0.22$ nm]; (e) TEM image of a Co_xB–SBH particle with a shell; and (f) HRTEM image of the Co_xB–SBH sample (inset: inverse FFT for the selected region with a partially ordered structure).

The presence of traces of carbon is also seen in the EDX spectra (Figure 9.6c). The oxygen-containing shell around the cobalt boride particles has been described in the literature and may consist of amorphous oxide compounds of boron [27] and cobalt [32], or of a CoB–O phase [26]. It can be suggested that during the TEM examination of Co_xB–SBH and Co_xB–AB the shell was destructed under the reducing action of the electron beam to form either small oxide crystallites of cobalt or the metal.

The inset in Figure 9.6f shows a high-resolution TEM image of a part of a cobalt boride particle in Co_xB–SBH after image filtering. The cobalt borides prepared in the medium of $NaBH_4$ were thought to be X-ray amorphous [24]. According to the TEM image, the structure of the cobalt boride shows ordered areas of ~1 nm. The interplanar spacing derived from this image is 0.24 nm, but its assignment to a particular phase is rather difficult because of the strong irregularity of the structure. It may correspond to a disordered Co_3B phase. Note that microcrystallites of metallic cobalt (fcc, face centered cubic) of a size of 0.19 nm were detected in a high-resolution TEM study of cobalt borides prepared in the system $CoCl_2 + NaBH_4$ [31]. It is obvious that the small region of the coherent scattering from the observed structure does not allow detection of the structural ordering by X-ray diffraction (XRD) method. A similar study of Co_xB–AB could not be performed because of the large size of its particles. It can be suggested that the partially ordered structure in the highly dispersed amorphous cobalt boride (Co_xB–SBH) is built by two-dimensional particles of hexagonal metallic cobalt linked together by boron atoms. An analogous conclusion has been made earlier in Ref. [33] describing flat nanoclusters of nickel of a thickness of 15–20 Å.

Thus, the samples Co_xB–SBH and Co_xB–AB prepared from $CoCl_2$ in solutions of $NaBH_4$ and NH_3BH_3, respectively, differ from each other in Co : B ratio, magnetic properties, particle size, and the degree of oxidation. The sample Co_xB–SBH is more dispersed and has a higher content of oxygen as compared with Co_xB–AB. The low content of oxygen in the Co_xB–AB sample is due to the formation of ten times larger particles of the cobalt boride, apparently because of the slower rate of its formation (Figure 9.1) as NH_3BH_3 is a milder reducing agent when compared with $NaBH_4$. The different ferromagnetic to oxidized phase ratios in Co_xB–SBH and Co_xB–AB allows one to understand the observed difference of their magnetic properties (Figure 9.4).

9.2.1
A Study of the Crystallization of Amorphous Cobalt Borides Forming in the Medium of Sodium Borohydride and Ammonia Borane

One of the traditional approaches to studying amorphous phases has been to study the processes of their crystallization, which are determined by the state of the initial phase. Thus, methods of the thermal analysis allow observation of the heat evolution in the processes of crystallization. Figure 9.7 shows DTA curves for different rates of heating (from 2.5 to 15 °C min^{-1}) for Co_xB–SBH and Co_xB–AB. Both samples show an exothermic peak with a maximum at 444–487 °C (T_{max}). In the literature,

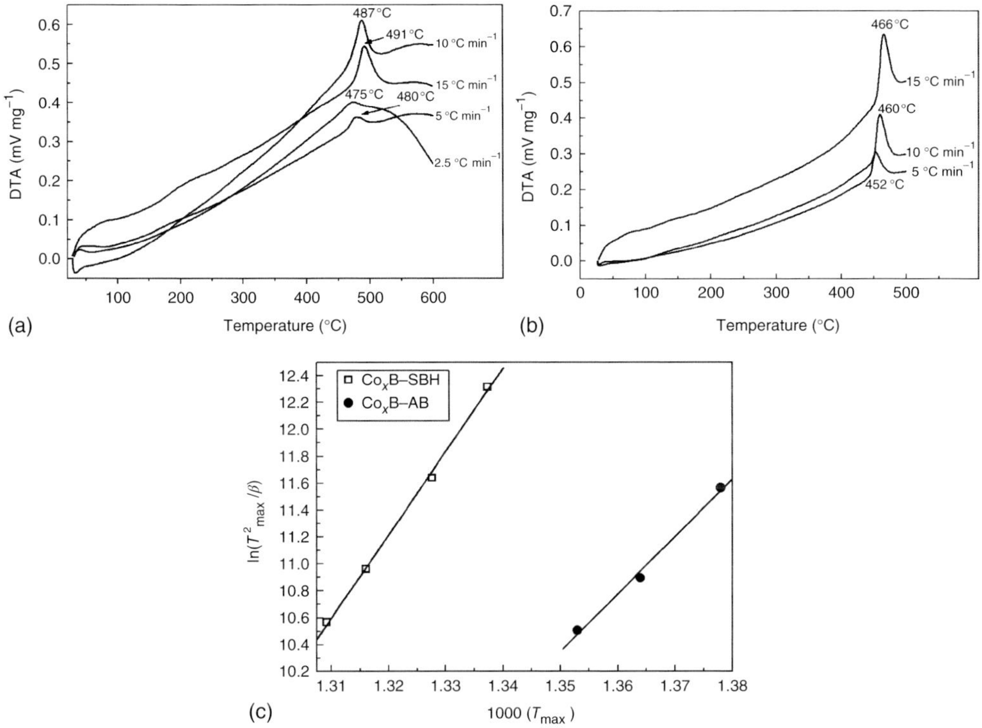

Figure 9.7 DTA profiles of (a) Co_xB–SBH and (b) Co_xB–AB recorded at a heating rate of 2.5, 5, 10, and 15 °C min^{-1}. (c) Kissinger plots for Co_xB–SBH and Co_xB–AB.

this peak was assigned either to the crystallization of amorphous cobalt boride [9, 34] and/or to cobalt boride transformation to metallic cobalt [26]. The effective activation energy of this phase transformation has been calculated by the Kissinger method (Equation 9.2), which is traditionally used to calculate the activation energy of crystallization (E_a) of amorphous boron-containing systems [35–38]. It was equal to 541 kJ mol^{-1} for Co_xB–SBH and to 355 kJ mol^{-1} for Co_xB–AB (Figure 9.7c), in agreement with the values given in the literature. For analogous systems, the value of E_a varies from 300 to 650 kJ mol^{-1} [35, 36]. It should be noted that at the same rates of heating Co_xB–AB shows lower values of T_{max} as compared with Co_xB–SBH (Figure 9.7a,b), although samples with larger particles usually show higher values of T_{max} [39]. In our case, such a behavior appears to be associated with differences in the state of the amorphous phases and the peculiar features of their crystallization processes.

Crystallization of amorphous cobalt borides described in the literature are rather complex and the crystallization products are varied. Thus, thermal treatment of cobalt borides in an inert atmosphere has been reported to lead to the formation of metallic cobalt and various cobalt borides (CoB, Co_2B, and Co_3B), both as individual compounds and as a mixture (Table 9.2) [24].

Table 9.2 Products of Co_xB crystallization in an inert atmosphere according to XRD analysis.

Literature	Heat treatment conditions	Temperature (°C)	XRD data
Corrias *et al.* [27]	2 h, Ar	200	Amorphous
		350	Amorphous
		450	Co, Co_3B
		600	Co, Co_2B
Corrias *et al.* [40]	10 °C min^{-1}, Ar	580	Co_2B, traces of Co_3B
Glavee *et al.* [25]	Rapid heating, holding for 2 h, rapid cooling, Ar	500	Co_2B
Lu *et al.* [6]	2 h, vacuum (10^{-6} Torr)	500	Co_2B, Co
Li *et al.* [34]	2 h, N_2	200	Amorphous
		400	Co, Co_2B, Co_3B
		500	Co, Co_2B, Co_3B
Wu *et al.* [41]	N_2	300	CoB, Co
		500	CoB, traces of Co
		700	Co
Lee *et al.* [42]	2 °C min^{-1}, 2 h, Ar	100	Amorphous
		200	Amorphous
		300	Co
		500	Co
Walter *et al.* [43]	N_2	200	Amorphous
		300	Amorphous
		400	Co_3B
Dai *et al.* [44]	2 °C min^{-1}, 2 h, Ar	200	Amorphous
		300	Co (15 nm)
		400	Co, Co_2B, Co_3B
		500	Co (30 nm), Co_2B, Co_3B
Huang *et al.* [45]	Ar	300	Amorphous
		500	Co, Co_2B, Co_3B
		700	Co, Co_2B, Co_3B
Kalidindi *et al.* [46]	12 h, Ar	500	Co, Co_2B
Patel *et al.* [47]	2 h, Ar, H_2	400	Amorphous
	2 h, O_2	400	Co
Ma *et al.* [48]	2 h, N_2	300	Amorphous
		500	Co_2B, Co_3B
	2 h, H_2	300	Co
Ma *et al.* [49]	1 h, Ar	300	Co, Co_2B, Co_3B
Kalidindi and Jagirdar [50]	12 h, Ar	350	Co, Co_2B

Analysis of the literature data shows that the boron content of Co_xB (determined by the conditions of their synthesis) and the conditions of thermal treatment are the key factors. The presence in Co_xB of oxygen-containing compounds should also be taken into consideration in data interpretation as it is known [18, 25] that at high temperatures in an inert atmosphere there takes place interaction of cobalt boride with cobalt borate in the solid state to form Co^0 by the reaction:

$$2Co_2B + Co(BO_2)_2 \rightarrow 5Co + 2B_2O_x \tag{9.5}$$

$$B_2O_x \xrightarrow{O_2} B_2O_3 \tag{9.6}$$

As mentioned here, in the case of $NaBH_4$, formation of cobalt borates is possible along with the cobalt boride [18, 25, 31]. Cobalt borates $Co(BO_2)_2$, $Co_3(BO_3)_2$ [51], and $Co_2B_2O_5$ [26] have been found among the crystallization products of amorphous borides formed in the interaction of sodium borohydride with a salt of cobalt in an aqueous solution.

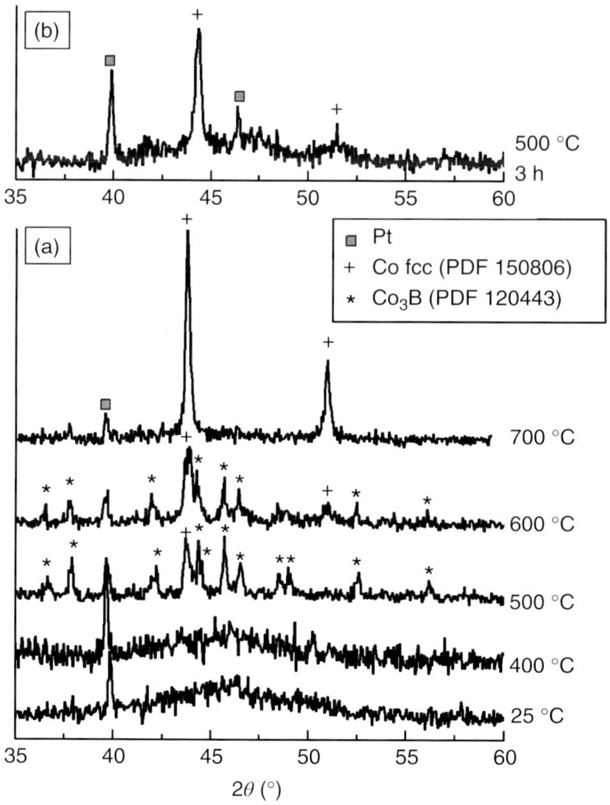

Figure 9.8 X-ray diffraction patterns of Co_xB–AB recorded (a) at 25 °C and after calcination at 400, 500, 600, and 700 °C in vacuum and (b) after calcination at 500 °C in Ar for 3 h.

In order not to consider the effect of oxygen-containing phases on the crystallization processes of amorphous cobalt borides, let us first consider the results of a high-temperature XRD study of a Co_xB–AB sample with a molar ratio of Co : B = 3 : 1 containing traces of oxygen (Figure 9.8). At temperatures from ambient up to 500 °C, Co_xB–AB remains X-ray amorphous. The diffraction pattern shows a broad line at $2\theta = 40–53°$, which has traditionally been assigned to amorphous cobalt borides [24]. At 500 °C, there takes place crystallization to form phases of Co_3B and metallic cobalt (fcc) with dispersions shown in Table 9.3. Estimated contents of the phases are 70 wt% Co_3B and 30 wt% Co^0, which is in agreement with the results presented in [52–54]. Further increase in temperature leads to an increase in Co^0 content (40 wt%), a decrease in the content of the cobalt boride phase Co_3B (60 wt%), and to formation of larger particles at the expense of sintering (Table 9.3). At 700 °C, the Co_3B phase transforms to metallic cobalt Co^0. Cobalt boride transformation to metallic cobalt with the rise of temperature to 700 °C has been observed rather often [24, 27, 41]. The disappearance of the Co_3B phase may be associated with its disproportionation as described in the literature [52–54]. Increased content of metallic cobalt was observed not only with the rise of temperature to 700 °C but also when the sample was held at a lower temperature for a longer time (Figure 9.8b).

The results of a high-temperature investigation of a sample of Co_xB–SBH with the molar ratio Co : B : O = 2 : 1 : 1 are shown in Figure 9.9a. Its crystallization occurs at a higher temperature (600 °C) than in the case of Co_xB–AB. It should be noted that the X-rays (Figures 9.8 and 9.9) and DTA (Figure 9.7c) values of the crystallization temperature are different from each other because of the different sensitivities of the methods and the different conditions of the experiments. The DTA data were obtained under helium, whereas the XRD study was performed in a vacuum. Nevertheless, both methods give consistent results. The Co_xB–SBH sample, which has a higher activation energy of crystallization, crystallizes at a higher temperature both in helium and under vacuum. However, in the case of Co_xB–SBH, no crystalline phases of cobalt borides were formed in spite of the

Table 9.3 Crystallite size data for the phases resulting from a progressive heating of Co_xB–AB and Co_xB–SBH at a rate of 10 K min^{-1}.

Sample	Heat treatment conditions	Phase composition	Crystallite size (nm)
Co_xB–AB	500 °C, vacuum	Co_3B	74
		Co	27
	600 °C, vacuum	Co_3B	87
		Co	22
	700 °C, vacuum	Co	63
	500 °C, Ar	Co	37
Co_xB–SBH	600 °C, vacuum	Co	10
	700 °C, vacuum	Co	79
	400 °C, air	Co	6.5

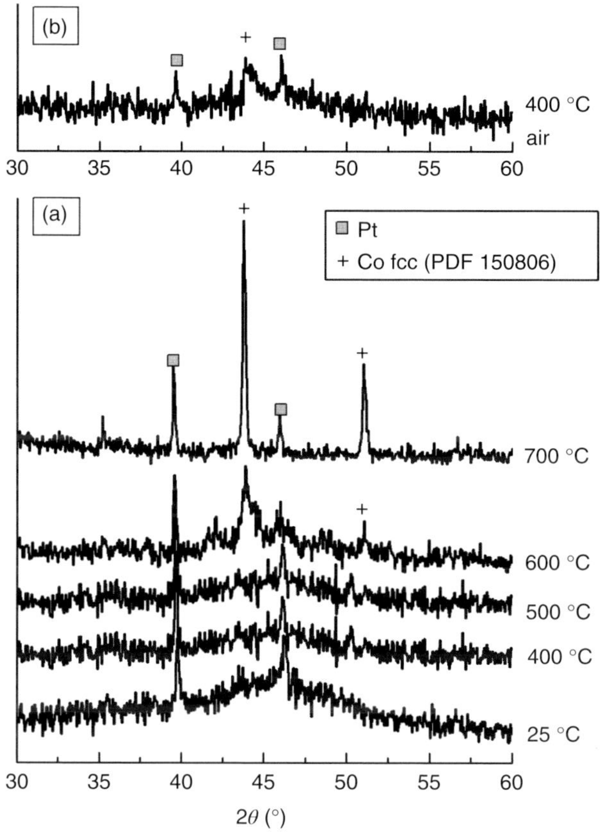

Figure 9.9 X-ray diffraction patterns of Co_xB–SBH recorded (a) at 25 °C and after calcination at 400, 500, 600, and 700 °C in vacuum and (b) after calcination at 400 °C in air.

fact that it has a higher content of boron than Co_xB–AB and, hence, a higher probability for the cobalt boride to become crystallized [55]. In our case, in the diffraction pattern, only reflections of metallic Co^0 (fcc) are observed. At 700 °C, the sample was fully crystallized to form Co^0, as indicated by the disappearance of the halo at 40–53°. The observed results can be explained by the presence of oxygen and the decreased content of boron in the cobalt boride phase as a result of the formation of borates, which was confirmed by FTIR spectroscopy (Figure 9.5). Calculations have shown that the content of oxygen was sufficient for the boride particles to be oxidized to metallic cobalt by Equation 9.5. Thermal treatment performed in the presence of oxygen of the air provides for a rapid formation of finely dispersed particles of Co^0 (6.5 nm, Table 9.3) already at 400 °C (Figure 9.9b). The obtained results are consistent with the data in Glavee *et al.* [18] and indicate that during the thermal treatment of amorphous cobalt borides in the presence of oxygen-containing compounds or oxygen gas Co_xB oxidizes to Co^0.

9.2.2
The Effect of the Reaction Medium on the State of Cobalt Boride Catalysts

Apart from catalytic activity, stable operation is also an important characteristic of a catalyst cobalt boride prepared by reduction of $CoCl_2$ in a solution of $NaBH_4$ *ex situ* (Co_xB–SBH), tested in the hydrolysis of an aqueous solution of $NaBH_4$ and in a solution of $NaBH_4$ stabilized by NaOH. Alkaline solutions of sodium borohydride have traditionally been used because of their high stability toward spontaneous hydrolysis in storage. In the tests of Co_xB–SBH in the aqueous solution of $NaBH_4$, there was a fall in hydrogen generation rate by about 35% during the first five cycles of hydrolysis followed by a rather stable generation of H_2 for at least another 11 cycles (Figure 9.10). The yield of hydrogen was practically unchanged from cycle to cycle. In the Co_xB–SBH tests in the alkaline solution of $NaBH_4$, the rate of H_2 generation in the first cycle (68.7 ml s^{-1} $g_{Co}{}^{-1}$) was slightly higher than in the aqueous solution (58.2 ml s^{-1} $g_{Co}{}^{-1}$). This result is in agreement with the literature data on the effect of alkali on the activity of analogous systems (Co_xB, prepared *in situ* from $CoCl_2$, Co_3O_4, and nanoparticles of Co [56, 57] and of the samples prepared *ex situ* from $CoCl_2$ [43, 58–60]). Further cycles showed a drastic fall in activity. In the fifth cycle, the activity amounted to 40% of the initial value and to 70% of the Co_xB–SBH activity in the fifth cycle of hydrolysis of the aqueous solution of $NaBH_4$. In the 12th cycle, the activity amounted to 6.5 and 13%, respectively. There also was a decrease in hydrogen yield by 10% toward the 12th cycle.

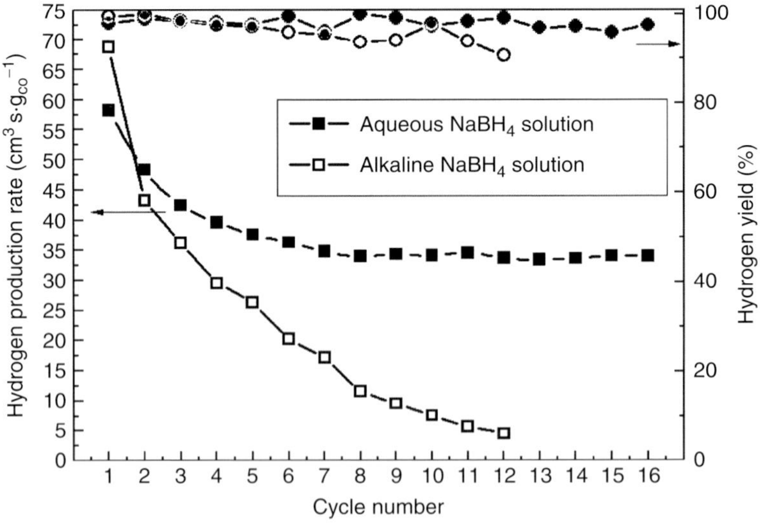

Figure 9.10 Generation rate and yield of hydrogen during stability test of Co_xB–SBH for hydrolysis of aqueous (0.12 M $NaBH_4$) and alkaline (0.12 M $NaBH_4$ and 0.04 M NaOH) solutions of $NaBH_4$ at 40 °C.

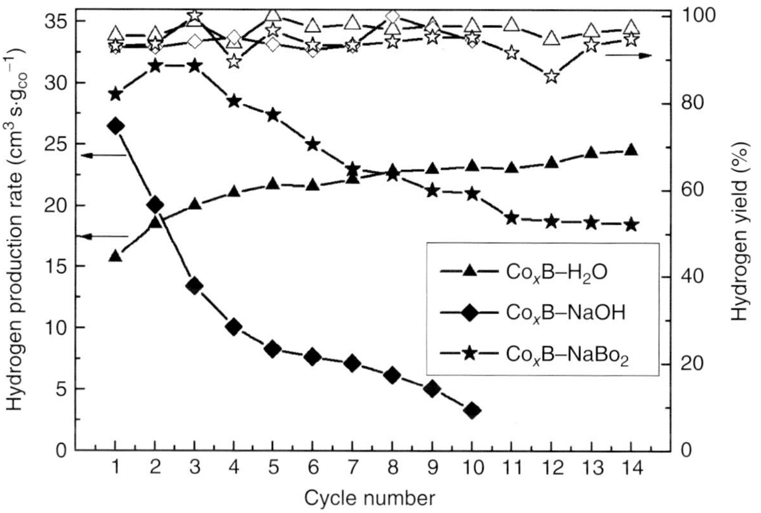

Figure 9.11 Generation rate and yield of hydrogen during stability tests of Co_xB-H_2O, $Co_xB-NaBO_2$, and $Co_xB-NaOH$ for hydrolysis of aqueous of $NaBH_4$ solutions (0.12 M) at 40 °C.

To get an insight into the reasons for the activity behavior of Co_xB-SBH in the hydrolysis of the aqueous and alkaline solutions of sodium borohydride, the influence of reaction medium components and pH value on the state of cobalt boride was studied. For this purpose, Co_xB-SBH was held in distilled water (pH = 6; sample Co_xB-H_2O), in a solution of the hydrolysis products formed in a 0.12 M aqueous solution of $NaBH_4$ (pH = 10.4; sample $Co_xB-NaBO_2$), and in a 5 wt% solution of NaOH (pH = 14; sample $Co_xB-NaOH$) for 5 h at 40 °C under stirring. The samples were then separated by filtration and dried at 110–130 °C.

For all sample Co_xB-H_2O, $Co_xB-NaBO_2$, $Co_xB-NaOH$ tested in the hydrolysis of aqueous solution of $NaBH_4$ (Figure 9.11), the initial rate of hydrogen generation was smaller than that of the starting Co_xB. In the case of Co_xB-H_2O, the activity was increasing with each cycle and toward the 14th cycle the activity amounted to 56% of that of the starting sample. In the case of $Co_xB-NaBO_2$, the rate of hydrogen generation increased in the first three cycles after which it was gradually decreasing. The $Co_xB-NaOH$ sample showed the strongest deactivation. The rate of hydrogen generation in the 10th cycle was only 12% of that in the 1st cycle. To explain these differences in catalytic properties, Co_xB-H_2O, $Co_xB-NaBO_2$, and $Co_xB-NaOH$ have been studied by chemical analysis, XRD, and FTIR spectroscopy.

The chemical analysis has shown (Table 9.4) that treatment of Co_xB-SBH in the alkaline solution leads to leaching of a considerable amount of boron (from 7.4 to 0.48 wt%), from the sample. In contrast to the X-ray amorphous Co_xB-SBH, the diffraction pattern of $Co_xB-NaOH$ shows peaks of $Co(OH)_2$ and metallic cobalt with an hcp (hexagonal close packed) and fcc structure (Figure 9.12). It is seen that the peaks of metallic cobalt are anisotropically broadened, which may be indicative of the presence of packing defects or of a microdomain structure with alternating

Table 9.4 Influence of cobalt boride treatment on its chemical analysis.

Sample	Content (wt%)			Co : B : O molar ratio
	Co	B	O[a]	
Co_xB–SBH	81.7	7.4	10.9	2 : 1 : 1
Co_xB–H_2O	66.3	2.67	31.03	1 : 0.2 : 1.7
Co_xB–$NaBO_2$	66.8	3.02	30.18	1 : 0.25 : 1.7
Co_xB–NaOH	86.1	0.48	13.42	1.7 : 0.05 : 1

[a] Oxygen content was calculated assuming that sample consists only of Co, B, and O elements.

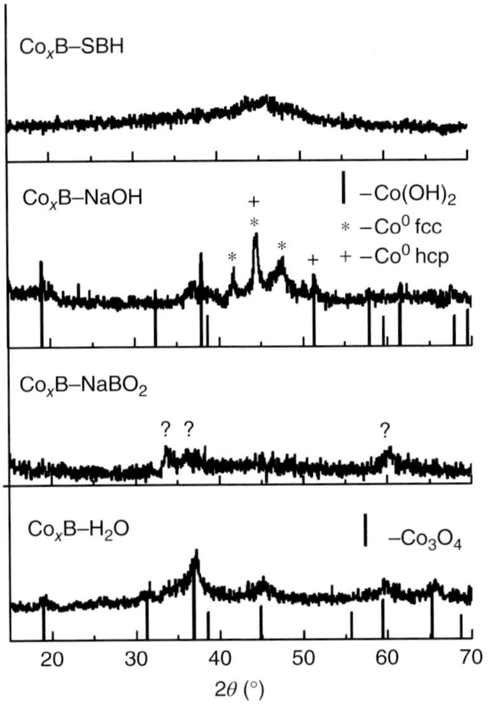

Figure 9.12 X-ray diffraction patterns of Co_xB–SBH, Co_xB–NaOH, Co_xB–$NaBO_2$, and Co_xB–H_2O registered at 25 °C.

hcp and fcc structures of Co. The formation of cobalt hydroxide has been reported for the interaction of salts of cobalt with a solution of $NaBH_4$ stabilized by an alkali [22–24] and Co^0 was found after a prolonged exhaustive washing of cobalt borides with hot water [6, 27].

Additional information on the changes in Co_xB–SBH composition under the action of NaOH can be obtained by FTIR spectroscopy, which allows

Table 9.5 B–O absorption bands for borates [28].

Frequency regions	Band assignments
1500 to 1200 cm^{-1}	$\nu_{as}(B_{(3)}-O)$ asymmetric stretching
1300 to 1150 cm^{-1}	$\delta(B-O-H)$ in-plane bending
1200 to 800 cm^{-1}	$\nu_{as}(B_{(4)}-O)$ asymmetric stretching
950 to 850 cm^{-1}	$\nu_s(B_{(3)}-O)$ symmetric stretching
890 to 740 cm^{-1}	$\nu_s(B_{(4)}-O)$ symmetric stretching
750 to 620 cm^{-1}	$\gamma(B_{(3)}-O)$ out-plane bending

characterization not only of the crystalline but also of the amorphous phase of the sample. A characteristic feature of the Co_xB–SBH sample is the presence of absorption bands at 1392, 1066, 857, and 704 cm^{-1}, characteristic of the B–O vibrations in borate anions (Table 9.5), which arise because of the presence in Co_xB–SBH of the amorphous phase of cobalt borate. Treatment of the cobalt boride by a solution of NaOH with pH = 14 leads to considerable changes in its IR spectrum (Figure 9.13). The absorption bands corresponding to cobalt borate are practically absent and new absorption bands appeared at 3623, 666, 582, 515, 435, and 315 cm^{-1}. According to [60, 61], the absorption bands at 3623, 515, and 315 cm^{-1} can be assigned to the absorption by cobalt hydroxide β-$Co(OH)_2$; those at 666 and 582 cm^{-1}, to Co_3O_4 [62]; and the absorption bands at 435 cm^{-1}, to CoO [63].

The presence of $Co(OH)_2$ is confirmed by XRD data (Figure 9.12). It should be noted that the intensity ratio of the absorption bands at 666 and 582 cm^{-1} does not correspond to that in the standard Co_3O_4 [64]. The relative intensity of the absorption band at 582 cm^{-1} in Co_xB–NaOH is considerably higher. It appears that the intensity of this absorption band has a contribution from the CoOOH phase absorbing at 584 cm^{-1} [65]. After 10 cycles of the test of Co_xB–NaOH in the hydrolysis of an aqueous solution of $NaBH_4$, its spectrum (Figure 9.13) shows, against a structureless background, absorption bands at 1370, 1240, 1155, and 1025 cm^{-1} of very small intensity probably belonging to the absorbed borate anions (Table 9.5). The structureless absorption over the entire spectral range under study with an edge at 500 cm^{-1} due to the absorption by free charge carriers (electrons) and the presence of the weak absorption bands indicate that it is cobalt that appears to be the main phase. This conclusion also follows from the change in physical properties. The sample under study has a characteristic metallic luster and shows considerably stronger magnetic properties than the initial Co_xB–SBH (Figure 9.14).

Thus, in a solution of NaOH with pH = 14, a considerable amount of boron is leached from the sample, which is accompanied by destruction of both the cobalt boride phase and the cobalt borate phase contained in Co_xB–SBH. This gives rise to crystalline phases of metallic Co^0 and $Co(OH)_2$, and amorphous oxides of cobalt CoO, Co_3O_4, and CoOOH. It is known that Co^0 and $Co(OH)_2$, which are the main

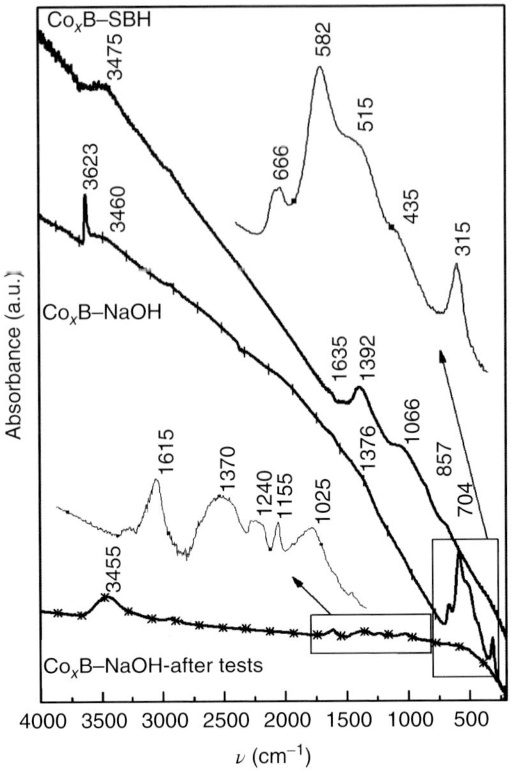

Figure 9.13 FTIR spectra of initial Co_xB–SBH, Co_xB–NaOH, and Co_xB–NaOH after its tests in 10 cycles of $NaBH_4$ hydrolysis.

phases, are characterized by a low activity in the hydrolysis of $NaBH_4$ [9, 22, 23, 29, 41, 66–68]. $Co(OH)_2$ and CoOOH (Figure 9.15) were also shown to have a rather low rate of their reduction by $NaBH_4$. Of the cobalt compounds formed, only Co_3O_4 and CoO (Figure 9.15) can undergo a sufficiently rapid reduction in the reaction medium with a formation of an active Co_xB phase.

It can, therefore, be suggested that the low content of boron in the sample and the presence of phases of a low-activity Co^0, $Co(OH)_2$, and CoOOH lead to a rapid loss in activity in stability tests (Figure 9.10). As was shown by FTIR spectroscopy (Figure 9.13), after the tests the Co_xB–NaOH sample largely consisted of metallic cobalt. Its stronger magnetic properties, as compared with Co_xB–SBH, may cause an enhanced aggregation and rapid deactivation.

Treatment of Co_xB–SBH by a less alkaline solution of sodium borate (the product of $NaBH_4$ hydrolysis) with pH = 10.4 increases the oxygen content of the sample from 10.9 to 30.18 wt%, with the boron content changing from 7.4 to 3.02 wt% (Table 9.4). The XRD pattern of Co_xB–$NaBO_2$ (Figure 9.12) shows a sharp peak with an interplanar space of 2.67 Å and two haloes with *d* of 2.46 and 1.54 Å, which so far could not be assigned to any known compounds of cobalt or sodium. A study

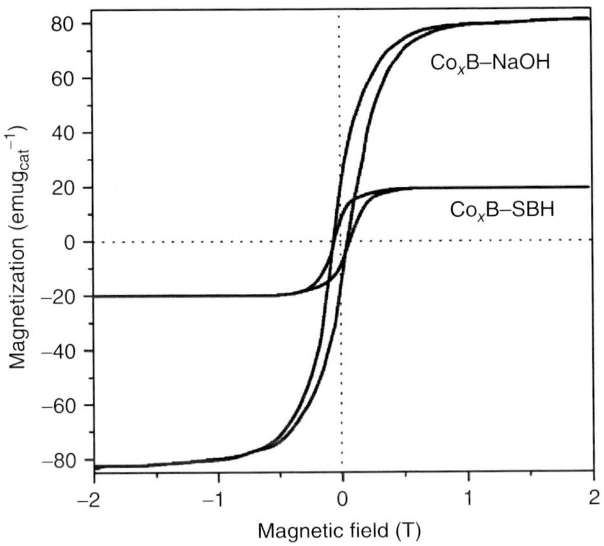

Figure 9.14 Magnetization curves of Co$_x$B–SBH and Co$_x$B–NaOH samples.

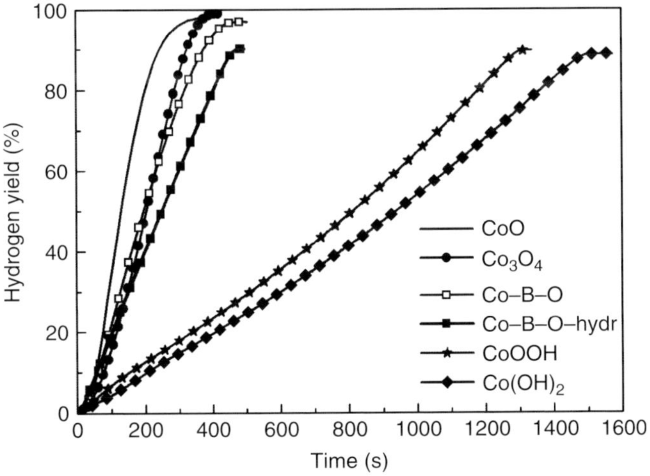

Figure 9.15 Hydrogen generation from aqueous solution of $NaBH_4$ (0.12 M) at 40 °C over amorphous cobalt borate (Co–B–O and Co–B–O–hydr) in comparison with that over Co_3O_4, CoO, CoOOH, and $Co(OH)_2$.

of this sample by IR spectroscopy indicated a change in the character of the B–O bond vibrations at 1500 to 1100, 1200 to 800, and 700 to 500 cm^{-1} as compared with the initial Co$_x$B–SBH (Figure 9.16). Similar IR spectra have also been found for asharites $Me_2B_2O_4(OH)_2 \cdot nH_2O$ where Me is Mg, Mn, and Fe [69]. It appears that cobalt borate becomes partially hydrolyzed during its interaction with water [70] to form a compound of cobalt with a structure closely resembling that of the

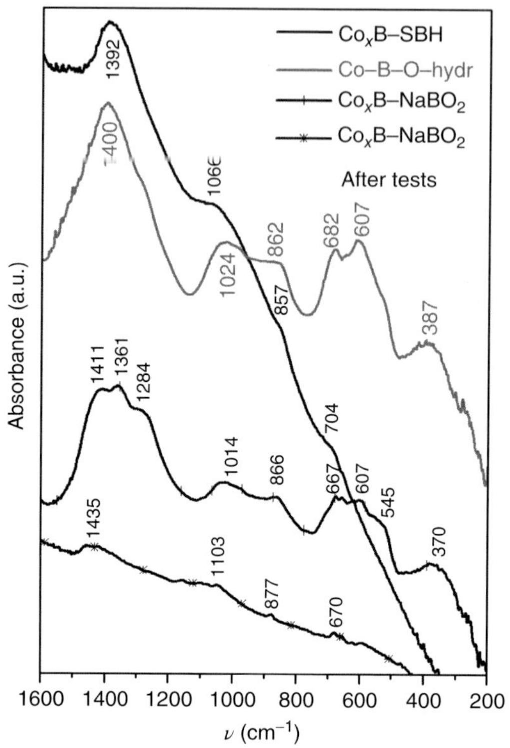

Figure 9.16 FTIR spectra of Co_xB–SBH, Co_xB–$NaBO_2$, Co–B–O–hydr, and Co_xB–$NaBO_2$ tested in 14 cycles of $NaBH_4$ hydrolysis.

ascharites (spectrum of Co–B–O–hydr on Figure 9.16). Besides, it is known that the state of borate anions in aqueous solutions is rather unstable and there may take place their association and polymerization, the different species being in a dynamic equilibrium.

After 14 cycles of the stability tests of Co_xB–$NaBO_2$ in the hydrolysis of $NaBH_4$, the FTIR spectrum showed absorption of a low intensity in the region of borate anion vibrations, indicating reduction of its cobalt borate compounds. It can be suggested that upon the contact of the cobalt boride with the alkaline solution of sodium borate with pH = 10.4 there takes place hydrolysis of the cobalt borate present on its surface. Hydrolyzed species of cobalt borate may also appear as a result of destruction of the cobalt boride phase as follows from a decreased boron content of the sample. As was shown on Figure 9.15, Co–B–O–hydr has a rather high reactivity. It can be suggested that the stable generation of hydrogen and even some increase in the rate of its generation in the first three cycles of the Co_xB–$NaBO_2$ stability tests (Figure 9.11) were due to the reduction of cobalt borates to form an active phase. Further gradual decrease in the Co_xB–$NaBO_2$ activity indicates occurrence of deactivation processes. Nevertheless, the extent of deactivation in the case of Co_xB–$NaBO_2$ is not so appreciable in contrast to the

Co_xB–NaOH sample in which no active phase of cobalt borate was detected. The reason for this may be found in the different nature of the oxidation processes in the strongly alkaline medium with pH = 14 and in the medium with pH = 10.4 containing borate ions, which promotes formation of more reactive phases of cobalt borates.

Co_xB–SBH kept in water showed a change in its chemical composition, which was practically the same as that of Co_xB–$NaBO_2$ (Table 9.4). The sample Co_xB–H_2O had an increased content of oxygen from 10.9 to 31.03 wt% and a decreased content of boron from 7.4 to 2.7 wt% as compared with the initial Co_xB–SBH. According to the IR data (Figure 9.17), the hydrolyzed form of cobalt borate (Co–B–O–hydr, Figure 9.16) also forms in the case Co_xB–H_2O, as in the case of Co_xB–$NaBO_2$. In the IR spectrum of Co_xB–H_2O, there appear absorption bands at 800 to 300 cm^{-1} belonging to the Co–O bond vibrations, which seem to overlap the absorption bands due to the cobalt borate. The strong absorption bands with maxima at 663 and 588 cm^{-1} are characteristic of Co_3O_4. XRD of Co_xB–SBH kept in water also shows broad peaks corresponding to the Co_3O_4 phase (Figure 9.12). Earlier, the Co_3O_4 phase was detected by XPS and X-ray analysis among the oxidation products of cobalt borides after they were thoroughly washed by water [9, 71, 72].

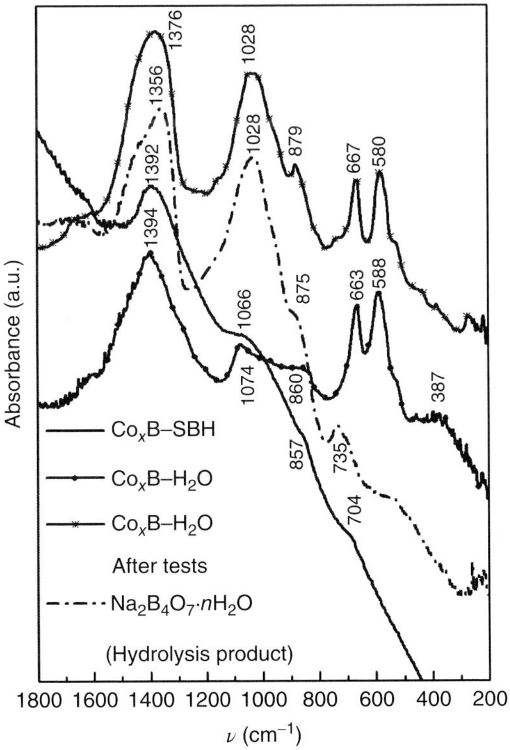

Figure 9.17 FTIR spectra of Co_xB–H_2O and Co_xB–H_2O tested in 14 cycles of $NaBH_4$ hydrolysis in comparison with those of Co_xB–SBH and $Na_2B_4O_7{\cdot}nH_2O$.

After the Co_xB-H_2O stability tests in 14 cycles of $NaBH_4$ hydrolysis, its IR spectrum showed an increased intensity of the absorption bands in the region of the asymmetric valence vibrations of the BO_4 group (1200 to 800 cm^{-1}). The absorption maximum of the BO_3 group was shifted from 1394 to 1376 cm^{-1}. The absorption band at 387 cm^{-1} due to the Co–O vibrations disappeared from the spectrum [73], indicating that there was a partial reduction of the oxygen-containing compounds of cobalt. Comparison of this spectrum with the spectrum of $Na_2B_4O_7 \cdot nH_2O$ prepared by evaporating the solution after the $NaBH_4$ hydrolysis shows that they are very close to each other (Figure 9.17). It can therefore be suggested that during the Co_xB-H_2O test there takes place reduction of cobalt borate phase and adsorption on the catalysts surface of the products of $NaBH_4$ hydrolysis. Thus, a prolonged contact of Co_xB–SBH with an aqueous medium leads to its oxidation to form crystalline Co_3O_4 and a partially hydrolyzed phase of cobalt borate. As mentioned, these oxygen-containing compounds of cobalt show high activity in the hydrolysis of $NaBH_4$ (Figure 9.15). It appears that it is their reduction that leads to the increased activity of Co_xB-H_2O in each cycle of the $NaBH_4$ hydrolysis (Figure 9.11). This means that as Co_xB-H_2O interacts with the hydride solution the content of the Co_xB phase increases as a result of a gradual reduction of the cobalt borate and cobalt oxide and hence this leads to an enhanced rate of hydrogen generation. This is also indicated by the results of the IR spectroscopy study of Co_xB-H_2O after its test. However, along with the activation of the sample, there also takes place adsorption of the reaction products on its surface and probably an oxidation of the active Co_xB phase (Figure 9.17) which hampers the growth in activity. It is the ratio of the Co_xB oxidation and reduction processes in the reaction medium that determines the observed rate of the $NaBH_4$ hydrolysis.

Summing up the obtained results, it can be concluded that the stability of the nanosized particles of cobalt borides in the hydrolysis of $NaBH_4$ depends on the composition of the reaction medium. A drastic deactivation of the particles is observed only in the $NaBH_4$ hydrolysis in an alkaline solution. In the hydrolysis in an aqueous solution, after the initial decrease of the activity a stable generation

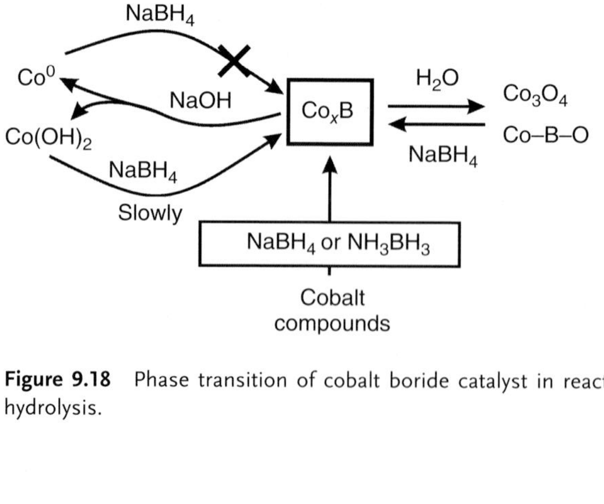

Figure 9.18 Phase transition of cobalt boride catalyst in reaction medium of $NaBH_4$ hydrolysis.

of hydrogen was observed. It was shown that the different behavior of Co_xB is associated with the influence of the composition of the reaction medium on the phase composition of the nanosized particles (Figure 9.18). Thus, in the absence of alkali in the reaction medium, there takes place oxidation of Co_xB to form phases of oxides and borates of cobalt. Their rapid reduction in the hydrolysis of the aqueous solution of $NaBH_4$ with the formation of the active phase of cobalt boride ensures a stable generation of hydrogen in the stability tests of the Co_xB–SBH catalyst. The high value of pH (14) upon the addition of NaOH leads to practically full leaching out of boron from Co_xB with the formation of phases of a low-activity Co^0, $Co(OH)_2$, CoOOH, and this is the reason for the poor stability of the catalyst in the hydrolysis of the alkaline solution of $NaBH_4$.

9.3 Conclusions

The low-temperature hydride synthesis has been used for the preparation of nanosized catalytically active cobalt borides for the hydrolysis of sodium borohydride and ammonia borane. The rate of formation, the activity in $NaBH_4$ and NH_3BH_3 hydrolysis, and the physicochemical properties of the prepared cobalt boride catalysts depend on the nature of the cobalt compounds and the hydride used. The composition and the structure of the cobalt borides have been studied by FTIR spectroscopy, high-resolution TEM with an EDX analysis and magnetic susceptibility measurements. The interaction of $CoCl_2$ with a solution of NH_3BH_3 leads to the formation of an amorphous phase of cobalt boride with a Co : B molar ratio of 3 : 1. The boride particles formed in NH_3BH_3 had the largest size of 150–500 nm and contained traces of oxygen compounds because of the low reducing power of the boride. In the case of $NaBH_4$, which is a stronger reducing agent, there was an instantaneous formation of more dispersed and more active particles of cobalt boride (20–60 nm) with a higher degree of oxidation (Co : B : O = 2 : 1 : 1).

The cobalt borides forming in the aqueous solutions of hydrides are X-ray amorphous. In the case of nanosized boride, there were areas with an ordered structure with a size of ~1 nm. Such a structure is different from that in the stoichiometric borides. TEM and FTIR data have shown the cobalt boride particle to consist of a ferromagnetic core of cobalt boride enclosed by a 2–3 nm thick shell containing oxides (CoO, Co_3O_4) and borate of cobalt, which form as a result of oxidation of the particle surface by the reaction medium. It was found that the presence of oxygen-containing compounds in the amorphous cobalt boride has a substantial influence on its crystallization. The crystalline phase of cobalt boride (Co_3B) was observed only in the sample prepared in the medium of ammonia borane. The presence of oxygen in the highly dispersed cobalt boride prepared from $CoCl_2$ by reduction with sodium borohydride leads to its oxidation to metallic cobalt upon heating.

The activity of the cobalt boride in the $NaBH_4$ hydrolysis drastically decreases as the hydrolysis goes on because of the presence of alkali added to stabilize sodium borohydride solution. At the same time, in the hydrolysis of a nonalkali aqueous solution of $NaBH_4$, after an initial decrease in activity there was a stable generation of hydrogen. These differences are explained by the influence of the reaction medium on the phase composition of the nanosized particles of cobalt borides. Thus, in the absence of alkali, Co_xB is oxidized to form cobalt oxide and cobalt borate phases. In the aqueous solution of $NaBH_4$, they undergo a rapid reduction to form the active phase of cobalt boride, which ensures a stable generation of hydrogen in stability tests. As a result of the high value of pH in solutions with NaOH, practically all boron is leached from the cobalt boride to form Co^0, $Co(OH)_2$, and CoOOH, which have a low activity. This explains the low stability of the catalyst in the hydrolysis of the alkaline solution of $NaBH_4$.

Thus, by varying the conditions of the hydride synthesis of catalysts it is possible to obtain nanosized particles of cobalt borides with a controllable catalytic activity in the hydrolysis of sodium borohydride and ammonia borane.

Acknowledgment

The authors thank their colleagues A.M. Ozerova, D.G. Kellerman, G.V. Odegova, V.I. Zaikovskii, O.A. Bulavchenko for their considerable contributions and the Russian Academy of Sciences (project No 24.54) and the Siberian Branch of the Russian Academy of Sciences (project No 32) for financial support.

References

1. Kellner, C.S. and Bell, A.T. (1982) *J. Catal.*, **75** (2), 251–261.
2. Cuenya, B.R. (2010) *Thin Solid Films*, **518** (12), 3127–3150.
3. Simagina, V.I., Netskina, O.V., Tayban, E.S., Komova, O.V., Grayfer, E.D., Ischenko, A.V., and Pazhetnov, E.M. (2010) *Appl. Catal. Genet.*, **379** (1–2), 87–94.
4. Simagina, V.I., Netskina, O.V., Komova, O.V., Odegova, G.V., Kochubei, D.I., and Ishchenko, A.V. (2008) *Kinet. Catal.*, **49** (4), 568–573.
5. Klabunde, K.J., Stark, J.V., Koper, O., Mohs, C., Khaleel, A., Glavee, G., Zhang, D., Sorensen, C.M., and Hadjipanayis, G.C. (1994) in *Nanophase Materials – Synthesis, Properties, Applications* (eds G.C. Hadjipanayis and R.W. Siegel), Kluwer Academic Publishers.
6. (a) Lu, J., Dreisinger, D.B., and Cooper, W.C. (1997) *Hydrometallurgy*, **45** (3), 305–322; (b) Glavee, G.N., Klabunde, K.J., Sorensen, C.M., and Hadjapanayis, G.C. (1992) *Langmuir*, **8** (3), 771–773.
7. Arzac, G.M., Rojas, T.C., and Fernández, A. (2012) *Appl. Catal. B Environ.*, **128**, 39–47.
8. Simagina, V.I., Storozhenko, P.A., Netskina, O.V., Komova, O.V., Odegova, G.V., Larichev, Y.V., Ishchenko, A.V., and Ozerova, A.M. (2008) *Catal. Today*, **138** (3–4), 253–259.
9. Cavaliere, S., Hannauer, J., Demirci, U.B., Akdim, O., and Miele, P. (2011) *Catal. Today*, **170** (1), 3–12.
10. Garron, A., Bennici, S., and Auroux, A. (2010) *Appl. Catal. Gen.*, **378** (1), 90–95.
11. Wang, L., Zhan, X., Yang, Z., and Ma Chin, C. (2011) *J. Chem. Eng.*, **19** (4), 693–697.
12. Santos, D.M.F. and Sequeira, C.A.C. (2011) *Renewable Sustainable Energy Rev.*, **15** (8), 3980–4001.

13. Orimo, S.I., Nakamori, Y., Eliseo, J.R., Zuttel, A., and Jensen, C.M. (2007) *Chem. Rev.*, **107** (10), 4111–4132.
14. Komova, O.V., Simagina, V.I., Netskina, O.V., Kellerman, D.G., Ishchenko, A.V., and Rudina, N.A. (2008) *Catal. Today*, **138** (3–4), 260–265.
15. Simagina, V.I., Komova, O.V., Ozerova, A.M., Netskina, O.V., Odegova, G.V., Kellerman, D.G., Bulavchenko, O.A., and Ishchenko, A.V. (2011) *Appl. Catal. Gen.*, **394** (1–2), 86–92.
16. Lin, H.K., Wang, C.B., Chiu, H.C., and Chien, S.H. (2003) *Catal. Lett.*, **86** (1–3), 63–68.
17. Bezemer, G.L., Radstake, P.B., Koot, V., van Dillen, A.J., Geus, J.W., and de Jong, K.P. (2006) *J. Catal.*, **237** (2), 291–302.
18. Glavee, G.N., Klabunde, K.J., Sorensen, C.M., and Hadjapanayis, G.C. (1992) *Langmuir*, **8** (3), 771–773.
19. Ozerova, A.M., Simagina, V.I., Komova, O.V., Netskina, O.V., Odegova, G.V., Bulavchenko, O.A., and Rudina, N.A. (2012) *J. Alloy Compd.*, **513**, 266–272.
20. Nagle, L.C. and Rohan, J.F. (2010) *ECS Trans.*, **24** (41), 13–25.
21. Amendola, S.C., Onnerud, P., Kelly, M.T., Petillo, P.J., Sharp-Goldman, S.L., and Binder, M. (1999) *J. Power Sources*, **84** (1), 130–133.
22. Liu, B.H. and Li, Q. (2008) *Int. J. Hydrogen Energy*, **33** (24), 7385–7391.
23. Wu, C., Bai, Y., Wu, F., Yi, B.L., and Zhang, H.M. (2010) *Int. J. Hydrogen Energy.*, **35** (7), 2675–2679.
24. Demirci, U.B. and Miele, P. (2010) *Phys. Chem. Chem. Phys.*, **12** (44), 14651–14665.
25. Glavee, G.N., Klabunde, K.J., Sorensen, C.M., and Hadjipanayis, G.C. (1993) *Langmuir*, **9** (1), 162–169.
26. Garron, A., Swierczynski, D., Bennici, S., and Auroux, A. (2009) *Int. J. Hydrogen Energy*, **34** (3), 1185–1199.
27. Corrias, A., Ennas, G., Licheri, G., Marongiu, G., and Paschina, G. (1990) *Chem. Mater.*, **2** (4), 363–366.
28. Jun, L., Shuping, X., and Shiyang, G. (1995) *Spectrochim. Acta*, **51A** (4), 519–532.
29. Patel, N., Guella, G., Kale, A., Miotello, A., Patton, B., Zanchetta, C., Mirenghi, L., and Rotolo, P. (2007) *Appl. Catal. Gen.*, **323**, 18–24.
30. Markova-Deneva, I. and Univ, J. (2010) *Chem. Technol. Metall.*, **45** (4), 351–378.
31. Arzac, G.M., Rojas, T.C., and Fernandez, A. (2011) *ChemCatChem.*, **3** (8), 1305–1313.
32. Andrieux, J., Swierczynski, D., Laversenne, L., Garron, A., Bennici, S., Goutaudier, C., Miele, P., Auroux, A., and Bonnetot, B. (2009) *Int. J. Hydrogen Energy*, **34** (2), 938–951.
33. Kozitsyna, N.Y., Troitskii, S.Y., Kryukova, G.N., Martens, M.V., Grushetskaya, M.A., Stolarov, I.P., Novgorodov, B.N., Kolomiychuk, V.N., Kochubey, D.I., Vargaftik, M.N., and Moiseev, I.I. (2003) *Mendeleev Commun.*, **13** (1), 1–2.
34. Li, H.X., Wu, Y.D., Luo, H.S., Wang, M.G., and Xu, Y.P. (2003) *J. Catal.*, **214** (1), 15–25.
35. Yuan, Z.Z., Chen, X.D., Xu, H., Qu, X.L., and Wang, B.X. (2006) *J. Alloy Compd.*, **422** (1–2), 109–115.
36. Lesz, S., Nowosielski, R., Kostrubiec, B., Stok losa, Z., and Achieve, J. (2006) *Mater. Manuf. Eng.*, **16** (1–2), 35–39.
37. Machizaud, F., Kuhnast, F.A., Flechon, J., Auguin, B., and Defresne, A. (1981) *J. Phys.*, **42** (1), 97–106.
38. Long, J. (2006) ProQuest LLC.
39. Santos, I.A., Eiras, J.A., Araujo, E.B., Sampaio, J.A., and Catunda, T. (2001) *J. Mater. Sci. Lett.*, **20**, 1815–1817.
40. Corrias, A., Ennas, G., Musinu, A., Marongiu, G., and Paschina, G. (1993) *Chem. Mater.*, **5** (12), 1722–1726.
41. Wu, C., Wu, F., Bai, Y., Yi, B.L., and Zhang, H.M. (2005) *Mater. Lett.*, **59** (14–15), 1748–1751.
42. Lee, J., Kong, K.Y., Jung, C.R., Cho, E., Yoon, S.P., Han, J., Lee, T.G., and Nam, S.W. (2007) *Catal. Today.*, **120** (3–4), 305–310.
43. Walter, J.C., Zurawski, A., Montgomery, D., Thornburg, M., and Revankar, S. (2008) *J. Power Sources*, **179** (1), 335–339.
44. Dai, H.B., Liang, Y., Wang, P., and Cheng, H.M. (2008) *J. Power Sources*, **177** (1), 17–23.

45. Huang, Y.Q., Wang, Y., Zhao, R.X., Shen, P.K., and Wei, Z.D. (2008) *Int. J. Hydrogen Energy*, **33** (23), 7110–7115.
46. Kalidindi, S.B., Indirani, M., and Jagirdar, B.R. (2008) *Inorg. Chem.*, **47** (16), 7424–7429.
47. Patel, N., Fernandes, R., Guella, G., Kale, A., Miotello, A., Patton, B., and Zanchetta, C. (2008) *J. Phys. Chem. C*, **112** (17), 6968–6976.
48. Ma, L., Peng, Q.R., and He, D.H. (2009) *Catal. Lett.*, **130** (1–2), 137–146.
49. Ma, H., W.Q, J., Zhao, J.Z., Liang, J., and Chen, J. (2009) *J. Alloy Compd.*, **474** (1–2), 584–589.
50. Kalidindi, S.B. and Jagirdar, B.R. (2010) *J. Indian Inst. Sci.*, **90**, 181–187.
51. Srivastava, A.K., Madhavi, S., White, T.J., and Ramanujan, R.V. (2005) *J. Mater. Chem.*, **15** (41), 4424–4428.
52. Khan, Y. (1986) *J. Non-Cryst. Solids*, **86** (1–2), 137–148.
53. Zern, A., Kleinschroth, I., Gonzalez, A., Hernando, A., and Kronmuller, H. (1999) *J. Appl. Phys.*, **85** (11), 7609–7615.
54. Takahashi, M., Kim, C.O., Koshimura, M., and Suzuki, T. (1978) *Jpn. J. Appl. Phys.*, **17** (10), 1911–1912.
55. Hasegawa, R. and Ray, R. (1979) *J. Appl. Phys.*, **50** (3), 1586–1588.
56. Damjanovic, L., Bennici, S., and Auroux, A. (2010) *J. Power Sources*, **195** (10), 3284–3292.
57. Damjanovic, L., Majchrzak, M., Bennici, S., and Auroux, A. (2011) *Int. J. Hydrogen Energy*, **36** (3), 1991–1997.
58. Jeong, S.U., Cho, E.A., Nam, S.W., Oh, I.H., Jung, U.H., and Kim, S.H. (2007) *Int. J. Hydrogen Energy*, **32** (12), 1749–1754.
59. Walter, J.C., Zurawski, A., Montgomery, D., Thornburg, M., and Revankar, S.T. (2008) *ECS Trans.*, **6** (17), 55–63.
60. Shieh, S.R. and Duffy, T.S. (2002) *Phys. Rev. B.*, **66** (13), 134301–8.
61. Lutz, H.D., Moller, H., and Schmidt, M. (1994) *J. Mol. Struct.*, **328**, 121–132.
62. Preudhomme, J. and Tarte, P. (1971) *Spectrochim. Acta, A*, **27** (9), 1817–1835.
63. Tirosh, E., Shemer, G., and Markovich, G. (2006) *Chem. Mater.*, **18** (2), 465–470.
64. Bhowmik, R.N. and Naresh, N. (2010) *Int. J. Eng. Sci. Technol.*, **2** (8), 40–52.
65. Tang, C.W., Wang, C.B., and Chien, S.H. (2008) *Thermochim Acta*, **473** (1–2), 68–73.
66. Akdim, O., Demirci, U.B., Muller, D., and Miele, P. (2009) *Int. J. Hydrogen Energy*, **34** (6), 2631–2637.
67. Jeong, S.U., Kim, R.K., Cho, E.A., Kim, H.J., Nam, S.W., Oh, I.H., Hong, S.A., and Kim, S.H. (2005) *J. Power Sources*, **144** (1), 129–134.
68. Liu, B.H., Li, Z.P., and Suda, S. (2006) *J. Alloy Compd.*, **415** (1–2), 288–293.
69. Suknev, V.S. and Brovkin, A.A. (1970) *J. Appl. Spectrosc.*, **12** (2), 248–254.
70. Keshan, A.D. (1955) Synthesis of Borates in Aqueous Solution and its Investigation, Riga (in USSR).
71. Liu, C.H., Chen, B.H., Hsueh, C.L., Ku, J.R., Tsau, F., and Hwang, K.J. (2009) *Appl. Catal. B.*, **91** (1–2), 368–379.
72. Delmas, J., Laversenne, L., Rougeaux, I., Capron, P., Garron, A., Bennici, S., Swierczynski, D., and Auroux, A. (2011) *Int. J. Hydrogen Energy*, **36** (3), 2145–2153.
73. Suknev, V.S. (1970) *J. Appl. Spectrosc.*, **12** (1), 76–79.

10
Reactive and Metastable Nanomaterials Prepared by Mechanical Milling

Edward L. Dreizin and Mirko Schoenitz

10.1
Introduction

During last several decades, mechanical milling has been extensively explored to prepare multiple advanced powder-like materials including mechanically alloyed and reactively milled compositions. Extensive reviews on preparation and characterization of such materials are available in the literature [1, 2]. While detailed mechanisms of material refinement at the nano or even atomic scale by simple mechanical interaction with the milling media remain elusive, an extensive database is currently available for the laboratory scale preparation of composite materials and alloys, including amorphous and nano-crystalline compounds [3–5]. More recently, mechanical milling was proposed as a viable method for preparation of reactive (or mechanically activated) materials [6–9]. Such materials are usually micron-sized powders; each particle is a fully dense composite of two or more components capable of a highly exothermic reaction. For example, a cross-sectioned image of an Al-CuO nanocomposite particle is shown in Figure 10.1. Inclusions of CuO with characteristic dimensions of about 100 nm are nearly uniformly dispersed in the matrix of Al. Upon thermal initiation, a redox reaction $2Al + 3CuO \rightarrow Al_2O_3 + 3Cu$ begins at all Al-CuO interfaces resulting in effectively volumetric self-heating of the entire composite particle. Because of the very developed reactive interface area, such materials may be comparable to or even more reactive than traditional energetic nanocomposites prepared by intimate mixing or self-assembly of nanopowders of the starting components [10–13].

The reactive interface structures and properties in the materials prepared by mechanical milling are different from those existing in the mixed nanopowders, most significantly because of the absence of natural passivating oxide layers on the surface of Al or other metals used as fuels. Further, there is no porosity inherent for the mixed nanopowders, resulting in a higher density and energy density and also in different mechanisms for propagation of combustion waves in the reactive composite prepared by mechanical milling. In addition to reactive nanocomposite materials, mechanical milling is used to prepare reactive mechanically alloyed

Metal Nanopowders: Production, Characterization, and Energetic Applications, First Edition.
Edited by Alexander Gromov and Ulrich Teipel.

Figure 10.1 A cross-sectioned particle of Al-CuO nanocomposite material prepared by arrested reactive milling.

powders. In such powders, the reactivity of one component can be assisted by a highly reactive additive; alternatively, alloys can be designed to undergo phase changes upon heating, which generate fresh surface capable of rapid reactions with ambient oxidizers. This chapter will discuss preparation, characterization, and properties of mechanically alloyed and reactive nanocomposite materials prepared by mechanical milling to date focusing on the work by the researchers from New Jersey Institute of Technology.

10.2 Mechanical Milling Equipment

Four different types of ball mills were used for preparation of reactive materials, including SPEX CertiPrep shaker mills (8000 series), a SPEX 6850 freezer mill, a Retsch PM400 planetary mill, and a Union Process 01HD attritor. Freezer and attritor mills are capable of milling at cryogenic temperatures. All mills have been customized and were used for preparation of reactive materials. Shaker and planetary mills are equipped with sets of hardened steel and zirconia jars. Planetary mill jars are equipped with customized lids enabling evacuating closed jars as well as adding liquid process control agent (PCA) without opening the sealed jars. Custom-built pressure-proof milling jars for the planetary mill allow systematic research of high-energy and gas-generating materials. In addition, the planetary mill is modified to include an air conditioner for the milling chamber and custom-designed heat sinks for the milling jars, both aimed to enable milling at reduced temperatures. Milling jars equipped with the heat sinks and with a customized lid as well as pressure-proof milling jars are shown in Figure 10.2.

Figure 10.2 Different milling jars used for preparation of reactive mechanically alloyed and nanocomposite powders.

In order to enable temperature monitoring, wired temperature sensors are used for the shaker mills, while wireless temperature sensors are attached to each milling jar of the planetary mill [14]. For the attritor mill, processing parameters including power and torque are monitored in real time enabling one to track milling progress non-disruptively.

10.3
Process Parameters

Common parameters varied for preparation of materials by mechanical milling include ball to powder mass ratio or charge ratio, C_R, material and dimensions of the milling balls, milling time, τ, intensity of milling often related to the rate of the milling jar or impeller agitation expressed in rpm, type and amount of the PCA, and milling temperature, T_m. General guidance for selection of C_R, rpm, and PCA is available from several reviews and books on mechanical alloying and reactive milling [5, 15].

For reactive and metastable materials described in this chapter, common C_R for the shaker mill varied from 5 to 10 with a powder charge of approximately 3–5 g per milling jar. For the planetary mill, the values of C_R varied from 3 to 6 with the powder charge variable from 20 to 50 g per milling jar. Much higher values of C_R are typical for attritor mills; in this work the powder charge per jar varied from 30 to 50 g with a C_R of approximately 30–40. The common PCA used for preparation of most mechanically alloyed powders was stearic acid (approximately 2 wt% of the powder charge) preventing cold welding, especially at early stages of milling. Stearic acid was also used in earlier experiments aimed to prepare reactive nanocomposite materials, such as nanocomposite thermites and boron-based nanocomposites. Liquid PCA, such as hexane and heptane were used commonly for preparation of reactive nanocomposite powders. Such liquid PCA also serves to prevent self-sustaining reaction in the milling jar and thus its use is important to ensure safe materials manufacturing. Nonpolar liquids are preferred since common solvents such as ethanol or acetone readily react with freshly formed aluminum surfaces. Prepared nanocomposite powders are preferably handled while under liquid PCA to minimize their exposure to environmental oxygen and humidity.

The temperature of the milling environment was found to be an important parameter affecting both the rate of material refinement and the final properties of the prepared materials. Conversely, the temperature of the milling jars is affected by energy dissipated in the interactions of the milling tools, which is in turn affected by the size, shape, and mechanical properties of the milled powder. The temperature as measured directly from the surface of the milling jar can therefore serve as an important indicator of the milling progress. For example, a set of temperature traces measured while preparing a mechanically alloyed Al-Mg powder in a planetary mill is presented in Figure 10.3. The traces from top to bottom reflect changes in the environment temperature, while the specific shape of the traces reflect variations in the milling jar temperature. The trace corresponding to the highest milling temperature is recorded without any additional cooling of the milling vial. Other traces show effect of venting and air-conditioning the milling compartment of the mill as well as effect of custom heat sinks on the processing temperature. The temperature traces exhibit characteristic peaks at early milling times signifying formation of metal flakes. After longer milling, flakes become thicker while breaking up laterally, eventually forming work hardened mechanically alloyed powders. It is observed that the flake formation is accelerated by reduced temperatures; however, the formation of equiaxial alloy particles is delayed. At the same time, it is observed that the solubility of Mg in Al is increased at reduced temperatures, so that the compositions of the prepared alloys are further away from the equilibrium intermetallic phases.

In experiments with reactive nanocomposite powders [16], it was observed that the uniformity of mixing between different phases can be improved and the scale of mixing can be refined if milling is performed at cryogenic temperatures. Effect of milling media density and PCA in a shaker mill was addressed in a study on Al-MoO_3 reactive nanocomposites [17]. Milling balls made of alumina, zirconia, steel, and tungsten carbide were used. Varying milling parameters lead to different scales of structural refinement in the nanocomposites. The crystallite size decreased with increase in the milling intensity caused by reduced amount of PCA or increased

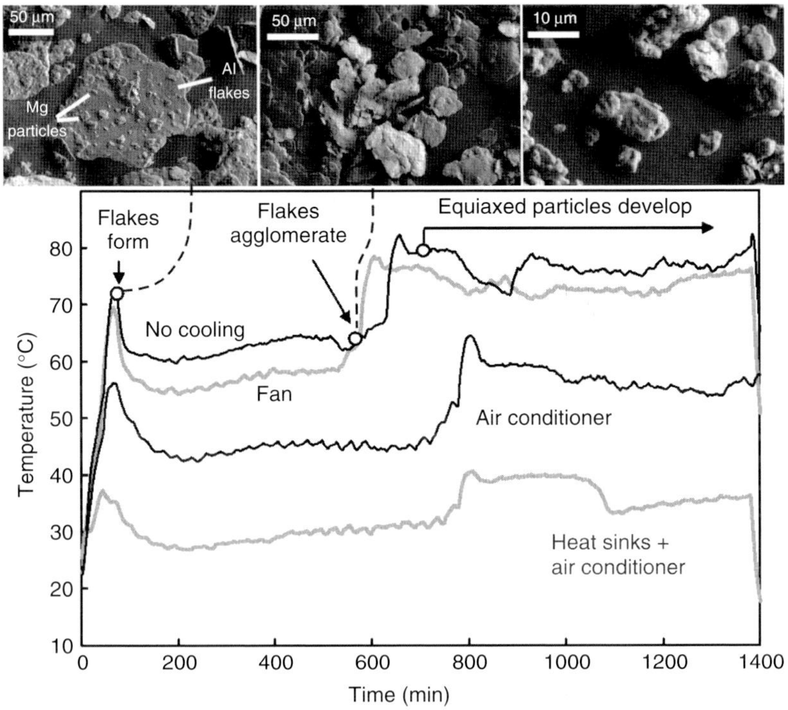

Figure 10.3 Effect of environment temperature on milling progress for a Al-30 at% Mg mechanically alloyed powder [14].

density of the milling media. However, increase in milling intensity also resulted in undesirable partial reaction between starting materials. Milling media with densities in the range of 5–8 $g\,cm^{-3}$ resulted in the highest structural refinement accompanied by the lowest degree of partial reaction. Powders produced with such milling media were most reactive. Based on those experiments, steel milling media was generally recommended for the optimized milling conditions required for manufacturing of reactive nanocomposite powders by mechanical milling.

A most useful guidance for the selection of appropriate milling parameters when preparing reactive materials using different milling devices derives from the concept of the milling dose introduced to relate milling progress to the energy transfer from the milling media to the powder. The milling dose, D_m is defined as the rate of energy transfer from the milling tools to the powder, E_d, multiplied by the milling time, τ, and divided by mass of the milled powder, m_p:

$$D_m = \frac{(E_d \tau)}{m_p} \tag{10.1}$$

The concept of the milling dose was considered for three milling devices: shaker, attritor, and planetary mills. In experiments aimed to develop and validate the approach, the same starting powders were used in each mill. The milling time, τ, for each mill was measured while the final product with the desired properties was

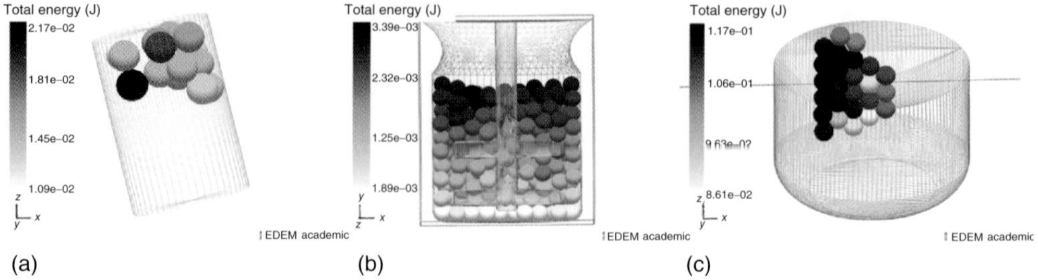

Figure 10.4 DEM descriptions of shaker, attritor, and planetary mills (left to right) [18].

prepared. Theoretical models of the three mills were set up using discrete element modeling (DEM) [18] and the energy dissipation rate, E_d, required for calculating the milling dose was computed. DEM descriptions for different types of mills are illustrated in Figure 10.4. For each device, the milling vials are designed in Creo Elements/Pro and incorporated into EDEM code by DEM Solutions. The motion patterns of the milling vials and impeller, for the attritor mill, are reproducing the experimental situation directly. The energy dissipation rate calculation used directly the information about the properties of the milling media, characteristics of the milling media interaction, such as restitution and friction coefficients, and the mill operating parameters, such as rpm. Specific experiments were designed and performed to quantify restitution [19] and friction coefficients [18].

Current research established the utility of the milling dose-based approach in adequately describing the milling synthesis of new materials. Specifically, for a correctly set up model, greater energy dissipation rate describes a higher intensity milling process resulting in a faster preparation of the desired material. An overall correlation between milling dose (both predicted and experimental) and the final properties of the prepared materials was also established [18–22].

10.4 Material Characterization

Partially and fully milled powder samples were preliminarily examined using a Phenom Tabletop Microscope by FEI Technologies Inc. (a scanning electron microscope (SEM) with a backscattered electron detector.) Morphologies and elemental compositions of the prepared powders were examined in detail using a LEO 1530 Field Emission SEM, equipped with an energy dispersive X-ray spectrometer (EDX). In addition to "as prepared" powders, samples were embedded in epoxy and cross-sectioned for analysis. Images were taken using backscattered electron imaging. Phase compositions of the samples were analyzed using X-ray diffraction (XRD) on a Phillips X'pert MRD (materials research diffractometer) powder diffractometer operated at 45 kV and 40 mA, using Cu-K_α radiation ($\lambda = 1.5438$ Å).

Particle size distributions of the prepared materials were measured with low-angle laser light scattering using a Beckman-Coulter LS230 Enhanced Particle

Analyzer. Powder suspensions for analysis were prepared in ethylene glycol or, for materials stable in water, in distilled water, and ultrasonicated during the measurement to minimize particle agglomeration.

Intermetallic, decomposition, and oxidation reactions were studied using Differential thermal analysis (DTA), differential scanning calorimetry (DSC) and thermogravimetric analysis (TGA). Two instruments were routinely used: a Netzsch Instruments STA 409 PC/4/H Luxx Simultaneous TG-DTA-DSC apparatus, equipped with an additional customized furnace enabling measurements in controlled water vapor environments, and a premium performance research grade TGA model Q5000 IR by TA Instruments capable of high accuracy measurements at extremely high heating rates, approaching 1000 $K\,min^{-1}$.

The thermal analyzers were temperature-calibrated using the known melting points of a set of certified high-purity metals. In addition, the TA Q5000IR was calibrated using the Curie points of alumel, nickel, and cobalt.

Measurements were performed in different gas mixtures. Typically, for intermetallic reactions the furnace was flushed with ultra-high purity argon (Matheson, 99.999%) at about 50 $ml\,min^{-1}$. For most oxidation experiments, argon was mixed 1 : 1 with oxygen (Matheson, 99.98%). Experiments were performed at various heating rates and the maximum temperature was 1495 °C for the STA 409 and 1200 °C for the Q5000 IR. Typically, after a heating ramp, samples were cooled to room temperature, and heated again to the same maximum temperature at the same heating rate. The second heating was then subtracted from the first to obtain the baseline for the measurement. TGA was performed in either instrument using sintered alumina sample carriers designed to maximize the exposed sample surface. Scanning calorimetry was typically performed in covered alumina sample pans. Processes at lower temperatures were also examined with better sensitivity using covered Pt/Rh sample pans which are lined with thin-walled alumina inserts.

In many oxidation experiments involving Al-rich powders, samples were blended with nanocrystalline α-Al_2O_3 powder (Inframat® Advanced Materials, crystal size ~40 nm, particle size ~150 nm). This was done to minimize the effect of coalescence of Al particles upon melting. A typical sample blend consisted of 25 wt% of a prepared composite and 75 wt% of nanocrystalline α-Al_2O_3. The powders were mixed in a hexane suspension using an ultrasonic bath for 1.5 h. Once blended, the samples were dried in vacuum and used for the oxidation experiments.

The custom water vapor furnace for the STA 409 was adapted to allow mixing of steam with other oxidizer gases. Oxidation of aluminum powders was analyzed in various mixtures of H_2O, O_2, and CO_2 where the whole range of gas compositions was covered [23–25].

10.5
Ignition and Combustion Experiments

Ignition was studied at heating rates ranging from 1000 to 50 000 $K\,s^{-1}$ in air using an electrically heated filament [26, 27], shown schematically in Figure 10.5a.

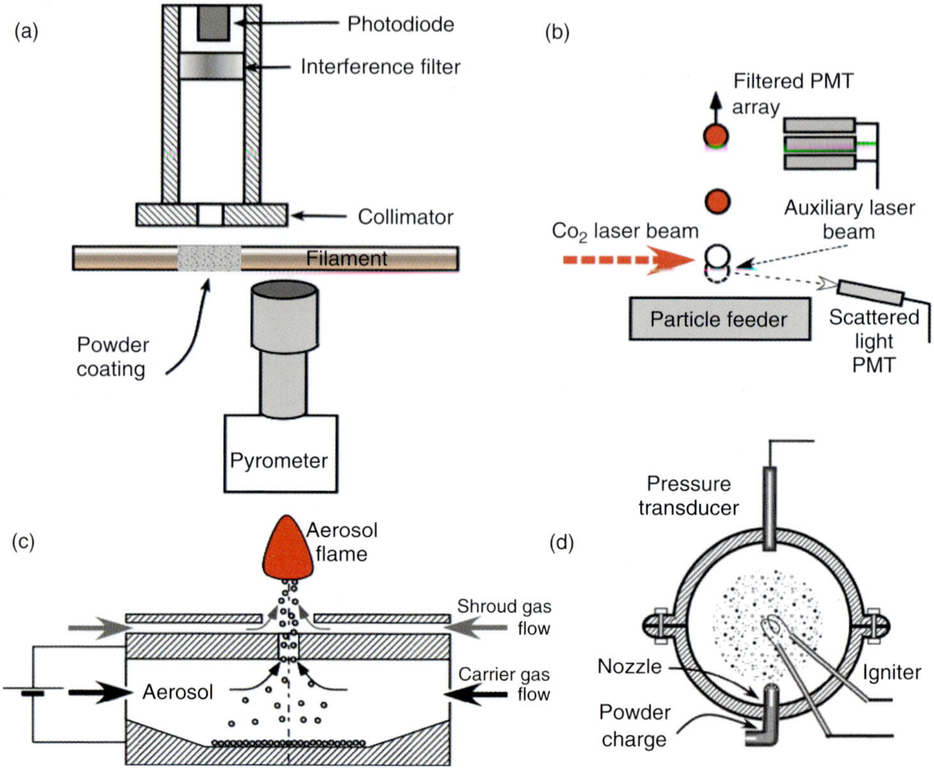

Figure 10.5 Heated filament setup used for powder ignition experiment (a); Laser ignited particle combustion setup for single particle combustion experiment (b); lifted laminar aerosol burner (c), and constant volume explosion setup for aerosol combustion experiment (d).

The heating rate of a nickel-chromium filament was controlled via a DC power supply coupled with a set of resistors, connected in series with the filament. A small amount of powder slurry with hexane as a liquid carrier was coated onto the filament. Hexane was dried prior to the experiment leaving a thin powder coating on the filament surface. Ignition was observed using both a silicon photodiode and a high-speed video camera. The temperature history of the heated filament was measured using a high-speed infrared pyrometer focused on the uncoated filament surface adjacent to the powder coating, giving effectively the environment temperature of the heated powder.

Burning characteristics of individual composite particles in air were investigated using a laser ignition apparatus (see a simplified diagram in Figure 10.5b), described in detail in Refs [28–30] Individual particles were fed by a laminar air flow into a CO_2 laser beam, where they were ignited. Before ignition, they crossed a beam of a 785 nm, 30 mW laser and the scattered light pulses were used to measure particle sizes in-situ. Four photomultiplier tubes (PMTs, Hamamatsu H3164-10), equipped with interference filters (486, 532, 568, and 589 nm), were used to detect the light emission from particle burning. These wavelengths were selected targeting primarily aluminum combustion details, so that 486 nm coincided with one of the strongest AlO emission lines. Conversely, other filters were selected to avoid strong AlO lines enabling measurements of the gray body emission produced by the burning particles. 532 and 589 nm filters were used for combustion temperature measurement.

The same particle feeder as used for laser ignition experiments was also employed to produce and study lifted laminar aerosol flames of the prepared powders [27, 31] and compare them to analogous flames for pure metal (Al and Mg) aerosols [32]. The particle feeder, as shown in Figure 10.5c, consists of a plate capacitor with a small hole in the top electrode [33]. The voltage applied to the capacitor controls the number density of the produced aerosol. The carrier gas flow controls the speed of the aerosol jet. The shroud flow together with the carrier gas flow determines the oxidizing environment. The aerosol number density is increased substantially compared to that used in the individual particle experiments. Thus, once ignited, the aerosol jet is capable of maintaining a flame. The speed of the flame propagation and flame temperature were measured. In addition, the particle burn times were inferred from processing the flame thickness measurements. The burn times were correlated with particle sizes using measured particle size distributions.

Combustion of the prepared composite materials was studied using a constant volume explosion (CVE) experiment [34–40]. A simplified diagram of the experiment setup is shown by Figure 10.5d. A 9.2 l explosion vessel was initially evacuated. Composites were aerosolized and introduced into the vessel with an air blast delivered from a high-pressure reservoir. After a short delay, aerosolized powder was ignited by an electrically heated wire placed in the center of the vessel. The pressure trace was recorded by a pressure transducer. Samples were tested in air with a constant powder load (4.65 g), corresponding to a fuel-rich condition ensuring reliable ignition for different materials. The combustion products were collected after the experiment, embedded in epoxy and cross-sectioned for SEM examination.

10.6 Starting Materials

Starting materials are commercially available powders of metals, oxides, or other compounds. For Al-based nanocomposite materials, −325 mesh, 99.9% by Atlantic Equipment Engineers is commonly used. Metal oxides used included MoO_3 (99.95% pure, by Alfa Aesar), CuO (99% pure, 1–5 μm, by Atlantic Equipment Engineers), Fe_2O_3, (99.5% by Alfa Aesar), and Bi_2O_3 (99.97% by Skylighter Inc.). Other metals used included powders of Mg, Ti, Zr, Li, Fe, Zn, B, and W. Oxidizers included Fe_3O_4, WO_3, and sodium nitrate, $NaNO_3$. For selected materials, elemental iodine has been used as a biologically active volatile ingredient. PTFE (polytetrafluoroethylene) $(C_2F_4)_n$ was used in composites as oxidizing agent, and to provide biocidal combustion products. Composites of aluminum and hydrocarbons (paraffins, polyethylene) have been prepared and examined as well. Requirements of the milling process for component purity and particle size are relatively relaxed so that commercial-grade micron-sized powders are typically sufficient.

10.7 Mechanically Alloyed and Metal–Metal Composite Powders

10.7.1 Preparation and Characterization

All alloys or metal–metal composite materials described here are based on aluminum as the most common metal used in energetic formulations. The goal of preparing reactive mechanically alloyed powders was to produce materials with shorter ignition delays compared to pure Al, but with combustion enthalpy and flame temperatures similar to those for pure Al powders. In addition, it was desirable to be able to adjust combustion dynamics of the mechanically alloyed particles, for example, accelerate the overall burn rate. Mechanically alloyed powders offer several potential ways to achieve the stated goals. Oxide film produced on particle surface of such alloyed powders may not be pure Al_2O_3, and thus is likely to be more permeable, enabling exothermic reactions to begin at lower temperatures. Mechanically alloyed powders are often metastable and undergo weakly exothermic intermetallic reactions upon heating. These reactions could accelerate the oxidation occurring simultaneously, considering that the rate of oxidation is a strong function of temperature. Furthermore, formation of new intermetallic phases is accompanied by formation of new crystallites and grain boundaries, which generates fresh surfaces prone to oxidation. Finally, some of the alloying additives may be selectively oxidizing sooner than Al, causing an additional heat release at lower temperatures and thus accelerating ignition.

Characteristic milling times vary from 1 to 8 h for preparing mechanical alloyed powders using a planetary mill at room temperatures with typical ball to powder mass ratio of 3–5 and rotation speed of 350 rpm [14, 19, 21, 41–43]. In some cases, however, longer milling times, up to 100 h were used [42]. Shorter milling times are used for preparation of metal–metal composite powders [44]. For some materials, intermetallic phases begin to form after only 30 min of milling, so even shorter milling times are needed if intermetallics are undesirable.

In most cases, stearic acid was used as a PCA preventing cold welding. Typical amount of stearic acid was 2 wt% of the powder mixture. Some materials were prepared with different PCA, for example, hexane or heptane added to the powder mixture.

Milling compromises the natural protective oxide layer (most importantly, alumina) on the surface of the prepared powder particles. Therefore, such particles can be pyrophoric if exposed to air or other oxidizer-enriched environment. As a general rule, aluminum-based mechanically alloyed powders should be passivated before handling in air. A simple passivating method is to leave the powders in a glovebox filled with Ar, where relatively small oxygen concentrations (of the order of 2–3%) are present. Current experience suggests that after a 24 h exposure to the glovebox environment with low oxygen concentration, mechanically alloyed powders can be handled in air. Prior to passivation (or instead of it, if subsequent handling of dry powders in air is not required), it is safe to store the powders under a nonpolar liquid hydrocarbon, such as hexane or heptane.

Characteristic shapes and sizes of the mechanically alloyed powders are illustrated in Figure 10.6. The shape of the milled powders typically evolves as laid out in

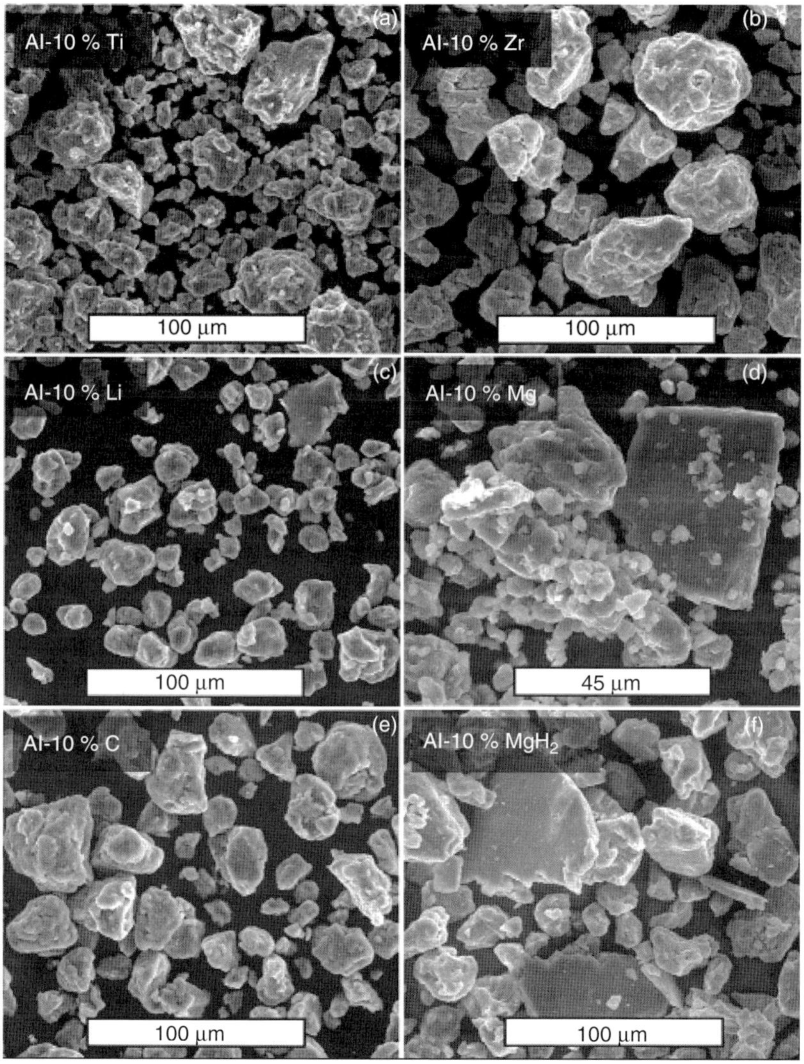

Figure 10.6 Representative SEM images of Al-based mechanically alloyed powders [31].

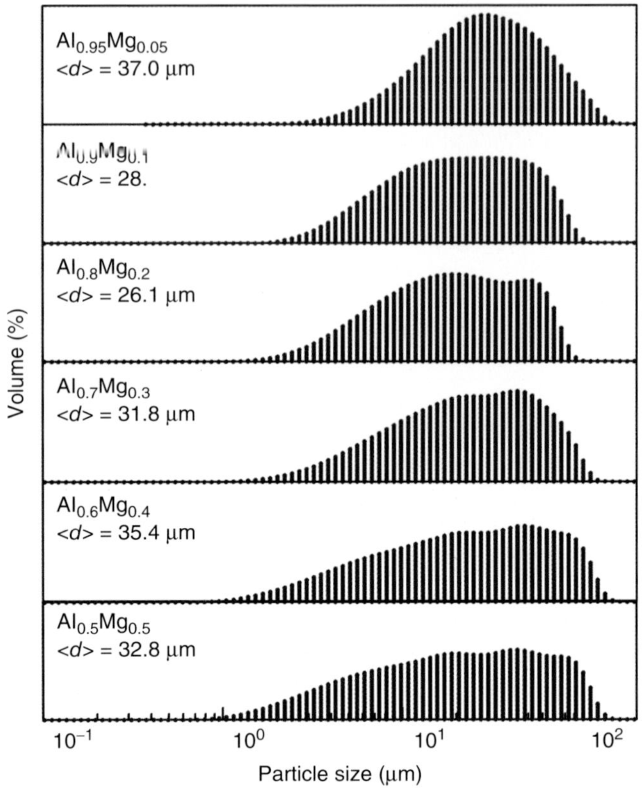

Figure 10.7 Particle size distributions of mechanically alloyed Al-Mg powders [45].

the "Process parameters" section above (cf. Figure 10.3). Early in the process, particles show flake or plate-like shapes, and the final, steady state shapes are rough but approximately equiaxed with typically wide size distributions. The steady state particle shapes do not depend significantly on the constituent metals.

The representative series of size distributions for mechanically alloyed Al-Mg powders are shown in Figure 10.7. The steady state size distribution can be influenced somewhat by the choice of milling ball size and PCA [3]. However, for Al-based materials obtained powder particle sizes were always in the 1–100 μm range; the sizes of mechanically alloyed particles did not depend on the size distributions of the starting component powders.

Depending on the milling conditions, the product powder can be classified as mechanically alloyed, or as metal–metal composite. Early in the milling process, the continuous cold-welding and re-fracturing of the component powders yields composite materials, in which the components retain their identity in terms of structure and composition. Backscattered electron images of selected cross-sections are shown in Figure 10.8. The distribution of the components relative to each other is determined by the properties of the component powders: more ductile components result in more lamellar structures.

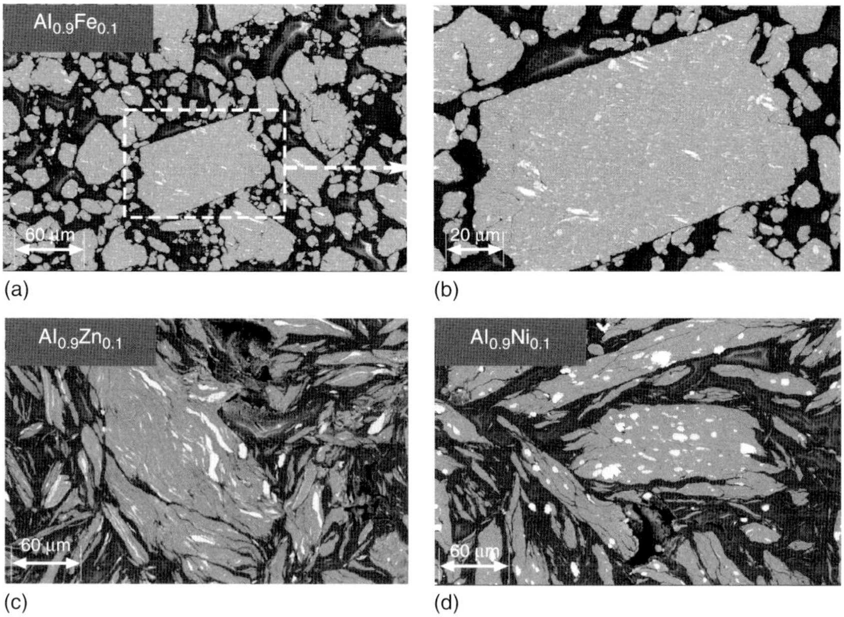

Figure 10.8 Selected metal-metal composite powders [44].

As milling progresses, the components become more and more refined until they can no longer be distinguished compositionally and structurally, forming mechanically alloyed powders. Cross-sections of such powders show simply homogeneous materials. The altered state of the material can however be recognized structurally. Figure 10.9 illustrates this change by showing XRD patterns of an Al-30 at% Mg powder as a function of milling time. At milling times of 3 h and below, Al remains identifiable, while Mg has disappeared, and weak peaks of $Al_{12}Mg_{17}$ are present. At longer milling times most Mg dissolves into *fcc* Al, forming a supersaturated solution that is recognized by a change in the

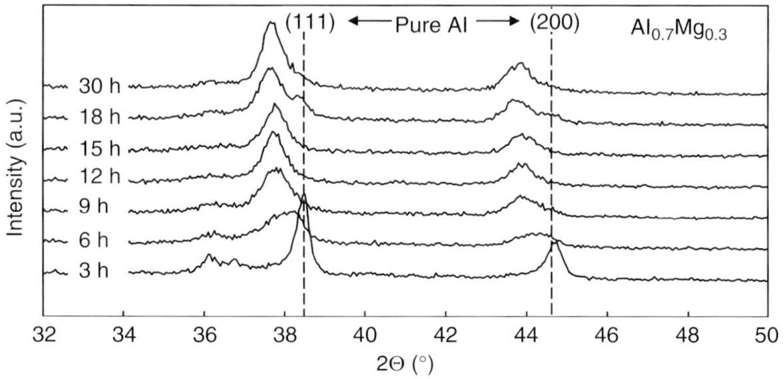

Figure 10.9 Evolution of Al-Mg powder during prolonged milling [45].

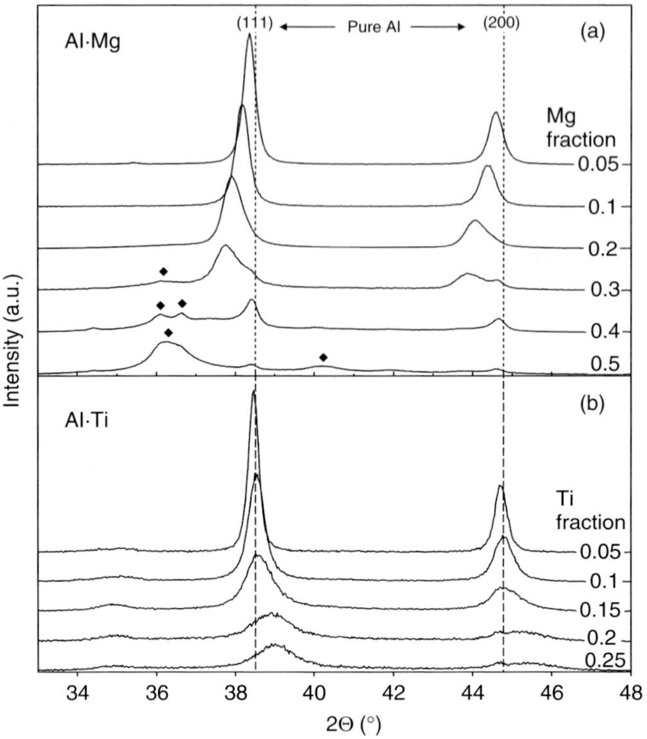

Figure 10.10 XRD patterns indicating steady state phase assemblages for an Al-Mg powder (12 h milling) [45] and an Al-Ti powder (15 h milling) [46].

fcc Al lattice parameter, and the corresponding diffraction peaks [45] The solid solution achieves steady state and no longer changes after 9–12 h. The extent of the solid solution does depend however on the bulk composition of the powder. Figure 10.10a shows XRD patterns of mechanically alloyed Al-Mg powders with varying bulk compositions that have achieved their steady state. Higher bulk Mg concentrations result in increased peak shift and therefore in solid solutions with correspondingly higher Mg concentrations. It should be noted, however, that a material balance showed that not all of the bulk Mg is contained in the solid solution, and therefore some Mg must exist in an X-ray amorphous disordered state [45]. At bulk Mg concentrations of 30 at% and above, a set of diffuse peaks is consistent with a disordered form of the equilibrium phase $Al_{12}Mg_{17}$.

Other systems may form metastable phases with defined compositions. Figure 10.10b shows XRD patterns of mechanically alloyed Al-Ti powders. At increasing bulk Ti concentrations, the XRD peaks shift away from the position of pure Al. A clear relation between the *fcc* Al lattice parameter and Ti dissolved in Al was not found, but the XRD patterns are consistent with the emergence of the disordered, metastable $L1_2$ phase of Al_3Ti [46]. This concept was substantiated by the sequence of phases ($L1_2 \rightarrow D0_{23} \rightarrow D0_{22}$) observed during relaxation of this metastable phase [46].

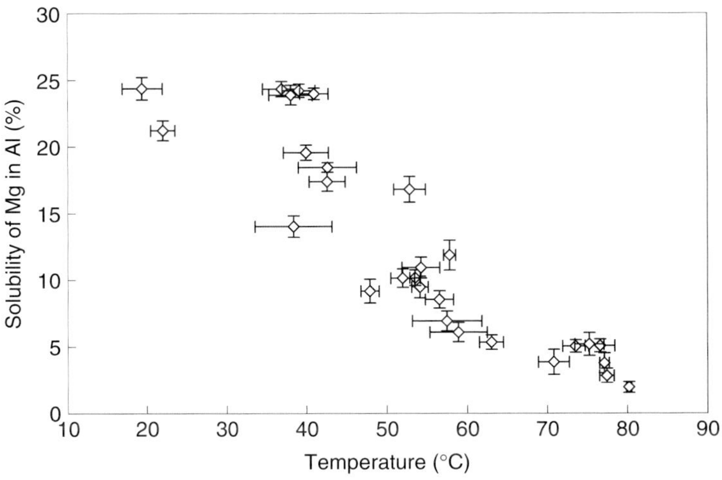

Figure 10.11 Degree of Mg dissolution into *fcc*-Al depending on the milling jar temperature [14].

At long milling times, complete amorphization may be achieved. For example, for a blend with Al 30 at% Li bulk composition, dissolution of Li in the *fcc* Al structure occurred after the initial flake stage, accompanied by the formation of δ-LiAl. At extended milling times the LiAl disappeared first, and after milling for 102 h the *fcc* phase disappeared as well [42].

As the steady state phase composition depends on the equilibrium between the mechanical work causing formation of metastable phases and annealing at the milling temperature causing the metastable phases to revert to stable phases, the extent to which metastable phases are formed during milling is a function of milling environment temperature. This effect is illustrated in Figure 10.11, where for mechanically alloyed Al-Mg powders, the environment temperature was varied [14]. With the use of custom heat sinks (cf. Figure 10.1), the resulting milling jar temperature varied between 20–30 and ~70–80 °C. At the higher temperatures, most of the Mg was present as $Al_{12}Mg_{17}$, and only small amounts were dissolved in *fcc* Al, while at the lower temperatures negligible amounts of $Al_{12}Mg_{17}$ were observed, and the concentration of Mg in *fcc* Al was highest.

When mechanically alloyed powders are annealed, successively less metastable phases are observed. Al-rich mechanical alloys in the Al-Mg system will first form the more Mg-rich $Al_{12}Mg_{17}$, until at higher temperatures the equilibrium phase assemblage of *fcc* Al and Al_3Mg_2 forms. Similarly, Al-rich Al-Ti mechanical alloys form more and more ordered polymorphs of Al_3Ti when annealed. These changes have generally small heat effects, but may lead to surface modifications that can be beneficial for oxidation and ignition [27].

Contamination of the milled powders may be a concern. Ceramic milling tools will introduce detritus in the milled material which causes dispersion strengthening, and therefore aids the formation of steady state particles, but can also diminish the amount of active Al. Similarly, metallic (steel) milling tools will introduce iron contamination in the system causing limited solution hardening. The observed levels are low – <1 at%, however. The PCA is another source of contamination. Stearic acid in particular will break down gradually during milling, and the milled powder will accumulate carbonaceous residue. This may lead to the formation of aluminum carbide during subsequent heating of the powder. In the case of mechanically alloyed Al-Ti, the Al carbide was observed to further react, forming Ti carbide at higher temperatures [47]. Whether any of the sources of contamination are of practical relevance should be decided on a case by case basis.

10.7.2
Thermal Analysis

Metastable states induced by milling relax upon heating, and can influence the oxidation (and ignition) of such materials in a number of ways. Changes can be determined by the altered bulk composition, by modification of the surface of the oxidizing aluminum, by the exothermicity released when metastable phases revert to their stable equivalents, or by exothermic reactions between the component metals. Selected examples are shown below.

Upon heating, all mechanically alloyed powders experience weakly exothermic phase changes resulting in relaxation of their metastable structures and formation of intermetallic phases. Such phase changes detected by DSC are illustrated for several prepared mechanically alloyed powders in Figure 10.12 [42, 45, 48]. Typically, such phase changes occur at low temperatures; for Al-based alloys the formation of intermetallic phases typically occurs well before Al melting point. As seen in Figure 10.12, it is also typical to observe several small exothermic peaks, corresponding to formation of various alloyed structures in the prepared materials. While the heat released in the intermetallic changes is typically small

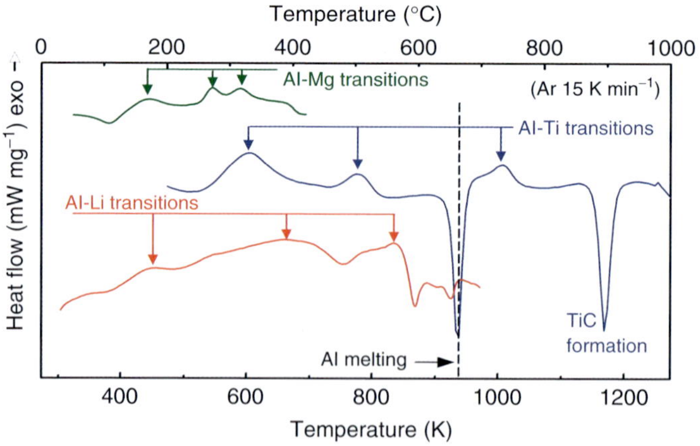

Figure 10.12 DSC traces for mechanically alloyed Al-Mg, Al-Ti, and Al-Li powders heated in Ar at 15 K/min [42, 45, 48].

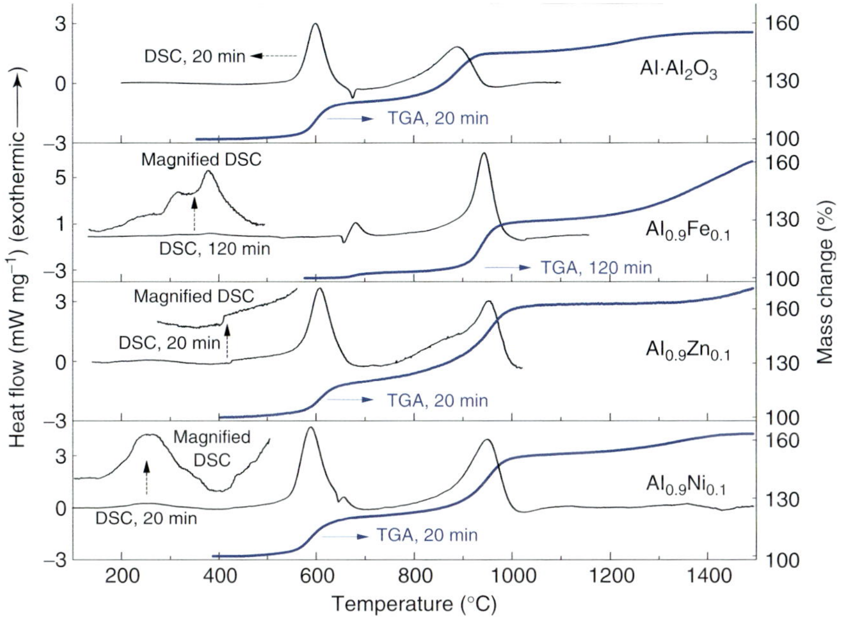

Figure 10.13 DSC and TGA results for oxidation of a set of Al-metal composite powders [44].

and insufficient to cause ignition of the mechanically alloyed powders, formation of fresh surface and grain boundaries caused by such transitions may be significant in altering the particle morphology and accelerating its oxidation. Thus, weakly exothermic, sub-solidus phase transformations illustrated in Figure 10.12 cannot be simply dismissed when analyzing ignition mechanisms of the mechanically alloyed powders.

Figure 10.13 shows the oxidation of a set of aluminum-metal composites milled for relatively short times [44]. Component metals are Fe, Zn, and Ni, respectively. The figure shows scanning calorimetry as well as thermogravimetric results. At the top, a composite material prepared from Al and alumina is shown to give a reference for the oxidation of pure aluminum with a particle shape and size distribution typical for milled powders. Milling times were generally short enough to avoid significant formation of any phases other than *fcc* Al and the respective other component. The DSC curves show various exothermic reactions occurring, generally below the onset of oxidation, similar to those shown in Figure 10.12. Oxidation occurs in stages. The earliest stage starts near 550 °C, below the melting point of Al at 660 °C. The second, and main oxidation step peaks around 900–1000 °C, and complete oxidation may not be observed until 1500 °C or higher. This behavior is consistent for all materials shown, that were milled for 20 min. The Al-Fe composite, milled for 120 min, begins oxidizing immediately after the Al melting point, and then proceeds to oxidize like the other cases. There is generally no overlap between the oxidation processes and the low-temperature reactions between the component metals.

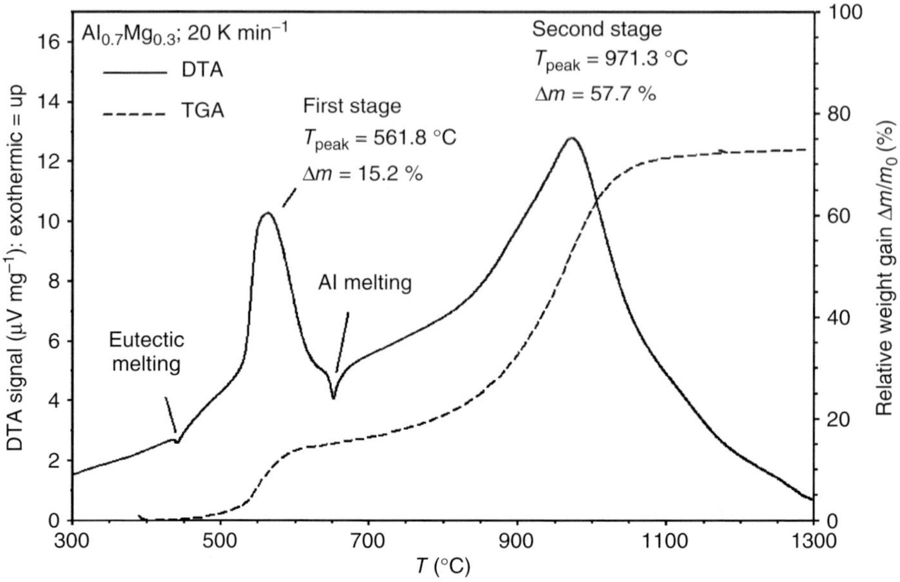

Figure 10.14 DTA and TGA results for oxidation of a mechanically alloyed Al-Mg powder in oxygen [49].

Figure 10.14 shows the oxidation of an Al-30 at% Mg mechanically alloyed powder (cf. also Figures 10.9 and 10.12) [49]. Here, melting transitions are observed for the Al-Al_3Mg_2 eutectic, and at the melting point of pure Al. Oxidation occurs in two major steps. The mass increase is observed before the Al melting point, but the majority of the mass increase occurs between 700 and 1100 °C. The relaxation of the metastable states in this mechanically alloyed powder occurs at a temperature below 400 °C, and does not appear to be connected to the oxidation process. Instead, the proportions of the mass gain steps are consistent with the sequential oxidation of first the Mg component, and later the remaining Al. This is supported by the observation of the Al melting point at 660 °C, while the liquidus for this composition should occur below 600 °C.

A different behavior is observed in mechanically alloyed Al-Ti powders, as illustrated in Figure 10.15 [48]. Like in the Al-Mg case, oxidation occurs in several steps; however the first step occurs above the Al melting point. It is less pronounced than in the case of Al-Mg, and it coincides with the final relaxation reaction of the metastable alloy, the transition from the $D0_{23}$ to the $D0_{22}$ phase of Al_3Ti at 723 °C [46, 47].

The endothermic transition seen near 900 °C indicates the formation of Al melt and TiC solids, therefore titanium is not removed from the alloy by selective oxidation [47].

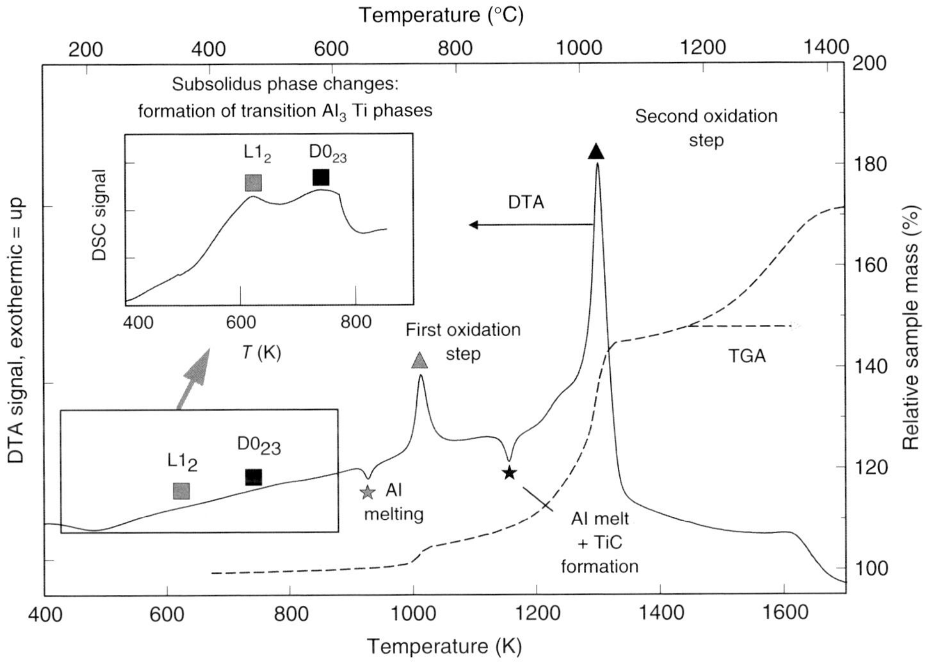

Figure 10.15 DTA (A) and TGA (B) results for a set of mechanically alloyed Al-Ti powders. Pure Al and Ti are shown for reference [48].

10.7.3 Heated Filament Ignition

In the heated filament experiments, traces of both photodiode, focused on the powder coating, and infrared pyrometer, focused on an uncoated filament surface, are measured simultaneously (cf. Figure 10.5a). To illustrate results obtained for different mechanically alloyed powders, the photodiode signal proportional to the optical emission intensity is plotted as a function of the filament temperature for selected experiments with different alloys. The temperature changes nearly linearly with time, the rate of temperature increase is adjustable by electrical settings. Results are illustrated in Figure 10.16 [26, 27, 50, 51]. For mechanically alloyed $Al_{0.7}Mg_{0.3}$ powder, optical emission shows a sharp spike at about 870 K, which is interpreted as ignition. The following decay and subsequent sharp increase in the emission signal are likely caused by the flame propagation through the entire length of the powder coating. The times of ignition cannot be readily identified if photodiode traces for $Al_{0.8}Ti_{0.2}$ and $Al_{0.7}Li_{0.3}$ alloys are plotted in a similar fashion. Such traces (not shown in Figure 10.17) exhibit continuous increase in the emission intensity with increasing temperature. To identify exothermic events leading to ignition for these two metals the time-derivatives of the photodiode signals were plotted, as also shown in Figure 10.16. For comparison, similar derivatives are shown for blank heated filaments for each respective case. For clarity, traces

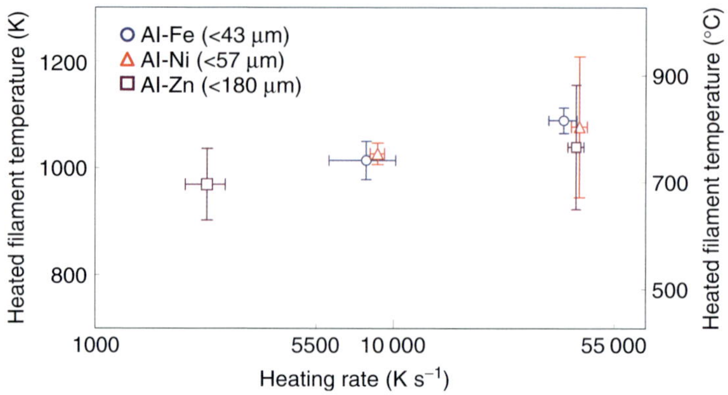

Figure 10.19 Effect of heating rate on ignition temperature for mechanically milled Al-metal composite powders [44].

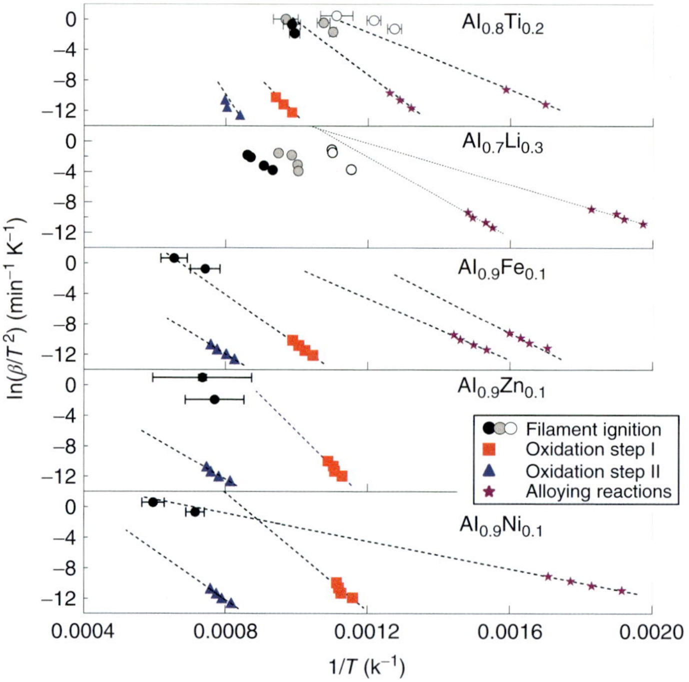

Figure 10.20 Kissinger plots [53] showing correlations between exothermic events observed in thermal analysis with ignition for mechanically alloyed Al-based materials [27, 42, 44, 50].

show locations of the exothermic sub-solidus reactions, typically, formation of different alloys occurring during heating of the ball milled powders; these reactions are typically detected in experiments performed both in argon and argon-oxygen environments. Red squares and blue triangles show initial oxidation reactions observed during heating these powders in argon-oxygen environments. Circles show ignition or exothermic events leading to ignition for different materials. For Al-Ti and Al-Li alloys, open, gray, and black circles show respectively first, second, and third exothermic events leading to ignition (cf. Figure 10.12), as established by processing their corresponding photodiode traces. Apparent correlations between sub-solidus alloying reactions and ignition are good for Al-Ti alloy and Al-Ni composite powders. For Al-Fe and Al-Zn, it is likely that the initial oxidation reactions serve as the ignition triggers. For Al-Li and Al-Mg alloys, it was observed that ignition is most likely caused by initial selective oxidation of Li and Mg, respectively.

10.7.4
Constant Volume Explosion

CVE tests were performed with several mechanically alloyed powders, including Al-Mg and Al-MgH_2 compositions [34]. Results are illustrated in Figures 10.21 and 10.22. Pressure traces in Figure 10.21 are shown for Al-Mg mechanically alloyed powders with different compositions compared to the blended Al and Mg powders with the same bulk compositions as for respective alloys. For all compositions, mechanically alloyed powders exhibit substantially greater rates of pressure rise and somewhat higher maximum pressure compared to the respective powder blends. It was suggested that in addition to early selective oxidation and ignition

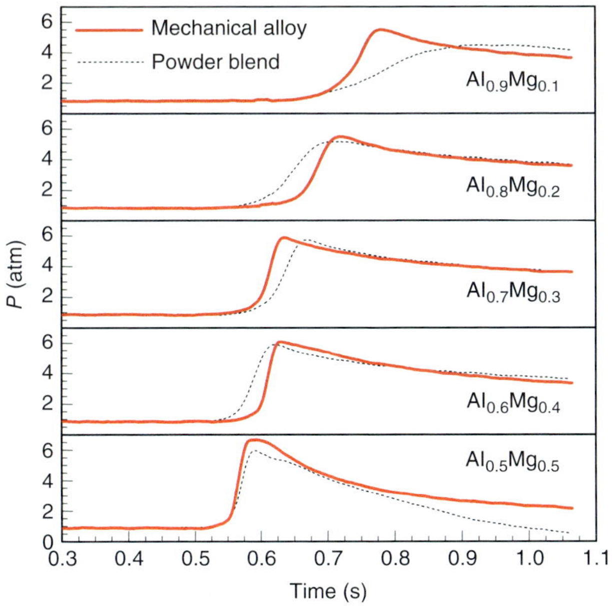

Figure 10.21 Pressure traces recorded in CVE experiments with mechanically alloyed Al-Mg powders burning in air. Reference traces are shown for Al/Mg powder blends with bulk compositions identical to those of alloys [34].

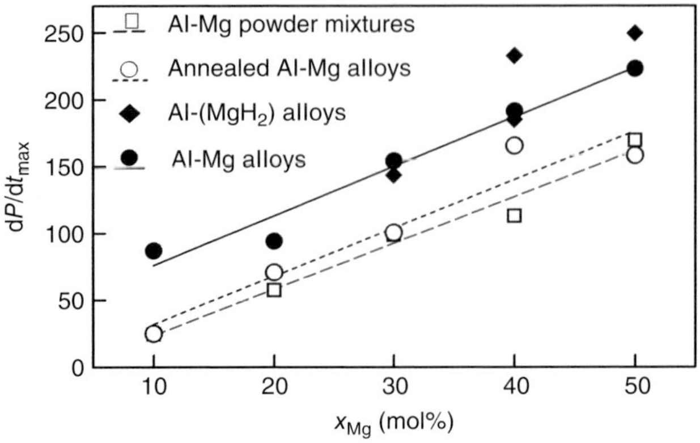

Figure 10.22 Maximum rates of pressure rise in CVE experiments with Al-Mg and Al-MgH_2 powders as a function of concentration of Mg [34].

of Mg (occurring in both blended and mechanically alloyed powders), exothermic subsolidus phase changes in the mechanically alloyed powders contribute to their enhanced ignition. To validate that hypothesis, mechanically alloyed powders were annealed at 663 K (390 °C) in vacuum before combustion experiments (see Figure 10.11 for reference). As shown in Figure 10.22, the rates of pressure rise for such annealed became lower and effectively the same as for the respective powder blends.

10.7.5 Lifted Laminar Flame (LLF) Experiments

Typical images of the LLF (lifted laminar flame) for metal powders are shown in Figure 10.23. The image on the left side is for pure Al aerosol and is taken with 125 μs exposure time, so that individual particle flames can be resolved. The image on the right shows combustion of mechanically alloyed $Al_{0.7}Mg_{0.3}$ powder; it is taken at a longer exposure time (10 ms), so that particle ignition events followed by the individual particle streaks are visible.

In addition to visualization of particle flames, LLF experiments produced two quantitative characteristics of powder combustion. Flame speed was determined once a stable laminar flame with a specific powder was established [31, 33]. The stable flame was achieved by varying the speed of the gas flow carrying the aerosolized powder. The flame speed was assumed to be equal to the absolute value of the velocity of particles entering the flame. The particle velocity was measured using a modulated laser sheet placed on the aerosol jet. An example of a photograph of an aerosol jet illuminated by such a laser sheet is shown in Figure 10.24. The length of the measured particle streaks at the height where the flame is established divided by the duration of the modulating pulse determines the powder velocity and thus the flame speed for the respective aerosol.

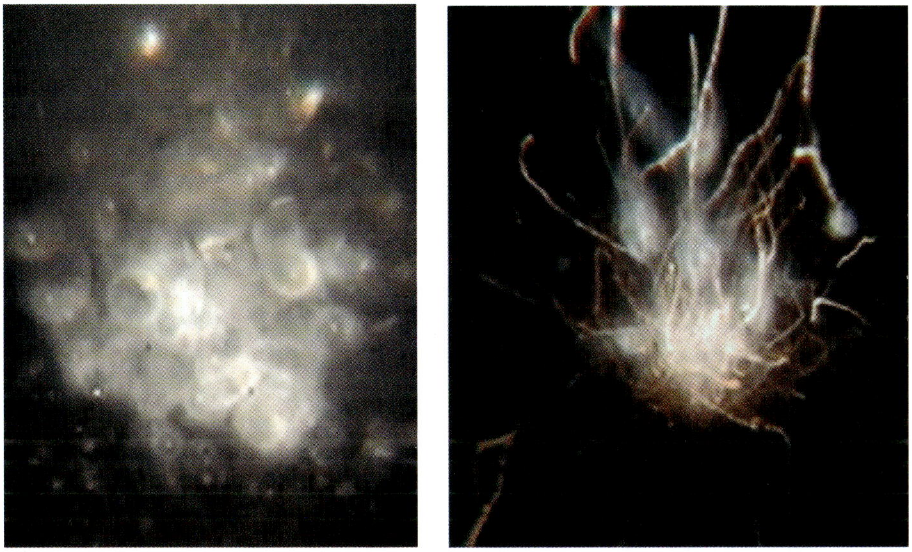

(a) (b)

Figure 10.23 Lifted laminar flames of aerosols of Al (left, exposure time 125 μs) [33] and mechanically alloyed $Al_{0.7}Mg_{0.3}$ powder (right, exposure time 10 ms) [51]. The width of the field of view is 4 mm for each image.

The second measurement gave the average burn rates for individual particles [32, 54]. The burn rates were obtained by matching the experimental and predicted flame radiation profiles, as illustrated in Figure 10.25. The experimental flame radiation profiles were obtained from the respective flame images. The predicted profiles used the experimental particle size distributions and an adjustable expression for the particle combustion time, τ, as a function of its diameter, d, raised to an exponent. Thus, the best match between the measured and calculated radiation profiles was found when both the exponent and respective constant, k (e.g., an evaporation constant for the traditional $\tau = kd^2$ expression for droplet combustion), were selected appropriately.

Results of the flame speed measurements for different mechanically alloyed Al-Mg powders are shown in Figure 10.26 [31]. It is interesting that for compositions with 20 and 30% Mg, the flame speeds are greater than for either pure Al or Mg. Flame speeds for different Al-based mechanically alloyed powders are shown in Figure 10.27 [31]. For all materials, substantial increase in the flame speed is observed compared to the pure Al. Thus, accelerated burn rates are generally anticipated for mechanically alloyed powders.

Burn rate comparisons based on the matching experimental and measured flame radiation profiles for different powders are shown in Figure 10.28. While the d^2 law was not the best choice for all powders, for consistent comparison of burn rates (or respective evaporation coefficients), the d^2 law was applied to reconstruct

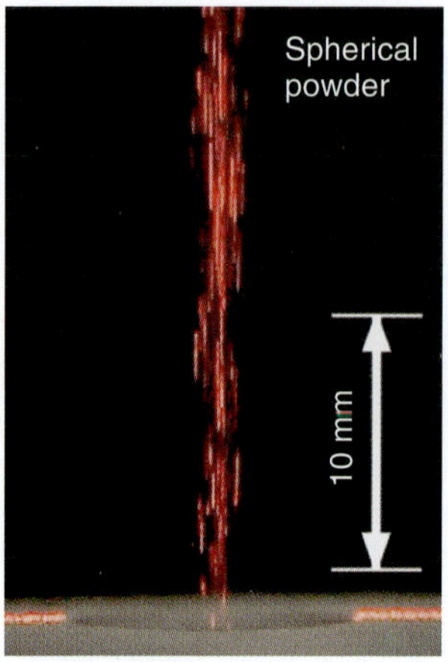

Figure 10.24 An aluminum aerosol jet illuminated by a modulated laser sheet. The powder is moving vertically up and is fed by electrostatic particle generator [33] (cf. Fig. 10.5c).

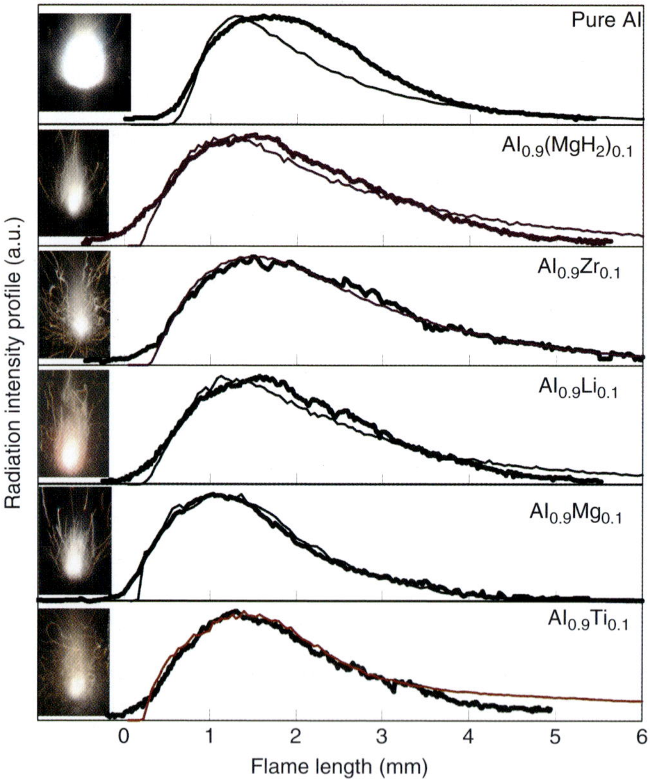

Figure 10.25 Experimental flame radiation profiles (bold lines) and radiation profiles calculated based on the powder particle size distributions (see text) for different mechanically alloyed powders. Flame images for respective alloys are also shown.

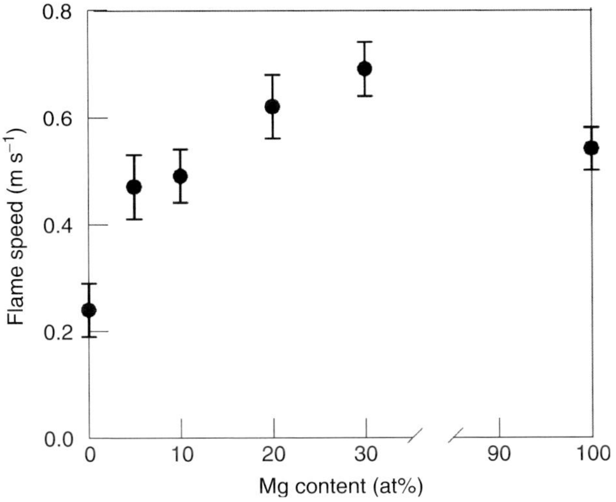

Figure 10.26 Lifted laminar flame speeds for mechanically alloyed Al-Mg powders [31].

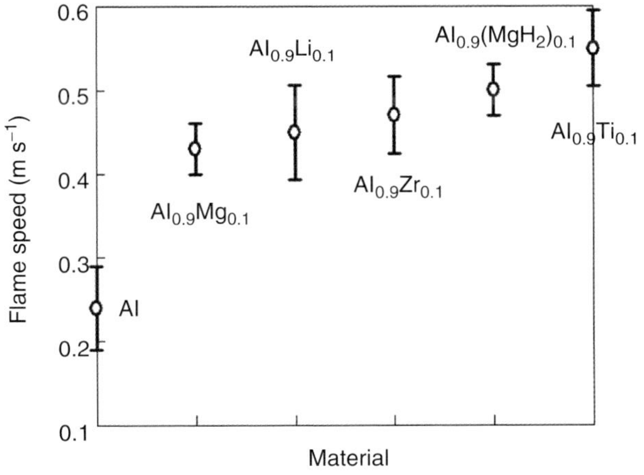

Figure 10.27 Lifted laminar flame speeds for different Al-based mechanically alloyed powders [31].

radiation profiles for all metals while preparing Figure 10.28. Results suggest that mechanically alloyed powders of Al-Mg and Al-Ti have greater burn rates than pure Al and even Mg powders. Thus, if such mechanically alloyed powders are prepared with size distributions matching that of currently used Al, they can replace Al as an additive to energetic formulations and yield greater burn rates. Further details on the burn rate assessments of different mechanically alloyed powders are available in the literature [32, 54].

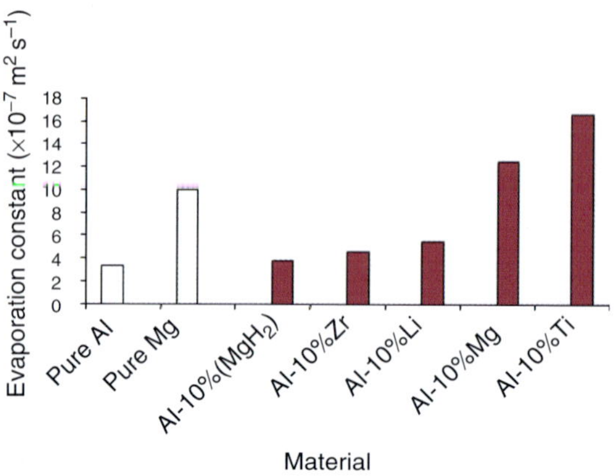

Figure 10.28 Evaporation constants, k, for particle combustion for different materials obtained by matching experimental radiation profile of lifted laminar flame of the respective aerosol with the radiation profile calculated using the corresponding particle size distributions and assuming that particle burn times, τ, are described as $\tau = kd^2$.

10.8 Reactive Nanocomposite Powders

Mechanical milling is suitable to prepare diverse reactive nanocomposite materials that are of interest for energetic applications. Development of such materials is motivated by several reasons:

- They can be used as ignition boosters for other reactive or energetic materials.
- For some applications (e.g., heaters or some pyrotechnics), reactive materials are desired that can initiate and/or react independently of the ambient air or other gaseous oxidizers.
- Adjusting the material composition enables one to tune the reaction rate and customize the material, for example, to generate permanent or condensable gases, reach and maintain specific reaction temperature, and so on.

Components for such composite materials can be metal fuels (e.g., aluminum or zirconium) and oxidizers, for example, Fe_2O_3, CuO, MoO_3, Bi_2O_3, and so on. Alternatively, metals can be combined with other metals (e.g., Al-Ni) or metalloids (e.g., B-Ti or B-Zr). Mechanical milling these components results in progressively more intimate mixtures with highly developed reactive interfaces. For preparation of all such reactive materials, the milling must stop before any exothermic reaction is mechanically initiated, hence their manufacturing method is referred to as *arrested reactive milling* or *ARM* [55].

The defining feature, common for all reactive nanocomposite materials prepared by ARM, is the fine scale on which reactive components are mixed with each other. This intimate mixing, illustrated in Figures 10.1 and 10.29, showing different types of such materials, generates a very well-developed interface between reactive

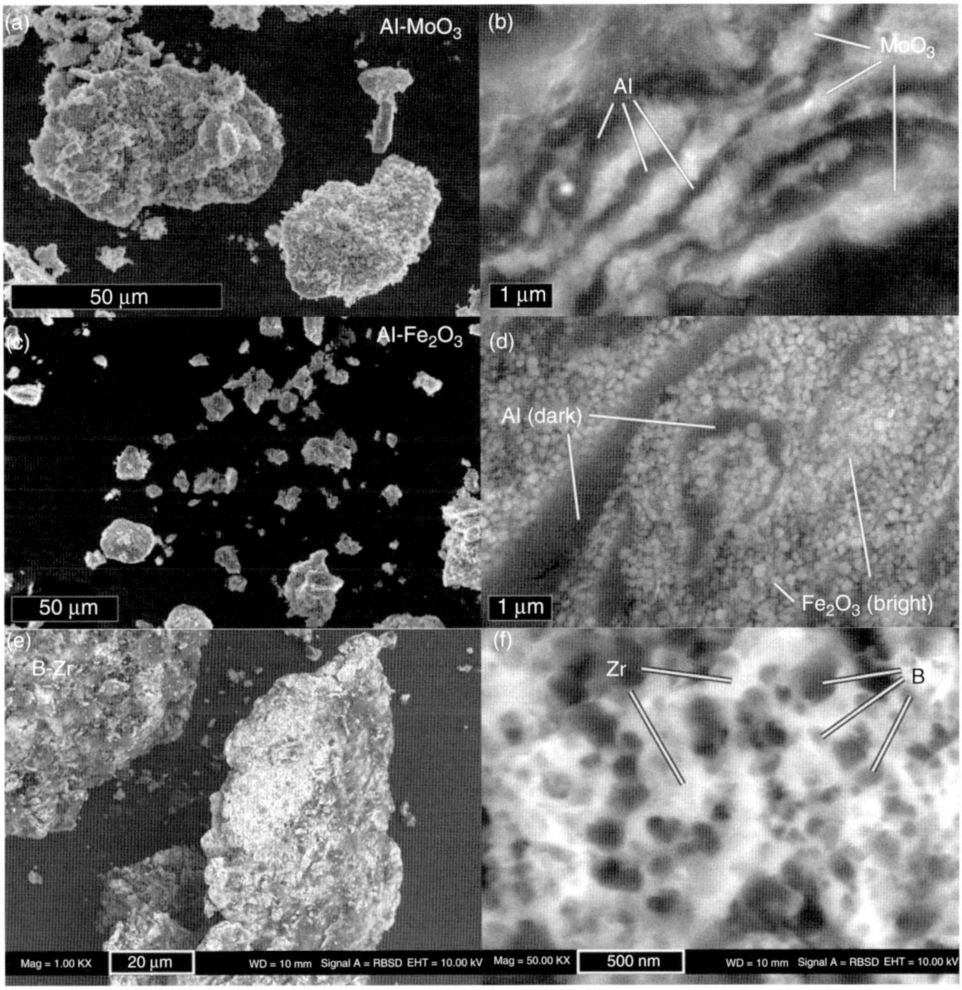

Figure 10.29 SEM images of reactive nano-composite powders and respective particle cross-sections (thermite cross-sections are from Ref. [59]).

components. The interface is formed at low temperatures typical for ball milling, so that the partially reacted intermediate layers between the reactive phases are thinner compared to similar layers in nanocomposite materials prepared at elevated temperatures, for example, by magnetron sputtering [56–58]. Thus, reactions occurring at fairly low temperatures, which are usually quite slow and do not result in any appreciable heat release for conventional, micron-scale materials, are accelerated and become significant for the fully-dense nanocomposite powders. Such reactions result in rapid and nearly volumetric heating and in modification of the material components triggered by various ignition stimuli. Both processes accelerate material ignition.

10.8.1
Preparation and Characterization

Preparation of the composite materials follows the same general procedure as outlined above. Milling conditions for reactive composites are typically "softer" than for mechanically alloyed powders, that is, milling times tend to be shorter, and charge ratios lower. Owing to these conditions, contamination of reactive nanocomposites by the milling media material is very minor.

Liquid PCAs are more frequently used. Because of its relatively low melting and boiling points, liquid PCA also serves as an endothermic heat sink so that if a reaction between the components happens to be triggered locally, in a collision between milling media, the generated heat dissipates quickly, quenching the reaction, and preventing the consumption of the whole powder load.

Composite morphologies shown in Figures 10.1 and 10.28 are typical. When both components are ductile, a layered structure is produced. When ductile and brittle components are mixed, as is the case of Al with Fe_2O_3, inclusions of the brittle component are embedded into the ductile matrix. For the case of B-Zr, B, and Zr represent the brittle and ductile components, respectively. The scale of mixing is reduced and its homogeneity is improved at longer milling times which can be achieved without initiating the self-sustaining reaction mechanically when sufficient amounts of liquid PCA are used. However, partial reaction between components becomes more and more significant, resulting in the formation of individual product particles, which later become mixed in with the rest of the material. An example of such a reacted particle detected in the prepared (and generally unreacted) material is shown in Figure 10.30. It is clear that locally, the temperature was substantially increased resulting in the formation of molten Al_2O_3 and Fe.

Formation of the nanocomposite structure and evolution of particle sizes and shapes at increasingly longer milling times is illustrated in Figure 10.31 for Al-CuO composites prepared using a shaker mill [59]. The phase contrast between Al and CuO-rich phases is well visible in the backscattered electron SEM images. For the starting mixture, (Figure 10.31, milling time 0 min) the bright CuO particles are well distinguished from gray Al and the dark epoxy background. After 2 min of dry milling, the Al particles are deformed, the CuO particles are reduced in size and become embedded in aluminum. Very large, dense agglomerates are formed. Most of the CuO particles are close to the surface of aluminum and located on or between the thick aluminum layers. After 16 min of milling with 1 ml of hexane, the particle sizes of Al and CuO reduce drastically. Some of the CuO particles are embedded in the Al matrix. However, unmixed particles of Al and CuO are still present. Both the particle sizes and the scale on which components are mixed vary widely. An increase in the milling time up to 60 min, was achieved without mechanically triggering the sample when 8 ml of hexane was added. Further reduction of the particle sizes and a more homogeneous mixing were thus achieved. Most of the particles are nanocomposites of CuO inclusions in the aluminum matrix. Shapes of nanocomposite particles vary from fairly *equiaxial* fine particles to thin flakes.

Figure 10.30 Nanocomposite $4Al \cdot Fe_2O_3$ powder prepared by ARM. Gray particles are composites of Al and Fe_2O_3; a dark particle in the center shows product of reaction occurring during milling.

However, as shown by an arrow, inclusions of metallic copper are also found, indicating that a small part of the material has reacted. Because the locally initiated reaction did not propagate through the entire sample, a useful compromise can be found between fine and uniform mixing and the maximum allowable extent of reaction during the ball milling. XRD is routinely used to characterize the compositions of the prepared composite powders and verify that the amounts of undesirable reaction products are small; in most acceptable materials such products are present in the amounts insufficient for detection from the powder XRD patterns.

In addition to the scale of mixing between components, an important parameter affecting behavior of the nanocomposite reactive materials is the particle size distribution of the prepared powder. The particle size distribution depends on the preparation conditions, such as milling duration, PCA used as well as on the bulk composition of the prepared material. For example, using large amounts of ductile Al results in formation of coarser particles as illustrated in Figure 10.32 for $Al\text{-}MoO_3$ composites [60]. Because of such effects, independent adjustment of material composition and particle size distribution is difficult.

10.8.2
Thermally Activated Reactions and their Mechanisms

Reactions in the prepared reactive nanocomposite materials were studied using thermo-analytical techniques (DSC/TGA) and microcalorimetry (using TAM III).

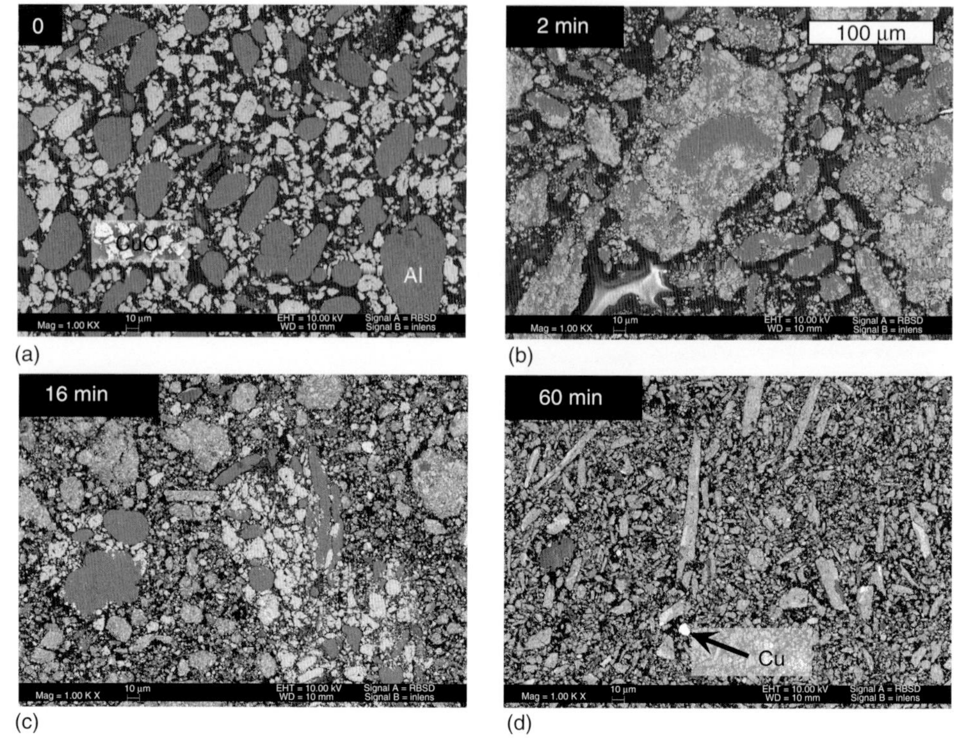

Figure 10.31 Evolution of particle sizes and morphologies for 2Al-3CuO composite prepared by different durations of milling; milling durations are shown in insets [60].

Characteristic DSC traces for several nanocomposite powders are shown in Figure 10.33 (data for Al·MoO_3 and Al·CuO are from [59–61], data for Al·Fe_2O_3 and Al·Bi_2O_3 are previously unpublished). A common feature for all shown DSC traces is the low-temperature onset of a broad exothermic hump beginning below 400 K. Respective exothermic processes, typically undetectable for reactive materials with coarser mixing scale between components, are significant for all nanocomposite materials prepared by ARM. For one of these materials, 2Al·3CuO, the reaction details for such low-temperature processes were studied by combining DSC measurements performed at different heating rates [39, 59] and microcalorimetry measurements using TAM III [62–64]. Results of microcalorimetry experiments are shown in Figure 10.34. Measurements were performed at several fixed temperatures, 303, 323, and 373 K (respectively, 30, 50, and 100 °C). Despite low temperatures, reaction rates, and heat effects were well quantified. The results were described in a multistep reaction model considering growth of and polymorphic phase changes in the Al_2O_3 layer separating Al from CuO [62]. Prior to discussing the reaction model, it is interesting to note that a second common feature among all DSC traces shown in Figure 10.33 is an exothermic peak occurring at about 870–880 K. According to the model discussed below (and developed based

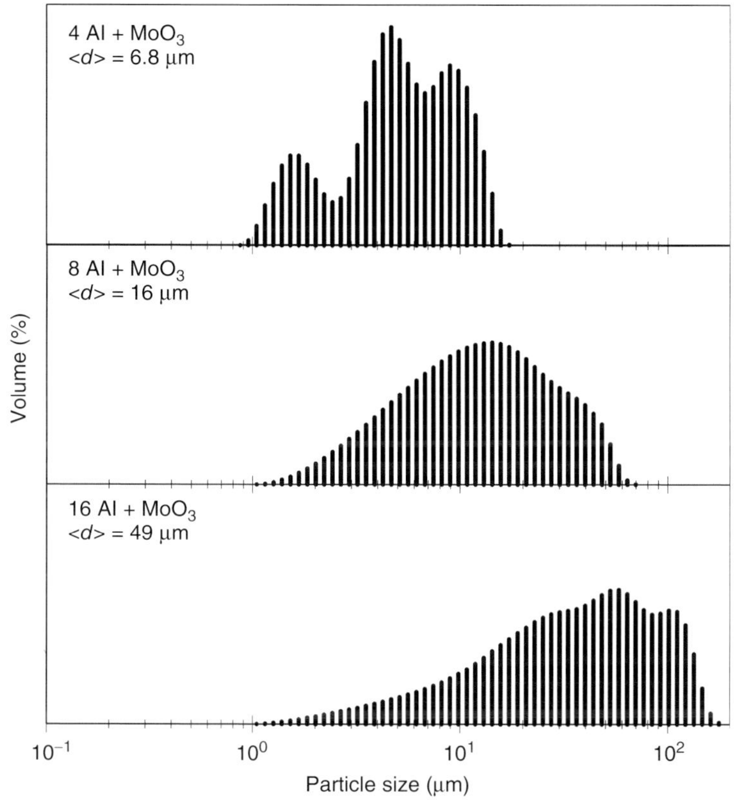

Figure 10.32 Particle size distributions for reactive Al-MoO_3 composites with varying compositions after 30 min of milling [61].

on detailed analysis of DSC and TAM III results for 2Al·3CuO), this exothermic peak is associated with formation of the first crystalline polymorph of alumina, γ-Al_2O_3 on the interface separating aluminum from the respective oxidizer. Thus, according to the model, this phase change and associated acceleration in the oxidation rate are predicted to occur for any oxidizer, which is consistent with the data shown in Figure 10.33.

To interpret the calorimetric measurements, a model for a reactive Al-CuO nanocomposite was developed that described the material as uniformly sized spherical inclusions of CuO in an Al matrix. Initially, CuO is separated from Al by a very thin amorphous aluminum oxide layer. Transport of reacting species through this layer defines the reaction rate; the transport is affected by temperature, growing layer thickness, and its diffusion resistance changing when polymorphic phase changes occur. It is assumed that the bulk reaction can be described as $2Al + 3CuO \rightarrow Al_2O_3 + 3Cu$.

As the reaction progresses, the amorphous aluminum oxide film thickens and, following conventional aluminum oxidation mechanism [65, 66], eventually

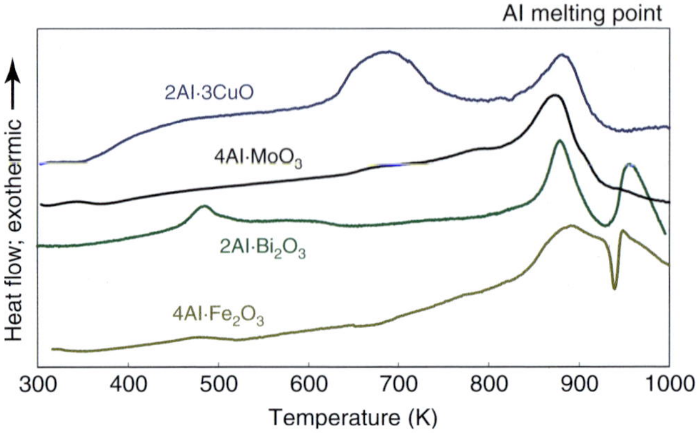

Figure 10.33 DSC traces for several nanocomposite thermite powders heated in Ar. Heating rate is 5 K/min [60–62].

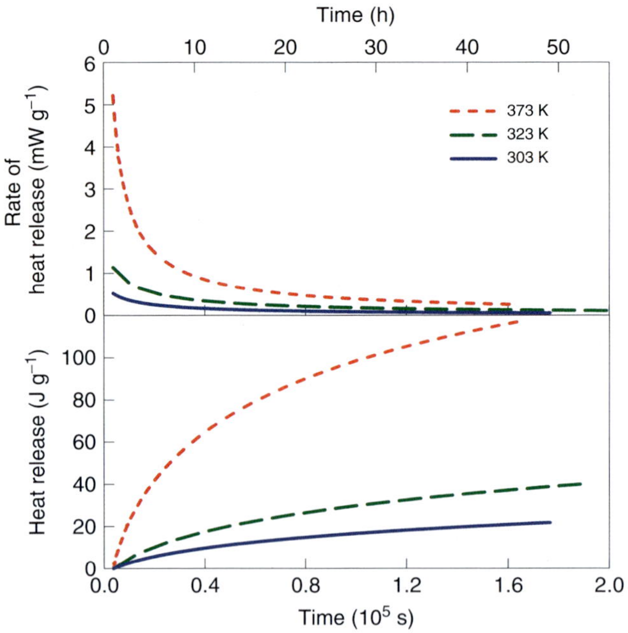

Figure 10.34 Heat release for 2Al•3CuO nanocomposite measured in microcalorimetry experiments [64].

transforms to γ-Al_2O_3. Further phase changes occurring at higher temperatures are neglected considering that the goal of this model is to describe phenomena leading to ignition, occurring at relatively low temperatures.

Several individual reaction steps are included in the model. The initial growth of the amorphous alumina is described using the Cabrera–Mott (CM) mechanism following Ref. [63]:

$$\frac{dm_{am}^{CM}}{dt} = 4\pi r_{am}^2 \rho_{am} k_0 \cdot \exp\left[-\frac{E_1}{RT} + \frac{E_2 \cdot r_{core}}{k_B T \cdot r_{am} \cdot h_{am}}\right] \quad (10.2)$$

where dm_{am}^{CM}/dt is the rate of mass increase of amorphous oxide due to this oxidation mechanism, T is temperature, R and k_B are respectively the universal gas and Boltzmann constants, ρ_{am}, r_{am}, and h_{am} are respectively the density, radius, and thickness of the amorphous oxide film, r_{core} is the radius of CuO inclusion, parameters k_0 and E_1 represent the pre-exponent and activation energy (these and other parameters used in the model are summarized in Table 10.1). In the CM mechanism, the mass transport is accelerated by an electric field produced by ions accumulated at the boundaries of the growing amorphous oxide layer; this effect is described by the parameter E_2. The effect of the electric field also depends on the thickness of the oxide layer h_{am}, and core (CuO inclusion) and shell (amorphous alumina) radii r_{core} and r_{am}, respectively. While the reaction continues, the thickness of the amorphous oxide layer, h_{am}, is increasing, so that the rate of reaction governed by the CM kinetics decreases. However, when the polymorphic phase transformation from amorphous to a higher density γ-Al_2O_3

Table 10.1 List of parameters used in the reaction kinetics model [62].

Parameter	Description	Value
E_1	Activation energy used in Cabrera–Mott equation	44.5 kJ mol^{-1}
E_{am}	Activation energy for diffusion-limited growth of amorphous oxide	105 kJ mol^{-1}
$E_{am\to\gamma}$	Activation energy for amorphous to γ-Al_2O_3 transformation	458 kJ mol^{-1}
E_γ	Activation energy for diffusion-limited growth of γ-Al_2O_3	210 kJ mol^{-1}
C_γ	Pre-exponent of γ-Al_2O_3 oxidation by diffusion	4.0784×10^{-3} kg m^{-1} s^{-1}
$F_{am\to\gamma}$	Pre-exponent for amorphous to γ-Al_2O_3 transformation equation	2×10^{-15} m s^{-1} K^{-1}
$K_{am\to\gamma}$	Exponential coefficient in amorphous to γ-Al_2O_3 transformation equation	1×10^{12} J mol^{-1} m^{-1}
c_{am}	Coefficient for reaction progress for growth of amorphous Al_2O_3	5
c_γ	Coefficient for reaction progress for growth of γ-Al_2O_3	8
b	Coefficient used in linear function of E_2	5.52×10^{-22} J nm K^{-1}
C_{am}	Pre-exponent for amorphous oxide growth by diffusion	5.098×10^{-8} kg ms^{-1}
h_0	Initial thickness of amorphous oxide layer	0.90 nm
h_γ^{min}	Minimum thickness of γ-Al_2O_3 producing diffusion resistance	3.47 nm

begins at higher temperatures, the thickness of the amorphous oxide may decrease causing respective acceleration of the CM reaction.

In parallel to the CM kinetics-controlled oxidation, the amorphous alumina layer can grow due to regular, thermally activated diffusion described as [67]:

$$\frac{dm_{am}^{Diff}}{dt} = f_{am}(\alpha_{am})\cdot C_{am} \exp\left(-\frac{E_{am}}{RT}\right)\left(\frac{1}{r_{core}} + \frac{1}{r_m}\right)^{-1} \tag{10.3}$$

where $f_{am}(\alpha_{am})$ is a progress function. The reaction progress α_{am} for the growth of amorphous oxide is defined based on the consumption of CuO:

$$\alpha_{am} = c_{am}\frac{\Delta m_{am}^{cum}}{m_{CuO}^{ini}} \tag{10.4}$$

where c_{am} is a constant, Δm_{CuO}^{cum} is the cumulative mass of CuO consumed in reaction, and m_{CuO}^{ini} is the initial mass of available CuO. The progress function $f_{am}(\alpha_{am})$ is described by an Avrami–Erofeev mechanism [68]:

$$f_{am}(\alpha_{am}) = 2(1 - \alpha_{am})[-\ln(1 - \alpha_{am})]^{1/2} \tag{10.5}$$

The progress function accounts for continuing consumption of CuO, resulting in a decreased reaction rate when substantial conversion from CuO to Cu occurs.

As the temperature increases and the thickness of the amorphous oxide grows, amorphous alumina begins to transform to γ-Al_2O_3. This phase change and the ensuing diffusion-controlled oxidative growth of γ-Al_2O_3 are described following an oxidation model for aluminum in oxygen [65–67]. The rate of the phase transformation is given by [65]:

$$\frac{dm_{am\to\gamma}^{tr}}{dt} = 4\pi r_{am}^2 F_{am\to\gamma} T\left[1 - \exp\left(-\frac{K_{am\to\gamma} h_{am}}{RT}\right)\right]\exp\left(-\frac{E_{am\to\gamma}}{RT}\right) \tag{10.6}$$

After alumina is converted to its γ-polymorph and its thickness reaches h_γ^{min}, the oxidation is described using the following Arrhenius-type expression:

$$\frac{dm_\gamma^{Diff}}{dt} = f_\gamma(\alpha_\gamma)C_\gamma \exp\left(-\frac{E_\gamma}{RT}\right)\left(\frac{1}{r_{core} + h_{am}} - \frac{1}{r_\gamma}\right)^{-1} \tag{10.7}$$

Similarly to Equations 10.3 and 10.7 contains a progress function $f_\gamma(\alpha_\gamma)$ describing the reaction progress for the oxidative growth of γ-Al_2O_3. As for the amorphous oxide, the reaction progress of the formation of γ-Al_2O_3 is defined based on consumption of CuO:

$$\alpha_\gamma = c_\gamma\frac{\Delta m_{CuO}^{cum} - \Delta m_{CuO}^{*}}{m_{CuO}^{ini}} \tag{10.8}$$

where c_γ is a constant, and Δm_{CuO}^{*} is the consumption of CuO at the moment when the oxidative growth of γ-Al_2O_3 begins. The reaction progress function in this case was described as a first order reaction:

$$f_r(\alpha_\gamma) = 1 - \alpha_\gamma \tag{10.9}$$

While more than one alumina polymorph can grow at the same time, in this model only one of the polymorphs, which has the greatest diffusion resistance,

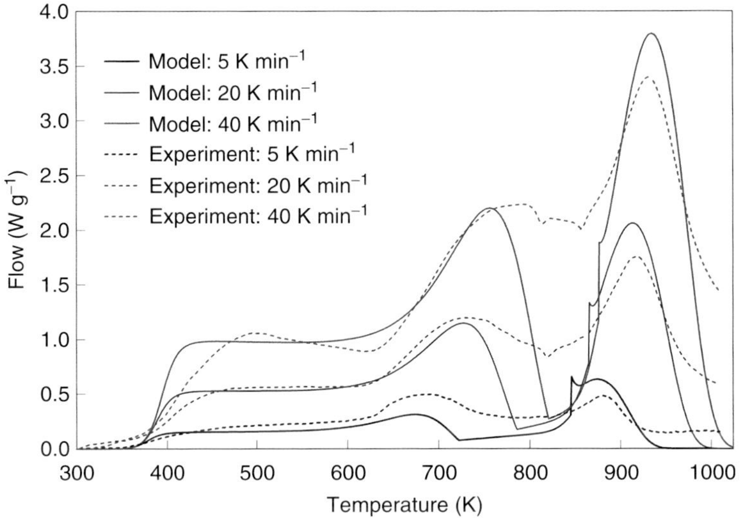

Figure 10.35 Measured and calculated DSC traces for a 2Al•3CuO reactive nanocomposite [63].

is allowed to grow due to oxidation at any given moment. Growth of an oxide polymorph as a result of phase change, for example, amorphous to γ-Al_2O_3, can occur simultaneously with the oxidative growth.

The correlation between experimental DSC traces recorded at different heating rates and predictions of the present model is illustrated in Figure 10.35. All main features observed in DSC signals are well represented in the model. Most importantly, the model is capable of predicting effect of heating rate on the peak shift observed in DSC measurements. Thus, the model is expected to be useful for prediction of behavior of reactive nanocomposite materials under different conditions, including very high heating rates typical of practical combustion configurations.

10.8.3 Ignition

Most ignition experiments used an electrically heated filament [16, 17, 39, 40, 59, 60, 69, 70], while initial experiments using individual nanocomposite particles ignited in the CO_2 laser beam have also been performed and interpreted [62]. In the heated filament tests, a very sharp emission spike unambiguously identifies ignition for reactive nanocomposite materials, unlike for mechanically alloyed powders discussed above. Examples of the photodiode traces recorded at different heating rates in experiments with 2Al·3CuO nanocomposite powders [59] are shown in Figure 10.36. Results are shown for two materials prepared using different milling times. The sample which was milled longer (60 min) ignites at

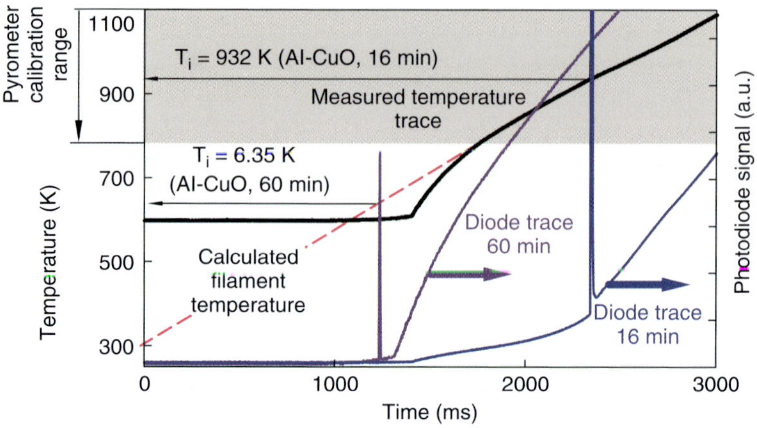

Figure 10.36 Photodiode and temperature traces recorded in heated filament ignition experiments with 2Al•3CuO nanocomposite powders [60].

a significantly lower temperature compared to the sample milled for only 16 min. This observation highlights the strong effect of refinement and scale of mixing achieved in a nanocomposite material on its measured ignition temperature. The same effect of reduced ignition temperature for samples with a better degree of compositional refinement was reported in Ref. [17] when the refinement was improved by increasing the amount of liquid PCA at otherwise similar milling conditions.

The effect of heating rate on ignition temperature can be readily identified in the heated filament ignition tests. However, in the range of heating rates of about 200–20 000 $K\,s^{-1}$ easily achieved experimentally, the changes observed in the measured ignition temperatures were relatively weak in most experiments with reactive nanocomposites. For example, results for Al·CuO and Al·MoO_3 nanocomposite thermites heated at different rates are shown in Figure 10.37. The data for Al·CuO combine experiments for composites with different metal rich compositions varied from 8Al·3CuO to 12Al·3CuO prepared with varied milling conditions [39] and include ignition tests in air and in argon. The data for 8Al·MoO_3 show results of ignition tests in air [69]. In an earlier study with Al·MoO_3 materials with varied Al/MoO_3 ratios, a slight, but reproducible decrease in the ignition temperature was observed when heating rates increased from about 4000 to 6000 $K\,s^{-1}$ [60]. Further increase in the heating rate resulted in a small increase in the ignition temperature. The ignition temperatures for all cases vary in relatively narrow ranges; it also appears from Figure 10.37 that the environment may have a more significant effect on the ignition temperature for Al·CuO than changes in the heating rate.

It is interesting to compare the overall range of ignition temperatures characteristic for different nanocomposite materials prepared by ARM to those of pure metals commonly used in energetic formulations, for example, aluminum and magnesium. This comparison is presented in Figure 10.38. Ignition temperatures

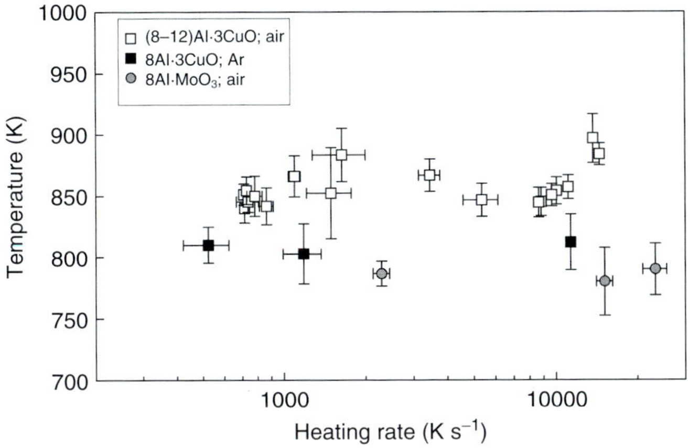

Figure 10.37 Ignition temperatures measured at different heating rates and in different environments for Al•CuO and Al•MoO_3 nanocomposite thermites [39, 70].

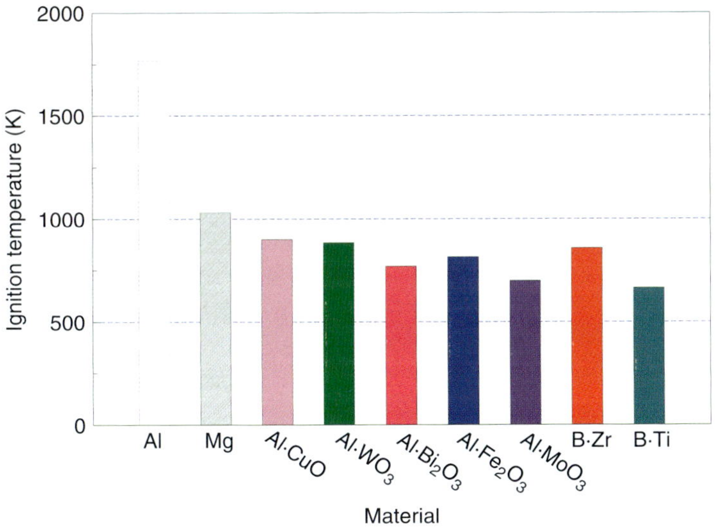

Figure 10.38 Ignition temperatures measured at a heating rate close to 2000 K/s for different nanocomposite materials. The data are compiled from different studies [16, 17, 39, 40, 60, 61, 72].

for different nanocomposite reactive materials are shown for the samples with reactive components mixed in stoichiometric proportions; all experimental data correspond to the heating rate of approximately 2000 K s^{-1}. The scale of mixing for all tested materials is close to 100 nm (i.e., the materials can be roughly approximated by a matrix of a more ductile component with 100 nm inclusions of a more brittle component).

Ignition temperatures for all nanocomposite powders are in the range of 650–900 K, whereas in the same experiment, Mg powder ignites when heated just above 1000 K and Al powder ignites when the temperature exceeds 1700 K.

When the heated filament ignition experiment for Al·CuO composites is described using the above reaction model [62], it becomes apparent that the temperature at which ignition is observed coincides with the transition from amorphous to γ-Al_2O_3. However, the model does not directly predict the temperature runaway, expected to occur upon ignition because it does not describe the release of gaseous oxygen accompanying ignition of nanocomposite thermites, similar to that described in Ref. [71]. The oxygen release causes dramatic changes in thermal conductivity of the powder layer coating and could thermally insulate the filament from the exothermically reacting powder thus enabling powder heating to occur faster than that of the filament. The effect of oxygen release on ignition of nanocomposite thermites is currently under investigation.

Experiments on ignition of individual nanocomposite Al·CuO particles heated in the CO_2 laser beam presented in Ref. [62] involve heating rates of the order of 10^6 K s^{-1}. In these experiments, laser power is increased in small increments until the particles passing the laser beam are observed to ignite. The minimum ignition energy is thus determined. This minimum energy can also be predicted modeling particle heating in a laser beam [72], while including the exothermic reactions occurring in nanocomposite thermites [62]. The calculated temperature history of an 8Al·3CuO particle heated in the laser beam with the experimental threshold ignition energy of 5.3 W is shown in Figure 10.39. The aluminum in the particle is observed to melt and the temperature runaway occurs soon after the melting is complete. The code continued to calculate the temperature history following the temperature runaway, shown as a dashed line in Figure 10.39; however, that part of the temperature history must be corrected considering combustion reactions that are not included in the present model. The temperature runaway starts to occur in the model at the laser power of 4.3 W, which is reasonably close to the experimental value of 5.3 W.

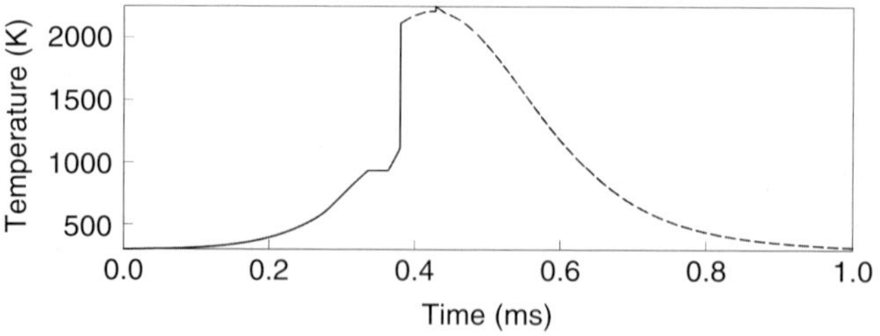

Figure 10.39 Calculated temperature history of an 8Al•3CuO particle heated in a laser beam [63].

10.8.4 Particle Combustion Dynamics

Experimental results on combustion dynamics of individual particles of Al·CuO and Al·MoO_3 nanocomposites prepared by ARM and ignited by a CO_2 laser beam have been reported recently [73]. The particles are passing through the laser beam where they ignite, and their combustion is monitored based on optical signatures. Characteristic emission traces for the two materials are shown in Figure 10.40, where reference trace for pure Al is also shown. For an Al·MoO_3 particle, the emission intensity increases slowly and small brightness oscillations are observed during this emission intensity ramp. After the emission peak, it rapidly decays; however, the particle remains luminous for an extended period of time. A magnified portion of the emission trace for Al·MoO_3 shows that the intensity changes noticeably, with the last sharp increase in emission most likely correlating with spearpoints observed in the particle streaks.

Nanocomposite Al·CuO particles ignited in the CO_2 laser exploded immediately upon ignition, which was most likely caused by the gaseous oxygen released during rapid decomposition of CuO. The emission trace shown in Figure 10.40 begins with a very sharp intensity increase, most likely associated with the explosion of the ignited particle. The following trace is most likely dominated by the emission from the largest produced fragment. A relatively slow decay in intensity does not directly lead to extinction; similar to Al·MoO_3, the low level luminosity is preserved for a while and an increase in emission is observed before the particle extinguishes.

A summary of the particle burn times as a function of their diameters is shown in Figure 10.41. For comparison, burn times for spherical Al powder are also shown. Details of data processing are discussed elsewhere [28–30, 74]. Despite large scatter among experimental points, it is clear that the burn times of the nanocomposite thermites are somewhat longer than those for Al particles of the same dimensions.

Average combustion temperatures measured for individual burning particles are shown in Figure 10.42. For Al·MoO_3, combustion temperatures are very similar to those of pure Al particles. For Al·CuO, combustion temperatures are somewhat lower.

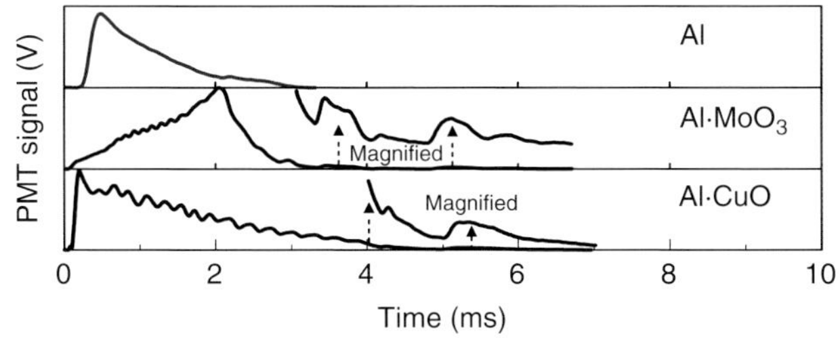

Figure 10.40 Optical emission produced by different material particles ignited in a CO_2 laser beam [75].

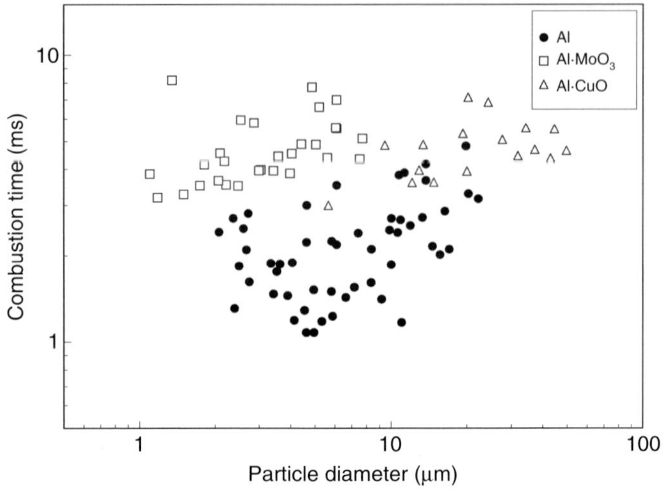

Figure 10.41 Combustion times measured for nanocomposite particles ignited in the CO_2 laser beam. Similar combustion times for spherical Al particles are also shown [75].

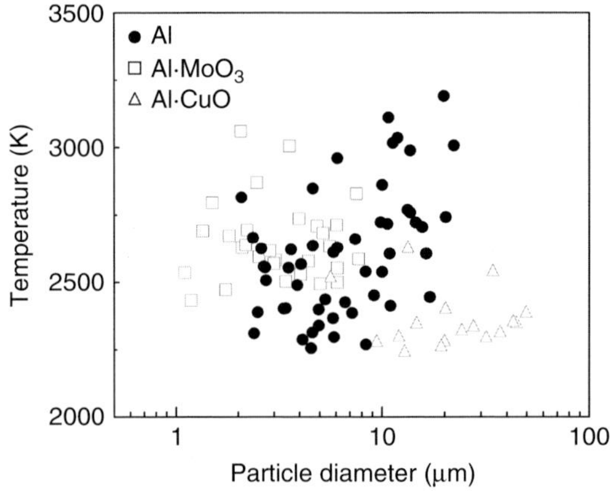

Figure 10.42 Average combustion temperatures measured for nanocomposite particles ignited in the CO_2 laser beam. Combustion temperatures for spherical Al particles are also shown [75].

10.8.5
Constant Volume Explosion

CVE experiments are used to determine maximum pressure and the rate of pressure rise caused by combustion of aerosolized powders. The maximum pressure is a useful gauge of the overall energy release whereas the rate of pressure rise is proportional to the flame propagation rate. Effects of ignition and combustion processes, addressed individually in the previously described experiments, are combined in affecting the pressure signatures recorded in CVE experiments.

Thus, CVE tests are more useful for evaluation of the overall potential of specific reactive materials than for detailed understanding of their ignition and combustion mechanisms.

Experiments comparing CVE pressure traces for reactive nanocomposite materials and for spherical Al powders, used as convenient reference materials are described in Refs [39, 40, 60, 75].

Effect of material composition on the measured CVE pressure traces is illustrated in Figure 10.43 [60] for a set of Al·MoO_3 nanocomposite thermites with different Al to MoO_3 ratios. It is observed that the best performing material has the composition of 8Al·MoO_3, which yields the strongest pressure pulse, effectively equal to that produced by pure Al powder, but a significantly greater rate of pressure rise. CVE experiments were also used to select which of the three nanocomposite powders, 8Al·MoO_3, 8Al·3CuO, and 2B·Ti, can serve as effective burn rate modifiers when used as an additive to pure Al [40]. The experiments were performed in gas mixtures containing 22% of O_2, 1.5–4.5% of CH_4, and the balance of N_2. It was observed that for all nanocomposite materials, the burn rates were improved while additives of 8Al·MoO_3 and 2B·Ti to micrometer-sized aluminum were found to be effective in increasing both the rate of pressure rise and maximum pressure. The best combination of the achieved burn rate and pressure was observed for 20 wt% of additive. Additives of nanocomposite 8Al·3CuO powder did not show appreciable improvements in aluminum combustion, most likely due to formation of an oxidation resistant $CuAl_2$ alloy detected in the combustion products. Formation of Al-Cu alloys was also found to be detrimental in reducing CVE combustion performance of various Al·CuO nanocomposite powders in Ref. [39].

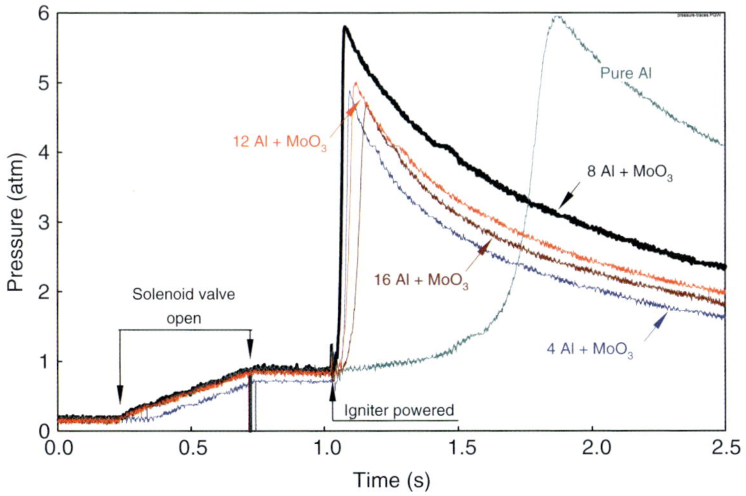

Figure 10.43 Pressure traces of Al and Al•MoO_3 nanocomposites with varying Al concentration recorded during CVE in a gas mixture comprising 22.5% O_2 and 77.5% N_2 [61].

Detailed comparisons of combustion of pure Al and 2B·Ti nanocomposite powder in CVE experiments performed in $N_2/O_2/CH_4$ gas mixtures are presented in Ref. [75]. Characteristic pressure traces are shown in Figure 10.44. The nanocomposite burns faster than Al, while the blended B and Ti powders with the same composition as the nanocomposite material burn slower.

Consistently with thermodynamic equilibrium calculations, experiments showed that the flame temperatures inferred from the measured pressures are higher for Al compared to both nanocomposite and blended powders of 2B·Ti. The maximum pressures for all three materials are shown as a function of gas composition in Figure 10.45. The combustion temperature for Al estimated from the pressures shown in Figure 10.45 and accounting for differences in the gas composition remained nearly constant at about 2560 K. The combustion temperature for the nanocomposite 2B·Ti increased from about 2180 to 2370 K as the methane concentration increased from 0 to 12%. At high methane concentration, the combustion temperatures obtained for nanocomposite and blended 2B·Ti powders were nearly the same.

The bulk burn rates inferred from the rates of pressure rise were consistently higher for the nanocomposite 2B·Ti powder, followed by Al and then by the blended 2B·Ti powder, as also shown in Figure 10.45. The difference in the burn rates became comparable to the scatter between experimental data points (quantified by error bars in Figure 10.45) when the concentration of methane increased to 12%.

These results were used to assess the efficiency of combustion for all the fuels by comparing predicted and experimental portions of combustion energy per unit mass of metal fuel used to produce the heated gaseous products. According to this assessment, nanocomposite boron-based fuels outperformed Al for all environments, with the difference increasing at higher methane concentrations.

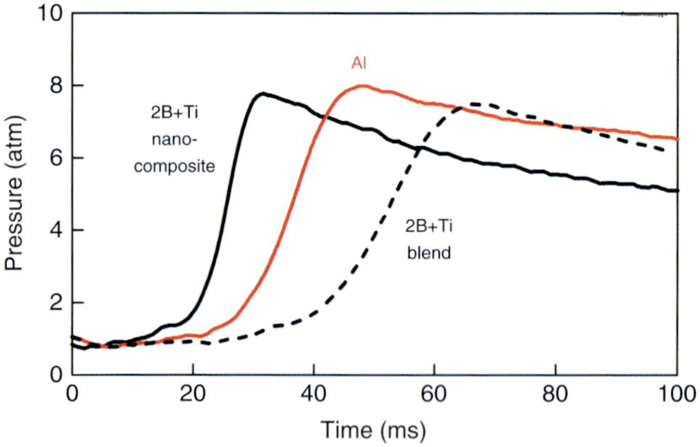

Figure 10.44 Examples of pressure traces recorded in constant pressure experiments with different fuels. All traces shown are for a 0.225 O_2 + 0.715 N_2 + 0.06 CH_4 gas mixture [77]. The fuel mass loads are selected based on the thermodynamic calculations to yield the highest flame temperature: Al: 2.64 g; 2B•Ti: 1.68 g.

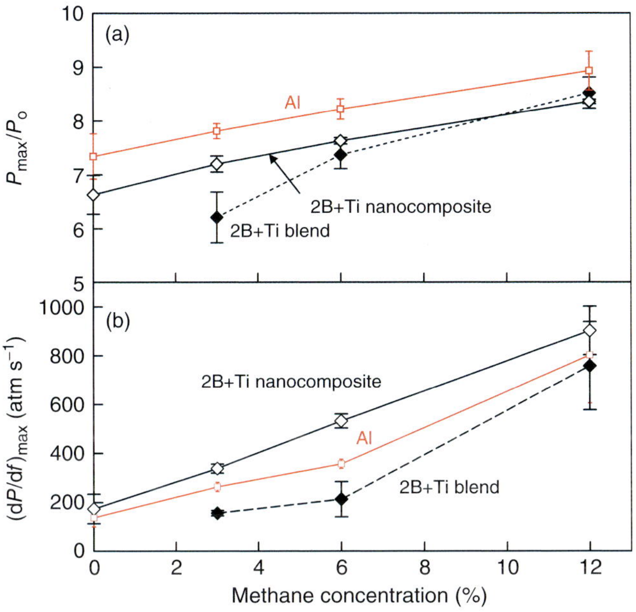

Figure 10.45 Maximum pressures and maximum rates of pressure rise measured in experiments with nanocomposite and blended 2B•Ti powders and reference spherical Al powder burning in different gas environments [77].

Nearly complete combustion was observed for both 2B·Ti fuels (nanocomposite and blended powders) at high methane concentration, when the highest rates of combustion were also observed.

10.8.6 Consolidated Samples: Mechanical and Reactive Properties

In some applications reactive structural components are desired, which could combine sound mechanical properties, such as strength and density, with ability to release chemical energy upon thermal or impact initiation. Several materials prepared by ARM were pressed to form bulk consolidated samples; mechanical and reactive properties of these samples were investigated and reported in Refs [76, 77].

Tensile fracture strength was used as the primary indicator of mechanical strength of the prepared samples. It was measured using the diametrical compression test designed for brittle materials [78]. Results of the tensile strength measurements for different prepared samples are shown in Figure 10.46. The tensile strength is plotted as a function of relative density (fraction of theoretical maximum density, TMD) for each material [76]. For comparison, experiments with consolidated pure Al powder and unmilled powder blends with the same

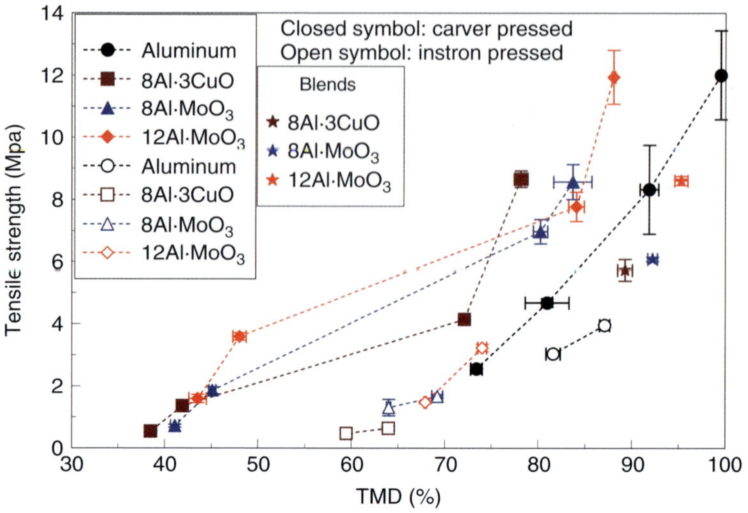

Figure 10.46 Tensile strength as a function of density for several materials and two rates of compaction [78].

bulk compositions as the ARM-prepared nanocomposite powders are shown. The results are shown for pellets of different materials consolidated using a Carver auto-series 25 ton press and an Instron 5567 universal testing machine. For the Carver press, the compression rate was kept constant using the built-in "medium" setting. The dwell time was set at 60 s. Compaction pressures used were in the range of 105–500 MPa. For the Instron device, a slower compression rate of 0.5 mm min^{-1} was used. Different relative densities were achieved applying different pressures; in the Instron, the applied pressure was limited to 176 MPa whereas in the Carver press, the maximum pressure was 496 MPa. It is observed that the relative density achieved with a slower compaction is increased (especially, accounting for pressure difference achieved for different compacted samples). Somewhat higher relative densities were achieved for the blended powders, in which aluminum was not work hardened, compared to the ARM-prepared nanocomposites. However, the tensile strength of the consolidated samples was similar for all materials.

Reactivity of the consolidated materials was tested using thermo-analytical measurements [76] and laser ignition tests [77]. DSC traces measured for consolidated and as-prepared powders were identical, suggesting that consolidation did not alter the reactivity of the nanocomposite material [76]. For the laser ignition tests, cylindrical pellets with diameters of 0.635 cm (1/4″) and thicknesses of 0.177–0.254 cm were prepared. The pellets were heated with a defocused CO_2 laser beam with the power varied in the range of 25–90 W. The pellets ignited, as illustrated in Figure 10.47, showing the temperature profile measured at the back surface of the pellet, whose front surface was laser heated. The time period while the laser remained on is also illustrated in Figure 10.47. The temperature ramp preceding

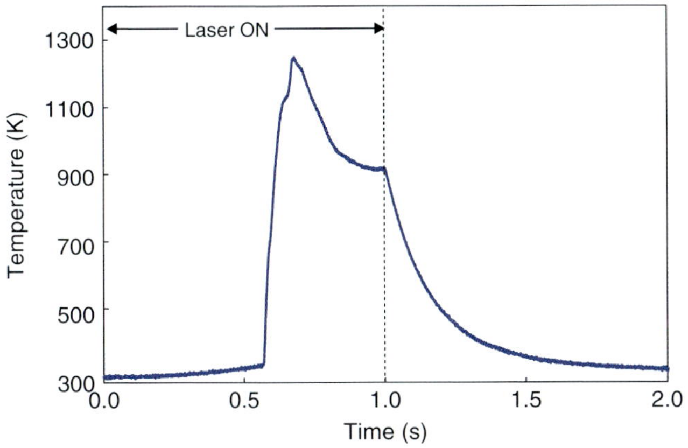

Figure 10.47 Temperature rise measured at the back of a 8Al•MoO_3 pellet with 98.7% TMD heated by a CO_2 laser beam [79].

the sharp increase at about 0.55 s is well described by a heat transfer model considering heat input from the laser beam propagating through the consolidated pellet due to its thermal conductivity. The ignited pellets disintegrated, so that the temperatures measured after ignition cannot be directly interpreted as pellet combustion temperatures. Ignition delays observed in experiments with different laser powers were successfully interpreted describing an exothermic reaction leading to ignition by a simple Arrhenius function [77].

10.9 Conclusions

Multiple reactive mechanically alloyed and nanocomposite materials were prepared by mechanical milling and characterized. It is observed that the nanocomposite materials ignite much more readily than pure Al powders commonly used as high energy density additives in energetic formulations. However, the individual particle burn rates and combustion temperatures of pure Al powders could not be exceeded. In cloud combustion, where ignition and combustion effects are combined, nanocomposite reactive materials can perform better than pure Al powders with comparable particle sizes; the improvement in performance depends on the oxidizing environment. Prepared nanocomposite powders can be consolidated, similar to powders of elemental metals. Powders pressed without heating retain their reactivity and can be used in reactive structures. The mechanisms of thermal initiation of the prepared nanocomposite materials include multi-step reactions involving growth and modification of the product layers separating reagents. In addition, in thermite materials decomposition of the oxidizer and stability of different intermediate suboxide phases should be accounted for.

References

1. Suryanarayana, C. and Koch, C.C. (2000) Nanocrystalline materials – current research and future directions. *Hyperfine Interact.*, **130** (1–4), 5–44.
2. Weeber, A.W. and Bakker, H. (1988) Amorphization by ball milling. A review. *Phys. B: Phys. Condens. Matter*, **153** (1–3), 93–135.
3. Suryanarayana, C. (2008) Recent developments in mechanical alloying. *Rev. Adv. Mater. Sci.*, **18** (3), 203–211.
4. Suryanarayana, C., Ivanov, E., and Boldyrev, V.V. (2001) The science and technology of mechanical alloying. *Mater. Sci. Eng. A*, **304–306** (1–2), 151–158.
5. Suryanarayana, C. (2001) Mechanical alloying and milling. *Prog. Mater. Sci.*, **46** (1–2), 1–184.
6. Dreizin, E.L. (2009) Metal-based reactive nanomaterials. *Prog. Energy Combust. Sci.*, **35** (2), 141–167.
7. Mukasyan, A.S. *et al.* (2011) Mechanical activation and gasless explosion: nanostructural aspects. *Chem. Eng. J.*, **174** (2–3), 677–686.
8. Rogachev, A.S. and Mukasyan, A.S. (2010) Combustion of heterogeneous nanostructural systems (review). *Combust. Explos. Shock Waves*, **46** (3), 243–266.
9. Korchagin, M.A. and Bokhonov, B.B. (2010) Combustion of mechanically activated 3Ti + 2BN mixtures. *Combust. Explos. Shock Waves*, **46** (2), 170–177.
10. Prentice, D., Pantoya, M.L., and Gash, A.E. (2006) Combustion wave speeds of sol-gel-synthesized tungsten trioxide and nano-aluminum: the effect of impurities on flame propagation. *Energy Fuels*, **20** (6), 2370–2376.
11. Bockmon, B.S. *et al.* (2005) Combustion velocities and propagation mechanisms of metastable interstitial composites. *J. Appl. Phys.*, **98** (6), 1–7.
12. Gash, A.E. *et al.* (2003) Nanostructured energetic materials with Sol-Gel methods. Materials Research Society Fall 2003 Meeting, Boston, MA.
13. Clapsaddle, B.J. *et al.* (2003) Synthesis and characterization of mixed metal oxide nanocomposite energetic materials. Materials Research Society Fall Meeting 2003, Boston, MA.
14. Umbrajkar, S.M. *et al.* (2005) Effect of temperature on synthesis and properties of aluminum-magnesium mechanical alloys. *J. Alloys Compd.*, **402** (1–2), 70–77.
15. Murty, B.S. and Ranganathan, S. (1998) Novel materials synthesis by mechanical alloying/milling. *Int. Mater. Rev.*, **43** (3), 101–141.
16. Badiola, C. *et al.* (2009) Nanocomposite thermite powders prepared by cryomilling. *J. Alloys Compd.*, **488** (1), 386–391.
17. Umbrajkar, S.M., Schoenitz, M., and Dreizin, E.L. (2006) Control of structural refinement and composition in Al-MoO_3 nanocomposites prepared by arrested reactive milling. *Propellants Explos. Pyrotech.*, **31** (5), 382–389.
18. Santhanam, P.R. and Dreizin, E.L. (2012) Predicting conditions for scaled-up manufacturing of materials prepared by ball milling. *Powder Technol.*, **201**, 401–411.
19. Ward, T.S. *et al.* (2005) A study of mechanical alloying processes using reactive milling and discrete element modeling. *Acta Mater.*, **53** (10), 2909–2918.
20. Ward, T.S., Chen, W., Schoenitz, M., Dreizin, E.L., and Dave, R. (2005) Nanocomposite energetic powders prepared by arrested reactive milling. 43rd: Meeting And Exhibit, Reno, NV.
21. Jiang, X. *et al.* (2009) Mechanical alloying and reactive milling in a high energy planetary mill. *J. Alloys Compd.*, **478** (1–2), 246–251.
22. Santhanam, P.R. and Dreizin, E.L. (2012) Real time indicators of material refinement in an attritor mill. *AIChE J.*, **59**, 1088–1095.
23. Schoenitz, M., Chen, C.M., and Dreizin, E.L. (2009) Oxidation of aluminum particles in the presence of water. *J. Phys. Chem. B*, **113** (15), 5136–5140.
24. Zhu, X., Schoenitz, M., and Dreizin, E.L. (2009) Aluminum powder oxidation

in CO_2 and mixed CO_2/O_2 environments. *J. Phys. Chem. C*, **113** (16), 6768–6773.

25. Zhu, X., Schoenitz, M., and Dreizin, E.L. (2010) Oxidation of aluminum particles in mixed CO_2/H_2O atmospheres. *J. Phys. Chem. C*, **114** (44), 18925–18930.
26. Ward, T.S. *et al.* (2006) Experimental methodology and heat transfer model for identification of ignition kinetics of powdered fuels. *Int. J. Heat Mass Transfer*, **49** (25–26), 4943–4954.
27. Shoshin, Y.L. *et al.* (2006) Ignition of aluminum-rich Al-Ti mechanical alloys in air. *Combust. Flame*, **144** (4), 688–697.
28. Gill, R.J., Mohan, S., and Dreizin, E.L. (2009) Sizing and burn time measurements of micron-sized metal powders. *Rev. Sci. Instrum.*, **80** (6), 064101-1–064101-7.
29. Gill, R.J., Badiola, C., and Dreizin, E.L. (2010) Combustion times and emission profiles of micron-sized aluminum particles burning in different environments. *Combust. Flame*, **157** (11), 2015–2023.
30. Badiola, C., Gill, R.J., and Dreizin, E.L. (2011) Combustion characteristics of micron-sized aluminum particles in oxygenated environments. *Combust. Flame*, **158** (10), 2064–2070.
31. Shoshin, Y. and Dreizin, E.L. (2004) Laminar lifted flame speed measurements for aerosols of metals and mechanical alloys. *AIAA J.*, **42** (7), 1416–1426.
32. Shoshin, Y. and Dreizin, E. (2003) Particle combustion rates in premixed flames of polydisperse metal – air aerosols. *Combust. Flame*, **133** (3), 275–287.
33. Shoshin, Y. and Dreizin, E. (2002) Production of well-controlled laminar aerosol jets and their application for studying aerosol combustion processes. *Aerosol Sci. Technol.*, **36** (9), 953–962.
34. Schoenitz, M., Dreizin, E.L., and Shtessel, E. (2003) Constant volume explosions of aerosols of metallic mechanical alloys and powder blends. *J. Propul. Power*, **19** (3), 405–412.
35. Santhanam, P.R. and Dreizin, E.L. (2010) Characteristics of metal combustion obtained from constant volume explosion experiments. 48th AIAA Aerospace Sciences Meeting Including the New Horizons Forum and Aerospace Exposition, Orlando, FL, January 4–7, 2010.
36. Trunov, M.A., Hoffmann, V., Schoenitz, M., and Dreizin, E. (2006) Combustion of boron-titanium nanocomposite powders in different environments. 42nd AIAA/ASME/SAE/ASEE Joint Propulsion Conference and Exhibit, Sacramento, CA, July 11, 2006.
37. Stamatis, D., Jiang, Z., Hoffmann, V., Schoenitz, M., and Dreizin, E. (2008) Fully dense, aluminum-rich Al-CuO nanocomposite powders for energetic formulations. 44th AIAA/ASME/SAE/ASEE Joint Propulsion Conference and Exhibit Hartford, CT, July 21, 2008.
38. Badiola, C., Schoenitz, M., and Dreizin, E.L. (2009) Mechanically alloyed Al-Ti powders prepared by mechanical milling at cryogenic temperatures. 45th AIAA/ASME/SAE/ASEE Joint Propulsion Conference and Exhibit, Denver, CO, 2009.
39. Stamatis, D. *et al.* (2009) Fully dense, aluminum-rich Al-CuO nanocomposite powders for energetic formulations. *Combust. Sci. Technol.*, **181** (1), 97–116.
40. Stamatis, D. *et al.* (2010) Aluminum burn rate modifiers based on reactive nanocomposite powders. *Propellants Explos. Pyrotech.*, **35** (3), 260–267.
41. Zhu, X., *et al.* (2005) Materials Research Society Fall Meeting. Reactive AI-Li powders prepared by mechanical alloying. Boston, MA.
42. Zhu, X., Schoenitz, M., and Dreizin, E.L. (2007) Mechanically alloyed Al-Li powders. *J. Alloys Compd.*, **432** (1–2), 111–115.
43. Jiang, X., Trunov, M., Schoenitz, M., Dave, R., and Dreizin, E. (2008) Mechanical alloying and reactive milling in a high energy planetary mill. 44th AIAA/ASME/SAE/ASEE Joint Propulsion Conference and Exhibit, Hartford, CT, July 21, 2008.
44. Aly, Y., Schoenitz, M., and Dreizin, E.L. (2011) Aluminum-metal reactive composites. *Combust. Sci. Technol.*, **183** (10), 1107–1132.

45. Schoenitz, M. and Dreizin, E.L. (2003) Structure and properties of Al-Mg mechanical alloys. *J. Mater. Res.*, **18** (8), 1827–1836.
46. Schoenitz, M., Zhu, X., and Dreizin, E.L. (2004) Mechanical alloys in the Al-rich part of the Al-Ti binary system. *J. Metastable Nanocryst. Mater.*, **20–21**, 455–461.
47. Schoenitz, M., Zhu, X., and Dreizin, E.L. (2005) Carbide formation in Al-Ti mechanical alloys. *Scripta Mater.*, **53** (9), 1095–1099.
48. Zhu, X., Schoenitz, M., and Dreizin, E.L. (2006) Oxidation of mechanically alloyed Al-rich Al-Ti powders. *Oxid. Metals*, **65** (5–6), 357–376.
49. Schoenitz, M. and Dreizin, E. (2004) Oxidation processes and phase changes in metastable Al-Mg alloys. *J. Propul. Power*, **20** (6), 1064–1068.
50. Zhu, X. (2006) *Mechanical Engineering*, New Jersey Institute of Technology, Newark, NJ, p. 173.
51. Shoshin, Y.L., Mudryy, R.S., and Dreizin, E.L. (2002) Preparation and characterization of energetic Al-Mg mechanical alloy powders. *Combust. Flame*, **128** (3), 259–269.
52. Mohan, S., Trunov, M.A., and Dreizin, E.L. (2003) Characterization of aluminum powder ignition, in *The 2003 Technical Meeting of the Eastern States Section of the Combustion Institute*, AIAA Aerospace Sciences Meeting and Exhibit, The Combustion Institute, University Park, PA, pp. 3001–3006.
53. Kissinger, H.E. (1957) Reaction kinetics in differential thermal analysis. *Anal. Chem.*, **29** (11), 1702–1706.
54. Shoshin, Y.L. and Dreizin, E.L. (2006) Particle combustion rates for mechanically alloyed Al-Ti and aluminum powders burning in air. *Combust. Flame*, **145** (4), 714–722.
55. Dreizin, E.L. and Schoenitz, M. (2009) Nano-composite energetic powders prepared by arrested reactive milling. US Patent 7,524,355.
56. Duckham, A. *et al.* (2004) Reactive nanostructured foil used as a heat source for joining titanium. *J. Appl. Phys.*, **96** (4), 2336–2342.
57. Blobaum, K.J. *et al.* (2003) Al/Ni formation reactions: characterization of the metastable Al9Ni2 phase and analysis of its formation. *Acta Mater.*, **51** (13), 3871–3884.
58. Gavens, A.J. *et al.* (2000) Effect of intermixing on self-propagating exothermic reactions in Al/Ni nanolaminate foils. *J. Appl. Phys.*, **87** (3), 1255–1263.
59. Umbrajkar, S.M., Schoenitz, M., and Dreizin, E.L. (2006) Exothermic reactions in Al-CuO nanocomposites. *Thermochim. Acta*, **451** (1–2), 34–43.
60. Umbrajkar, S.M. *et al.* (2008) Aluminum-rich Al-MoO_3 nanocomposite powders prepared by arrested reactive milling. *J. Propul. Power*, **24** (2), 192–198.
61. Umbrajkar, S.M. *et al.* (2008) On problems of iso conversion data processing for reactions in Al-rich Al-MoO_3 thermites. *Thermochim. Acta*, **477** (1–2), 1–6.
62. Stamatis, D., Ermoline, A., and Dreizin, E.L. (2012) A Multi-step reaction model for ignition of fully-dense Al•CuO nanocomposite powders. *Combust. Theor. Modell.*, **16** (6), 1011–1028.
63. Ermoline, A., Schoenitz, M., and Dreizin, E.L. (2011) Reactions leading to ignition in fully dense nanocomposite Al-oxide systems. *Combust. Flame*, **158** (6), 1076–1083.
64. Ermoline, A., Stamatis, D., and Dreizin, E.L. (2012) Low-temperature exothermic reactions in fully dense Al-CuO nanocomposite powders. *Thermochim. Acta*, **527**, 52–58.
65. Trunov, M.A., Schoenitz, M., and Dreizin, E.L. (2006) Effect of polymorphic phase transformations in alumina layer on ignition of aluminium particles. *Combust. Theor. Modell.*, **10** (4), 603–623.
66. Trunov, M.A. *et al.* (2005) Effect of polymorphic phase transformations in Al_2O_3 film on oxidation kinetics of aluminum powders. *Combus. Flame*, **140** (4), 310–318.
67. Schoenitz, M. *et al.* (2010) Oxidation of aluminum powders at high heating rates. *Thermochim. Acta*, **507–508**, 115–122.
68. Vyazovkin, S. *et al.* (2011) ICTAC kinetics committee recommendations for

performing kinetic computations on thermal analysis data. *Thermochim. Acta*, **520** (1–2), 1–19.
69. Stamatis, D., Dreizin, E.L., and Higa, K. (2011) Thermal initiation of Al-MoO_3 nanocomposite materials prepared by different methods. *J. Propul. Power*, **27** (5), 1079–1087.
70. Schoenitz, M., Umbrajkar, S., and Dreizin, E.L. (2007) Kinetic analysis of thermite reactions in Al-MoO_3 nanocomposites. *J. Propul. Power*, **23** (4), 683–687.
71. Zhou, L. *et al.* (2010) Time-resolved mass spectrometry of the exothermic reaction between nanoaluminum and metal oxides: the role of oxygen release. *J. Phys. Chem. C*, **114** (33), 14269–14275.
72. Mohan, S., Trunov, M.A., and Dreizin, E.L. (2008) Heating and ignition of metallic particles by a CO_2 laser. *J. Propul. Power*, **24** (2), 199–205.
73. Dreizin, E.L. *et al.* (2011) Combustion dynamics of particles of metal-based reactive materials. *Int. J. Energetic Mater. Chem. Propul.*, **10** (4), 297–319.
74. Badiola, C. and Dreizin, E.L. (2012) On weak effect of particle size on its burn time for micron-sized aluminum powders. *Combust. Sci. Technol.*, **179**, 10.
75. Trunov, M.A. *et al.* (2008) Combustion of boron-titanium nanocomposite powders in different environments. *J. Propul. Power*, **24** (2), 184–191.
76. Stamatis, D. *et al.* (2011) Consolidation and mechanical properties of reactive nanocomposite powders. *Powder Technol.*, **208** (3), 637–642.
77. Stamatis, D. and Dreizin, E.L. (2011) Thermal initiation of consolidated nanocomposite thermites. *Combust. Flame*, **158** (8), 1631–1637.
78. Amorós, J.L. *et al.* (2008) Green strength testing of pressed compacts: an analysis of the different methods. *J. Eur. Ceram. Soc.*, **28** (4), 701–710.
79. Schoenitz, M., Ward, T.S., and Dreizin, E.L. (2005) Fully dense nano-composite energetic powders prepared by arrested reactive milling. 30th International Symposium on Combustion, Chicago, IL.
80. Badiola, C., Schoenitz, M., Zhu, X., and Dreizin, E.L. (2008) Synthesis and characterization of aluminum-rich nanocomposite powders at cryogenic temperatures. AIChE Annual Meeting, Philadelphia, PA, 2008.

11
Characterizing Metal Particle Combustion *In Situ*: Non-equilibrium Diagnostics

Michelle Pantoya, Keerti Kappagantula, and Cory Farley

11.1
Introduction

When considering ignition and combustion of solid fuel and oxidizer powders from a fundamental perspective, an energy threshold, defined here as apparent activation energy, must be achieved in order for a reaction to initiate. Once ignited and in order for the reaction to self-propagate, the heat generated by the reaction, called *heat of combustion*, must be sufficiently high such that enough energy is transferred to the adjacent particles to elevate them to a state beyond their activation energy. It has been shown that as the fuel particle size decreases the apparent activation energy for a mixture is reduced such that less energy is required to sustain a self-propagating reaction as compared to the same composition with larger particle size reactants [1]. In fact, smaller fuel particle size results in increased reactivity in general; and, has been characterized in diverse ways: from faster flame speeds [2] to shorter ignition delay times [3] to greater energy generation [4].

The focus of this chapter is on aluminum (Al)-fueled composites. These composites are sometimes referred to in the literature as pyrotechnics, thermites, reactive materials, reactive composites, or metastable interstitial composites (MICs). Figure 11.1 shows a scanning electron micrograph (SEM) image of Al nanopowder combined with molybdenum trioxide (MoO_3) nanopowder. It is noted that the MoO_3 particles have a length exceeding the nanoscale, but particles with at least one dimension on the nanoscale are defined as nanoparticles. When the diameter of the Al particle reduces to 40–120 nm (in contrast to traditional 10–100 μm), their reactivity drastically increases. Flame rates reach $1\,km\,s^{-1}$ for reactive nanocomposites (e.g., $Al + MoO_3$ and $Al + I_2O_5$) in contrast to $1\,m\,s^{-1}$ for micron-size Al particles [2, 5], ignition delay times are reduced by three orders of magnitude [3], and the apparent activation energy decreases to the point of igniting the aluminum reaction even in the solid phase at temperatures well below Al melting point at 660 °C [6]. These observations have spurred researchers to consider new and different mechanisms for Al combustion. In the first section of this chapter, a brief review of Al reaction mechanisms that includes classical diffusion theories applied to micron scale

Metal Nanopowders: Production, Characterization, and Energetic Applications, First Edition.
Edited by Alexander Gromov and Ulrich Teipel.

Figure 11.1 (a,b) Aluminum spherical-shaped fuel particles combined with MoO_3 rectangular shaped oxidizer particles to form an energetic nanocomposite. (Images were taken at 52KX magnification and 5.0 kV using an SEM at Los Alamos National Laboratory.)

Al particles as well as new theories proposed for nanometric Al particles is presented.

Resolving these new theories tailored to nanoparticle reactions is difficult because observing reactions at nanometric spatial resolution coupled with picosecond time scales is experimentally challenging. However, each mechanistic theory produces predictions for macroscopic behavior that can be directly checked in experiments using currently available techniques. Describing these techniques is the objective of this chapter.

The first technique discussed is measuring flame propagation rate. The flame propagation rate, or sometimes referred to as the *flame speed*, depends on a number of parameters, such as particle size distribution, shape and mass density of the powder, diameter and length of the sample tube, oxidizer, gas flows, radiation losses, ignition kinetics, and others that are difficult to model. When designing experiments, the goal is to fix all parameters except for one. For example, reduce the Al average particle size. In this way, the influence of one variable can be ascertained and knowledge of the reaction kinetics associated with that single varied parameter provides insight into the mechanism controlling reactivity.

Finally, flame temperature is also discussed. For fast reactions that produce high temperatures capturing flame temperature data is challenging. One approach is to couple point source measurements from a multichannel radiometer to high speed infrared (IR) diagnostics. This new approach enables a two-dimensional transient measurements even for large variations in temperature (e.g., up to 3000 K).

Both of these diagnostics will be described following a brief review of (i) ignition and combustion fundamentals applied to solid materials and (ii) aluminum reaction mechanisms. This information provides researchers an understanding of experiments that can be used with diagnostics to understand reaction behaviors of any energetic composite.

11.2 Ignition and Combustion of Solid Materials

11.2.1 Ignition

The first step for solid combustion is the initiation of the chemical reaction achieved when a certain energy threshold for the reactants has been met. Svante Arrhenius studied this energy threshold and coined the term *activation energy*, E_a and identified the relationship between E_a and reaction rate, $k(T)$ seen in Equation 11.1. In this equation, A is a pre-exponential factor, R is the gas constant, and T is absolute temperature.

$$k(T) = A \exp\left(-\frac{E_\mathrm{a}}{RT}\right) \tag{11.1}$$

Most solid combustion reactions are exothermic and energy released is greater than the activation energy plus dissipated energy to the surroundings. If the energy transferred to the neighboring reactants is greater than the activation energy, then the chemical reaction initiates, and the combustion of the particles propagates along the powder [7].

The activation and initiation of a combustion reaction is called *ignition*. Glassman describes two types of ignition: thermal and spark ignition [8]. As long as the rate of heat generation is greater than the rate of heat dissipation, the ensuing energy buildup could ignite a combustion reaction. In 1935, Semenov modeled this behavior by combining an energy balance with the Arrhenius form of the chemical rate equation to develop a time constant comparison of the generation and dissipation energy terms as seen in Equation 11.2.

$$\left(\frac{\tau_\mathrm{r}}{\tau_\mathrm{l}}\right) = \frac{e(Q/c_\mathrm{v})}{RT_\mathrm{o}^2/E_\mathrm{a}} \tag{11.2}$$

where τ_r and τ_l are the time constants for the generation and dissipation, respectively, Q is the thermal energy release of the chemical reaction, R and c_v are the ideal gas constant and specific heats, and e is the number e. This relationship has been modeled for Al reacting with a metal oxide to show the required size of a hot spot to initiate thermal ignition [9]. Thermal ignition of micron-Al composite energetic materials usually requires on the order of 100 J over a time scale of seconds, whereas nano-Al formulations require only 1 J over millisecond [3]. Both laser heating and resistive heating wires are common forms of thermal ignition.

For a more controlled ignition, a trigger is often used. Spark ignition is popular because it is reliable and safe and experiments show that significantly less energy is required to initiate an Al-based composite with a spark than thermal stimuli (millijoule compared to joule) [10]. Most discussion on spark ignition centers on the term *minimum ignition energy*, which describes the smallest energy impulse that can ignite a given mixture under given circumstances.

Very few studies on electrostatic ignition sensitivity of Al particles have been reported in the literature and all of which were performed for the electrostatic discharge (ESD) of energy into a sample rather than by pouring induced interparticle transport (e.g., electrostatic sensitivity that can occur from pouring a powder sample). An interesting finding from studying the literature on electrostatic ignition of powders is that a paradox exists regarding the electrical properties of the powder and the corresponding electrostatic ignition behavior. Glor [11] studied dust particles that had accumulated a charge through interparticle transport. He found that as the powder's electrical conductivity *decreased*, so did the minimum ignition energy; in other words, materials with a decreased conductivity are more readily ignited by ESD [11]. In contrast, Foley *et al.*'s [12] study on Al-CuO showed that *increasing* electrical conductivity actually decreased the minimum ignition energy of the mixture. A striking difference in these two studies is the way in which the electrical stimuli were introduced to the sample. For Glor [11] electrostatic charge accumulated within the sample, while in Foley *et al.* [12], the electrostatic charge was discharged into the sample. This paradox poses new research questions that have potential for impactful development in this field.

According to a review in 1966, ignition theories are classified into three major groups, (i) gas-phase, (ii) heterogeneous, and (iii) solid-phase [13]. In the case of aluminum combustion, most reactions fall in the latter two groups, with nanometer scale aluminum particles having been shown to react in the solid phase. Reaction rates for these reactions are dependent on the chemical reaction rate constant, $k(T)$, and how strongly the oxidizer adsorbs to the surface of the fuel [7]. If the oxidizer is weakly adsorbed, then the reaction rate, R, is proportional to the concentration of oxidizer [O].

$$R \propto k(T)[\mathrm{O}] \tag{11.3}$$

Second, if the oxidizer is strongly adsorbed, then R will depend mainly on $k(T)$.

$$R \propto k(T) \tag{11.4}$$

The remaining scenario involves a case where the combustion products P compete with the oxidizer for adsorption sites on the surface of the fuel.

$$R \propto k(T)[\mathrm{O}]/[P] \tag{11.5}$$

11.2.2
Propagation

After the flame has been ignited, the chemical reaction generates enough energy to drive the surrounding reactants to their activation energy and ignition. The "ignition and energy spread" cycle repeats with the surrounding reactants. The cyclic repetition of (i) activation, (ii) ignition, and (iii) energy transfer describes the concept of flame propagation, which is displayed graphically in Figure 11.2.

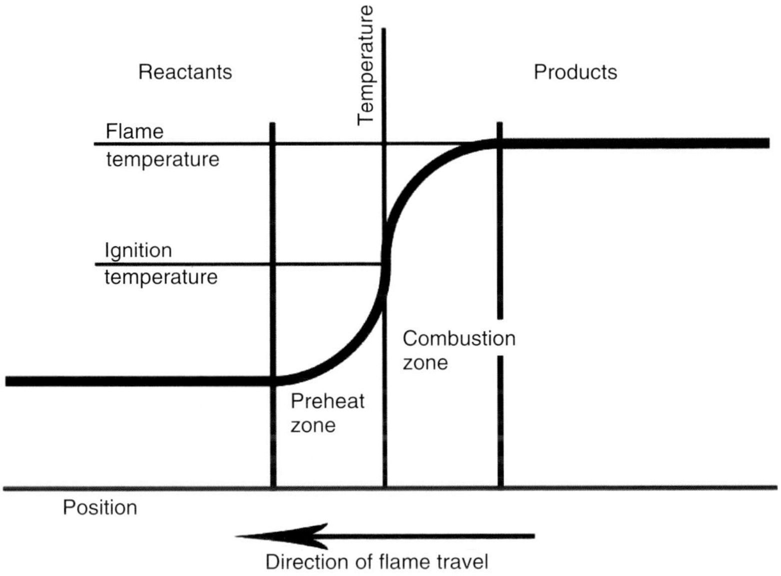

Figure 11.2 Concept sketch of temperature profile across reaction zone of the heterogeneous composite.

While these are diffusion reactions, there is no single expression relating flame speed to any combination of properties for these solid combustion reactions. Much of the theory reported is based on gas-phase diffusion reactions pertaining to hydrocarbon combustion [7], or solid non-gas-generating diffusion reactions [14]. The turbulence, kinetics, and multiphase nature of Al composite reactions are still not well understood and predictive models for powder flame speeds have therefore not been developed.

To understand the mechanics of solid combustion, a brief discussion on the combustion of carbon is presented. Unlike aluminum, carbon is not encased in any sort of protective shell; therefore, when burned in air, the surface of the carbon particle comes into direct contact with oxygen. Owing to this intimate contact, the burning of a solid carbon particle can be analogously modeled to a burning fuel droplet.

The derivation begins with a first law balance for the surface of the particle in a one-dimensional Cartesian system (Equation 11.6).

$$\sum \dot{Q}_{CV} - \sum \dot{W}_{CV} + \sum_{in} \dot{m}(h + \text{ke} + \text{pe}) - \sum_{out} \dot{m}(h + \text{ke} + \text{pe}) = \frac{dE_{CV}}{dt} \quad (11.6)$$

In Equation 11.6, $\dot{Q}_{CV}\,\dot{W}_{CV}$ are the heat and work boundary transfer rates for the control volume; $(h + \text{ke} + \text{pe})$ are the enthalpy, kinetic, and potential energy transfers associated with a given mass flow rate $\dot{m}$; and E_{CV} is the total energy of the control volume. Assuming that (i) burning is steady state, (ii) potential and kinetic energies are negligible, (iii) $\dot{W}_{CV}$ is negligible, and (iv) bulk mass ($\dot{m}$) is a steady, single flow, then, Equation 11.6 simplifies to Equation 11.7 where (″) denotes a flux over a given area A.

$$[\dot{Q}''A]_x - [\dot{Q}''A]_{x+\Delta x} + [A\dot{m}''h]_x - [A\dot{m}''h]_{x+\Delta x} = 0 \tag{11.7}$$

Dividing Equation 11.7 by $(-A\Delta x)$ yields Equation 11.8.

$$\frac{\dot{Q}''_{x+\Delta x} - \dot{Q}''_x}{\Delta x} + \frac{\dot{m}''h_{x+\Delta x} - \dot{m}''h_x}{\Delta x} = 0 \tag{11.8}$$

Then, take the limit as $\Delta x \to 0$ to derive Equation 11.9.

$$\frac{\mathrm{d}\dot{Q}''}{\mathrm{d}x} + \dot{m}''\frac{\mathrm{d}h}{\mathrm{d}x} = 0 \tag{11.9}$$

From here, we need to take a detailed look at the heat transfer, $\dot{Q}''$. Until now, we have been treating the mass flow, $\dot{m}''$, through the control volume as if it was a single component, steady flow. However, this is a reacting mixture and each species in the flow is changing as combustion takes place inside the control volume. In order to account for this, it is common practice to include the deviations of each species diffusion rates into the heat transfer term. Therefore, the heat transfer term includes both Fourier conduction and the enthalpies of species deviating from the bulk flow when diffusing through the carbon particle as seen in Equation 11.10.

$$\dot{Q}'' = -\lambda\frac{\mathrm{d}T}{\mathrm{d}x} + \sum \dot{m}''_{\mathrm{i,diff}} h_i \tag{11.10}$$

The term, $\dot{m}''_{\mathrm{i,diff}}$, is mass diffusion deviating from bulk values with corresponding enthalpy, h_i and λ is the thermal conductivity. Recall from Fick's law of mass diffusion (Equation 11.11).

$$\dot{m}''_{\mathrm{i,diff}} = -\rho D\frac{\mathrm{d}Y_i}{\mathrm{d}x} \tag{11.11}$$

Substituting Equation 11.10 into Equation 11.11 yields Equation 11.12. Assuming all the species have a similar diffusivity, D.

$$\dot{Q}'' = -\lambda\frac{\mathrm{d}T}{\mathrm{d}x} - \rho D \sum \frac{\mathrm{d}Y_i}{\mathrm{d}x} h_i \tag{11.12}$$

From here, it can be shown that the Shvab–Zeldovich energy equation for spherical geometry is Equation 11.13.

$$\frac{1}{r^2}\frac{\mathrm{d}}{\mathrm{d}r}\left[r^2\left(\rho v_{\mathrm{r}}\int c_{\mathrm{p}}\mathrm{d}T - \rho D\frac{\partial \int c_{\mathrm{p}}\mathrm{d}T}{\partial r}\right)\right] = -\sum h^{\mathrm{o}}_{f,i}\dot{m}''' \tag{11.13}$$

In order to get to this equation, the following assumptions have been made: (i) steady state; (ii) no kinetic or potential energy effects; (iii) no control volume work (i.e., shaft or piston); (iv) no radiation; and (v) $\mathrm{Le} = \alpha/D = 1$. Assuming a quasi-steady burning, $\dot{m}_r$ is a constant with respect to r; therefore

$$\frac{\mathrm{d}(\rho v_r r^2)}{\mathrm{d}r} = 0 \tag{11.14}$$

Recall that ν_r is bulk flow velocity. Thus, with constant properties, Equation 11.13 can be rewritten as Equation 11.15.

$$\frac{d}{dr}\left(r^2\frac{dT}{dr}\right) = \frac{\dot{m}c_{pg}}{4\pi\lambda}\frac{dT}{dr} \quad (11.15)$$

This equation assumes no reactions (evaporation only, $\sum h^{o}_{f,i}\dot{m}'''_i = 0$) at the surface of the droplet. From here, it is straightforward to arrive at the diffusion droplet lifetime equation.

$$t_d = \frac{D_o^2}{K} \quad (11.16)$$

Here, t_d is total burn time, D_o is initial diameter, and K is an evaporation constant calculated from material properties and temperature differences. Now, burn time can be examined for a kinetically driven reaction. The flux of fuel at the droplet surface can be described as Equation 11.17.

$$\dot{m}''_f = \frac{\dot{m}_f}{4\pi r^2} = \rho_s\frac{dr_s}{dt} \quad (11.17)$$

The subscript s denotes the particle surface, while the flux of oxidizer can be described by Equation 11.18.

$$\dot{m}''_o = \frac{\dot{m}''_f}{i} = k_s[O]_s \quad (11.18)$$

Here, i is the mass stoichiometric index and k_s is the reaction rate constant and $[O]_s$ is the concentration of oxidizer on the surface. However, $[O]_s$ is not readily known. However, if the reaction is kinetically limited, it can be assumed that the concentration on the surface is equal to the free stream concentration for oxidizer, or

$$[O]_s = [O]_\infty \quad (11.19)$$

Combining equations yields Equation 11.20.

$$\dot{m}''_f = \frac{\dot{m}_f}{4\pi r^2} = \rho_s\frac{dr_s}{dt} = ik_s[O]_s = ik_s[O]_\infty \quad (11.20)$$

which simplifies to Equation 11.21.

$$\frac{dr_s}{dt} = \frac{ik_s[O]_\infty}{\rho_s} \quad (11.21)$$

Separate the variables, then integrate and solve for time.

$$t_k = \frac{\rho_s D_o}{2i\rho[O]_\infty k_s} \quad (11.22)$$

Therefore, in kinetic-limited reactions, the burn time is proportional to the droplet diameter or Equation 11.23.

$$t_k \propto D_o^1 \quad (11.23)$$

11.2.3 Flame Speeds

Flame speed is typically defined for gas-phase reactions as "the velocity at which the unburned gases move through the combustion wave in a direction normal to the wave surface" [8]. For example, in a flame tube, the flow rate that holds the flame stationary can be directly related to the flame speed. However, in multiphase media with varying particle geometries, the concept of flame speed becomes less clear. For multiphase combustion involving solid-phase reactants, the flame speed is defined here as the velocity at which the flame front consumes and propagates through the unburned particles.

Flame speed experiments for solid particle combustion can either consist of measuring the burn time for a single particle or burn rate for powders. Early experiments for calculating the burn time of solid particles involved dropping them into a flat flame, and in 1963, Friedman and Maček [15] conducted this experiment with aluminum particles. The light traces from the burning particles were recorded and used to estimate burn times that correlated to $D^{1.5}$ power for aluminum particles from 10 to 74 μm. Recall that for purely diffusion driven reactions, the burn time should be $\propto D^2$. For purely kinetic reactions, the correlation would be $\propto D^1$. So, a reaction of $\propto D^{1<x<2}$ would have a mixture of diffusion and kinetic elements in the combustion. More recently, burn time experiments for single particles have been conducted in shock tubes. Beckstead [16], in 2005, collected data from a variety of experiments for particles ranging from 20 to 800 μm and found a correlation of $D^{1.8}$. Lynch *et al.* [17] correlated 3–11 μm particles at $D^{0.3}$. Finally in 2011, Gill *et al.* [18] used a CO_2 laser to ignite 10–14 μm particles and correlated them to $D^{0.8-0.9}$. The reaction kinetics involved causing the burn times to fall out of the range from kinetic to diffusion controlled are not fully understood at this time.

The powder flame speed measurements typically fall into three categories: aerosol, open tray, and closed tube. Aerosols containing solid particles are extensively studied in an effort to understand dust explosions. Solid particle aerosols behave similarly to the premixed flame previously discussed [19]. Bunsen burner style experiments with a constant flow are difficult to execute owing to the difficulty in mixing the solid particles into the flow. Instead, the powder is typically dispersed throughout a volume and then ignited. For volumes such as an open duct, the flame position is measured with optical high-speed imaging. For spherical closed volumes, the flame speed can be related to pressure traces from pressure transducers within the vessel [20] or derived from optical techniques [21].

For combustion of masses of nonaerosolized powders, flame speed measurements are typically divided into open and closed measurements. For closed tube measurements, powder is loaded into a tube in such a way as to maintain a constant overall bulk density and ignited by a spark, hot wire, or other stimuli [22, 23]. Optical readings can be taken by either high-speed imaging or photodiodes. High-speed cameras give more information to help determine the nature of the combustion and make accurate flame speed estimates. However, in many cases, flame speeds are so high (i.e., >1000 m s^{-1}) that a camera with the resolution and frame rate required to get reliable information may prove cost prohibitive [2]. Photodiodes give similar information at a much lower cost. A photodiode works by converting light into an electrical signal. An oscilloscope triggers on a certain voltage met when a threshold of light signifying combustion strikes the diode. The drawback to photodiodes is that only a few set positions are monitored and therefore only average speed data may be inferred during analysis.

Instead of photodiodes, pressure transducers can be used to help locate the passing flame front. Heat from the flame and gaseous products increases the pressure as the flame progresses. In 2007, Bockmon *et al.* [2] designed an experimental setup that used both optical and pressure information by combining photodiodes and pressure transducers. In their experiments, flame speed exhibited a similar trend with pressure history.

Here, high-speed video imaging was used for the flame speed measurements in an open tray. Loose powders were loaded into an open tray and ignited. The ensuing videos were analyzed and calibrated. Flame speed was determined by combining the distance information with the frame rate as a timing device. Drawbacks to this method, aside from the camera cost, include spatial effects if the camera is not perfectly perpendicular to the sample. Also, with an optical camera, the luminescent front is assumed to travel with the flame at the same velocity. For highly convective flows, the hot product gases may be pushed ahead of the flame at a temperature capable of radiating visible light. The hot gases may increase the luminescence to a point where oversaturation leads to systemic error in flame speed calculations.

11.3 Aluminum Reaction Mechanisms

Before the thrust toward nanoparticles, numerous models detailed ignition and combustion of micron Al particles. Early studies describe particle burning using a droplet "d^n" model, in which d represents the droplet diameter and the exponent n was determined by the formation of an "oxide cap" [24]. More refined models that considered diffusion of an oxidizer into the particle interior through the oxide shell and phase transitions in the oxide shell [25–29] were then developed.

The key to understanding the reactivity of Al resides in recognizing the role of the passivation shell in this core–shell particle. Figure 11.3 shows a transition electron microscope (TEM) image of a single Al particle. In solid-phase reaction models [24, 26, 29], the aluminum oxide shell (Al_2O_3) grows in an oxidizing environment. For example, Brooks *et al.* [24] modeled both steady and unsteady combustion of Al. According to their model, while the Al particle burns, the residue of an oxide cap remains, while the particle size gets smaller. The accumulation of aluminum oxide on the particle surface reduces the reaction rate as a result of limiting the space for Al evaporation [24]. On the other hand, a shrinking core model is described by Levenspiel [28] for spherical nanometric Al oxidation. According to this model, gaseous reactant (i.e., oxygen) diffuses through the shell to the unreacted core.

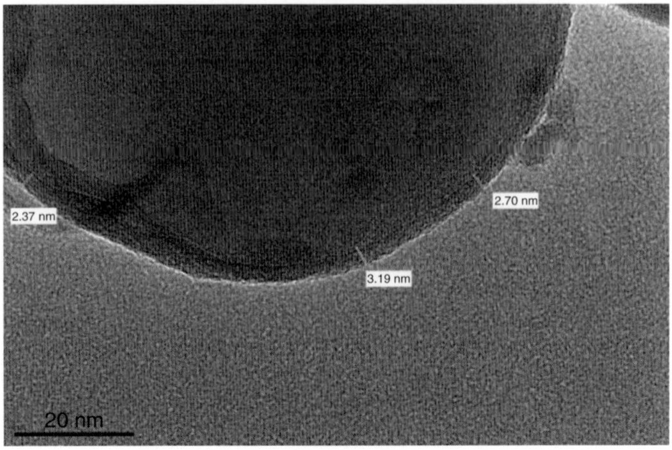

Figure 11.3 Aluminum particle imaged with a TEM showing the distinct core–shell structure associated with Al and its passivation shell, Al_2O_3. Note the slight variation in shell thickness along the edge of the particle. (Image captured with a TEM at Texas Tech University.)

At the core, a chemical reaction of gaseous reactant with solid Al takes place. The reaction front moves radially inward and replaces the unreacted core with an entirely reacted outer layer [28]. In other work, Henry *et al.* [30] suggested that Al_2O_3 prevents Al from igniting until the Al_2O_3 melts. When the Al_2O_3 melts, isolated islands form around the Al core. The Al surface is exposed to oxidizer and the particle ignites [30]. According to the diffusion theory, the reaction is controlled by diffusion of atoms through an oxide shell [12, 31]. The growth of an oxide film is described using a typical Arrhenius-type term. The burning time for diffusion (t) follows the law $t = d^n/K$, where $1.5 < n \leq 2$ and K is the burning rate constant. Divergence from the law is attributed to oxide accumulation on the surface [24].

Additional studies on Al combustion further elucidated its reaction mechanism. Poletaev *et al.* [27] studied the combustion zone of a steady-state laminar diffusion flame of Al particles of diameter 4.8 μm with atomic and molecular emission spectra. The result showed that metal particles burn individually to form microflames. This is an interesting study because it showed that individual microflames comprise a macroscopic powder burn and is a good model relating single particle burning to a macroscopic collection of particles burning. More recently, Huang *et al.* [32] studied the combustion of Al particle dust in a laminar air flow theoretically for a range of particle sizes extending from nanoscale to micron scale. In their study, it was assumed that ultrafine particles are initially in a liquid phase and then evaporate and react with an oxidizer in gas phase. This assumption led to flame speed calculations that agreed with experimental data. When the particle size is reduced from micron to nanoscale, flame speeds increase. For micron and larger particles, droplet surface regression follow $d^{-0.92}$ law but for nanoparticles droplet regression follow $d^{-0.52}$ law [32].

There have been at least three reports of advanced diffusion-based models describing nanoparticle oxidation. Zachariah and coworkers [25, 26] developed a model based on modification of the diffusion coefficients for metal and oxygen across an oxide shell nanoparticle through introduction of a convective force driven by pressure gradients. Another approach taken by Trunov and coworkers [29] describes oxidation in terms of a series of polymorphic phase transitions within the alumina shell. The model emphasizes the thickness-dependent multiple-phase transitions in alumina during particle heating and predicts complex heating and oxidation kinetics. This model was derived for relatively slow oxidation rates and with experiments involving micron-scale particles such that nanoparticle behavior could be extrapolated [29]. A third model is based on a mechanochemistry approach and is applicable only at high heating rates [23–25, 48–50]. The theory describes heating, phase change, and rapid pressure buildup within the fuel particle core. The passivation shell bursts, causing cavitation and dispersion of molten Al that then oxidizes. The mechanism has been coined the melt dispersion mechanism because it results in a burstlike dispersion of molten fuel, promoting accelerated oxidation [23–25, 48–50].

All of these models provide inspiration for designing experiments that assess the validity of these theories. The remaining portion of this chapter is devoted to characterization strategies that are currently used to study solid energetic material composite combustion.

11.4
The Flame Tube

The flame tube has become a bench mark diagnostic for examining flame propagation of nano aluminum (nm Al) powders combined with a solid oxidizer. The first study demonstrating this diagnostic was published in 2002 by Bockmon *et al.* [2] and was used as a tool to characterize the combustion loose powders of nm Al-MoO_3. For this reason, the flame tube has also been referred to as the *Bockmon tube*. Before the first reporting of the flame tube, this diagnostic was not useful because micron-scale Al particles do not propagate combustion waves well, or at all, in this semiconfined tube. However, with the advent of nanotechnology, nm Al particles are significantly more ignition sensitive [1] such that less energy is required to ignite [3] (and therefore propagate) a reaction.

Flame tube experiments usually start with the process of homogeneous mixing of reactants. The solid reactant particles are suspended in a solution of hexane or isopropanol and then a Misonix Sonicator 3000 probe vibrating at ultrasonic speeds breaks up agglomerates and provides homogeneous mixing of the reactants. The liquid is evaporated using a rotary evaporator or simply by pouring the mixture into a Pyrex container placed on a hot plate at a low temperature to evaporate the liquid. After evaporation, the powder is reclaimed for experimentation.

The flame tube can be described as a long transparent tube (usually 10 cm in length) made out of Acrylic™ or Lexan™ with a constant inside diameter ranging from 3 to 5 mm. Son *et al.* [36] examined the sensitivity of the tube inside diameter on the flame propagation speed and found that for sufficiently small tubes, diameters larger than roughly 4 mm do not appreciably influence flame propagation speeds. The tubes are filled with roughly 200–300 mg of powder mixture and placed onto a vibrating block for approximately 5 s to reduce density gradients, after which more powder is added. This process results in a bulk density expressed as a percentage of the theoretical maximum density (TMD). The TMD is calculated as the ratio of the actual mass of reactants divided by sample volume to the weighted average of pure solid densities of each reactant in the matrix. When the powder is loaded into a tube as described here, the bulk density can range anywhere from 3 to 10% TMD and are considered loose powders (as opposed to a consolidated pellet). Variations in powder bulk density can lead to variations in flame speed [2] and care should therefore be taken to ensure that the overall bulk density is limited to a discrete range.

Figure 11.4 shows the experimental test setup including flame tube, test block, and sensors. The tube may be machined with sensor ports for monitoring pressure and/or optical light intensity at various axial locations along the tube. These measurements, coupled with high-speed imaging of the reaction propagation, provide information about the pressure profile as well as the propagation speed of the luminous reaction front. The propagation speed is typically too fast to allow accurate and simultaneous temperature measurements.

The assembled test block is placed into a stainless steel combustion chamber. Ignition is typically by an electric match or hot wire (thermal ignition) and the tube can be open at one end while the ignition source seals the other end. The combustion chamber has an acrylic window (139.7 × 25.4 mm dimensions) that enables observation of the reaction with a high-speed video camera. The chamber is connected to a vacuum pump to set the chamber to a desired pressure. A pressure gauge on the top of the chamber measures pressure. Also, the chamber

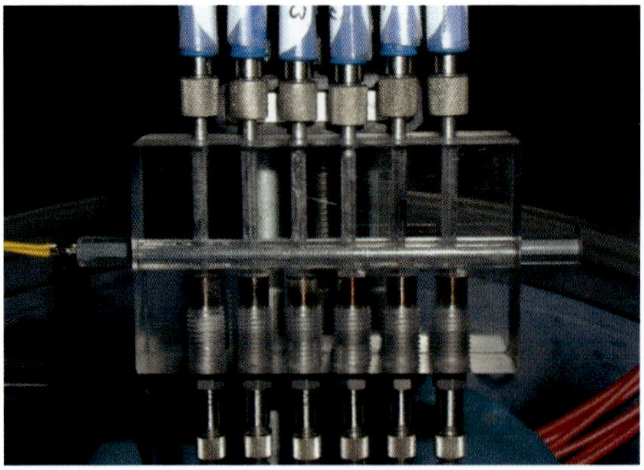

Figure 11.4 Photograph of assembled test block that includes flame tube and sensor ports. (The image was captured at Texas Tech University Energetic Materials Combustion Laboratory.)

Table 11.1 Flame speeds from flame tube measurements characterizing the combustion of nm Al particles combined with a solid oxidizer.

Composition	Al size (nm)	Active Al content (%)	Equivalence ratio	%TMD	Velocity ($m\,s^{-1}$)	Peak pressure (MPa)	References
$Al + MoO_3$	79	81	3.6	9	1025	n/a	[36]
$Al + MoO_3$	80	80	1.35	10	989	17.9	[2]
$Al + MoO_3$	80	88	1.4	11	950	18.6	[37]
$Al + CuO$	80	88	1.1	6	802	13.1	[37]
$Al + WO_3$	80	88	1.9	9	925	26.9	[37]
$Al + Bi_2O_3$	80	88	1.3	17	646	53.4	[37]
$Al + Bi_2O_3$	80	88	1.3	47	560	39.3	[37]
$Al + Fe_2O_3$	52	75	1.4	7	617	n/a	[41]
$Al + Fe_2O_3$	52	75	1.4	7	8	n/a	[41]
$Al + Fe_2O_3$	50	70	n/a	n/a	25	n/a	[39]
$Al + NiO$	50	75	2.5	7	194	1.68	[42]
Al + PTFE[a]	50	75	1.39	9	837	14.37	[43]
$Al + I_2O_5$	80	80	1.25	7	1035	28	[5]

[a] PTFE = polytetrafluoroethylene (DuPont manufactured Zonyl™ fluoropolymer).

was designed for environmental controls such as gas composition variations (e.g., air or argon) and humidity controls. Several researchers have used the flame tube in inert (argon) [37], vacuum [38] as well as high-pressure environments [39]. The high-speed camera is aligned perpendicular to the direction of propagation and records flame position as a function of time within the tube.

Table 11.1 summarizes flame speeds from much of the literature that incorporate a flame tube measurement for characterizing the combustion of nm Al particles combined with a solid oxidizer. Some physical interpretations from Table 11.1 reveal that reaction propagation falls into mainly three categories determined by the behavior of the oxidizer upon nano Al ignition. Studies show that metallic oxidizers that decompose or gasify at temperatures below the adiabatic flame temperature for the reaction result in higher flame propagation rates [40]. In Table 11.1, this case holds for oxidizers such as CuO, MoO_3, WO_3, and Bi_2O_3. Metal oxidizers that decompose or gasify above the adiabatic flame temperature for the reaction likely initiate in the condensed phase, which promotes slower propagation speeds. In Table 11.1, this case holds for Fe_2O_3 and NiO. Finally, a third category may be associated with reducing agents that fall under the halogen family such as fluorine and iodine that typically decompose at temperatures well below the melting temperature of Al and act as aggressive oxidizers partly due to their relatively high electronegativity.

Recent research also reveals that when Al reacts with a halogenated oxidizer (i.e., either a fluoropolymer-like polytetrafluoroethylene or certain iodine oxide species that decompose below the ignition temperature of Al) the halogen reacts exothermically with the alumina passivation shell surrounding Al particles [44–46].

Many thermal equilibrium studies point to specific interactions between fluorine and alumina as well as iodine and alumina that lead to preignition reactions because they occur just before aluminum ignition [44–46]. In this way, the alumina shell serves to provide a site of exothermic activity in contrast to simply acting as a diffusion barrier inert to the combustion process as with the metal oxides.

11.5 Flame Temperature

11.5.1 Background

Despite the wide amount of research investigating Al thermites, surprisingly little is known about the temperature of the flame front and spatial distribution of temperature generated during combustion. Most researchers use thermoequilibrium software to predict reaction temperature, and Fischer and Grubelich [47] tabulated these values for hundreds of different formulations. These calculated predictions provide a qualitative comparison of candidate formulations for a particular application. However, they are not quantitative indicators of thermal behavior because they assume equilibrium conditions under either a constant pressure or volume condition with no energy loss. These calculations do not represent the high heating rates of real-time combustion or consider the extent of reaction completion.

Temperature measurements of energetic materials employ both invasive and noninvasive techniques. Most common invasive temperature measurement techniques make use of various types of thermocouples or probe thermometers that are embedded into the reaction zone or semi-invasive techniques that require information about the emission properties of certain materials that change with temperature. A temperature difference generates a voltage in the thermocouple (due to the Seebeck effect), which is then converted into temperature that can be recorded. Although thermocouples are an inexpensive and simple method to measure temperatures, the accuracy of the temperature measurement is questionable because of the invasive nature of the technique. With the use of thermocouples, the true temperature profile cannot be captured as the very presence of the thermocouple disturbs the temperature field and the mass transport processes. In many situations, the time scale of the energetic reactions is smaller than the response time of the thermocouples. In addition, thermocouples can measure temperatures at limited number of locations and cannot provide the temperature field of a target. The multiphase fluid dynamic and chemical processes that occur in the reaction zone of a thermite system are complex. Owing to the high temperatures and pressures within the combustion products, direct *in situ* measurement of the thermodynamic properties of the multiphase products is difficult.

Among the noninvasive techniques, researchers have applied absorption-emission spectrometry of various chemical species, fiber optic thermometry, XRD (X-ray diffraction), acoustic pyrometry, CARS (Coherent anti-Stokes Raman scattering) spectroscopy, and NMR (nuclear magnetic resonance) methods. Radiation thermometry belongs to the noninvasive class of temperature measurement techniques. One of its greatest advantages is to reliably follow rapid temperature changes in a fast-evolving system as a function of time. Owing to its noninvasive nature, it can be used in a harsh environment where high temperature, pressure, corrosion, and contaminants would prohibit the use of invasive forms of temperature measurement. It is thus highly suitable for measuring the temperature of reaction zone in a combustion environment.

Thermal imaging cameras (TICs), commonly referred to as *IR* cameras, are noncontact temperature measurement devices that determine apparent temperature from the IR radiation emitted. They employ pyroelectric detectors that measure the radiation in the field of interest by sensing changes in the polarity of the detector material due to exposure to incoming radiation and translate that into temperature using Planck's law coupled with the emissivity of the material.

IR thermometry has been used to measure temperatures in the highly evolving, harsh environment of combustion reactions as it has the capability to record and calculate the temperature of a required area in which the reaction is taking place, thus generating a unique temperature field as a function of time. Although there is a balance between the resolution of the temperature field and the area that can be monitored by the camera, with the development of sophisticated diagnostics for IR radiometry such as high resolution IR cameras, this technique has become one of the favored noninvasive measurement methods.

However, one of the drawbacks of IR cameras is that calculations of temperatures require an estimation of emissivity of the reacting materials. Energetic material reactions include continuous phase changes in the reacting particulate media. Owing to this, coupled with the added variations in the fluid field in which the reaction is taking place, emittance will vary throughout the regions of the thermite reaction. In addition, estimation of emissivity of the combustion products is very complex because there are multiple components of the reaction emitting and possessing different surface finish characteristics. Thus, IR thermometry is a useful temperature measurement technique either for point source or the spatial distribution of temperature *only if* there is a good estimate of the emissivity associated with the reacting media.

One way to solve the emissivity estimation problem is by using multiwavelength pyrometers, also known as *ratio pyrometers*. These instruments measure the emitted radiant flux from the same target area at two or more different, but close wavelengths. The ratios of the signals detected at the different wavelengths by the photodetectors are calculated and the temperature of the target area is deduced from this ratio. The underlying assumption is that the emissivity of the target area is independent of the wavelengths, which is true for gray surfaces. It has been shown that the majority of the condensed products of combustion reaction are usually gray emitters over a relatively narrow range of wavelengths [48]. Thus, dual or multiwavelength pyrometry is an apt technique to measure the temperature of a target area by eliminating the emissivity determination problem.

This very advantage has been exploited by researchers, who have used this multiwavelength pyrometry in determining the temperature of combustion systems [49, 50]. Multiwavelength pyrometers, however, can determine the temperature as a function of time and not as a function of space. More specifically, determination of a temperature field is possible employing IR thermometry but not possible with multiwavelength pyrometry. A larger target area would supply more radiant flux to a multiwavelength pyrometer, but the instrument would still yield only one set of temperature values. In thermite systems, not confined to a small volume, there will be several temperature gradients spatially.

This chapter introduces a unique measurement technique that can determine the spatial distribution of temperature for a multiphase fast-reacting thermite as a function of time. This technique employs IR thermometry to determine the various thermal zones in the reaction area and a multiwavelength pyrometer to identify the highest temperature of the combustion reaction. These two sets of data are then integrated to correctly identify the emissivity of the thermite and generate an accurate temperature field of the area of interest during the reaction.

11.5.2
Radiometer Setup

Figure 11.5 shows the setup for the temperature measurement experiments using the radiometer. A powder thermite sample is placed on a pedestal and a nichrome wire initiated the reaction. The pedestal was surrounded by thermal shields for safety. A five-channel radiometer was used for point-source temperature measurements. The radiometer was positioned 25 cm from the sample. A sapphire window (selected because of its transparency to majority of radiation in the IR region) was placed between the sample and the radiometer in order to protect the radiometer lens from the combustion debris.

On ignition, the thermal radiance emitted from the exothermic reaction is detected by the radiometer's primary lens and passes through five beam splitters

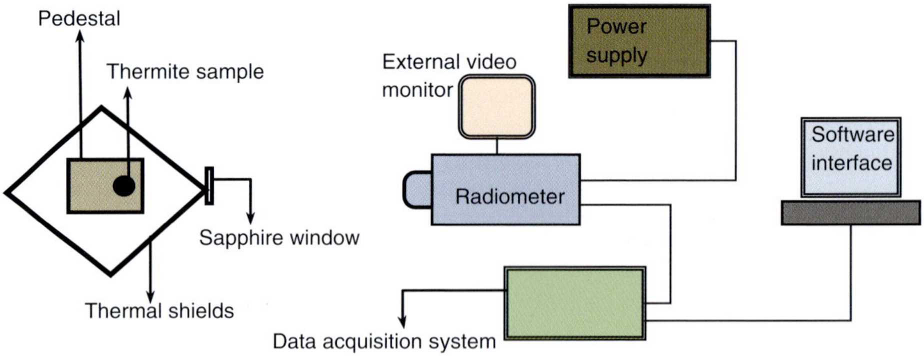

Figure 11.5 Schematic of the radiometer apparatus used for measuring sample temperature.

housed behind the lens to create six identical images. One of these images is sensed by an in-built IR video camera housed behind the beam splitters connected to an external video monitor, which is used for aligning the radiometer with the sample. The other five images pass through narrow bandpass filters and are sensed by five photodetectors. The signal generated by each photodetector corresponds to light intensity of a different part of a light spectrum ranging from about 900 to 4000 nm. The signals are then conditioned by amplifiers before being digitized and stored by the data acquisition system.

11.5.3 Infrared Setup

Figure 11.6 is a schematic for obtaining thermal images of energetic material combustion. The powder sample is placed over a nichrome wire (which provided thermal stimulus for reaction initiation) attached to an acrylic plate that is painted black. The plate is placed in a volume formed by four shields: three on the three sides of the plate and one on top as shown in Figure 11.5. A 14-bit, high-speed midwave (3–5 μm) IR camera with an indium antimonide (In-Sb) detector, procured from FLIR Systems, is placed at a distance of 68 cm from the fourth unconfined side of the volume. The IR camera records at a sample rate of 402 Hz with an integration time of 0.28 μs. A Mikron M300 blackbody with an effective emissivity of +0.995, temperature range of 200–1150 °C and an accuracy of 0.25% of the reading ±1 K was used to calibrate the IR camera. Light emitted by the sample on its ignition was intercepted by the camera in the form of analog/digital (AD) count. This data was converted into thermal radiance values (R) by the IR camera software and imaged by comparing it to the blackbody calibration. The IR camera recorded the thermal event from the thermite reaction as a series of thermal images.

11.5.4 Linking Radiometer and IR Data for a Spatial Distribution of Temperature

For each test conducted, the software provided with the IR camera was used to identify the thermal image with the maximum amount of radiance and mark

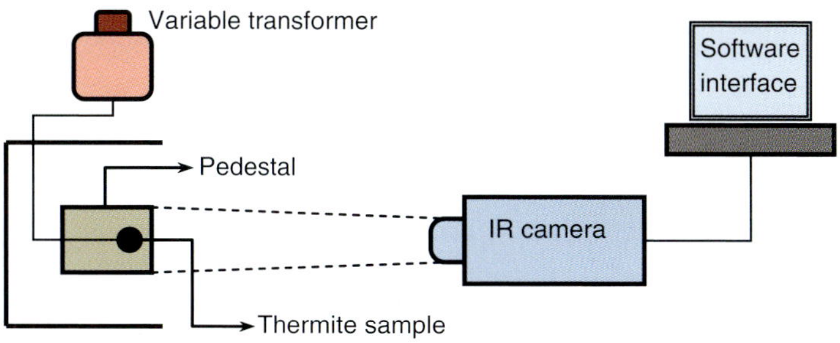

Figure 11.6 Schematic of the thermal imaging apparatus.

an area of interest within each frame for further analysis. The thermal images boxed an area with length of about 8 cm and width of 3.5 cm. From each of these identified thermal images, the maximum thermal radiance value is identified. The point in the thermal image corresponding to this radiance was located. This point was also the location of the maximum temperature of the thermite as the point of maximum radiance is also the point that has the maximum temperature, according to the theory of radiometry. Thus, the temperature of this point was identified as T_{max} (measured by the radiometer) for the corresponding reaction. Therefore, a thermal image of the thermite reaction with the maximum temperature value and its location was obtained at the instant when it recorded its maximum thermal radiance. However, the temperature values elsewhere on the image were still unknown. To determine these, the emissivity (ε) of the material needed to be determined at that instant of time when the thermite reached its maximum temperature (T_{max}). In order to do so, the Planck's equation for band emission (with $T = T_{max}$) was solved for the bandwidth of 3–5 μm, as that is the operating range for the IR camera.

Each location in the thermal image recorded at the instant of T_{max} had a unique radiance value, which, when combined with the ε using Planck's equation, yielded the temperature of that particular location. Thus, the temperature at each location

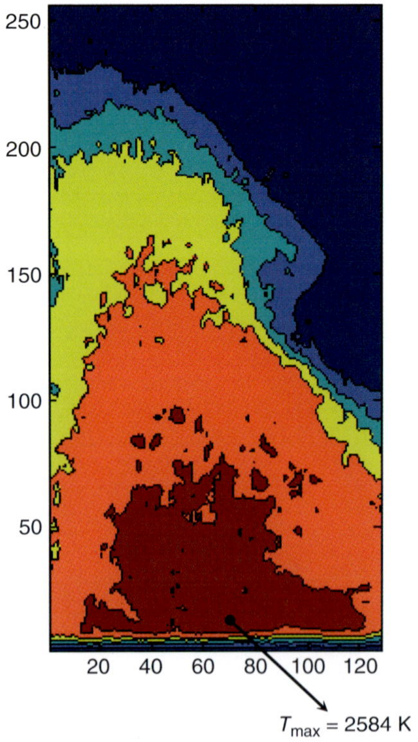

Figure 11.7 Thermal image for Al/PTFE reaction with temperature gradations.

was determined, which determined the complete temperature field of the reaction at the instant when it reached its maximum temperature.

Figure 11.7 shows the thermal image of Al/PTFE (polytetrafluoroethylene) reaction at the frame when it reaches T_{max} and the spatial distribution of temperature at that instant. In the thermal image, the lowest temperature is 750 K or below and every location at this temperature is in blue. The areas with the higher temperature in the thermal image are in red. The contour thresholds in the image are 500 K degree gradients. The location of P_{max} is noted and the maximum temperature is labeled in the image. The radiance values recorded by the IR camera at each point of the thermal image and the corresponding temperature values are provided as supplementary information.

11.6 Conclusions

Energetic materials involving particle composites of fuel and oxidizer present challenging opportunities for examining and characterizing their reaction behaviors. This chapter reviews fundamental ignition and propagation behaviors inherent in solid combustion systems and specific to aluminum-fueled composites. Also, this chapter highlights two fundamental diagnostic techniques that provide a means to quantify flame propagation rates and spatial distributions of temperature as a function of time. These laboratory scale diagnostics can be used safely and inexpensively to observe combustion behaviors and isolate influential parameters controlling overall reaction kinetics.

Acknowledgment

The authors are grateful for support from the Army Research Office under contract number W911NF-11-1-0439 and helpful conversations with our program manager, Dr. Ralph Anthenien.

References

1. Hunt, E.M. and Pantoya, M.L. (2005) Ignition dynamics and activation energies of metallic thermites: from nano- to micron-scale particulate composites. *J. Appl. Phys.*, **98** (3), 034909.
2. Bockmon, B., Pantoya, M., Son, S., Asay, B., and Mang, J. (2005) Combustion velocities and propagation mechanisms of metastable interstitial composites. *J. Appl. Phys.*, **98**, 064903.
3. Granier, J. and Pantoya, M. (2004) Laser ignition of nanocomposite thermites. *Combust. Flame*, **138**, 373–383.
4. Crane, C.A., Pantoya, M.L., and Dunn, J. (2010) Evaluating energy transfer from energetic materials to steel substrates. *Int. J. Thermal Sci.*, **49** (10), 1877–1885.
5. Russell, R., Bless, S., and Pantoya, M. (2011) Impact driven thermite reactions with iodine pentoxide and silver oxide. *J. Energetic Mater.*, **29** (2), 175–192.

6. Sanders, V.E., Asay, B.W., Foley, T.J., Tappan, B.C., Pacheco, A.N., and Son, S.F. (2007) *J. Propul. Power*, **23**, 707.
7. Turns, S.R. (2012) *An Introduction to Combustion: Concepts and Applications Third Edition*, 3rd edn, McGraw Hill, New York.
8. Glassman, I. and Yetter, R.A. (2008) *Combustion*, 4th edn, Elsevier, San Diego, CA.
9. Granier, J., Mullen, T., and Pantoya, M.L. (2003) Non-uniform laser ignition in energetic materials. *Combust. Sci. Technol.*, **175** (11), 1929–1951.
10. Weir, C., Pantoya, M.L., Ramachandran, G., Dallas, T., Prentice, D.J., and Daniels, M.A. (2013) Electrostatic discharge sensitivity and electrical conductivity of composite energetic materials. *J. Electrostat.*, **71** (1), 77–83.
11. Glor, M. (1985) Hazards due to electrostatic charging of powders. *J. Electrostat.*, **16**, 175–191.
12. Foley, T., Pacheco, A., Malchi, J., Yetter, R., and Higa, K. (2007) Development of nanothermite composites with variable electrostatic discharge ignition thresholds. *Propellants Explos. Pyrotech.*, **32**, 431–434.
13. Goyne, C.P., Stalker, R.J., Paull, A., Brescianini, C.P., Bradley, H., Dehority, G.L., Ibiricu, M., and Price, E.W. (1966) Theory of ignition of solid propellants. *AIAA J.*, **4** (7), 1153–1181.
14. Armstrong, R. (1990) Models for gasless combustion in layered materials and random media. *Combust. Sci. Technol.*, **71**, 155–174.
15. Friedman, R. and Maček, A. (1963) Combustion studies of single aluminum particles. *Symp. (Int.) Combust.*, **9** (1), 703–712.
16. Beckstead, M.W. (2005) Correlating aluminum burning times. *Combust. Explos. Shock Waves*, **41** (5), 533–546.
17. Lynch, P., Krier, H., and Glumac, N. (2009) A correlation for burn time of aluminum particles in the transition regime. *Proc. Combust. Inst.*, **32** (2), 1887–1893.
18. Gill, R.J., Badiola, C., and Dreizin, E.L. (2010) Combustion times and emission profiles of micron-sized aluminum particles burning in different environments. *Combust. Flame*, **157** (11), 2015–2023.
19. Proust, C. and Veyssiere, B. (1988) Fundamental properties of flames propagating in starch dust-air Mixtures. *Combust. Sci. Technol.*, **62** (4–6), 149–172.
20. Pu, Y.K., Jia, F., Wang, S.F., and Skjold, T. (2007) Determination of the maximum effective burning velocity of dust–air mixtures in constant volume combustion. *J. Loss Prev. Process Ind.*, **20** (4–6), 462–469.
21. Bradley, D., Chen, Z., and Swithenbank, J.R. (1989) Burning rates in turbulent fine dust-air explosions. *Symp. (Int.) Combust.*, **22** (1), 1767–1775.
22. Weismiller, M., Malchi, J., Yetter, R.A., and Foley, T.J. (2009) Dependence of flame propagation on pressure and pressurizing gas for an Al/CuO nanoscale thermite. *Proc. Combust. Inst.*, **32** (2), 1895–1903.
23. Martirosyan, K.S., Wang, L., and Luss, D. (2009) Novel nanoenergetic system based on iodine pentoxide. *Chem. Phys. Lett.*, **483** (1–3), 107–110.
24. Brooks, K.P. and Beckstead, M.W. (1995) Dynamics of aluminum combustion. *J. Propul. Power*, **11** (4), 775–780.
25. Park, K., Lee, D., Rai, A., Mukherjee, D., and Zachariah, M.R. (2005) Size-resolved kinetic measurements of aluminum nanoparticle oxidation with single particle mass spectrometry. *J. Phys. Chem. B*, **109** (15), 7290–7299.
26. Rai, A., Park, K., Zhou, L., and Zachariah, M.R. (2006) Understanding the mechanism of aluminum nanoparticle oxidation. *Combust. Theor. Modell.*, **10** (5), 843–859.
27. Poletaev, N.I. and Florko, A.V. (2008) Spectral studies of the gas component of an aluminum dust flame. *Combust. Explos. Shock Waves*, **44** (4), 437–443.
28. Levenspiel, O. (1999) *Chemical Reaction Engineering*, 3rd edn, John Wiley & Sons, Inc., New York.
29. Trunov, M.A., Schoenitz, M., and Dreizin, E.L. (2006) Effect of polymorphic phase transformations in alumina layer on ignition of aluminum particles. *Combust. Theor. Model.*, **10** (4), 603–623.

30. Henry, B.M., Dinelli, F., Zhao, K.-Y., Grovenor, C.R.M., Kolosov, O.V., Briggs, G.A.D., Roberts, A.P., Kumar, R.S., and Howson, R.P. (1999) Microstructural study of transparent metal oxide gas barrier films. *Thin Solid Films*, **355**, 500–505.
31. Rode, H., Hlavacek, V., Viljoen, H.J., and Gatica, J.E. (1992) Combustion of metallic powders: a phenomenological model for initiation of combustion. *Combust. Sci. Technol.*, **88**, 153.
32. Huang, Y., Risha, G.A., Yang, V., and Yetter, R.A. (2009) Effect of particle size on combustion of aluminum particle dust in air. *Combust. Flame*, **156**, 5–13.
33. Levitas, V.I., Pantoya, M.L., and Dikici, B. (2008) Melt dispersion versus diffusive oxidation mechanism for aluminum nanoparticles: critical experiments and controlling parameters. *Appl. Phys. Lett.*, **92**, 011921.
34. Levitas, V.I., Asay, B.W., Son, S.F., and Pantoya, M.L. (2007) Mechanochemical mechanism for fast reaction of metastable intermolecular composites based on dispersion of liquid metal. *J. Appl. Phys.*, **101** (8), 083524.
35. Levitas, V.I., Asay, B.W., Son, S.F., and Pantoya, M.L. (2006) Melt dispersion mechanism for fast reaction of nanothermites. *Appl. Phys. Lett.*, **89** (7), 071909.
36. Son, S.F., Asay, B.W., Foley, T.J., Yetter, R.A., Wu, M.H., and Risha, G.A. (2007) Combustion of nanoscale Al/MoO_3 thermite in microchannels. *J. Propul. Power*, **23** (4), 715–721.
37. Sanders, V.E., Asay, B.W., Foley, T.J., Tappan, B.C., Pacheco, A.N., and Son, S.F. (2007) Reaction propagation of four nanoscale energetic composites (Al/MoO_3, Al/WO_3, Al/CuO and Bi_2O_3). *J. Propul. Power*, **23** (4), 707–714.
38. Asay, B.W., Son, S.F., Busse, J.R., and Oschwald, D.M. (2004) Ignition characteristics of metastable intermolecular composites. *Propellants Explos. Pyrotech.*, **29** (4), 216–219.
39. Weismiller, M.R., Malchi, J.Y., Yetter, R.A., and Foley, T.J. (2009) Dependence of flame propagation on pressure and pressurizing gas for an Al/CuO nanoscale thermite. *Proc. Combust. Inst.*, **32**, 1895–1903.
40. Sullivan, K. and Zachariah, M.R. (2010) Simultaneous pressure and optical measurements of nanoaluminum thermites: investigating the reaction mechanism. *J. Propul. Power*, **26** (3), 467–472.
41. Plantier, K.B., Pantoya, M.L., and Gash, A.E. (2005) Combustion wave speeds of nanocomposite Al/Fe_2O_3: the effects of Fe_2O_3 particle synthesis technique. *Combust. Flame*, **140** (4), 299–309.
42. Dean, S.W., Pantoya, M.L., Gash, A.E., Stacy, S.C., and Hope-Weeks, L. (2010) Enhanced convective heat transfer from non-gas generating nanoscale thermite reactions. *J. Heat Transfer*, **132** (11), 5.
43. Watson, K.W., Pantoya, M.L., and Levitas, V.I. (2008) Fast reactions with nano and micron aluminum a study on oxidation versus fluorination. *Combust. Flame*, **155** (4), 619–634.
44. Osborne, D.T. and Pantoya, M.L. (2007) Effect of aluminum particle size on the thermal degradation of Al/teflon mixtures. *Combust. Sci. Technol.*, **179** (8), 1467–1480.
45. Dean, S.W. and Pantoya, M.L. (2009) The influence of alumina passivation on nano-Al/Teflon reactions. *Thermochim. Acta*, **493** (1–2), 109–110.
46. Farley, C. and Pantoya, M.L. (2010) Reaction kinetics of nanometric aluminum and iodine pentoxide. *J. Therm. Anal. Calorim.*, **102** (2), 609–613.
47. Fischer, S.H. and Grubelich, M.C. (1998) Theoretical energy release of thermites, intermetallics and combustible metals. 24th International Pyrotechnics Seminar, Monterey, CA, SAND98-1176C.
48. Goroshin, S., Frost, D., and Levine, J. (2006) Optical pyrometry of fireballs of metalized explosives. *Propellants Explos. Pyrotech.*, **31**, 169–181.
49. Tarasov, M.D., Tolshmyakov, A.I., Kuznetsov, F.O., Petrushin, O.N., Petushkov, V.S., Savelev, Y.A., Tarakanov, M.Y., and Tilkunov, V.A. (ed.), Frank, A.M., and Paisley, D.L. (1996) Detonation and shock wave front temperature measurement by two-wave pyrometer based on optical fibers. Proceeding of the SPIE, 22nd International

Congress on High-Speed Photography and Photonics, Santa Fe, NM, Vol. 2869, pp. 891 890.

50. Lefrancois, A., Grouffal, J.Y., Bouinot, P., and Mencacci, S. (2002) Temperature and pressure measurements comparison of the aluminized emulsion explosives detonation front and product expansion. Conference Proceedings of 12th International Detonation Symposium, San Diego, CA, 2002.

12
Characterization and Combustion of Aluminum Nanopowders in Energetic Systems

Luigi T. De Luca, Luciano Galfetti, Filippo Maggi, Giovanni Colombo, Christian Paravan, Alice Reina, Stefano Dossi, Marco Fassina, and Andrea Sossi

12.1
Fuels in Energetic Systems: Introduction and Literature Survey

During the past decade or so, progress in technologies made available a new class of energetic materials, nanosized metal powders, characterized by very small size and thus very high specific surface. The different chemical and physical properties of nanopowders, with respect to the corresponding bulk materials, make them appealing for many industrial applications. In particular, nanosized aluminum (nAl) is broadly exploited to improve the performance of energetic systems (ES)s [1, 2], promote sintering processes for oxide ceramics [3], produce aluminum hydroxide fibers by hydrolysis [4], provide hydrogen sources for hydrogen energetics [5], and support the synthesis of nitride ceramics [6].

In energetic applications (propellants, pyrotechnics, explosives, and combustion synthesis), Al is widely used because of its high combustion enthalpy, easy availability, low toxicity, and good stability [2]. Thanks to its high specific surface, nAl allows incrementing burning rate and combustion efficiency with respect to ES containing micron-sized aluminum (μAl) [7].

In spite of more than a decade of investigations reported in the open literature about nAl use in ES [1, 2, 7–9], the topic is in continuous development because of the potential and new perspectives offered by these systems [10]. Techniques currently used to produce and passivate nAl particles are discussed in Section 12.3.

Some basic definitions are recalled in this introductory section because they are used throughout the chapter. At least for propulsion applications, the gravimetric specific impulse I_s is among the most important parameters of merit. It is traditionally (but improperly) as the thrust $\mathcal{T}$ per unit weight flow rate of burned propellant $I_s = \mathcal{T}/\dot{w}$. Under the ideal assumptions of a nonreactive and calorically perfect gaseous mixture expanding through a gas-dynamic nozzle (frozen expansion model), one finds $I_s \sim \sqrt{T_{cc}/\overline{\overline{\mathcal{M}}}}$, where T_{cc} represents the temperature and $\overline{\overline{\mathcal{M}}}$ the molar mass of the gas at the exit of the combustion chamber. The hypothesis of thermochemical equilibrium is, in general, assumed for the computation of

Metal Nanopowders: Production, Characterization, and Energetic Applications, First Edition.
Edited by Alexander Gromov and Ulrich Teipel.

combustion chamber properties. That is, in this simplified picture, the analysis of a specific impulse translates into the investigation of reactant effects on combustion temperature and molar mass of the reaction products.

For system-level considerations, it is important to monitor the volumetric (or density) specific impulse as well, defined as $I_v = I_s \cdot < \rho_p >$, where $< \rho_p >$ represents the mean propellant density. A large propellant density is always convenient for flying systems; in some cases, it is a dominating factor. An instance of I_v reduction against an increment of I_s can be found in solid propellants, when some types of metal hydrides are used to replace regular aluminum (for details, see Section 12.2).

12.1.1
An Overall Introduction to Energetic Systems

Metal powders have been investigated as high-energy fuels since the beginning of rocket history. Solid rocket propellants commonly contain Al powder, thanks to the high energy release from the oxidation of Al to Al_2O_3. Owing to the large negative enthalpies of formation of both liquid and solid Al_2O_3, respectively, −387.3 and −400.5 kcal mol^{-1} [11], the oxidation of metallic Al to either of these products is accompanied by a heat release of more than 7 kcal g^{-1} of Al. The temperature increase due to Al oxidation and the reduction of H_2O and CO_2, respectively, to H_2 and CO results in a specific impulse improvement of around 10%; Al high density (2.70 in terms of specific gravity) allows obtaining a significant gain in propellant mass fraction, approximately estimated at 15%. A consequence of Al addition, even in low mass fractions, is the presence of condensed combustion products (CCPs); condensed particles are not only effective in damping acoustic rocket motor instabilities but they also reduce the delivered specific impulse and cause nozzle erosion. In two-phase flows, the largest particles are not able to follow the stream lines and the droplet diameter becomes one of the important factors affecting the droplet's entrapment, showing the need to control the alumina particle size. Two-phase flow losses are peculiar aspects of the combustion processes influenced by Al powders. Moreover, in rocket motors that burn aluminized propellants, a significant amount of Al_2O_3 slag can accumulate in the rear chamber region close to the nozzle. This alumina accumulation may have important consequences on specific impulse, thermal insulation behavior, and thrust vectoring, and constitutes an overload that might endanger the propulsive mission. An example is given by space shuttle boosters which, in static tests of the high-performance motor version [12], produced a slag weight in the range 426–1407 kg.

Al does not vaporize at the typical burning surface temperature of AP/HTPB (ammonium perchlorate/hydroxyl-terminated polybutadiene) propellants, around 900–950 K. Many experimental investigations [13–15] show that Al particles adhere to the burning surface temporarily: here, they form aggregates, which later burn and detach from the propellant surface subjected to a very complex convective flow. For composite aluminized propellants, surface oxidation of Al can occur, after melting at 933 K, in the oxide layer cracks resulting from the different phase change expansions of Al and Al_2O_3. This oxidation provides an opportunity for concentrations of Al droplets, confined in their oxide layer, to sinter together into bigger aggregates. At the melting temperature of the oxide layer (2320 K), the Al surface tension converts the aggregate to one or more droplets, and the surface tension of the insoluble oxide causes it to retract to one or more lobes on the Al droplet surface. The breakdown of the oxide layer leads to a rapid increase in Al oxidation and temperature increase of the droplet. The combustion-condensed products of aluminized propellants consist of agglomerates of hundreds and even thousands of microns, as well as of smaller particles [16] that have sizes on the order of 1 μm. The size of the captured particles in a typical solid propellant combustion process results from the combustion of large particles, larger than those originally contained in the grain; they come from an agglomeration process. Because of their size, their combustion time are longer than those of the loaded Al particles and influences combustion efficiency and motor performance.

Combustion processes of conventional μAl powders develop relatively far from the propellant surface and do not significantly contribute to propellant burning rate. A great deal of attention was recently focused on ultrafine energetic particles, particularly Al nanoparticles, because of the significant increase in propellant burning rates, shorter ignition delays, and shorter agglomerate burning times. Under the assumption of the D^2-law for particle burning, it was observed that a reduction factor of 10 in the particle diameter results in a reduction of 100 in burning time [17]. Nanoparticles typically enhance the linear burning rate of aluminized solid propellants by 100% under standard operating conditions. Nonequilibrium processes, often used to produce ultrafine nanopowders, are probably responsible for the unusual combustion behavior of these powders. It has been demonstrated that the particle size has an effect on the reactivity of Al powders; the well-known Alex powder produced in Russia, with diameters in the range of 100–200 nm and Brunauer, Emmett, and Teller (BET) surface areas of approximately $12\,m^2\,g^{-1}$, displays an unusual behavior in thermal tests [18, 19] showing a strong exothermic peak in air at temperatures that are 100–200 K below the Al melting point. This phenomenon, the result of oxidation of particles, only occurs with nanoscale powders. The observed low-temperature exothermic behavior was related to a "structural energy" due to defects in the crystal lattice; however, successive studies did not find evidence for this internal energy [20].

nAl particles are very sensitive to aging problems. For this reason, organic coatings were investigated to test their effectiveness in preventing the oxidation of nAl powders. The palmitic acid coating, used in L-Alex powder, forms an effective barrier to hydrolysis and only minor degradation of the metal, after accelerated aging lasted 40 days, could be detected [21]. The passivation coating of palmitic acid, chemically bound to the Al surface via a carboxylate linkage, degrades under prolonged exposure to hot, humid conditions. The coating is retained on the metal surface and provides excellent protection against the oxidation. Further, L-Alex incorporated into HTPB binder systems shows good compatibility.

The increased amount of Al_2O_3 oxide coating, which reduces the achievable specific impulse and hinders propellant manufacturing because of the fine particle loading, prompted the development of propellants with a mixing of conventional μAl and fine Al (with bimodal size Al distributions) [22]. Poor mechanical properties and high content of Al_2O_3 require further investigation.

12.1.2
Experimental Investigations on Micro and Nano Energetic Additives

The thermal behavior of μAl and nAl powders can be investigated using differential scanning calorimetry (*DSC*), simultaneous thermogravimetry–differential thermal analysis (*TG–DTA*), and more advanced systems for the identification of emitted vapors and the understanding of transformation mechanisms, given by the coupling of a gas analyzer (*GC*, gas chromatograph; *QMS*, quadrupole mass spectrometer; *FTIR*, Fourier transformation infrared spectrometer, etc.) to a TGA (thermogravimetric analysis) system. Thermal analyses can also be performed using an accelerating rate calorimetry (*ARC*), which is an adiabatic calorimeter suitable for thermal and pressure hazard parameters, including onset temperature, adiabatic temperature increase, pressure generation rate, time to maximum rate, temperature of no return, and self-accelerating decomposition temperature [23].

One of the most important properties of a powder is its specific surface area, expressed in square meter per gram. The *BET* technique (from the authors Brunauer, Emmett, and Teller) uses the principle of the physical inert gas adsorption (typically nitrogen) on a solid surface to varying of the relationship between the partial pressure of nitrogen and its vapor pressure to the temperature of liquid nitrogen.

Electron microscopy is used to examine structures over many orders of magnitude, from macroscopic to atomic scale, allowing a study of structure–property relationships. In addition, with increasing resolution of modern instruments, the study of shape, dimension, morphology, and defects has improved considerably in depth and detail. In this context, scanning electron microscopy (*SEM*) and transmission electron microscopy (*TEM*) are used for the purpose; in the second case, *TEM* is both conventional transmission electron microscopy (*CTEM*) and high-resolution electron microscopy (*HREM*). In the *SEM*, specimens may be observed directly with little or no surface preparation, over useful magnifications ranging from approximately 20 to 100k×. Powder preparation for *TEM*, instead, is rather complicated, owing to the necessity of obtaining films about 100 nm thick by means of an ultramicrotome. To ensure that the sample will withstand the cutting process, without suffering extensive modification or damage, it is necessary to provide it with extra rigidity. This is usually obtained by a process known as *embedding*, which involves introducing the sample into a suitable medium such as an epoxy resin and then polymerizing it by gentle heating and/or by irradiation with ultraviolet light. Additives are often employed to speed up the polymerization process and to vary plasticity and hardness of the end product in readiness for sectioning. The sections thus produced are placed on a copper folding grid of 100 × 200 meshes. In order to reduce electrical charging of the samples (Al plus resin), ultramicrotomed films are coated with a thin evaporated carbon film before their observation in the electron microscope at room temperature [24].

XPS is used, in addition to sputtering of Ar^+ at 4 keV, to obtain chemical depth profiles that give information on the surface passivation layers and allow chemical species to be quantified. The sputtering rate is estimated, so that the time scale can be converted into a depth scale. This surface analysis explores a volume with a relatively large surface area (hundreds of microns) but a very thin thickness (about 10 nm). By using sputtering, hundreds of nanometers in depth can be investigated. All the elements are detected in the spectrum, provided that they are over the instrument sensitivity, except hydrogen. To characterize the surface chemistry, atomic absorption spectrometry (AAS) and Auger electron spectrometry (AES) can also be used. The first technique is based on the absorption of optical radiation (light) by free atoms in the gaseous state and allows the quantitative determination of chemical elements; the second, based on the Auger effect, and related to the analysis of energetic electrons emitted from an excited atom, is a technique for probing the chemical and compositional surface that is well developed in metallurgy.

XRD spectra are obtained to disclose the crystalline phases and their relative abundance. A qualitative analysis can be carried out by means of the standard powder method. In addition, a full profile-fitting procedure, based on the Rietveld method [25], allows accurate quantitative analysis to be performed and complete structural refinements of each identified phase to be obtained. Furthermore, by means of the investigation of the intrinsic broadening of diffraction peaks, accurate values of coherent scattering-domain lengths can also been obtained.

TG–DTA and *DSC* results show that the reactivity of Al powders in air depends on the particle size and the history of the sample. According to Kwok *et al.* [26], an early oxidation of Al nanopowders can be observed between 700 and 870 K, while from an *ARC* experiment performed on a mixture of Al/H_2O, exothermic activity is detected at 305 K because of the high reactiveness of Al with water. Aging of Al was investigated for samples exposed to humid atmosphere at room temperature for more than 47 days, showing that Al completely reacted in 74 days. A value of approximately 13% mass of Al_2O_3 was obtained from the AAS analysis for Alex. Following the same authors, *XPS* and *AES* results indicate that most of the Al on the surface is the Al^{3+} species, mixed with oxygen, carbon, and nitrogen; a thickness of about 10 nm was estimated for the oxide layer, and near the surface of Alex an appreciable AlN amount was also detected. Outgassing studies, performed using a *TG–DTA–FTIR–MS* technique, show desorption of H_2O and CO_2 from Al nanopowders.

12.1.3
Theoretical/Numerical Investigations on Energetic Additives

Typical combustion products of composite propellants are H_2O, CO_2, CO, HCl, and N_2, but the low mass fraction of molecular oxygen contained in the products stresses the role of H_2O and CO_2, which are the primary oxidizing species for Al combustion.

The combustion of Al is similar to that of a liquid fuel droplet. The reaction is limited by the diffusion of the oxidizer toward the particle surface and the vaporized metal away from the particle surface, which are responsible for the development of a flame zone at a distance from the particle surface where the vaporized metal and the oxidizer mix and burn. Al vaporization is due to the heat conduction from the flame back to the surface. Al combustion produces a condensed phase that can diffuse back to the particle surface; the condensation on the surface forms an oxide cap and provides additional energy that can increase the particle combustion rate. But the oxide cap can also reduce the surface area available for combustion and radiate heat away from the particle. Both of these effects would slow the reaction as burning of the droplet progresses [27].

Data are usually correlated assuming the burning time to be proportional to the diameter of the original particle squared. In general, burning time is proportional to the diameter raised to a power between 1.5 and 2.0. Different explanations for the change in exponent have been proposed. The most notable model is that of Beckstead [28], who formulated a comprehensive model for the burning behavior of spherical Al particles. His assumption treats the Al droplet as a rigid sphere, and forces the formation of an oxide cap on the side of the droplet. His calculated results compared fairly well with experimental results, both in diffusion flame structure and diameter dependence on the burning rate. Beckstead [17] found a D^n power law, with $n = 1.8–2.0$, suitable over the 10–100 μm sized particles. The applicability of this law was not established for nanosized particles, because chemical kinetics becomes the dominant mechanism rather than diffusion just as for μAl particle combustion.

Bucher *et al.* [29] investigated the species close to the burning Al particles in CO_2, N_2O, and CO as well as in mixtures of O_2, N_2, and argon at atmospheric pressure. Under the tested conditions, AlO was identified as a primary intermediate gas-phase product species.

Sarou-Kanian *et al.* [30] investigated the combustion of large Al particles (on the order of 3 mm). In a CO gas environment, carbon can deposit on the particle surface, which suppresses the rate of Al vapor departure from the particle surface as well as rate heterogeneous combustion. They found that a small amount of water vapor or hydrogen in a CO_2 gas mixture can cause the oxide shell to recede, but it does not recede in a dry CO_2 environment. They also observed that C can dissolve into the liquid Al, eventually forming a carbon shell on the molten Al droplet surface, which can prevent further oxidation of the Al particle. Burned particles were cross sectioned and examined through X-ray diffraction (XRD) techniques; Al, Al_2O_3, Al_4C_3, and Al_4O_4C were discovered in the interior of the particles, indicating that carbon dissolution occurred during combustion. However, the effects of carbon dissolution in a molten Al droplet may not produce a very significant effect in the case of metallized solid propellant combustion, owing to the relatively low concentrations of CO_2 and water vapor in combustion products. Bazyn *et al.* [31] studied the combustion of 10 μm sized Al particles in the reflected region of a shock tube and found that the CO_2

was slightly more effective than H_2O as an oxidizer. They also found pressure exponents of −0.9, 0.3, and 0.2 for O_2, CO_2, and H_2O on particle burning time. Diffusion-controlled combustion theory predicts no pressure dependence for oxidizing species, but as particle sizes become smaller, the diffusion length and time scales are reduced and kinetic-controlled combustion becomes increasingly prevalent. The strong pressure exponent for O_2 is indicative of kinetic-controlled combustion mechanisms, whereas the low values of pressure exponents for H_2O and CO_2 are indicative of diffusion-controlled combustion mechanisms. Their results also show that CO_2 is about twice as effective as H_2O and half as effective as O_2, nearly the opposite trend reported by Beckstead for much larger μAl particles.

In recent years, interest and availability of nanoscale materials have increased significantly; application of nAl particles in chemical propulsion required the investigation of their combustion processes. Al flakes (16 μm in diameter and 200 nm in thickness) have been applied to energetic materials for chemical propulsion; during combustion processes of metallized energetic explosive materials, Al flakes could be dispersed at very high speed into a hot gas environment, in which the time of dispersion could be so short that the particles retain their nonspherical shapes. The times required for the diffusion of the oxidizer to the particle surface are decreased significantly, if compared with larger Al spheres, because of their nanosized thickness. Therefore, the gas-phase diffusion time is significantly shorter than the conventional case, leading to a chemical-kinetic-controlled situation. The model simulation should include heterogeneous reaction rates. A detailed combustion model must include both gas-phase and heterogeneous reaction chemistry if the model is to be valid for both large and small Al particles. A model of Al particle combustion should also have to take into account the oxide growth.

Recently, Houim and Kuo [32] formulated a comprehensive model of nonspherical Al particle ignition and combustion; the model simulates many physicochemical processes associated with particle deformation, interaction with surrounding fluid motion, mass evaporation, energy transfer, and chemical reactions. The model is based on the volume of fluid multiphase approach; it can calculate oxide cap formation without assuming a definite oxide lobe shape or end position. The interfacial treatment can account for droplet surface tension, condensed-phase material heat up as well as boiling, condensation, and heterogeneous chemical reactions. Numerical results obtained from this generalized multiphase combustion model have demonstrated the capability of solving the detailed species mass fraction distribution over Al particles of different shapes. Numerical results also showed the time evolution of molten Al particles into various shapes depending on their initial geometric configurations. Calculated results also demonstrated the importance of particle spin rates in particle morphology. At high spin rates, a disk-shaped Al flake can break apart into multiple fragments; this phenomenon is important because the exposed surface area can vary on account of the dynamic interaction of the surrounding fluid with the particle.

12.1.4 Thermites

Thermites are a class of pyrotechnic compositions of metal and metal/nonmetal oxide that can produce strong, exothermic redox reactions [33, 34]. A thermite reaction is written as

$$\mathrm{Me + XO \rightarrow MeO + X}, \quad \Delta H_{rx} > 0$$

with Me and O being, respectively, a metal and oxygen, X either a metal or a nonmetal, and ΔH_{rx} the heat of reaction. A detailed list of possible thermite adiabatic flame temperatures, T_f, and reaction enthalpies, ΔH_{rx}, is reported in [35]. A brief summary of these data is given in Table 12.1. Aluminum is commonly used as a metal in thermite reactions because of its relatively low melting point associated with a high boiling temperature and its easy availability (reduced cost). The low melting point of Al is an advantage because of the crucial role played by reactant diffusion in thermite reaction: the liquid phase yields enhanced diffusion with respect to the solid state. High metal boiling points yield higher T_f, as this parameter is limited by the reactants' vaporization temperatures [35]. Thermite reactions can deliver ΔH_{rx} higher than those of conventional explosives, although the energy release of the latter is faster [36]. Industrial uses of thermites include railroad track welding [37], torches for atmospheric/underwater cutting, creation of *in situ* alumina liners in pipes [34, 35] and ES. In particular, $Al + MoO_3$ thermites are considered as candidates for substitution of lead-based composites in fire gun primers [38, 39].

12.1.4.1 Nanocomposite Thermites

Thermite formulations based on nanosized particles are also called *nanothermites* or *superthermites*. In this section, a brief survey of open literature studies on

Table 12.1 Adiabatic flame temperatures and reaction enthalpies for Al, Mg, and B thermites [35].

Formulation	T_f (K)	ΔH_{rx} (cal g^{-1})	ΔH_{rx} (cal cm^{-3})
$2Al + B_2O_3$	2327	780.7	1971
$2Al + 3CuO$	2843	974.1	4976
$2Al + Fe_2O_3$	3135	945.4	3947
$8Al + Fe_2O_4$	3135	878.8	3747
$2Al + MoO_3$	3253	1124	4279
$2B + CuO$	2843	738.0	4182
$2B + Fe_2O_3$	2065	590.1	2751
$8B + Fe_2O_4$	1903	530.1	2462
$3Mg + B_2O_3$	3873	2134	1195
$Mg + CuO$	6502	1102	4336
$3Mg + Fe_2O_3$	3135	1110	3579
$4Mg + Fe_2O_4$	3135	1033	3383

nanocomposite thermites is given. Performance of a thermite formulation depends on several factors such as component particle sizes, equivalence ratio, fuel/oxidizer particle size distributions, and actual density (loose powders/pressed pellets).

Nanocomposite thermites exhibit a reduced ignition delay with respect to conventional, coarser counterparts thanks to their enhanced reactivity demonstrated by DSC and laser ignition tests [40]. Pantoya and Granier [40] investigated the behavior of $Al + MoO_3$ in DSC (argon atmosphere, heating rate of $10\,K\,min^{-1}$) for two different Al particle sizes. The tested μAl (10–14 μm) provides an exothermic reaction for $T = 1232.15$ K, after Al melting. However, nAl (40 nm) presents a marked exothermic peak for $T = 749.15$ K. These results agree with ignition tests conducted by laser heating on pressed pellets ($\rho_{actual} = 38\%$ of TMD) under ambient pressure. The enhanced reactivity of nanosized particles cannot strictly be related to the reduced melting temperature of nAl [40]. Significant reduction in the melting temperature of nanosized metal with respect to the bulk material requires particle diameters ≤30 nm [40, 41]. Although these finer particles may be present, their mass fraction is estimated to be too low to significantly influence the powder behavior [40–43]. As discussed by Moore *et al.* [44], addition of nAl to μAl can reduce ignition delay of thermite formulations (Figure 12.1). Similar results were presented by Popenko *et al.* [45] for the behavior of mixtures on nAl and μAl powders.

For a given thermite formulation, ignition delay and burn rate depend on the equivalence ratio of the formulation, Φ. Experiments reveal that fuel-rich formulations exhibit shortened ignition delay and higher burning rate [40, 46, 47]. Some authors report similar findings could be due to the influence of test environment oxidizers (as O_2, CO_2, and H_2O). Asay *et al.* [48] conducted combustion tests on low-density samples of nAl (44 nm) + MoO_3 (15.5 nm) at ambient pressure and under rough vacuum (~3.3 Pa). The burning rate result was not significantly different between atmospheric and vacuum tests. This suggests that environment oxidizers do not exert a marked influence on thermite combustion. Sanders *et al.* [47] experimentally investigated the burn rate of nAl (80 nm) with different metal

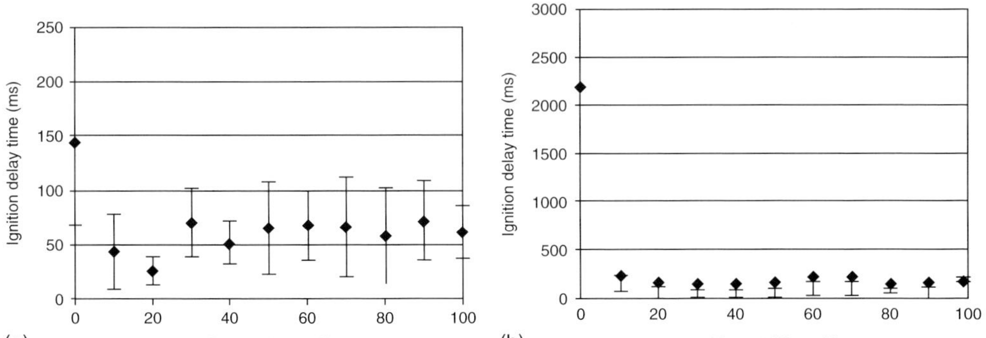

Figure 12.1 Ignition delay of $Al + MoO_3$ ($\Phi = 1.3$) under laser radiation (50 W) for (a) 4 μm Al + nAl and (b) 20 μm Al + nAl. The mean particle size of nAl is 80 nm [35].

oxides, including MoO_3 and CuO. Equivalence ratios ranging from 0.8 to 1.8 were considered in the investigation. Results show that maximum burning rate is usually associated with fuel-rich conditions. For nAl + CuO (~20 nm, rods), the maximum burning rate is achieved for $\Phi \sim 1$.

A possible explanation for this behavior is suggested by Sanders *et al.* considering thermochemistry analysis: for all the tested formulations except nAl + CuO, the gaseous reaction product amount has a maximum for $\Phi > 1$. Under these circumstances, both Al and the oxidizers are in liquid or gaseous phase, thus providing an effective contribution to the convective heat transfer ruling thermite combustion at low densities. The Al reaction with CuO requires a temperature close to the stoichiometric value to maximize the gas mass fraction (that is, $\Phi \sim 1$). These results testify the importance of convective heat transfer in thermite burning, especially for loose powder/low-density samples. Increasing sample density yields a more effective conductive heat transfer contribution and a reduction in convection. Dedicated tests conducted by Asay *et al.* [48] on Al + MoO_3 thermite showed that thermal radiation heat transfer exerts no significant influence on the reaction propagation speed.

Detailed investigation of the combustion of nAl + MoO_3 composites was conducted by Granier and Pantoya [40, 46, 49]. The standard μAl powders and their nanosized counterparts were tested considering different sample densities. The conventional μAl had particle sizes ranging from 3 to 20 μm, while nAl was characterized by particle sizes in the range of 30–100 nm. An overview of the results achieved in tests comparing μAl and nAl particles in thermite formulations is shown in Figure 12.2. The first remarkable difference between conventional and

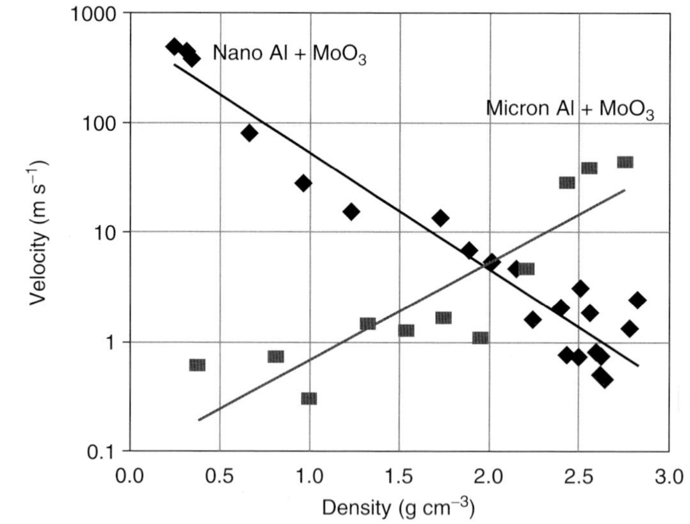

Figure 12.2 Combustion velocity as a function of Al particle size for Al + MoO_3 ($\Phi = 1.2$) [40]. Particle sizes range from 10 to 14 μm for micron-sized formulations and from 76 to 80 nm for the nanocomposites. The investigated actual densities extend from 6.5% to 73% of TMD.

nanocomposite thermites is given by the effect of sample compaction (i.e., actual density) on the burning rate. According to Figure 12.2, increasing the sample density yields burning rate increases for micron-sized formulations, while the opposite trend characterizes the behavior of nanothermites.

As pointed out by Pantoya *et al.* [49], this could be related to different concurring phenomena ruling the combustion of micron- and nanosized thermites. By increasing the tested sample compaction, thermal conduction and thermal diffusivity of the strand increase for both coarser thermites and nAl-containing formulations. For micron-sized thermites, Al and MoO_3 diffusion path reduction and the Al_2O_3 shell damage induced by the pressing of the pellets can enhance diffusion of the oxidizer toward the metal fuel, thus yielding to the combustion wave propagation velocity increase for increasing densities. The nanothermites' behavior can be related to the *melt dispersion mechanism* recently proposed by Levitas *et al.* [49–52]. According to this mechanochemical process, fast heating ($>10^6\ K\,s^{-1}$) yields intense internal stresses in the nanosized particles because of the differences in thermal expansion coefficients between the Al and the Al_2O_3. When Al melting point is reached, a strong strain is generated by the volume change of the nAl particle core. When the ratio between nAl particle radius and oxide layer thickness is lower than a critical value ($R/\delta < 19$), this volumetric strain induces fracturing of the Al_2O_3 shell. Disruption of the external Al_2O_3 shell yields an unbalanced pressure wave. This leads to an unloading wave in the liquid Al droplet that, in turn, creates a tensile pressure. Thanks to the latter, Al droplets are dispersed into the surrounding environment with a high speed ($\sim 100\ m\,s^{-1}$) and the particle combustion is no more limited by diffusion. For compacted nanothermites, the melt dispersion mechanism could be hindered by possible oxide shell damage induced by the pressing and disruption of the spherical particle shape. The reduced resistance of the damaged oxide shell could impede the pressure buildup into the particle core induced by Al melting. Similarly, changes in particle morphology can create zones with reduced tensile strength. The strong compaction of the particles reduces the free space for melt aluminum dispersion, thus limiting/canceling the combustion wave velocity enhancement. In spite of the absence of a direct, experimental confirmation of the melt dispersion mechanism, this is currently the only possible explanation for the behavior of nanothermites and some of the peculiar behaviors of nAl [49].

12.1.5 Explosives

Explosive materials are reactive substances in gaseous, liquid, or solid state, containing a great amount of potential energy that can be released in a very short time through exothermic reactions. Substances that detonate are said to be *high explosives* and substances that deflagrate *low explosives*. While high explosives can reach a pressure peak of 300 000 bar, the peak of a deflagrating or low explosive does not usually exceed 4000 bar.

Explosive materials may also be classified by their sensitivity, which relates to the required heat, or shock or friction intensity for the material ignition or detonation. *Primary explosives* are materials extremely sensitive to stimuli such as heat, impact, friction, static electricity, or electromagnetic radiation; a relatively small amount of energy is required for initiation. *Secondary* or *tertiary explosives* are materials relatively insensitive to these parameters; examples of secondary explosives include 2,4,6-trinitrotoluene (TNT) and cyclo-1,3,5-trimethylene-2,4,6-trinitramine (RDX).

The energy available from a standard quantity of explosive is defined as the *explosive strength*; it can be referred to as unit weight (weight strength) or unit volume (bulk strength).

Stability is a fundamental property of explosive materials; it is the ability of an explosive to be stored without deterioration. For explosives, stability commonly relates to ease of detonation, concerned with kinetics, which is responsible for the decomposition rate. Groups such as *nitro* ($-NO_2$), *nitrate* ($-ONO_2$), and *azide* ($-N_3$), are known to be intrinsically labile; these groups exhibit high sensitivity to flame or mechanical shock. Stability is also affected by the temperature of storage. The rate of decomposition of explosives increases at higher temperatures. All standard military explosives may be considered to have a high degree of stability in the temperature range from 263 up to 308 K, but each is characterized by a temperature at which its stability is reduced and its decomposition rate rapidly increases. At temperatures approximately above 343 K, most explosives become unstable. Other parameters affecting stability are sunlight and electrical discharge. Many explosive compounds containing nitrogen groups rapidly decompose, under the sun's ultraviolet rays, affecting their stability. Static or other electrical discharge may cause a reaction, even detonation; note that electrostatic or spark sensitivity to initiation is common in a number of explosives, and, for this reason, for safe handling of explosives a proper electrical grounding is required.

The oldest explosive, blackpowder or gunpowder, is made of potassium or sodium nitrate, sulfur, and charcoal. It is a deflagrating explosive, nowadays mainly used in fireworks and fuses. The first of the modern explosive materials is *nitroglycerine*, which is a detonating material. It is a liquid, extremely sensitive substance, usually mixed with other absorbent substances to give dry, stable explosives. Straight dynamites contain nitroglycerine, sodium or potassium nitrate, and combustible materials such as flour or wood pulp, in a ratio such that the combustible material is fully oxidized by the nitrate. Another important explosive material is *ammonium nitrate*. It is used in mixtures with other explosives, such as nitroglycerine, or with combustible materials. These mixtures have a low explosion temperature and a low sensitivity; for this reason, they are very safe to handle, in spite of the fact that ammonium nitrate is probably responsible for the most serious accidental explosions in the history of catastrophic events caused by explosive materials.

Most commercial explosives are mixtures of explosive and nonexplosive substances. The fundamental requirement of such mixtures is the necessary explosive powder that they should provide. Fuel and oxidant are contained in all explosives, sometimes in the same molecule; if the fuel/oxidant ratio does not correspond to the correct stoichiometry, the produced combustion products will contain nitric oxide and nitrogen dioxide for an oxygen excess, and carbon monoxide for an oxygen deficiency. Addition of ammonium nitrate can give the needed oxygen required. The fuel may be represented by a wide range of materials, including organic compounds or metals, in compositions where oxygen is in excess. To reduce the sensitivity of the explosive, to improve the long-term stability or its manufacture, a lot of other compounds may be added. Explosives in common use belong to one of the following four classes:

1) Nitroglycerine gelatins ($\rho = 1.4{-}1.6\,g\,cm^{-3}$)
2) Nitroglycerine semigelatins and powders ($\rho = 0.7{-}1.3\,g\,cm^{-3}$)
3) TNT powders ($\rho \approx 1\,g\,cm^{-3}$)
4) Ammonium nitrate powders ($\rho \approx 1\,g\,cm^{-3}$).

Al powders are a common ingredient in energetic materials. They are used to increase the energy and raise the flame temperature in rocket propellants; they are also added to explosives to enhance air blast, raise reaction temperatures, create incendiary effects, and increase bubble energies in underwater weapons. In explosives, it is generally assumed that combustion of Al particles occurs behind the reaction front, during the expansion of the gaseous detonation products. Particles act as inert ingredients and do not participate in the reaction zone.

It was previously stressed that nAl powders were known to react more rapidly than conventional μAl grades in propellant and explosive compositions. In explosives, it is not clear if nAl particles are small enough to react faster than micrometric particles and, therefore, to participate in the detonation reaction zone. Improvements using nanopowders in explosives have been investigated in [18, 53]; other research activities demonstrated some improvements in the velocity of detonation (VoD) of ADN/Al blends using nAl [54]. Tulis *et al.* [55] also measured an increase in the VoD of fuel-oxidizer blends (lactose–Al–AP) when Alex was compared to Al flakes. However, other reports [56–58] showed either no improvement in performance or even a decrease in performance when nAl was used. The reaction mechanism of nAl in explosives is not yet clear. Many researchers have reported on a large number of different compositions with mixed results.

The potential of nAl powders in explosive compositions was extensively studied by the Defence Research and Development Canada-Valcartier (DRDC-V) and the Defence Science and Technology Organisation (DSTO). They developed different plastic-bonded explosives (PBXs) and TNT-based formulations to compare ultra-fine and conventional μAl powders [59]. For PBX compositions, no significant differences were observed between formulations containing μAl and nAl powders. Improvements were observed in the VoD and in plate dent depths using nAl for mixes of TNT and Al. Aquarium tests were run on TNT/Al blends to confirm the improvement observed, to evaluate the relative performance of the different sizes of Al, and to qualify the behavior of Al in such an explosive. The nAl chosen for their experimental study was Alex® with particle size in the 100–200 nm range. The μAl powder used was Valimet H-15 (average particle size 12 μm); Valimet H-2 (spherical, 2 μm); and MDX-75 from Alcan (spherical, 21 μm) were also tested. The reference Al used by DSTO had an average particle size of 17 μm.

To perform the study, three different types of explosives were tested: cast-cured PBXs, TNT-based melt-cast explosives, and an ammonium nitrate fuel oil (ANFO) formulation. The PBXs were selected to confirm the results available in the literature on HTPB/RDX/Al PBXs. The compositions were mixtures of HTPB, HMX, and Al. Cylinders of 25.4 and 38.1 mm in diameter and 228 mm in length were cast.

Experimental results point out that the addition of nAl Alex has very little effect on VoD, thus confirming previous studies on AP- and RDX-based PBX explosives. Conversely, the heat of detonation shows a reverse trend, with a small increase going from Al H-15 to Alex. When Al particle size decreases, more Al reacts in the system, contributing to the detonation energy.

For the ANFO formulations, crushed AN was sieved (150 mesh or 105 μm) and the passing fraction was blended with fuel oil and 10% of Al (either Alex or H-15). The final blend was oxygen balanced by adjusting the fuel oil. The final composition was 87.5% AN, 2.5% fuel oil, and 10% Al. To adjust the final explosive density and increase the sensitivity, microballoons (2–5%) were added.

Experimental results for the aluminized ANFO mixes, when μAl is replaced by Alex, show again a small decrease in VoD and an increase in the heat of detonation. All the mixtures of the families presented in this section are oxygen deficient; although ANFO explosives are characterized by a balanced oxygen content, this does not help improve the performance.

The largest changes with Al type is shown by mixtures with only TNT and Al. Table 12.2 presents the performance; mixes with Alex always show higher VoD

Table 12.2 Performance of TNT/Al explosives [59].

Al (%)	Al type	Density ($g\ cm^{-3}$)	Charge diameter (mm)	Detonation velocity ($m\ s^{-1}$)	Heat of detonation ($kJ\ kg^{-1}$)
10	H-15	1.67	25.4	5718	5307
10	Alex	1.67	25.4	6508	5479
10	H-15	1.64	31.8	6554	—
10	Alex	1.68	31.8	6638	—
20	H-15	1.74	25.4	6182	—
20	Alex	1.74	25.4	6598	—
20	H-15	1.72	31.8	6456	—
20	Alex	1.74	31.8	6724	—
30	MDX-75	1.82	25.4	5740	6184
30	H-15	1.82	25.4	6228	6519
30	H-2	1.82	25.4	6483	6660
30	Alex	1.82	25.4	6628	6749
30	MDX-75	1.80	31.8	6400	—
30	H-15	1.80	31.8	6428	—
30	H-2	1.82	31.8	6707	—
30	Alex	1.84	31.8	6708	—
30	H-15	1.81	57.8	6591	—
30	Alex	1.84	57.8	6742	—

compared to those added with μAl H-15. The difference in performance, which is larger at small charge diameters, decreases as the diameter increases. All these charges are close to the critical diameter, estimated to be 27.4 mm for cast TNT [60]. Results in Table 12.2 show the effect of the Al particle size on the performance of TNT/Al mixtures: smaller Al particles determine a higher VoD in unconfined charges. However, note that at a diameter of 31.8, mixes with Al H-2 give the same VoD as that obtained with Alex. The increase in performance is therefore not limited to nAl. The enhancement from nAl has been previously reported in [58]. Increases in the brisance in TNT/Alex and TNT/RDX/Alex blends were measured by Dorsett *et al.* [61].

Underwater detonations are investigated in aquarium tests, useful to study the early-time formation of the water shock wave and the gas bubble of detonation products. The detonation front within the solid explosive can be differentiated from the expanding products because the water acts to contain the detonation products. Dorsett *et al.* [61] demonstrated that Alex reacted earlier and faster than μAl in tritonal-type compositions, measuring a peak at 8 μs lasting 17 μs for Alex, compared to a peak at 25 μs lasting for more than 100 μs for a 17 μm Al particle size. This early reaction can be responsible for the larger heat of detonation as well as the increased brisance.

In conclusion, nAl powders did not increase the VoD of aluminized PBXs based on an inert binder. When Alex replaced μAl, the performance (VoD) of PBXs based on an energetic binder was decreased. In ANFO compositions at 10% Al, Alex also reduces the VoD, but increases the heat of detonation. Alex increases both VoD and heat of detonation in TNT/Al mixes; the increase of VoD is more pronounced at small charge diameters, close to the critical diameter. In aquarium tests, expansion of gases appears faster for tritonal-type formulations containing nAl. Alex is shown to decrease significantly the critical diameter of tritonal-type mixes. There is no increase in air-blast performance by nAl; a reduction of air-blast pressure is measured for inert PBXs and melt-cast explosives.

Examples of new materials added with nanopowders have been shown earlier, but a lot of research activity has to be done in order to understand and control the effects of nanopowders in explosive materials. Delivery of energy from explosives and propellants represents a strategic and challenging area for the present and for the future.

12.1.6
A Short Historical Survey of SPLab Contributions

The Space Propulsion Laboratory (SPLab) of Politecnico di Milano investigated in the past 10 years a lot of propellant formulations for solid rocket applications, mainly of the family AP/Al/HTPB, and fuel formulations, mainly HTPB- and paraffin-based, for hybrid rocket systems including nanopowder additives.

Solid propellants are analyzed in terms of steady burning rate, usually measured up to 70 bar, using an automated image technique from video recording. Details of the composition are investigated, linked to the coarse and fine AP fractions and to the associated change in pressure sensitivity. The corresponding flame structure is analyzed by a high-speed and high-resolution video recording system, which allows investigating in detail the CCPs taking place on the burning surface as a result of the aggregation and agglomeration processes. The word agglomeration is reserved for the spherical drops of liquid metal in combustion with an oxide cap, while the word aggregation is reserved for the partially oxidized objects of irregular shape typically seen as precursors of agglomeration [62]. Agglomeration always implies a loss of the initial particle individuality, while aggregation may keep some remnant of the initial particle individuality. Optical measurements of the agglomerate diameters detaching from the burning surface can be performed or movies can be analyzed using dedicated software. Ignition delay is measured under radiant conditions; a CO_2 laser beam is used to ignite propellant strands, investigating the effect of radiant flux and pressure. Other studies involve the pressure deflagration limit (PDL), the metal powder dispersion, and analyses of the solid combustion residues.

Tests performed at SPLab with a variety of nanopowders show [63–66], in general, faster burning rates, quicker ignition, and reduced agglomeration but also problems with manufacturing and mechanical properties. High-speed and high-resolution video system techniques, such as SEM, X-ray photoelectron spectroscopy (XPS), and XRD analyses, allow characterizing the condensed products, showing the important effects of the oxidizer grain size on Al agglomeration phenomena at the propellant burning surface. Larger AP grain sizes lead to a significant increase in the agglomeration processes and in the agglomerate size. The nAl metal fuel, with respect to the micro-sized fuel, promotes a more intense heat release near the burning surface, which favors higher steady burning rates and reduces agglomerate sizes. Aggregation and agglomeration phenomena are also affected by the different propellant microstructure for nanometric Al propellants, leading to a peculiar form of aggregates. Visual analyses performed at SPLab point out for AP/HTPB formulations the emission, from the propellant microstructure over the combustion surface, of aluminized solid aggregates whose different structure is shown in Figure 12.3.

These aggregates emanate from the propellant, grow on the burning surface, accumulate, and protrude into the gas phase until their detachment. The possibility exists for nAl formulations of in-depth metallic networks forming during propellant manufacturing. This distinctive difference with respect to micro-Al can sensibly influence agglomeration effects.

SPLab research activities are also carried out for the development of solid fuel formulations for hybrid propulsion systems. Several formulations, HTPB-based or paraffin-based, added with different families of energetic additives, are extensively investigated.

Experimental tests are performed in a two-dimensional (2D)-radial microburner. The main aim of the facility is the relative grading of different fuel formulations under a variety of operating conditions. Thanks to its original design, the implemented experimental rig enables a continuous visualization of the head end of the tested strand during the combustion. Thus, r_f data are reduced by a continuous

Figure 12.3 Aggregation flakes issuing from the burning surface of an AP-based nAl propellant (5 bar).

tracking of the central port diameter during the combustion. The test rig enables the combustion of cylindrical strands with single central port perforation. Solid fuel grains are accommodated in stainless steel cases with $L_p = 30$ mm. The relatively small size of the tested strand enables execution of a high number of tests at reduced costs.

An optical, time-resolved technique for regression rate measurement, developed and validated at SPLab [67, 68], is currently in use. The ballistics of HTPB in GOX (gaseous oxygen) under quasi-steady and forced transient operating conditions was investigated by a time-resolved, optical technique. Under quasi-steady operating conditions, burning tests exhibited a marked regression rate at the beginning of the run and then a monotonic decrease of this parameter as combustion proceeds. This behavior is caught by a phenomenological, numerical approach aiming at regression rate simulation, developed from a model available in the literature. Under forced transient operating conditions, with oxidizer mass flow rate throttling down till strand shutdown, solid fuel exhibited a non-strictly monotonic ballistic response with the regression rate decreasing with a faint oscillating behavior [69].

Other research activities were performed using a 2D slab rig [70, 71]. The average regression rate of several fuel formulations was measured at a reference condition, corresponding to 150 kg m^{-2} s^{-1} oxidizer mass flux and 1.5 bar operating pressure. Tests were performed in double-slab configuration, with pure oxygen as oxidizer. The results of the ballistic characterization for different paraffin-based compositions are shown in Figure 12.4; the regression rate increase with respect to pure HTPB, measured at the same reference oxidizer mass flux, can be appreciated.

The regression rate percentage increase with respect to the reference baseline formulation (pure HTPB) ranges from +57% for HTPB filled with MgH_2, up

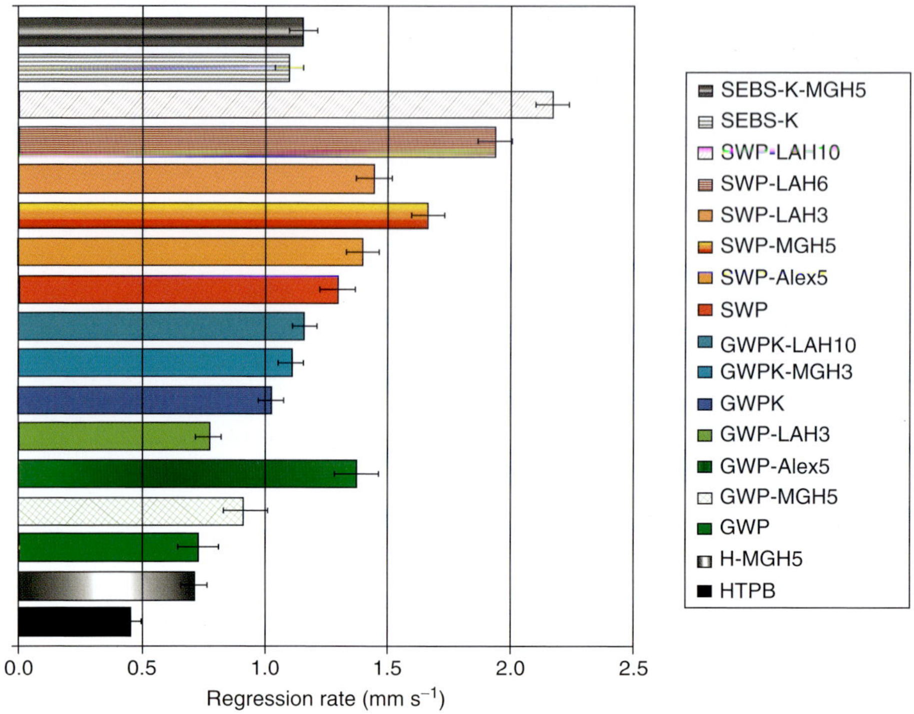

Figure 12.4 Regression rate comparison for the investigated fuel formulations. Oxygen mass flux: 150 kg m^{-2} s^{-1}. Operating pressure: 1.5 bar. GW: gel wax; GWP: gel wax strengthened with a poly-urethane foam; GWPK: gel wax strengthened with a poly-urethane foam and added with kerosene; SWP: solid wax strengthened with a poly-urethane foam; SEBS: styrene-ethylene-butylene-styrene block copolymer to improve paraffin-based fuels mechanical properties: MGH: magnesium hydride (3 and 5%); LAH: lithium aluminum hydride (3, 6 and 10%).

to +378% for SWP filled with $LiAlH_4$. Intermediate values are obtained for the GWP-based fuels (from +61% up to +202%) and for GWPK-based fuels (from +126% up to +144%). The highest regression rate enhancement, under the investigated operating conditions, is obtained with $LiAlH_4$ addition in SWP binder (up to +378% compared to pure HTPB). Results obtained suggest that kerosene addition is effective in enhancing GW-based fuels' r_f, by decreasing their viscosity and thus increasing their tendency to entrainment.

The same experimental rig assisted the development and validation of a simulation code specialized in hybrid rocket flame investigation. The code, based on the OpenFOAM framework library, is capable of steady-state as well as transient investigation of the turbulent reacting boundary layer, typical of this type of diffusive flame [72]. Some investigations were conducted on the combustion behavior of aluminum agglomerates in this type of flow, showing that in several cases, the residence time and the flow field conditions do not allow proper metal drop combustion [73].

12.1.7
Concluding Remarks on Energetic Additives

In energetic applications, such as propellants, pyrotechnics, and explosives, Al is widely used because of its high combustion enthalpy, easy availability, low toxicity, and good stability. Al powders are used to increase the energy and raise the flame temperature in rocket propellants; they are also added to explosives to enhance air blast, raise reaction temperatures, create incendiary effects, and increase bubble energies in underwater weapons. Combustion processes of conventional μAl powders develop relatively far from the propellant surface and do not significantly contribute to propellant burning rate. A great deal of attention was recently focused on ultrafine energetic particles, especially nanosized. These powders, characterized by very small size and consequently very high specific surface, are appealing because of their different chemical and physical properties, compared to the corresponding bulk or microsized materials. In particular, nAl is broadly exploited to improve the performance incrementing burning rate and combustion efficiency of ES, leading to shorter ignition delays and shorter agglomerate burning times with respect to ES containing μAl.

12.2
Thermochemical Performance of Energetic Additives

12.2.1
Ideal Performance Analysis of Metal Fuels

The use of energetic additives in solid propellants and hybrid solid fuels is of interest for several reasons. The increase in performance parameters represents one of the main drivers. In this sense, gravimetric as well as volumetric (or density) specific impulses are significant parameters for rocket propulsion purposes, being related to mass and volume savings at system level. A large propellant density is always convenient for flying systems. An instance of I_v reduction against an increment of I_s can be found in solid propellants, when some types of metal hydrides are used to replace regular aluminum [74]. On the other hand, additives can be adopted to improve other aspects, such as the efficiency of the combustion process. This is the case of nAl when it is added to hybrid rocket fuels: a specific impulse may be depressed but the mass burning rate increases [75, 76].

The scope of performance analysis consists of investigating the combustion parameters in connection with the propellant formulation (oxidizer, fuel, ingredient ratio), chamber pressure, and gas-dynamic expansion condition. Approaching this problem with thermochemistry allows for a simplified estimate of the performance without the use of complex kinetic models. On the other hand, restrictive hypotheses of equilibrium for chemical reactions, combustion chamber of infinite section area, ideal expansion, and many more assumptions lay beneath these results [77]. An ideal performance level represents a zero-loss characterization of a formulation, useful for cross comparison among different potential additives. However, the availability of small-scale and full-scale tests as well as the numerical investigations of rocket firings led to the definition of models or empirical correlations, capable of supplying realistic quantifications of performance detriments, identifying the contributions of different loss sources [78].

In order to obtain more accurate information on specific impulse, the hypothesis of frozen chemistry in the nozzle (recalled in Section 12.1) should be dropped. The complexity of chemical evolution is better approximated by assuming the so-called shifting equilibrium hypothesis, where each section of the nozzle is in local equilibrium according to the fluid conditions. It follows that the expanding mixture keeps on changing during expansion, thus influencing the specific impulse. Both for combustion chamber reactions and nozzle chemical evolution, the identification of the equilibrium state can take advantage of the thermodynamic potentials and, in particular, of the minimization of the Gibbs free energy [79]. Software codes using such a principle are capable of computing the main parameters of combustion as well as the expansion for reacting mixtures on the basis of thermodynamic properties of the initial composition. Results presented in this chapter are obtained using the NASA CEA software [80, 81]. We stress the fact that considerations mentioned here depend primarily on ideal thermochemical evaluations, where reactants are ideally mixed together and abstain from the real physical process of combustion (turbulent reacting boundary layer in hybrid rockets and diffusive flame in composite solid propellants).

Some standard parameters are fixed hereafter.

- Combustion chamber is infinite and pressure is set to 70 bar.
- Expansion is in vacuum, using a nozzle with exit-to-throat area ratio $\varepsilon = 40$.
- Unless stated, reactants are stored at ambient temperature.
- The oxidizer for hybrids is liquid oxygen stored at the boiling point (liquid at 90 K, $\Delta H = -13\,\text{kJ mol}^{-1}$), while AP is used for solids (NH_4ClO_4, $\Delta H_f^0 = -295.8\,\text{kJ mol}^{-1}$) [82].
- An HTPB binder is assumed, using the following thermochemical data [83]: $C_{7.075}H_{10.65}O_{0.223}H_{0.063}$, $\Delta H_f^0 = -58\,\text{kJ mol}^{-1}$.

The list of fuel additives tested in this chapter is reported in Table 12.3. The choice covered mostly metals that can be of interest for space propulsion. Aluminum, with a high metal content (namely, micrometric powder size), represents the standard. Boron is interesting because of its enthalpy of combustion. Magnesium is a metal which burns easily and, if properly deposited on boron particles, can promote their ignition [84]. Zirconium has a very high density and might be interesting in terms of volumetric specific impulse. In addition to metals, AlH_3 is chosen as a reference, being the best propulsion ingredient out of the family of metal hydrides [85].

12.2.2
Solid Propellant Optimal Formulations

Table 12.4 summarizes the results found for formulations that maximize the specific impulse of solid propellants. The columns report relevant results for

Table 12.3 List of fuel additives and relevant properties [74, 86].

Additive	Notes	ρ_{add} (g cm^{-3})
Al	Aluminum	2.70
Mg	Magnesium	1.74
B	Amorphous boron	2.37
MgB (5/95)	Boron coated with magnesium (respectively, 5% and 95% by mass), treated as a mixture of metals	2.33
MgB (20/80)	Boron coated with magnesium (respectively, 20% and 80% by mass), treated as a mixture of metals	2.21
AlH_3	Aluminum hydride, $\Delta H_f^0 = -11.4\,kJ\,mol^{-1}$	1.48
Zr	Zirconium	6.52

Table 12.4 Performance analysis of optimal propellant formulations.

Additive	AP/additive/HTPB	I_s (s)	T_f (K)	ξ_{cc}
	$<\rho_p>$ (g cm^{-3})	I_v (kg s m^{-3})	$\overline{\overline{\mathcal{M}}}$ (g mol^{-1})	ξ_t
Al	66/22/12	318.3	3581	0.35
	1.82	5.78E+05	27.1	0.37
Mg	62/25/13	309.7	3220	0.30
	1.66	5.14E+05	24.3	0.30
B	85/14/1	306.9	3568	0.00
	1.98	6.07E+05	31.6	0.00
MgB (5/95)	84/14/2	306.6	3537	0.00
	1.95	5.98E+05	31.1	0.00
MgB (20/80)	83/14/3	306.0	3502	0.00
	1.92	5.87E+05	31.0	0.00
AlH_3	64/30/6	346.4	3489	0.44
	1.68	5.81E+05	22.2	0.46
No additive baseline	90/0/10	297.9	3018	0.00
	1.75	5.22E+05	27.0	0.00

Formulations are expressed in mass percent over the whole composition. CCP content is in terms of mass fraction.

gravimetric and volumetric specific impulses, theoretical propellant density $<\rho_p>$, featured combustion chamber conditions (namely, adiabatic flame temperature T_{fl} and molar mass of the combustion products M_m), and mass fraction of the CCPs in the combustion chamber and in the nozzle throat. A baseline without additives is also assumed for a matter of comparison. Under the stated conditions, the use of pure aluminum as metal fuel can lead to an I_s increment of around +6%. The rest of the fuel additives does about +3%, except for zirconium which does not improve any performance and is not considered in the table. However, none of the metals can compete with aluminum hydride: AlH_3 yields about +16% of I_s with respect to the baseline.

All metal additives increment both adiabatic flame temperature and molar mass, whereas the hydride does the opposite because of the release of hydrogen. Considering the volumetric specific impulse, boron-based compounds take advantage of the high additive density and can compete with the value granted by Al and its hydride, despite a lower specific impulse. Some considerations should be added on CCPs. Thermochemical predictions are not capable of reproducing the complex phenomena, leading to aggregation–agglomeration processes. However, the predicted quantity is an indication of the losses that can be found during nozzle expansion. In this particular case, a zero-level CCP indicates the potential absence of two-phase flow losses for the specific propellant, as for B-based additives. Intermediate results are obtained for Mg-based propellants, where no CCPs are detected in the combustion chamber but oxides condense early during the expansion. As expected, Al-containing propellants (either pure metal or hydride) release a consistent fraction of Al_2O_3 in liquid or solid form, depending on the local temperature.

12.2.3
Hybrid Rocket Performance Analysis

Moving ahead to the analysis of hybrid solid fuels based on an HTPB elastomer with a dispersed additive, performance gain expectations are mostly unattended for I_s. All metal additives do not lead to improvements with respect to the baseline. Only AlH_3 leads to an increment limited to +6%. Given the more favorable O/F conditions if compared to solids, adiabatic flame temperature in hybrid rockets is higher and under the investigated conditions almost reaches 3900 K. Full oxidation of metals as Al and B is particularly demanding of oxidizing species; the drawback resides in the increment of the molar mass of the combustion products, which depresses the attainable specific impulse.

Table 12.5 reports only optimal compositions for tested ingredients. All metals are not mentioned as their addition into a hybrid fuel leads to specific impulse decrement and their optimal composition consists of a fuel made by plain HTPB. As

Table 12.5 Performance analysis of optimal hybrid fuel formulations.

Additive	O/F f_{ADD}	$<\rho_F>$ (g cm^{-3}) $<\rho_p>$ (g cm^{-3})	I_s (s) I_v (kg s m^{-3})	T_f (K) $\overline{\overline{\mathcal{M}}}$ (g mol^{-1})	ξ_{cc} ξ_t
AlH_3	0.90	1.28	376.9	3921	0.45
	0.74	1.21	4.56E+05	21.8	0.48
No additive baseline	2.40	0.92	354.6	3746	0.00
	0.00	1.07	3.78E+05	24.7	0.00

Additive and CCP content are expressed in terms of mass fraction.

Table 12.6 Performance analysis of optimal hybrid fuel formulations containing 10% of additive in the fuel.

Additive	O/F	$<\rho_F>$ (g cm^{-3})	I_s (s)	T_f (K)	ξ_{cc}
	f_{ADD}	$<\rho_p>$ (g cm^{-3})	I_v (kg s m^{-3})	$\overline{\mathcal{M}}$ (g mol^{-1})	ξ_t
Al	2.7	0.98	346.9	3813	0.04
	0.10	1.09	3.79E+05	26.7	0.05
Mg	2.1	0.97	352.1	3714	0.00
	0.10	1.08	3.79E+05	25.1	0.00
B	2.3	0.98	354.2	3822	0.00
	0.10	1.09	3.85E+05	25.3	0.00
MgB (5/95)	2.3	0.98	354.0	3818	0.00
	0.10	1.09	3.84E+05	25.3	0.00
MgB (20/80)	2.2	0.98	353.6	3793	0.00
	0.10	1.08	3.83E+05	25.0	0.00
AlH_3	2.8	0.96	348.9	3794	0.04
	0.10	1.09	3.79E+05	26.3	0.04
Zr	2.2	1.01	351.3	3780	0.04
	0.10	1.10	3.85E+05	25.4	0.04
No additive baseline	2.4	0.92	354.6	3746	0.00
	0.0	1.07	3.78E+05	24.7	0.00

in the case of solid propellants, the columns report relevant results for gravimetric and volumetric specific impulses (I_s and I_v), expected fuel and average propellant density (respectively, $<\rho_F>$ and $<\rho_p>$), combustion chamber conditions (namely, adiabatic flame temperature T_{fl} and molar mass of the combustion products M_m), and mass fraction of the CCPs.

It appears that the use of metal additives in hybrid rockets should be considered for reasons other than for specific impulse increase. Attention can be posed, for example, on the increment of fuel density and volumetric specific impulse or, again, on specific topics related to combustion issues.

Table 12.6 reports a series of suboptimal formulations containing 10% by mass of additive in the solid fuel. We adopt here the term *suboptimal* to address a formulation which does not represent the best possible propellant formulation but, for some reasons, the optimization follows some constraints. Specific reference to this kind of analysis can be found in [74]. Typical examples consist of the necessity to grant a minimum binder quantity for propellant mechanical consistency or the adoption of additives for burning rate enhancement.

Under the constraint of 10% additive, all metals depress the ideal gravimetric specific impulse of the baseline but increase the volumetric one, thanks to density improvement. The lowest performance is granted by Al-based compounds, both metal and its hydride. In all the tested cases, adiabatic flame temperature is about 200–300 K warmer than a solid propellant, although with production of condensed products. Mg and B fuels lead to a lower decrement with respect to Al, depending on reduced molar mass of the products. It should be noted that, under these constraints, a worsening effect is obtained also for AlH_3.

12.2.4
Oxidizing Species in Hybrid Rocket Nozzles

One important consideration for hybrid rocket applications is the presence of a high fraction of oxidizing species in the combustion products. In this respect, chemical erosion of the nozzle walls may become critical, especially when graphite is used. The amount of oxidizing species can be mapped through thermochemical computations. An indicative concentration of oxidizing species X_{OX}, useful for cross comparison between different operating conditions, can be obtained by summing up the contribution of main oxygen-carrier molecules, properly weighting their strength. According to Altman [87], a good indication is given by Equation 12.1. The symbol X_i indicates the molar fraction of the ith component at the throat section under the assumption of an expansion in shifting equilibrium.

$$X_{OX} = 2X_{O_2} + X_O + X_{H_2O} + X_{CO_2} \tag{12.1}$$

Table 12.7 reports the results for hybrid formulations with 10% of metal additive, already mentioned in Table 12.6. A couple of representative solid propellant compositions are mentioned as well. Under these operating conditions, the main oxidizing species predicted by thermochemistry consist of water vapor and carbon dioxide. Some leftovers of molecular and atomic oxygen are present as well. Looking at the overall concentration, all hybrid and nonmetallized solid rockets feature similar results. In this respect, the exhaust gases of the solid propellant are poorer in oxygen but richer in H_2O. However, the scenario changes completely when an aluminized propellant is assumed. Being a relatively fuel-rich mixture,

Table 12.7 Oxidizing species in the throat of the nozzle of hybrid rocket (molar fractions).

Additive	X_{O_2}	X_O	X_{H_2O}	X_{CO_2}	X_{OX}
Al	0.08	0.03	0.26	0.21	0.57
Mg	0.02	0.01	0.27	0.17	0.47
B	0.02	0.02	0.24	0.15	0.42
MgB (5/95)	0.02	0.02	0.24	0.15	0.42
MgB (20/80)	0.02	0.01	0.24	0.14	0.41
AlH3	0.08	0.03	0.28	0.20	0.59
Zr	0.03	0.02	0.28	0.18	0.50
Baseline	0.03	0.01	0.29	0.19	0.51
Solid no Al	0.01	0.00	0.44	0.16	0.61
Solid Al	0.00	0.00	0.07	0.01	0.08

X_{OX} is a weighted sum of the main oxidizing species. The last two lines of the table report two reference conditions of solid propellant, for comparison.

the presence of a metal suppresses most of the oxidizing capability of gas products. Conversely, this effect does not occur for hybrid rockets that benefit from the excess of oxygen.

12.2.5 Active Aluminum Content and Performance Detriment

Computations performed so far assume that the additives are 100% pure. In the case of Al, this is not true because metal passivation covers the external surface with a coat of Al_2O_3. Generally speaking, this layer of alumina is caused by the interaction of the metal with the atmospheric oxygen and has a thickness of about 3 nm, quite independent of the size of the particle itself [88]. For micrometric powders, the assumption that particles are fully made by active metal still holds, supported by the fact that specific surface is fairly limited (on the order of 1 $m^2\,g^{-1}$ or even less) and oxide layer mass results to be negligible with respect to the bulk of metal. On the contrary, when the size of Al particles is decreased in the nanometric range (for propulsion interests, 100 nm or below), the amount of oxide may not be negligible in contrast to the mass of the particle. That is, specific impulse can suffer from a reduction as part of the metal fuel is introduced in the propellant formulation already oxidized. A quantification of this effect is shown in Figure 12.5, where both a solid propellant (a) and a hybrid fuel (b) are supposed to embed partially degraded aluminum. Aluminized formulations in Table 12.4 and Table 12.6 are taken as reference.

As solid propellants take advantage of Al addition and, according to thermochemistry, its optimal fraction is 22% by mass, the reduction of active metal content generates a consistent depression of specific impulse, which reaches −3% for 85% metal purity and −10% for 50% aluminum content. Different results can be observed for hybrid rocket performance, whose dependence on active metal content is weaker (just −1% for a 50% metal purity), in consideration of the limited effect of

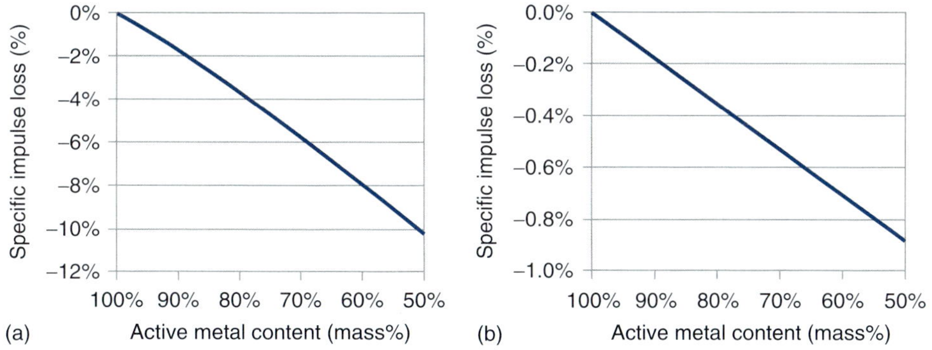

Figure 12.5 Specific impulse loss and active aluminum content. Conditions: 70 bar in the combustion and expansion ratio 40. (a) Solid propellant, AP/Al/HTPB = 66/22/12 by mass and (b) hybrid propellant, O/F = 2.7, 10% additive by mass.

metal on the performance of that specific composition. For the reasons mentioned, the use of metal nanoparticles must be carefully considered for solid propellants. The adoption of a supply featured by a low metal content may lead to undesirable depression of the specific impulse, thus jeopardizing the benefits from its use in terms of lower level of agglomeration. This fact paves the way for the application of mixed micro–nano metal blends for a proper compromise [89]. Conversely, hybrid rockets loaded with a limited mass fraction of metal are quite insensitive on this parameter, while the regression rate may benefit from nanometric aluminum addition to the solid fuel [9]. However, these considerations cannot be addressed by means of thermochemistry. Proper reference experimental data should assist in this choice.

12.2.6 Two-Phase Losses

The presence of CCPs in the reacting mixture comes along with a series of problems that may depress the delivered specific impulse as well as generate nozzle issues. The agglomeration process at the burning surface depends on a complex interaction of events related to propellant composition and ballistics, microstructure, and combustion conditions [66, 90, 91]. Before their interaction with the nozzle, such products are subjected to modifications while residing in the combustion chamber [92, 93]. Once in the nozzle, particles of oxide and molten metal may collide and accumulate on the walls, contributing to throat erosion or metal accumulation. In any case, a two-phase expansion occurs. This event is known to cause a detriment of ideal performance losses as metal particles are released in the atmosphere with a lower speed and higher temperature than the trailing expanding gas (velocity and thermal lag) [94].

Performance detriment has been the subject of thorough investigations for a long time and depends on several parameters such as the size of the expanding particles, the throat diameter, and the quantity of the condensed products for the specific formulation [78, 92, 94]. There is not a unified approach to the problem; rather, the foundation of the knowledge for each research group or propulsion company resides in its own experience and experimental capability. In 1986, an AGARD working group published a report where different methods of loss evaluation were contrasted with each other in a round robin test series [78]. According to the methodology recommended by the above-mentioned document, the share of two-phase losses Δ_{2P} is obtained by the empirical expression

$$\Delta_{2P} = C_3 \frac{\xi_{cc}^{C_4} D_p^{C_5}}{p_{cc}^{0.15} \varepsilon^{0.08} D_t^{C_6}} \tag{12.2}$$

In this formula, ξ_{cc} is the quantity of CCP (expressed in moles per 100 g of mixture), D_p consists of the agglomerate mean size in microns, D_t is the throat diameter in inches, ε represents the nozzle area ratio, and p_{cc} is the combustion chamber pressure in pounds per square inch. The coefficients C_3 through C_6 are

Table 12.8 Coefficients for BPD evaluation method for two-phase flow loss prediction valid for $\xi_{cc} > 0.09$.

$D_t < 1$	$1 \leq D_t \leq 2$	$D_t > 2$	D_p range
$C_3 = 9$	$C_3 = 9$	$C_3 = 13.4$	For $D_p < 4\,\mu m$
		$C_4 = 0.5$	
		$C_5 = 0.8$	
		$C_6 = 0.8$	
$C_4 = 0.5$	$C_4 = 0.5$	$C_3 = 10.2$	For $4 \leq D_p \leq 8\,\mu m$
		$C_4 = 0.5$	
$C_5 = 1$	$C_5 = 1$	$C_5 = 0.8$	
		$C_6 = 0.4$	
$C_6 = 1$	$C_6 = 0.8$	$C_3 = 7.58$	For $D_p > 8\,\mu m$
		$C_4 = 0.5$	
		$C_5 = 0.8$	
		$C_6 = 0.33$	

empirical parameters properly chosen from Table 12.8, depending on nozzle throat diameter and CCP mass-mean size.

Information on particle size can be obtained by direct collection and analysis of combustion residues in the plume of rocket or by modeling, resulting in a complex dependence on rocket parameters. Hermsen, on the basis of a wide database, which gathered experimental tests performed across almost two decades, developed several empirical correlations for CCP size prediction. Equation 12.3 represents an evolution of the standard formula included in the SPP computer code and mentioned in the AGARD working group report [78]. This is meant to be the interim candidate to be adopted, as suggested by Hermsen in 1981 [92]

$$D_p = 3.6304 D_t^{0.2932}(1 - \exp(-0.0008163\xi_{cc}p_{cc}\tau)) \tag{12.3}$$

The formula gives back the mass-mean size in microns D_p of agglomerates exiting the nozzle of a solid propellant rocket. As originally discussed by Hermsen, the main dependence on throat diameter D_t (in inches) is correlated to the velocity gradients in the nozzle, while the term $\xi_{cc}p_{cc}\tau$ (respectively, moles of CCPs per 100 g of propellant, the combustion pressure in pounds per square inch, the mean residence time in milliseconds) is meant to represent a limitation on particle size due to growth mechanism in the combustion chamber. It should be noted that the exponential term plays an important role only when the product $\xi_{cc}p_{cc}\tau$ takes a low value, which, in general, happens in subscale motors (low τ).

A simpler empirical correlation, valid in general for very large τ, was also included in the embodiment of the Bombrini Parodi Delfino (BPD) (now Avio) method as

$$D_p = 3.39 D_t^{0.4692} \tag{12.4}$$

The correlation does depend only on throat size and on correlated velocity gradients, neglecting information on growth phenomena in the combustion chamber.

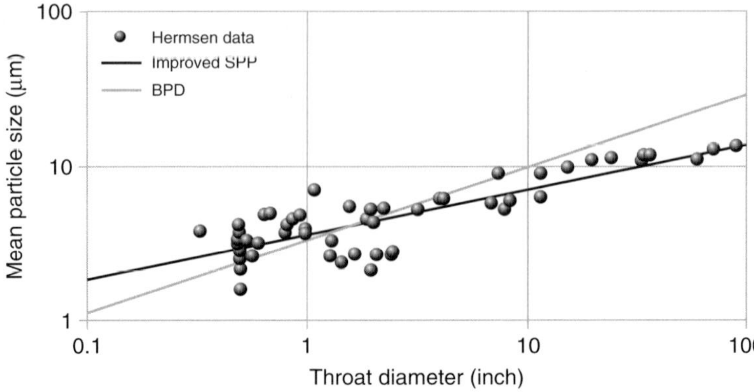

Figure 12.6 CCP mass-mean size out of the nozzle as a function of the throat diameter. Comparison of improved SPP correlation, BPD correlation, and Hermsen's experimental data [92], for $p_{cc} = 700$ psi, $\tau = 70$ ms, and $\xi = 0.315$.

A comparison between the two approaches under a specific reference condition is given in Figure 12.6, remembering that experimental data from Hermsen served the fitting of the improved SPP curve [94]. Experimental data used to fit BPD correlation were not available. In both cases, it is clear that the size of exhausted particles is above 10 μm only for large motor sizes.

Having in mind the information on the size of CCPs at the exit of the nozzle, the BPD method is now applied to a set of parametric test cases for the evaluation of Δ_{2P}. From the analysis of the formula itself, it appears that combustion chamber pressure and nozzle area ratio feature a weak dependence over the resulting losses, while the main drivers are represented by the throat diameter, amount, and size of CCP. Moreover, particle size is somehow coupled with other parameters.

Two representative conditions are selected, for a solid and a hybrid rocket. For the former family, the standard AP/Al/HTPB formulation reported in the first line of Table 12.4 is adopted, while, for the latter, the aluminized formulation reported in Table 12.6 is under consideration. Combustion pressure is set to 70 bar (1015 psi) and nozzle area ratio to 40. It should be noted that tables report the quantity of CCP in mass fraction, whereas the BPD model requires the number of moles per 100 g. In this case, the conversion is quite easy because all condensed products are represented by Al_2O_3 only.

The parametric analysis performed on throat diameter, varying CCP mass-mean size over a realistic range, shows that, in general, rocket motors with small nozzles suffer from important specific impulse losses. However, in these situations, the expected CCP size is relatively small, somehow limiting the losses. On the contrary, the increase of the nozzle diameter progressively reduces the losses, leading to a saturation effect, contained in the range 3–5% for solid propellants (Figure 12.7a). The same formulas applied to hybrid rocket motors show about the same trend, scaled down toward reduced losses, with a saturation effect around 2.5% (Figure 12.7b). This is reasonable considering that the amount of CCP

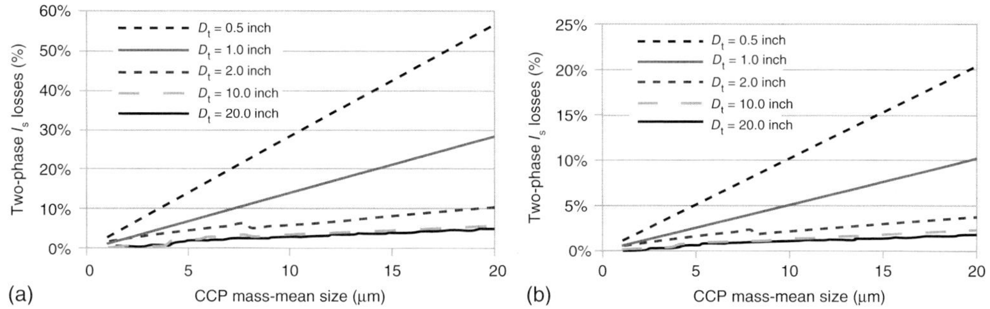

Figure 12.7 Two-phase losses as a function of CCP mass-mean size. Operating conditions: $p_{cc} = 70$ bar and $\varepsilon = 40$. (a) Solid propellant, AP/Al/HTPB = 66/22/12 by mass and (b) hybrid propellant, O/F = 2.7, 10% additive by mass.

produced is lower, by virtue of a reduced metal load. The reader may also note a couple of nonphysical steps in the plots, caused by the piece-wise definition of the fitting constants.

Finally, while solid rocket loss analysis is well assessed, these formulas should be handled with care for hybrid rockets and used only as an indication. This consideration derives from two main reasons. First, the BPD procedure was derived from empirical analysis of solid rocket firings and quantifies I_s losses in terms of percentage of the ideal rocket performance, instead of finding a value in seconds. *De facto* this makes the two-phase loss quantification dependent on the type of rocket, which, in principle, should not be the case. Second, the level of CCP for metallized hybrid rockets may be small enough to fall outside the range of confidence of the BPD model, as in this case. Specific investigation of hybrid rocket technology is still lacking on this topic and should be addressed in future research activities.

12.2.7
Concluding Remarks on Theoretical Performance

Among the possible additives for solid and hybrid propulsion, aluminum has always represented a reference from the performance point of view because of its high enthalpy of combustion. In solid propellants, the use of such an additive accomplishes a specific impulse increment along with propellant density gains, although CCPs are generated. Owing to two-phase flow expansion, the delivered specific impulse is reduced. For large rocket boosters, this decrement is quantified in the order of 2–5%, and may be increased by a higher amount of CCP, larger agglomerate size, and smaller throat area.

For hybrid rockets, it has been shown that I_s does not get any benefits from aluminum use. However, it is shown in other parts of this chapter that, for example, regression rate is increased by metal additives in the shape of nanometric powder.

The use of nanometric ingredients should be considered carefully. The improvement of powder reactivity is, in general, accomplished by its size reduction.

Nanoaluminum leads, in general, to faster burning energetic materials (both solid propellants and hybrid fuels) as well as smaller agglomerates with consequent reduced specific impulse losses. Nevertheless, a smaller fraction of active aluminum usually accompanies the reduction of mean particle size, inducing a consequent reduction of ideal specific impulse. Especially for solid propellants, the loss introduced by the reduction of the active metal content may overcome the expected gain for two-phase flow improvement. For this reason, the right compromise between reactants and metal content is crucial and requires a careful evaluation connecting experimental outcomes and analytical computations. As mentioned later, this can be obtained even by coating the powders with protecting or reacting layers that prevent the metal core from further oxidation or enhance its reactivity.

12.3 Nanosized Powder Characterization

12.3.1 Introduction

Various techniques were developed to obtain nanosized particles, such as plasma condensation or high-energy mechanical milling, but the most promising one is the electrical explosion of wires (EEW) [95]. This method is characterized by a process duration of 1–5 μs, plasma temperature $\sim 10^4$ K, average cooling rate of particles up to 10^9 K s^{-1}, and specific energy entering the wire >1 MJ kg^{-1} [7, 95]. However, a full theoretical explanation of the processes of stabilization and aging of coated metal nanoparticles is still missing [96, 97].

The minute nAl particles, without any protective layers, tend to significantly oxidize, and in the worst case to self-ignite, owing to chemical reactions triggered by contact with air and atmospheric moisture [98]. For this reason, the surface of industrial nAl is normally passivated by a slow stream of dry gas (Ar + 0.1 vol% air) for ALEX^{TM} or by stearic acid for L-ALEX^{TM} [99–102].

The passivation layer on the surface of nAl particles determines their reactivity, stability, ignition delay, and so on [98–102]. Properties of several particles are described in some of the recent papers on nAl, but they are far from large-scale implementation [103].

Together with desirable properties, nanosized particles are also characterized by different surface interactions with respect to bulk materials, resulting in difficulties in handling because of the tendency to form clusters that hinders the complete exploitation of nanosized materials potential [2]. Moreover, the percentage of metal content is lower with respect to coarser powders and tends to decrease fast if particles are exposed to air. This problem is even more serious if we think about the progressive reduction of metal content occurring for standard uncoated powders when decreasing from micrometric to nanometric size.

For these reasons, in addition to passivation, particles are also treated with additional coatings of materials suitable to improve their dispersion in the basic matrix and enhance mechanical properties [104]. A coating of protective reagents can further be applied to slow down the aging process, in order to maintain high metal content and to increase energy release during combustion [100, 103].

In this section, various characterization techniques applicable to nanosized particles are presented and an overview of results obtained for selected nAl is discussed. Particle visualization using TEM or SEM techniques, BET specific surface area measurements, and volumetric analyses for metallic aluminum determination are considered. An experimental investigation of particle reactivity is carried out, both at a slow heating rate using a DSC–TGA setup and at a fast heating rate with two different dedicated setups.

In order to fully understand the behavior of particles used in rocket propulsion, not only a characterization of the original ingredients is necessary but also the collection and analysis of CCPs. During combustion, at or near the surface, sintered aluminized structures grow and develop into aggregates and then the agglomerates burn in an environment constituted of binder and oxidizer degradation products with high heat release.

Al combustion may start at the surface but often extends far from it (distributed flame) through mainly a diffusion process. This fact produces condensed products in the combustion chamber, which are able to reduce acoustic instabilities (particles with size of several micrometers damp low frequencies ~500 Hz and particles below 1 μm damp high frequencies ~4000 Hz), but, on the other hand, they also cause thermal, velocity, and phase lags during expansion in the nozzle [105].

The analysis of residues is important to understand the combustion efficiency of Al and also to evaluate losses due to the two-phase flow of rocket motors, so it is possible to make more accurate predictions of system performance.

12.3.2 Facilities Used for Nanosized Powder Analyses

The apparatus used for SEM analyses was JEOL JSM7600f, provided with a detector for transmitted electrons (TED, transmission electron diffraction). The BET specific surface measurement was carried out by a QUANTACHROME NOVA 2200 E, while metal particle density was calculated by the gas picometer QUANTACHROME ULTRAPICNOMETER 1000. The device used for DSC/TGA analyses was a NETZSCH STA 449F3 JUPITER, implemented with a heating rate of $10\,\mathrm{K\,min^{-1}}$ under environment conditions. The X-ray diffractometer used for the analyses of CCPs was a PHILIPS X'PERT PRO θ/2θ. Being apparatus of commercial nature, they are not discussed further.

12.3.3 Tested nAl Powders: Production, Coating, and Properties

12.3.3.1 Production of nAl Particles

Two powders with nominal particle size of 100 and 50 nm were produced using EEW technique by Advanced Powder Technology LLC, Tomsk, Russia [106], in

Table 12.9 Production parameters of nAl.

Nominal particle diameter[a] (nm)	Electric capacitance (μF)	Wire diameter (mm)	Wire length (mm)	Argon pressure (MPa)	Voltage at EEW (kV)
100	1.6	0.32	100	0.2	30
50	0.8	0.25	80	0.2	30

[a] Declared by powder supplier.

an argon atmosphere. The diameter and length of the exploded wire (0.32 and 100 mm for the 100 nm samples; 0.25 and 80 mm for the 50 nm samples) are the main difference in the production parameters implemented for the two nAl batches (Table 12.9).

All tested powders, except the industrial L-ALEX™ (hereinafter L-ALEX$_{NPS}$), were passivated in a controlled atmosphere by a dry air stream (Ar + 0.1% air) and therefore they had a passivation layer composed of aluminum oxide [95, 100, 101]. Only L-ALEX$_{NPS}$ was obtained by immersing nAl in a proper stearic acid solution immediately after production.

12.3.3.2 **Coating of nAl Particles**

Air-passivated (APS) nAl were manufactured at laboratory scale (~100 g each) for the required coating process. Coating was performed using different chemicals: octadecanoic acid, hexadecanoic acid, 1H,1H-perfluoro-1-undecanol (10 : 1 fluorotelomer alcohol), ester from esterification of 1H,1H-perfluoro-1-undecanol with furan-2,5-dione, and Fluorel™ (a fluoroelastomer by 3M® [107]) + ester from esterification of 1H,1H-perfluoro-1-undecanol with furan-2,5-dione. For each type of chemical applied, the fraction of coating was chosen around 5–10 mass%, with respect to the initial dry nAl mass, to provide a complete coating of the particle surface.

Coating of L-ALEX$_{APS}$ or P-ALEX$_{APS}$ samples was obtained by suspending ALEX™ powder in a hydrocarbon solution of octadecanoic or hexadecanoic acid. The suspension was then placed in an ultrasonic dive mixer for 25 min. After that, the mixture was filtered through a polypropylene membrane and then exposed to fast heating up to 323 K for final evaporation of the solvent. Coating of F-ALEX$_{APS,A/E}$ samples followed a similar procedure characterized by a solution of telomer alcohol (subscript A) or its ester (subscript E) in isopropyl alcohol, filtration through polypropylene membrane, and final evaporation of the residual solvent by fast heating up to 323 K. Preparation of VF-ALEX$_{APS,F}$ samples followed a slightly different procedure: Fluorel™ and ester were solved in dimethyl sulfoxide; the solution was then added to ALEX™ and the resulting suspension was put into isopropyl alcohol before being filtrated by polypropylene membrane and exposed to fast heating. For VF-ALEX$_{APS,E}$ samples, ethyl acetate was used to solve the polymer; ALEX^{TM} was then put into the solution and the resulting suspension let to evaporate under ambient conditions.

The goal of ultrasonication is a better cohesion of the dissolved reagents with the particle surface and the destruction of particle clusters. The duration of the ultrasonication process (25 min) was selected on the basis of previous experimental results and literature data [108].

Application of hydrocarbon and fluorohydrocarbon coating over preliminary passivated metal particles is a less hazardous process than only passivation [109] and it is expected to provide easier interaction between coating and polymeric binder [110].

A summarizing prospect of the nomenclature of the different tested nAl samples, together with details of the coating procedure carried out for each chemical, is reported in Table 12.10.

12.3.3.3 Morphology and Internal Structure of nAl Particles

A qualitative analysis of SEM images, reported in Figure 12.8 for a 50 and 100 nm uncoated ALEX^{TM}, shows the presence of spherical particles characterized by a regular smooth surface and a broad range size, as already observed in [7, 98–102]. Micrometric and submicrometric objects are evident for both powders as well as particle tendency to stick together forming cluster structures. In this respect, it is not clear from the SEM images if clustering is caused by particle cold cohesion, because of short-range forces, or by some sintering processes during powder production.

A deep analysis using TED (Figure 12.9), allows revealing an inner powder structure of physical connections between particles (Figure 12.9a,c,d). Connection necks are visible both in coated and uncoated powders, but it is important to notice that this behavior is not generalized to all particles, confirming that clustering is caused by several phenomena; in the case of coated powders, such connections are clearly recognizable. This might be caused by the high viscosity of the coatings, which acts as a glue among particles as suggested in [98, 102]. The TED technique can be also applied to the study of particle oxide shell and coating. The aluminum oxide layer is well defined for all the powders reported in Figure 12.9 and is surrounded by a further external shell, made of different materials depending on the powder, only for coated ones (Figure 12.9b–d).

12.3.3.4 BET Area and Aluminum Content of nAl Particles

The specific surface area of several nAl powders, as measured by the BET technique, is reported in Table 12.11. As expected, the specific surface area is strongly influenced by the coating type and particle size. This can be explained looking both at the methodology used for the measurement and at the coating itself. In fact, the BET technique is based on the nitrogen adsorption at ambient temperature; however, the presence of a superficial coating plays a dual role, increasing average particle size and, depending on its nature, influencing cluster formation. Generally speaking, it is observable that, for a given powder family, a reduction of the average size causes a strong gain in terms of specific surface area; passing from a nominal size of 100 to 50 nm, increment is about +33% for ALEX^{TM} family and +58%

Table 12.10 Passivation and coating conditions of nAl.

Powder	Applied coating	Chemical formula of the applied coating	Preparation of of coating	Coating procedure	Sonication of mixture (min)	Separation of powder from suspension
ALEX™	None	Air ($N_2 + O_2$)	Not coated after passivation			
L-ALEX$_{NPS}$	Stearic acid	$CH_3(CH_2)_{16}COOH$	Not coated after passivation			
L-ALEX$_{APS}$			Fatty acid solved in saturated hydrocarbons	ALEX™ added to the solution	20	Filtration through polypropylene membrane + evaporation of residual solvent by fast heating
P-ALEX$_{APS}$	Palmitic acid	$CH_3(CH_2)_{14}COOH$				
F-ALEX$_{APS,A}$	Telomer alcohol	$H(CF_2CF_2)_5CH_2OH$	Telomer alcohol solved in isopropyl alcohol			
VF-ALEX$_{APS,F}$	Fluorel™ and ester from esterification of telomer alcohol with maleic anhydride	Vinylidenefluoride-hexafluoropropylene (70 : 30) copolymer + ester	Fluorel™ and ester solved in dimethyl sulfoxide	Isopropyl alcohol and ALEX™ added to the solution	25	Evaporation of solvent at ambient conditions
VF-ALEX$_{APS,E}$			Fluorel™ and ester solved in ethyl acetate	ALEX™ added to the solution		

All powders are passivated by air, except L-ALEX$_{NPS}$, passivated by stearic acid.

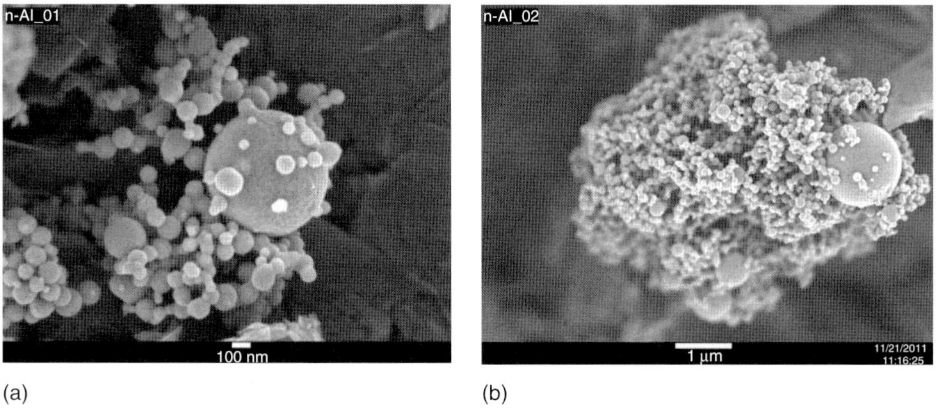

Figure 12.8 SEM images of ALEX™ uncoated powders. (a) 50 nm ALEX™ (50 000×) and (b) 100 nm ALEX™ (15 000×).

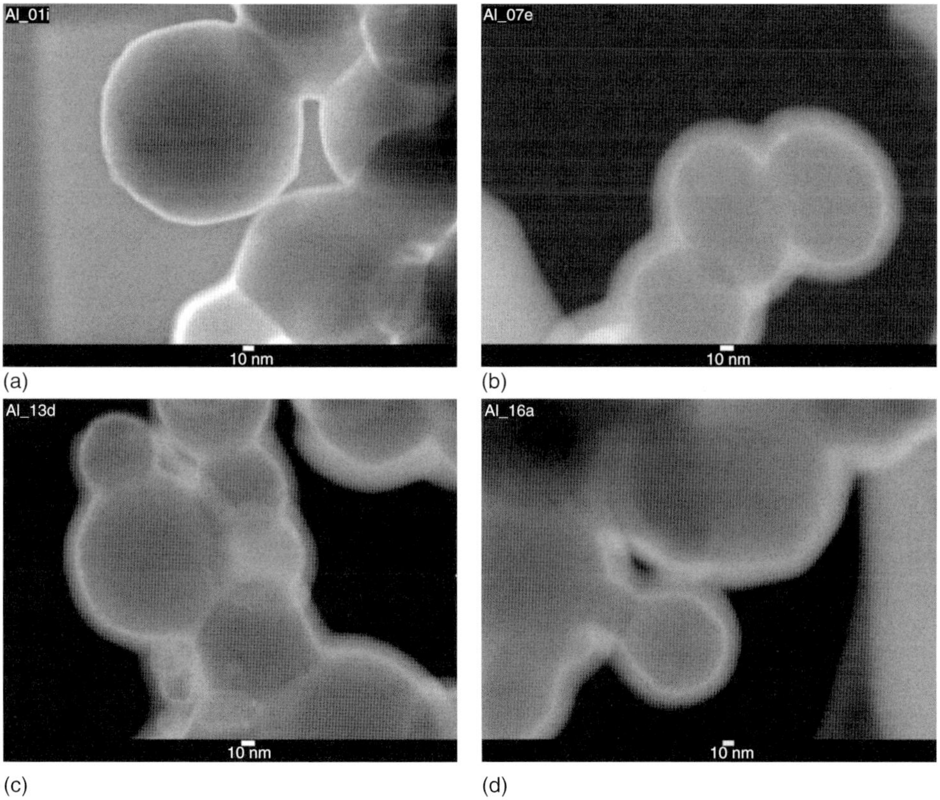

Figure 12.9 TED images of several coated and uncoated ALEX powders. (a) 100 nm ALEX™ (300 000×), (b) 100 nm L-ALEX$_{NPS}$ (350 000×), (c) 100 nm VF-ALEX$_{APS,F}$ (350 000×), and (d) 100 nm F-ALEX$_{APS,A}$ (300 000×).

Table 12.11 BET specific surface area and metallic Al content of nAl powders.

Powder abbreviation		BET specific surface, S_{sp} ($m^2 g^{-1}$)	Mean surface particle diameter[a], a_s (nm)	Metallic Al content[b], C_{Al} (% mass)
100 nm	ALEX™	11.8	202	89
	L-$ALEX_{APS}$	9.1	267	70
	P-$ALEX_{APS}$	Not available	Not available	79
	F-$ALEX_{APS,A}$	11.3	205	88
	VF-$ALEX_{APS,F}$	6.9	345	78
Average for 100 nm batch		*9.8*	*255*	*81*
50 nm	ALEX™	15.7	153	90
	L-$ALEX_{NPS}$	11.3	228	89
	P-$ALEX_{APS}$	13.9	173	71
	VF-$ALEX_{APS,F}$	10.9	215	78
Average for 50 nm batch		*13.0*	*192*	*82*

[a] For spherical particles: $a_S = 6/S_{sp}\rho_{Al}$.
[b] Evaluated after 30 min of reaction time in 5% NaOH aqueous solution.

for VF-$ALEX_{APS}$ family. It is then easy to understand that an increase of the average particle size related to the thickness of the coating leads to a reduction of the powder specific surface area. This problem worsens also using substances and coating procedures that favor the formation of large clusters (also in the micrometric range), such as fluorinated compounds.

Metal content analysis was performed through a volumetric method. This technique is based on the measure of hydrogen volume evolved by reaction of aluminum and water [111], according to the following equation

$$Al + 3H_2O \rightarrow Al(OH)_3 + (3/2)H_2 + 459.1\,kJ$$

A solution of water and NaOH is used to remove the oxide shell, allowing contact between aluminum and water. After a few hours, the process reaches the regime condition and the evolved hydrogen is measured through a graduated device. A simple relation allows computing the Al mass fraction.

Results are reported in Table 12.11. With the exception of the 50 nm ALEX™ powder, notice that the nAl metal content is always below 90%. This fact is quite interesting considering that the typical value of metal content for a standard μAl is over 99%.

Focusing attention on coated powders, the reduction of the active aluminum fraction is seen to depend on the material used. This result is not a surprise because, when applying a coating to a standard powder, the relative percentage of aluminum is obviously reduced, at least because of the mass addition. ALEX™, L-$ALEX_{NPS}$, and F-$ALEX_{APS,A}$ families show about the same metal content, while the VF-$ALEX_{APS,F}$ family is characterized by the lowest active aluminum mass fraction. However, it is important to remember that VF-$ALEX_{APS,F}$ features the most difficult coating procedure as well as a significant presence of very large clusters; it may then be supposed that a larger amount of residual "coating substance" is trapped among the particles with respect to the other samples.

The average powder size does not seem to have a clear influence on metal content. Although this fact can be explained with the coating process for the VF-ALEX$_{APS,F}$, overall this result is rather unexpected for the uncoated family. In fact, assuming a constant oxide shell thickness, a drop in particle size leads to a reduction of the aluminum content because of the increment of alumina "relative weight." A possible explanation for this result can be searched in the real powder size that can be different with respect to the one declared by the supplier.

The considerable difference in active Al content (C_{Al}) between uncoated ALEX^{TM}, industrial stearic acid-passivated L-ALEX$_{NPS}$ samples, and fluoropolymer-coated VF-ALEX$_{APS,F}$ is justified by the presence of a double shell structure, consisting of a passivating oxide layer plus a coating layer (Figure 12.9), as found also in [100, 102]. For the tested powders, organic coating application in average reduces S_{sp} by ~18% and decreases C_{Al} by 9% compared to uncoated nAl.

12.3.4
DSC/TGA Slow Heating Rate Reactivity

The reactivity parameters, as obtained under slow heating rate in DSC/TGA conventional apparatus, are defined by the curves illustrated in Figure 12.10. Samples were analyzed up to 1500 K with a heating rate of 10 K min^{-1}.

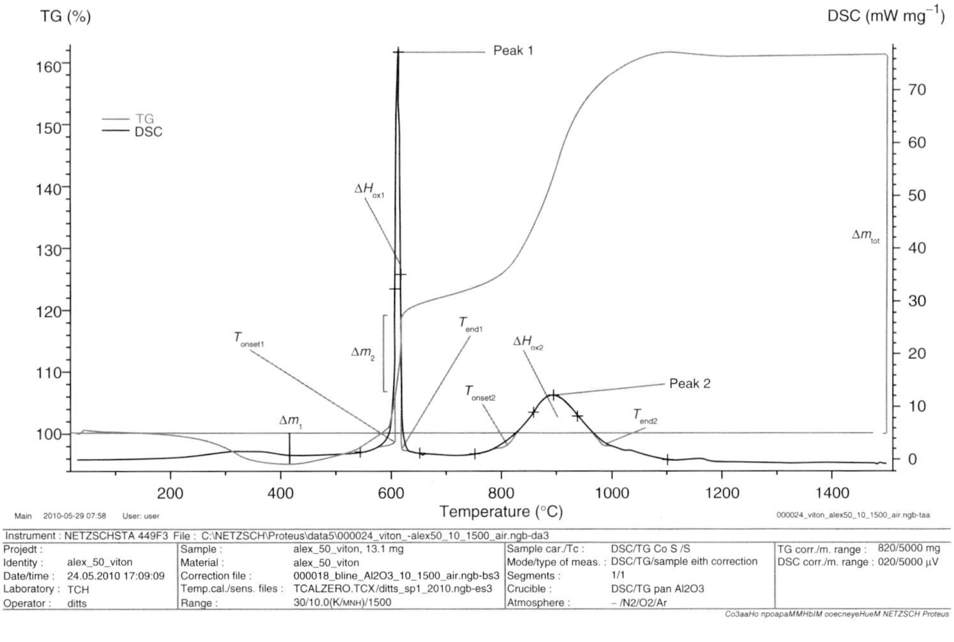

Figure 12.10 Reactivity parameters of 50 nm VF-ALEX obtained from DSC–TGA curves in air.

The first Al oxidation peak is clearly recognized and appears before the Al melting point (933 K). The second less intensive Al oxidation is observed above the melting point (~1173 K). The presence of two well-marked maxima of a positive heat specific flux (peak 1 and peak 2) testifies that the oxidation of nAl occurs in two stages (temperature zones). This two-stage characteristic may be related to any of the following phenomena:

- The periodic cracking of the oxide shell covering the particle metal core [112].
- The formation of AlN phase and its after-oxidation [113].
- The separate oxidation of different particle fractions (the finest particles are oxidized before metal core melting, while the larger ones react after melting in the same temperature range as the conventional micron-sized powders [1]).

The loss of mass before the first oxidation peak (Δm_1) may be explained by several reasons:

- evaporation of the adsorbed water on nAl particles;
- evaporation of residual solvent;
- adsorbed gases desorption;
- decomposition of the coating/passivating layer.

After Δm_1 at ~400 °C (673 K), a mass gain in each TGA curve (up to Δm_{tot}) is observed. The mass gain is related to the oxidation by air. The slope of the TGA curve decreases between the first and the second oxidation stages, and then the mass curve stabilizes toward the maximum. It is reasonable to compare the two studied batches of nAl in accordance with their particle diameter, as done hereafter.

12.3.4.1 **Nonisothermal Oxidation of 50 nm Powder**

The 50 nm VF-ALEX$_{APS,F}$ has the highest T_{onset1}: fluorohydrocarbons delay Al oxidation in air by ~20 K differently from hydrocarbons and Al_2O_3 inert layer (Table 12.4). The few nanometers of Al_2O_3 film thickness cannot be a thermal protection for Al core, as aluminum oxide has a rather high thermal conductivity (30 W m^{-1} K^{-1}). The ΔH_{ox1} of the first oxidation peak (Figure 12.10), as well as the sum of ΔH_{ox1} and ΔH_{ox2}, are the highest for the 50 nm industrial L-ALEX^{TM} sample (8540 and 12 160 J g^{-1}, respectively). This fact can be explained by the absence of the oxide layer and thus the high active Al content for this sample (Table 12.4). The presence of a double-shell structure surrounding the particle (passivating oxide layer plus coating layer) reduces the heat release of the coated powders with respect to the uncoated ALEX^{TM} and L-ALEX$_{NPS}$ samples. The first oxidation stage is the greatest for 50 nm ALEX^{TM} sample (ratio $\Delta H_{ox1}/\Delta H_{ox2}$ is maximal). The 50 nm VF-ALEX$_{APS,F}$ sample has the lowest enthalpy release, probably due to the fluoroelastomer coating performing a protective action over the particle core. The energy release ΔH_{ox2} exceeds the value of ΔH_{ox1} for the 50 nm VF-ALEX$_{APS,F}$ sample. The degree of transformation α_{1000} is nearly the same for 50 nm ALEX^{TM}, 50 nm L-ALEX$_{NPS}$, and 50 nm VF-ALEX$_{APS,F}$ powders with the exception of Al-4-50 sample (112%).

Evidently, this is the result of underestimated C_{Al} content for this sample [98] as well as for the L-ALEX$_{APS}$ one. Palmitic and stearic acids forms a soap-like film on the nAl particles during volumetric analyses in NaOH solution

$$CnH(2n+1)COOH + NaOH \rightarrow CnH(2n+1)COONa$$

12.3.4.2 **Nonisothermal Oxidation of 100 nm Powder**

The value of the T_{onset1} parameter is the highest for 100 nm VF-ALEX$_{APS,F}$ sample (888 K) compared to 846–860 K for the other samples (the same rule as for the 50 nm batch). Thus, thicker fluoroelastomer layers (6% mass for 100 nm VF-ALEX$_{APS}$) definitely shift the process of Al core oxidation to a higher temperature by ~20 K, on average. Moreover, the sample exhibits the lowest mass increase ($+\Delta m_{tot}$) by oxidation. In addition, the mass loss (Δm_1) for this sample is higher than for the other powders. Probably, fluoroelastomer coatings perform a protective action on the metal particle. This will provide more favorable aging characteristics of the powders at standard temperature and pressure. The initial mass decrease of 100 nm F-ALEX$_{APS}$ is the same as for 100 nm ALEX^{TM}. This testifies the low efficiency of the coating procedure. The values of $\Delta H_{ox1} + \Delta H_{ox2}$ are on the same level for all of the 100 nm samples but 100 nm VF-ALEX$_{APS}$. This latter powder exhibits the lowest total heat release in the investigated conditions.

12.3.4.3 **Passivation/Coating Efficiency**

Stearic acid-passivated L-ALEX$_{NPS}$ oxidizes very fast and strongly, even more than APS ALEX^{TM}. This is quite promising for the possible energetic applications of coated nAl [114].

Passivated and fluorohydrocarbon-coated nAl (F-ALEX$_{APS}$ sample) as well as the stearic acid-passivated nAl (L-ALEX$_{NPS}$) feature high and fast energy release by low-rate nonisothermal oxidation in air (Table 12.12). On the contrary, a fluoroelastomer coating (F-ALEX$_{APS}$) protects the particle metal core from oxidation, even at high temperatures. The result is a poor conversion of Al into Al_2O_3 even at 1000 °C (1273 K): an incomplete oxidation of Al occurs even at high temperatures, probably due to the presence of protective residual products from the elastomer decomposition. The alcohol-coated nAl behaves similarly to the uncoated one. The characteristic of DSC and TGA curves is typical for all samples with the exception of VF-ALEX$_{APS}$ samples: the peak temperature for the second oxidation stage is higher for the others (~900 °C vs ~800 °C, ~1173 K vs ~1073 K). The value of ΔH_{ox2} is higher for the 50 nm VF-ALEX$_{APS}$ sample than for the others (Table 12.12). Probably this is connected with the intensive reaction of melted Al with the decomposition products of the fluorocarbon coating [115]. However, the low ignition temperatures of coatings (see Table 12.13 and Table 12.14) imply their complete gasification at temperatures much lower than 1000 °C (1273 K), and, thus, only the reaction of Al with air is representative in the high temperature range.

Table 12.12 Reactivity parameters of nAl under nonisothermal oxidation in air.

Reactivity parameter[a]	Parameter value								
	100 ALEX^{TM}	100 L-ALEX$_{APS}$	100 P-ALEX$_{APS}$	100 F-ALEX$_{APS,A}$	100 VF-ALEX$_{APS}$	50 ALEX^{TM}	50 L-ALEX$_{NPS}$	50 P-ALEX$_{APS}$	50 VF-ALEX$_{APS}$
T_{onset1} (K)	852	860	846	851	888	852	855	860	885
$T_{ox\ peak1}$ (K)	876	868	880	884	895	873	877	865	893
Δm_1 (%)	1	3	3	1	6	1	5	3	5
ΔH_{ox1} (J g^{-1})	6 667	5 489	6 538	6 938	4 171	8 062	8 540	6 542	4 649
ΔH_{ox2} (J g^{-1})	4 365	4 223	4 539	4 679	3 919	4 052	3 620	4 348	5 866
$\Delta H_{ox1}/\Delta H_{ox2}$	1.5	1.3	1.5	1.5	1.1	2.0	2.4	1.5	0.8
$\Delta H_{ox1} + \Delta H_{ox2}$ (J g^{-1})	11 032	9 712	11 077	11 617	8 090	12 114	12 160	10 890	10 515
Δm_{tot} (up to 1273 K) (%)	66	65	66	69	53	70	70	71	62
α_{1000}[b] (up to 1273 K) (%)	83	104	94	88	76	87	88	112	89

[a] See Figure 12.10.
[b] Degree of transformation $\alpha(Al \rightarrow Al_2O_3) = +\Delta m_{tot}/(C_{Al} \cdot 0.89)$.
[c] C_{Al} (% mass) – metal content in the samples (see Table 12.11).

12.3.5
High Heating Rate Reactivity

Because the kinetic response of energetic ingredients is strongly sensitive to heating rates, an additional approach was taken to investigate the reactivity of nAl powders under operating conditions closer to the actual ones. An easy way to proceed consists in assessing the ignition temperature of the bulk powders under high heating rates. This analysis was carried out by two different techniques, using the facilities schematically represented in Figure 12.11.

Main differences concern the method used to heat up the powders till ignition as well as the sample holder. While a laser beam allows heating powder samples up to 150 K s^{-1}, a hot wire technique can reach heating rates up to 300 K s^{-1}.

12.3.5.1 nAl Powder Ignition Experimental Setup

Reactivity of powders under high heating rates was evaluated as follows. In the hot wire technique (Figure 12.11a), a few milligrams of metal powder are placed on a NiCr wire mounted on a ceramic support. In the laser technique (Figure 12.11b), tests are conducted by placing about 10 mg of metal powder on a thin steel plate that forms the upper part of the removable sample holder.

The sample holder is located in a 2 l steel vessel where pressure can be controlled and different kinds of oxidizer used. For tests carried out under subatmospheric conditions or using an oxidizer different from air, vacuum is first created in the chamber using a pump and then the selected oxidizer is injected slowly to avoid particle dispersion till the operating pressure is reached.

The sample holder is heated up by an analog power supply for the hot wire technique and by a continuous CO_2 source for the laser technique. In both instances, the temperature of the sample under test is monitored during the entire experiment using a 50 μm Pt–Pt/10%Rh thermocouple dipped into the metal particles. The thermocouple is connected to a scope and its signal is checked during the whole test duration. When combustion starts, the powder temperature rapidly increases. The ignition instant is located where the temperature–time curve features a sharp change in the slope.

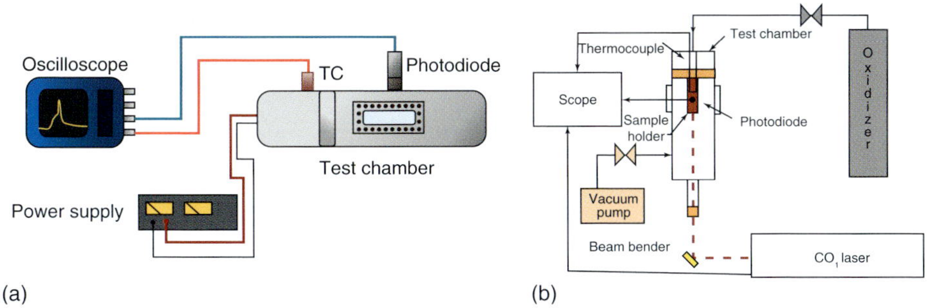

Figure 12.11 Setup used for test on powders reactivity under high heating rate. (a) Hot wire technique and (b) laser technique.

At the same time, a photodiode signal is collected as well. Again, ignition generates a change in the slope because of the fast intensification of the sample brightness. When using the laser technique, a signal from an infrared sensor is also collected to monitor the laser beam history.

The ignition temperature values obtained by thermocouples at each pressure are statistically handled assuming a Student's t distribution. The mean value, considered to be the best esteem for the real value, is calculated for each set of tests and the uncertainty, related to a level of confidence of 95%, is evaluated too. Therefore, the ignition temperature at each pressure is given as

$$T_{\text{ign}} = T_{\text{m}} \pm t\frac{\sigma}{\sqrt{N}} \tag{12.5}$$

where T_{ign} is the ignition temperature; T_{m} is the mean value; t is the Student's parameter, whose value can be found in tables; σ is the standard deviation; and N is the number of tests performed.

12.3.5.2 nAl Powder Ignition Representative Results

Results obtained from tests of nAl powders, in air at 1 bar, in the laser beam technique are summarized in Table 12.13. For each powder, at least 10 valid tests are considered.

The achieved experimental results point out the following facts.

- the ignition temperature of nanosized ALEX^{TM} is lower than the melting point of not only aluminum oxide but also metallic Al, thus confirming the higher reactivity of nAl particles with respect to μAl;
- comparing uncoated ALEX^{TM} of different size, one can notice a slight reduction of the ignition temperature, but this trend is obscured by the measurement uncertainty;
- the same remark is valid for VF-ALEX$_{\text{APS,F}}$: smaller particles are characterized by a slightly lower ignition temperature, but the difference remains within the measurement uncertainty;

Table 12.13 Ignition temperature of nAl powders in air at ambient pressure (laser beam technique).

Powder abbreviation		Average ignition temperature, T_{ign} (K)	Interval of confidence (K)
100 nm	ALEX^{TM}	875	17
	L-ALEX$_{\text{APS}}$	823	48
	L-ALEX$_{\text{NPS}}$	777	49
	F-ALEX$_{\text{APS,A}}$	910	37
	VF-ALEX$_{\text{APS,F}}$	732	27
50 nm	ALEX^{TM}	865	15
	L-ALEX$_{\text{NPS}}$	791	51
	F-ALEX$_{\text{APS,E}}$	892	9
	VF-ALEX$_{\text{APS,F}}$	716	18

- the same remark is valid for F-ALEX$_{APS,A}$: smaller particles are characterized by a slightly lower ignition temperature, but the difference remains within the measurement interval of confidence;
- a quite confusing trend emerges from the results obtained by testing L-ALEX, because of an unusually large measurement uncertainty as well as a double passivation technique. Within this large uncertainty, the lack of an aluminum oxide passivating layer makes L-ALEX$_{NPS}$ more easily ignitable than L-ALEX$_{APS}$.

The presence of coating sensibly reduces ignition temperature for all of the powders, with the notable exception of the fluorotelomer-coated F-ALEX$_{APS,A}$. The latter powder presents preburning characteristics similar to those of ALEX^{TM} (Table 12.11). In particular, the high S_{sp} and C_{Al} of F-ALEX$_{APS,A}$ contrast with the values achieved for the other coated powders. In this respect, further investigations are needed.

Apart from the ignition temperature results, from the same tests it is possible to verify the two-stage combustion process already identified with DSC/TGA analyses. By recording the combustion using a high-speed and high-resolution camera, it is possible to see all the features of the phenomenon: after the ignition, red waves propagate on the surface of the sample (first stage). During this stage, the laser is turned off. After a few seconds, one can notice that a second energetic reaction appears spontaneously: starting from the center a white light spreads over the sample. This second stage is explained with the production of AlN [116].

For some of the powders under examination, ignition temperatures were determined also at 300 K s^{-1} using the hot wire technique and the results are summarized in Table 12.14.

The results obtained by the hot wire technique (Table 12.14) are in broad agreement with those obtained by the laser beam technique (Table 12.13):

- Increasing the heating rate, T_{ign} decreases for uncoated ALEX^{TM}, while no clear effect is noticeable for coated powders.
- For both heating rates, reducing the size of uncoated powders implies a reduced T_{ign} (higher reactivity).
- For both heating rates, particle coating implies a reduced T_{ign} (higher reactivity) with respect to the corresponding uncoated powders.

Table 12.14 Ignition temperature of nAl powders in air at ambient pressure (hot wire technique).

Powder abbreviation		Average ignition temperature, T_{ign} (K)	Interval of confidence, (K)
100 nm	ALEX^{TM}	820	12
	ALEX^{TM} (3 years aged)	767	14
	L-ALEX$_{NPS}$	771	16
	VF-ALEX$_{APS,F}$	713	20
50 nm	ALEX^{TM}	779	11
	ALEX^{TM} (3 years aged)	761	13

- Both aged ALEX™ show a reduced T_{ign} with respect to the fresh powder batches.
- The small T_{ign} difference between 50 and 100 nm ALEX™ obtained when testing fresh powders through the laser technique is confirmed, after 3 years, by the hot wire technique.

The strong T_{ign} difference observed for ALEX™ 50 nm, in Table 12.13 and Table 12.14, is probably imputable to the use of two different production lots, requiring an improvement of powder manufacture and selection technique. The same explanation can be used to justify the discrepancy between old and fresh powders in terms of temperature variation passing from 50 to 100 nm ALEX™.

12.3.6 CCP Collection by Strand Burner

First, it is necessary to clarify that residuals produced by a strand burner or a comparable technique present different features with respect to the ones obtained by small- and full-scale motors, mainly due to the presence of the nozzle and the processes of coagulation and breakups in the flow evolution.

Figure 12.12 shows the experimental rig used to collect CCP. The apparatus is composed of a closed vessel containing the igniter, which consists of an electrically heated Ni–Cr alloy; the quenching pool, filled with an inert liquid (e.g., tetrachloroethylene, C_2Cl_4); and a pressure controller to set the operating condition for each run. Tests are performed under a controlled pressure, over the range 1–30 bar, in a nitrogen environment. Test samples are parallelepipeds (4 mm × 4 mm × 40 mm) inhibited on the lateral surfaces. CCPs of larger size (>1 μm) are collected in the quenching pool and then heat treated to obtain the complete evaporation of the inert liquid, while smoke oxide particles (SOPs, particles with size <1 μm) are collected through a cellulose filter.

When collecting CCPs by a strand burner, it is necessary to quench the samples using a proper media (liquid or gaseous) in order not to modify their chemical properties.

12.3.6.1 Condensed Combustion Product Analysis

A brief introduction regarding the different phases of alumina (Al_2O_3) is in order. During oxidation in pure oxygen, the particles undergo various changes. Aluminum particles are coated by a natural layer of amorphous alumina with a thickness between 0.5 and 5 nm, stable at ambient temperature. Increasing temperature, once the layer thickness grows over a critical value, amorphous alumina become metastable and a transition to γ-alumina takes place. For temperatures between 1223 and 1523 K, this layer undergoes further transitions to α-alumina, the thermodynamically stable crystal of aluminum oxide at each temperature. This transition can take place directly from γ-alumina to α-alumina or via the intermediate phases δ- and θ-alumina (Figure 12.13) [118].

The influence of the quenching liquid nature is underlined in Figure 12.14 (H_2O) and in Figure 12.15 (C_2Cl_4). Using H_2O, an interaction occurs between

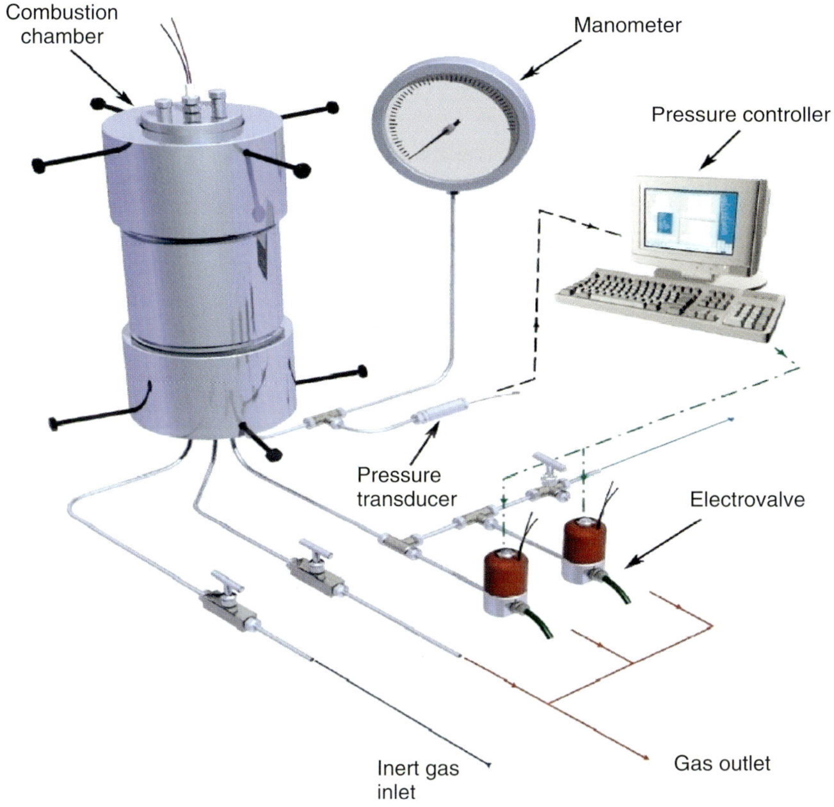

Figure 12.12 Experimental rig present at SPLab to collect condensed combustion products [117].

liquid and residues, which modifies their chemical properties with the formation of $Al(OH)_3$, while quenching with C_2Cl_4 (a fully inert liquid) involves no modifications.

Crystallographic analyses allow a better understanding of CCP properties and thus more accurate prediction models. XRD tests of the CCP generated by the same propellant formulation over the range 1–30 bar show an important presence of metastable structure (Table 12.15) [117].

Results show the pressure effect on combustion efficiency and also a strong presence of γ-alumina at every pressure: the amount of α-alumina decreases, increasing pressure. Fine particles, especially at higher pressure, do not show the presence of Al and are mainly constituted by γ-alumina.

Depending on the working conditions and propellant formulation [120], condensed phase particles ejected from the burning surface may have initial sizes in the range 10 to 200–800 μm with a trimodal distribution [121] covering a wide grain size distribution range (Figure 12.16).

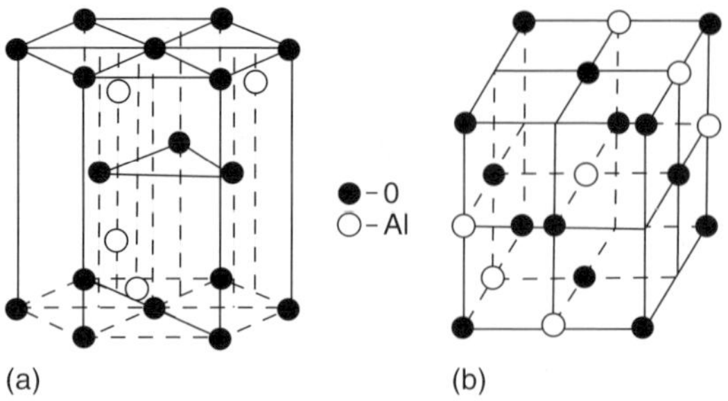

Figure 12.13 Diagram of alumina structure. (a) Alpha phase (corundum) and (b) gamma phase (spinel). (Reproduced with permission from [119] copyright © 1992 The American Institute of Aeronautics and Astronautics, Inc.)

Figure 12.14 (a,b) Visualization of coarse residues at 10 bar quenched in H_2O (10k×, 50k×) [117].

Table 12.15 XRD results on coarse residuals.

P (bar)	Al_2O_3/Al^0	Al^0	γAl_2O_3	δAl_2O_3	θAl_2O_3	αAl_2O_3	$AlCl_3(H_2O)_6$	$FeCl_2(H_2O)_2$	αNH_4Cl	NH_4ClO
1	0.2	78.5	—	1.7	9.1	6.2	2.4	1.0	0.3	0.8
5	2.1	32.1	32.1	10.1	18.9	5.8	—	—	1.0	—
10	4.8	17.2	37.6	11.4	30.4	3.4	—	—	—	—
20	7.8	11.4	39.2	22.3	22.7	4.4	—	—	—	—
30	8.8	10.2	37.6	26.4	23.6	2.2	—	—	—	—

Experimental campaigns conducted to understand the effects of nAl on the CCP structures and morphologies of solid propellants have shown a flame closer to the burning surface with respect to the corresponding µAl-based formulations. Thus, the aggregation/agglomeration process leads to the formation of smaller CCP. Indeed, spherical 30 µm Al, a very common fuel in aerospace application, presents

(a) (b)

Figure 12.15 (a,b) Visualization of coarse residues at 10 bar quenched in C_2Cl_4 (10k×, 50k×) [117].

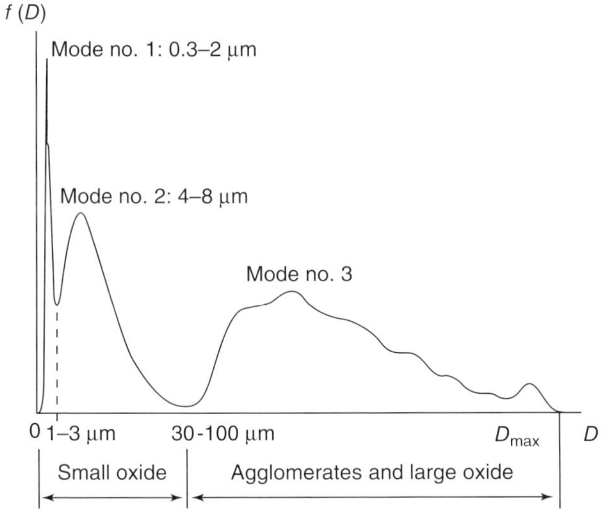

Figure 12.16 Qualitative character of the size distribution of condensed combustion products of aluminized propellants [121]. (Reproduced with permission from [121] copyright © 2000 Springer.)

residues with an average size of 13 μm, but propellants loaded with nAl (with size of 0.15 μm) develop residuals with a size around 6 μm (Figure 12.17) [122].

According to [121], the following mechanisms of formation/development of particles can be identified.

1) Condensation and coagulation in the vicinity of a particle being the mechanism that delivers the mass of the oxide particle (range: 2–6 μm).
2) Removal of the oxide film produced by heterogeneous reaction from the particle upon transition to vapor-phase burning with the formation of a single oxide particle ($D \leq 15$ μm).

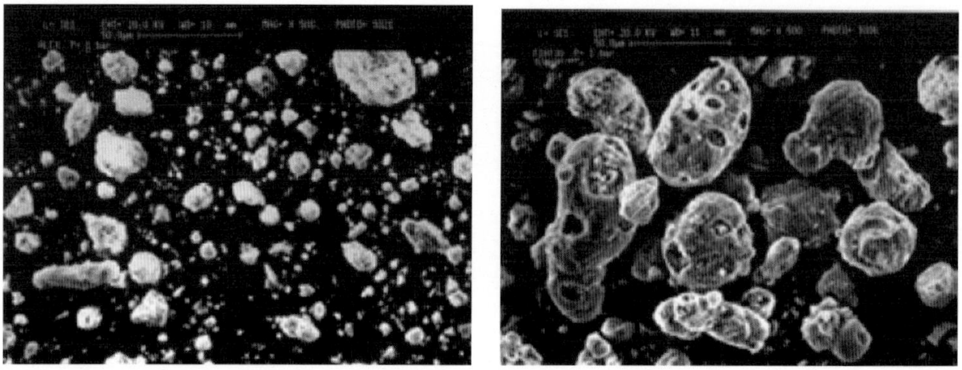

Figure 12.17 SEM micrograph of condensed combustion residues of spherical aluminum 30 μm (a) and nAl 0.15 μm (b). Magnification: 500× [122].

3) Process similar to no. 2 including the effects of fragmentation ($D \sim 2$–$3\,\mu m$).
4) Vapor-phase burning in the proximity of the particle ($D \sim 1\,\mu m$).
5) Condensation of the products of vapor-phase burning on the particle surface ($D \leq 3$–$4\,\mu m$).
6) Similar process to no. 5 but on a metal particle incompletely burned due to lack of oxidizing species ($D \leq 5\,\mu m$).

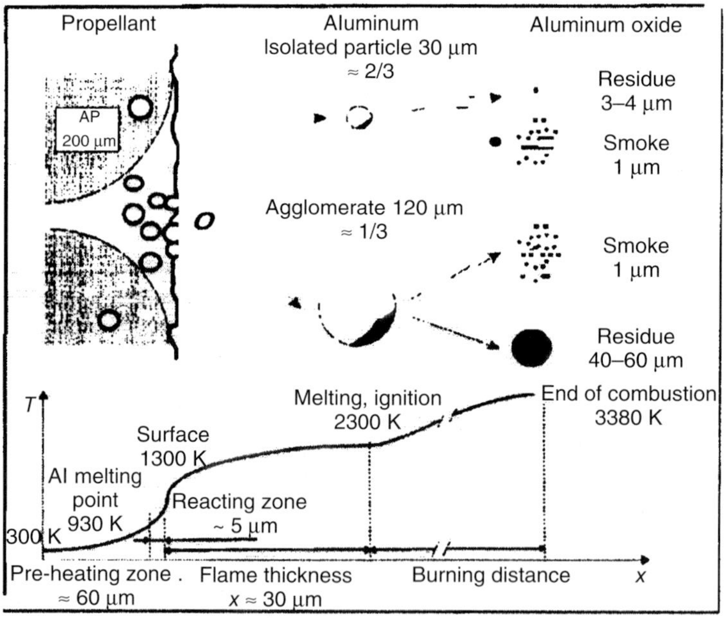

Figure 12.18 Heterogeneous vapor-phase combustion of agglomerates [123].

7) Formation of coarse oxide particle as a result of complete burnout of agglomerates (white oxide).
8) Formation of oxide particles 20–25 μm as a result of burnout of agglomerate fragments (Figure 12.18).

However, different experimental campaigns show that part (slightly < 50%) of the condensed products present a very fine grain size distribution, around 1 μm [117, 120].

An important consideration concerns the presence of the nozzle. In fact, the presence of the nozzle, in comparison with the strand burner, heavily modifies the grain size distribution pattern of CCP. Different numerical models and experimental works were conducted to analyze the Al_2O_3 particle in rocket plume with the aim of accurately predicting two-phase flow losses. Hermsen [92] compared several previous experimental works (Figure 12.19) to understand which parameters affect more the growth of oxide particles and stated that D_t (throat diameter), P (chamber pressure), ξ (Al_2O_3 concentration in chamber), and τ (average residence time in combustion chamber) are the main parameters that govern the process of growth–coagulation–fragmentation of CCP.

Data in Figure 12.20, obtained from several analyses [120–123, 92], agree in defining the range of alumina particle size enclosed in a short interval around 10 μm.

Analyses on alumina crystallography collected from rocket plumes show the presence of both γ-alumina and α-alumina (64–93% by mass of γ-alumina). In general, experimental evidence points out that smaller particles are preferably found in the γ-phase and larger particles in the α-phase [124].

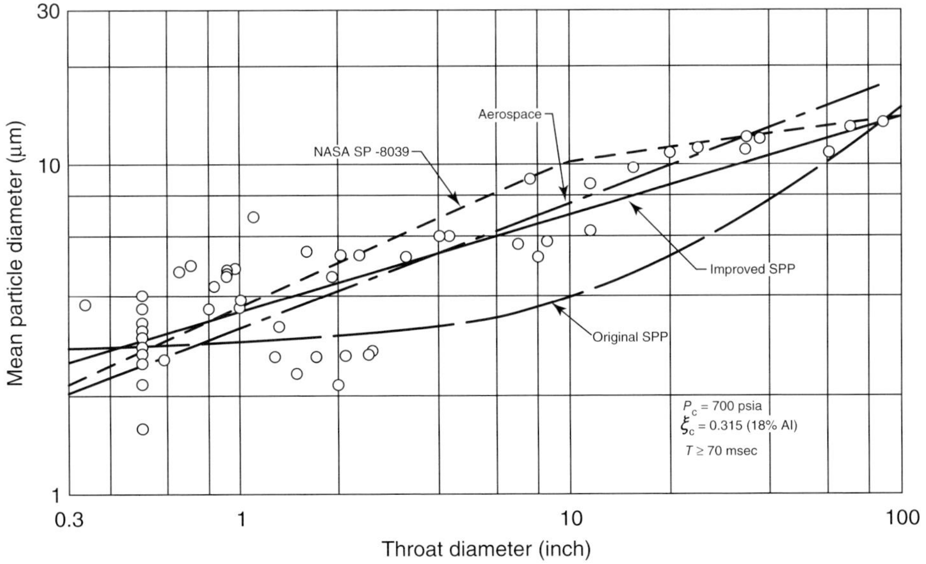

Figure 12.19 Comparing particle diameter calculated at nozzle throat and D_{43} at nozzle exit [117]. See also Figure 12.6.

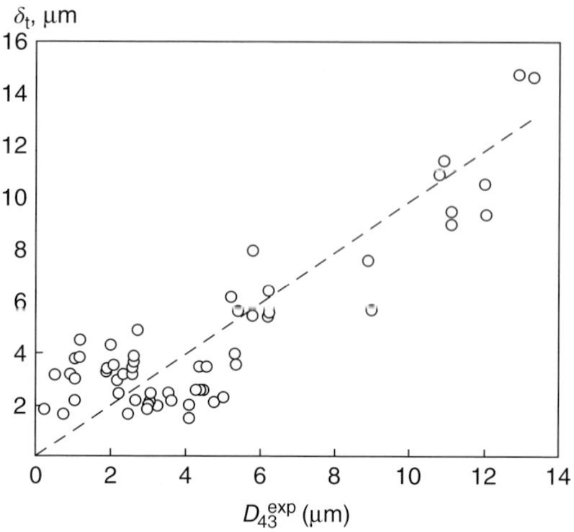

Figure 12.20 Comparison of particle size correlation. (Reproduced with permission from [92] copyright © 1981 The American Institute of Aeronautics and Astronautics, Inc.)

12.3.7 Concluding Remarks on Powder Characterization

Several organic reagents of different nature were applied to nAl produced by the EEW method: octadecanoic and hexadecanoic acid, telomer alcohol, Fluorel^{TM}, and ester. The difference between single- and double-shell-structured particles is mostly connected with the formation of thick layers on particles and different chemical origins of the coating reagents. Fatty acid has almost no effects on the slow nonisothermal oxidation behavior in air, while fluoroelastomers shift the oxidation to higher temperatures (by ~20 K for the first oxidation stage and by ~273 K for the second oxidation stage). The 50 nm batches of particles exhibit higher reactivity than the 100 nm counterpart. This is testified by lower $T_{ox\ peak1}$ and higher energy release ΔH_{ox1} in the first oxidation peak. A remarkable exception is represented by the fluorotelomer-coated 50 nm VF-ALEX$_{APS}$, for which ΔH_{ox2} exceeds heat release during the first oxidation peak ΔH_{ox1}. Studies on combustion of nAl particles with fluoroelastomer coatings are in progress [114].

12.4 Mechanical and Rheological Behavior with Nanopowders

12.4.1 Solid Propellants and Fuels: Mechanical and Rheological Behavior

Elastomers are polymeric materials characterized by viscoelastic behavior [125, 126]. The binders used in solid propellants and the solid matrices of hybrid fuel formulations are typically based on cured elastomers or polymers. In particular, polyurethane materials deserve particular attention thanks to their overall characteristics (easy manufacturing and loading with powdered energetic additives/oxidizers) [127, 128]. Different nonconventional binders can be considered because of their higher compatibility with exotic additives as metal hydrides [129, 130], or for the study of entrainment phenomena increasing the solid fuel regression rate [130–133]. Rheological and mechanical properties of ES depend on the original binder matrix and additive features. Uncured polymers loaded with powders behave like suspensions, while the final, solid ES generally present rubberlike behavior. Additive characteristics influencing the ES behavior are [125, 126]

- volume fraction, φ;
- mean particle size and size distribution;
- dispersion uniformity;
- mode (monomodal, multimodal);
- surface characteristics and morphology (e.g., coating presence);
- shape factor.

In this section, the rheological behavior of suspensions and the mechanical properties of cured solid ES are discussed, focusing on formulations loaded with nAl powders. Cured HTPB is the most commonly used binder for solid propellants/fuels thanks to its excellent overall mechanical properties [127, 128]. As for every polymer, cured HTPB properties depend on resin characteristics, detailed formulation of the final product, and manufacturing procedure [125].

Data concerning the uncured HTPB resin and the final cured HTPB formulation used in SPLab studies are reported in Table 12.16 and Table 12.17. The reported data yield $-NCO/-OH = 1.04$.

Details of the manufacturing procedure for SPLab HTPB binder and metal loaded formulations are reported in [68, 114, 134–137]. Manufacturing of the metallized solid fuels is conducted under controlled operating conditions. During mixing, the compound is processed under vacuum and at a temperature of 333 K. After nanopowder addition to the fuel matrix, the compound is ultrasonicated all along the mixing phase.

Table 12.16 Characteristics of the SPLab HTPB-R45 HTLO resin.

Property	Symbol	Units	Value
Mean molar mass	M_m	$kg\,mol^{-1}$	2.8
Hydroxyl number	OH_v	$mequiv_{KOH}\,g^{-1}$	0.84
Hydroxyl functionality	OH_{fun}	—	2.352
Specific gravity at 296 K	ρ	$kg\,m^{-3}$	901
Viscosity at 296 K	$\eta\ (T = 296\ K)$	mPa s	8000
Viscosity at 303 K	$\eta\ (T = 303\ K)$	mPa s	5000

Table 12.17 SPLab HTPB binder composition.

Ingredient	Mass fraction, % wrt total mass	Density, ρ (kg m^{-3})
HTPB-R45 HTLO	79.2	901
IPDI	7.70	1060
DOA	13.1	920
HTPB binder	100	~ 920

Excess addition of dibutyltin diacetate (curing catalyst) not reported.

12.4.2 Viscoelastic Behavior

The shear stress is related to the deformation rate by the following equation

$$\tau(t) = \mu(\dot{\gamma}) \cdot \dot{\gamma} \tag{12.6}$$

Newtonian systems are characterized by constant values of $\mu(\dot{\gamma})$, while non-Newtonian behavior characterizes systems whose viscosity depends on $\dot{\gamma}$ [125, 126, 138, 139]. When an elastic material is subject to a sinusoidal stress, it reacts with a sinusoidal strain. For such a material, no phase shift exists between the acting stimulus and its reaction ($\delta = 0°$). A perfectly viscous material behaves in a different way. When it is subject to a sinusoidal stress, the resulting strain exhibits a sinusoidal behavior, but a phase shift angle $\delta = 90°$ exists between the stimulus and the corresponding reaction. Viscoelastic materials respond to sinusoidal stresses with a strain exhibiting both viscous and elastic behaviors, thus presenting a strain phase shift angle $0° < \delta < 90°$ [125, 126], as shown in Figure 12.21. Thus, for this kind of materials, $\gamma(t)$ can be written as

$$\gamma(t) = \hat{\gamma} \cdot \sin(\omega \cdot t) \tag{12.7}$$

If $\gamma(t)$ is small enough that it lies in linear viscoelastic range, the sinusoidal stress will result

$$\tau(t) = \hat{\tau} \cdot \sin[\omega \cdot t + \delta(\omega)] \tag{12.8}$$

Introducing the complex shear modulus, $G^*(\omega) = \tau(\omega)/\gamma(\omega) = G'(\omega) + iG''(\omega)$, one finds

$$\tau(t) = \hat{\gamma} \cdot |G^*(\omega)| \cdot \sin[\omega \cdot t + \delta(\omega)] \tag{12.9}$$

The complex shear modulus $G^*(\omega)$ represents both the components of the material response to the applied deformation. The storage modulus, $G'(\omega)$, accounts for the elastic component, while the loss modulus, $G''(\omega)$, is related to the viscous part of the material response. The loss factor $\tan[\delta(\omega)] = G''(\omega)/G'(\omega)$ is used in order to evaluate the relative weight of the elastic behavior of the viscoelastic

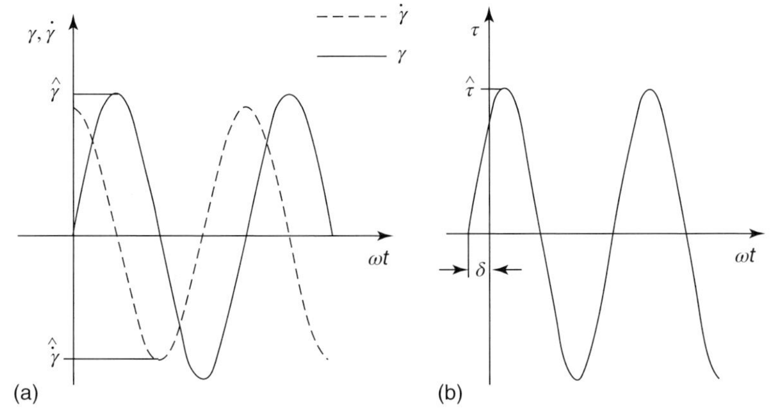

Figure 12.21 Deformation and shear rate profile: (a) traces in oscillatory shear conditions and (b) shear stress profile of a viscoelastic material. (Reproduced with permission from [138] copyright © 2001 Wiley.)

material response [125, 126, 138]. The presented treatment in terms of Equation 12.7, Equation 12.8, and Equation 12.9 can easily be extended to σ, ε, and $E^*(\omega) = E'(\omega) + iE''(\omega)$.

The loaded material response to a dynamic mechanical stress can exhibit peculiar features, as extensively discussed by Nielsen [125, 139], Smallwood [140], Fletcher and Gent [141], Payne [142–144], and as recently discussed by Wang [145] and Frolich *et al.* [146]. The typical behavior of a viscoelastic loaded material under dynamic deformation is reported in Figure 12.22, where strain-independent and strain-dependent components of the material response are shown. As originally proposed by Payne, the strain-independent contributions to the complex modulus are due

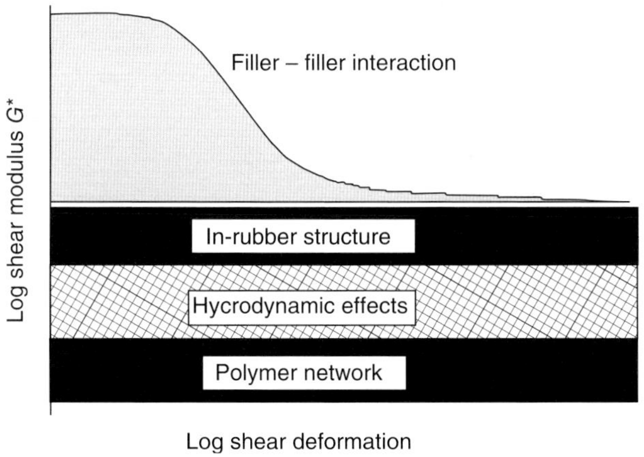

Figure 12.22 Complex modulus behavior [$G^*(\gamma)$] for a loaded rubberlike material. For nonloaded materials, the network contribution is the only possible response. (Reproduced with permission from [146] copyright © 2005 Elsevier.)

to polymer network (in turn depending on material nature/crosslink density), hydrodynamic effect (due to the viscoelastic material flow and additive particle presence), and in-rubber structure (polymer-filler physical and chemical linkages). The shear thinning component of complex modulus is due to the particle–particle interactions [142, 143, 146]. For small deformations, the close particle interactions can provide a significantly enhanced complex modulus. Increasing deformations yield increases in the interparticle distances and possible breakage of clusters. In turn, this reduces the complex modulus of the tested material for increasing deformations [146].

12.4.3 Additive Dispersion

Nanosized particles exhibit high S_{sp}, increasing their reactivity with respect to coarser, conventional micron-sized counterparts. Moreover, the size reduction of nanopowders can yield peculiar effects as enhanced sensitivity to van der Waals forces and Brownian motion [147]. These features affect the ES rheological and mechanical behaviors during manufacturing and combustion. In particular, the tendency of nanoparticles to form clusters in the storage/manufacturing phases must be contrasted in order to disperse the additive within the binder matrix down to the nanoscale. The most explored dispersion strategy is based on ultrasonication of particle suspension [126, 134–137, 148]. In this technique, the application of an ultrasound radiation source of intense power induces cavitation into the suspending fluid medium. Bubbles are created and grow till explosion. The process leading to bubble explosions yields significant temperature increases in short time intervals. The following cooling phase is so fast that clustering is hindered because of the decreased intermolecular and chemical bond forces [147]. In the open literature, frequencies used for ultrasound irradiation of suspensions extends from 5 kHz to 1 MHz. Lee and Mielewski [149] report that frequencies in the range 15–40 kHz can grant adequate dispersion for nanosized silica with nominal size down to 5 nm. A relationship between sonication duration time and the achieved dispersion degree is given by Gedanken [150]. Different techniques can be used to deliver the ultrasonic radiation to the suspension. In their work on nanosized additives for HTPB-based solid fuels, Risha *et al.* [148, 151] used an ultrasonic bath to sonicate the suspensions after mechanical mixing. Under the investigated conditions, an ultrasonication of 10 min yielded uniform additive dispersion and reduced clusters, as shown in Figure 12.23.

In order to encapsulate nanosized metal oxides with butyl rubber, Wang *et al.* [152] achieved high dispersed materials using a sonicating horn directly immersed in the particle suspension. Caneba *et al.* [153] proposed a combined approach using a sonicating horn immersed in a suspension placed in an ultrasonic bath. Although sonication is necessary to obtain a high degree of dispersion, it is not always sufficient to completely distribute nanoparticles in the suspending medium: additives are often used to obtain a stable and uniform dispersion [150, 154–156].

Figure 12.23 SEM of HTPB + 13% ALEX^{TM} highlighting nAl dispersion within cured matrix after ultrasonication [148].

12.4.4
Rheology of Suspensions

Detailed surveys of the rheological behavior of additive suspensions into fluid media are given by Nielsen [139], Russell [147], and Metzner [157]. Rheological behavior of energetic materials is extensively discussed by Teipel *et al.* [126], while the flow behavior of solid propellants was investigated by Miller *et al.* [158], Muthiah *et al.* [159], and Wildemuth and Williams [160].

When dealing with rheology of suspensions, the main parameter of interest is the relative viscosity. This is defined at a given shear rate (or shear stress) as

$$\eta_{rel} = \left.\frac{\eta_S}{\eta_{SM}}\right|_{\dot{\gamma}} \tag{12.10}$$

The suspension η_{rel} depends on several factors, including the suspending fluid viscosity affecting the polymer–particle hydrodynamic forces [126, 157].

Polymeric liquids loaded with additives exhibit increased viscosity, as shown in Figure 12.24. In spite of this, quantitative observations on η_{rel} depend on several factors and can yield not unique results [157]. An overview of literature relationships for η_{rel} evaluation is reported in Table 12.18. The Einstein equation fits to experimental data for very dilute suspensions ($\varphi \leq 0.02$), characterized by an almost linear η_{rel} increase [139, 157]. The Einstein coefficient k_E turns out to be 2.5 for well-dispersed spheres, but it can be evaluated with fair accuracy also for nonspherical particle dispersions. In particular, for agglomerates of rigid spheres, the following equation holds:

$$k_E = \frac{2.5}{\varphi_A} = \frac{2.5(V_S + V_L)}{V_S} \tag{12.11}$$

Thus, clustering of particles yields increased k_E values. In turn, this is related to enhanced viscosity of the considered suspension [139]. The oft-quoted Mooney

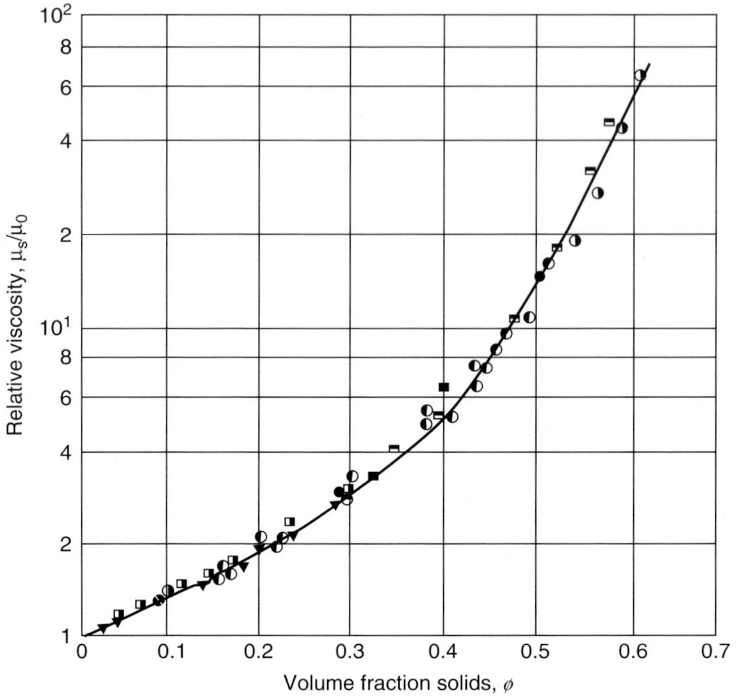

Figure 12.24 Relative viscosity of suspension of solid spheres, note monotonic μ_{rel} growth for increasing φ. (Reproduced with permission from [157] copyright © 1985 AIP Publishing LLC.)

Table 12.18 Overview of η_{rel} expressions for suspensions of rigid spherical particles [126, 157].

Author	Relative viscosity, μ_{rel} (see Equation 12.5)	Notes
Einstein	$1 + k_E\varphi$	$\varphi \leq 0.02$, $k_E = 2.5$ (for dispersed spheres)
Mooney	$\exp[k_E\ \varphi/(1 - \varphi/\varphi_{MAX})]$	$k_E = 2.5$ (for dispersed spheres)
Maron and Pierce	$(1 - \varphi/K)^{-2}$	$K = 0.68$ (typical for smooth spheres) [139]
Dougherty and Krieger	$(1 - \varphi/\varphi_{MAX})^{-[\eta]\cdot\varphi_{MAX}}$	$[\eta] = 2.5$ (typical for spheres)
Thomas	$1 + 2.5\ \varphi + 10.05\ \varphi^2 + 0.00273\exp(16.6\ \varphi)$	$0 \leq \varphi \leq 0.6$
Frankel and Acrivos	$9/8\cdot(1 - \varphi/\varphi_{MAX})^{\frac{1}{3}} / \left[1 - \left(1 - \varphi/\varphi_{MAX}\right)^{\frac{1}{3}}\right]$	High consistency for high φ
Chong, Christiansen, and Baer	$[1 - 0.25(\varphi/\varphi_{MAX})]^2/[1 - (\varphi/\varphi_{MAX})]^2$	—

equation and some of the relationships presented in Table 12.18 consider φ_{MAX} as a parameter for data fitting. The latter is the maximum filler volumetric fraction. These quantities differ because of particle–particle interactions and packing difficulties. Clustering reduces the φ_{MAX} value with respect to the case of a well-dispersed suspension. Thus, according to these equations, for clustered particle suspensions, η_{rel} is increased.

Teipel and Forther-Barth [138] investigated the rheological behavior of ALEX^{TM} suspensions in paraffin-oil and liquid HTPB. The tested ALEX^{TM} has $S_{sp} = 11.8\,m^2\,g^{-1}$ (as evaluated by gas adsorption) and was dispersed into the suspensions by stirring and sonicating the formulations. A summary of the Teipel and Forther-Barth results is reported in Figure 12.25, where the relative viscosities of the tested suspensions are plotted versus $\dot{\gamma}$ and φ. The tested pure paraffin oil is a relatively low-viscosity medium (η (293 K) = 198 mPa s) and presents a Newtonian behavior. Adding ALEX^{TM}, the resulting suspension exhibits a non-Newtonian, shear thinning behavior. In particular, Teipel and Forther-Barth noticed a marked η_{rel} increase for $\dot{\gamma} < 1$ Hz when testing suspensions with $\varphi \geq 21.53\%$ at $T = 293$ K. Under the investigated conditions, for $\dot{\gamma} > 10$ Hz, the relative viscosity of the suspensions decreases because of the hydrodynamic forces induced by the flow.

System-flow-induced structuring effects are well clarified by Figure 12.25b, comparing the η_{rel} of paraffin oil + ALEX^{TM} suspensions for two limiting shear rates. The HTPB resin tested by Teipel and Forther-Barth has η (293 K) = 6300 mPa s. The viscosity of the HTPB resin is so high that it induces hydrodynamic effects, yielding a Newtonian behavior over the whole investigated range. Similar results were achieved by Mary *et al.* [109] in their experimental activity on the suspension of HTPB + ALEX^{TM} with nAl volumetric fraction up to 10%.

In their work, Mary *et al.* evaluated the effects on η_{rel} of particle surface wettability for different nAl coatings. A shear thinning behavior is achieved by their tests on a suspension of polyethylene-coated ALEX^{TM} in HTPB.

This result underlines the importance of the high S_{sp} and surface characteristics of the nanosized particles. As observed by Popenko *et al.* [161], a significant increase in η_{rel} of ALEX^{TM} suspensions is achieved in ES for additive mass fractions >5%.

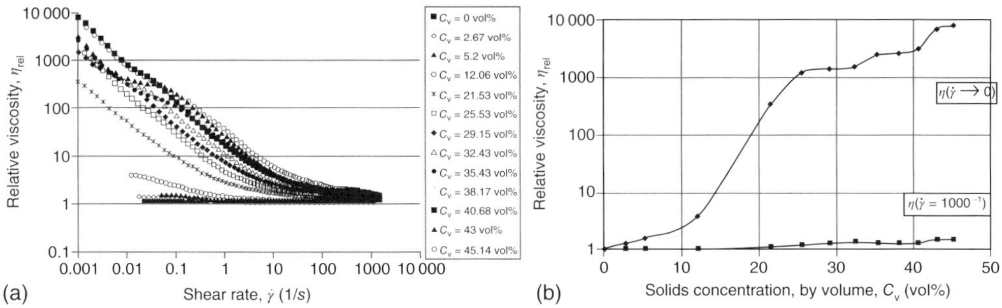

Figure 12.25 Relative viscosity of suspension of ALEX^{TM} in paraffin oil with $T = 293$ K [138]: (a) as a function of $\dot{\gamma}$ and (b) as a function of φ.

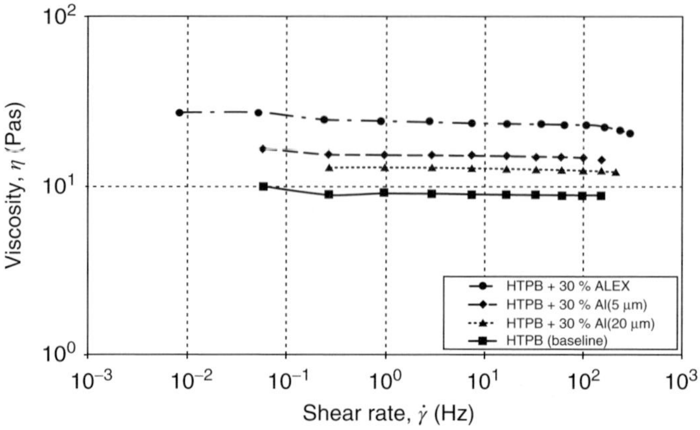

Figure 12.26 Viscosity of suspension of different Al powders in HTPB as a function of shear rate for $T = 293$ K [162]. Newtonian behavior over the whole investigated range.

Gromov *et al.* [162] investigated the behavior of μAl and nAl powders suspended in HTPB and GAP (glycidyl azide polymer). A summary of their results is presented in Figure 12.26 and Figure 12.27. Considering an Al mass fraction of 30% ($\varphi = 12.5\%$), HTPB + ALEX^{TM} exhibits an enhanced viscosity with respect to pure HTPB + μAl suspensions. This behavior, explained by S_{sp} of suspended particles and the hydrodynamic effect, characterizes the whole investigated $\dot{\gamma}$ and T ranges. Temperature effect on suspension rheology for HTPB + ALEX^{TM} is shown in Figure 12.27. As can be seen by the presented data, when temperature is increased, suspension viscosity is decreased, but its behavior remains Newtonian under the investigated conditions. The rheological behavior of GAP-based systems appears similar to the HTPB counterpart, although some difference can be noted

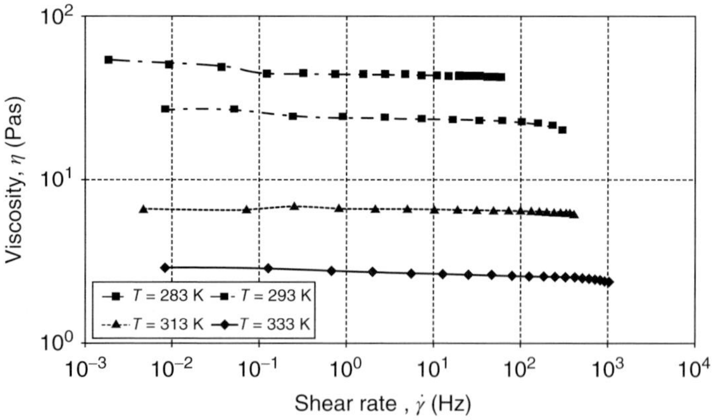

Figure 12.27 Viscosity of suspension of HTPB + ALEX^{TM} as a function of shear rate for different T [162]. Additive mass fraction is 30%, corresponding to $\varphi = 12.5\%$. Newtonian behavior over the whole investigated range.

for $\dot{\gamma} > 100$ Hz. Under these conditions, Gromov *et al.* noticed a shear thinning behavior common to pure GAP as well as to GAP + Al suspensions. This behavior is attributed to some system break-up within the suspending polymer.

12.4.5
Aging Effects

Aging can alter the mechanical characteristics and the combustion behavior of ES [163–169]. This is mainly due to different concurring phenomena:

- Oxidative crosslinking (polymer chain scission followed by oxidative attack);
- Migration and depletion of formulation ingredients (e.g., plasticizer, powder additives);
- Thermal-induced stresses and phase transitions due to temperature changes.

Because aging phenomena involve chemical reactions and diffusion of formulation ingredients, the common method for the investigation of these processes is ES storage under temperatures higher than the ambient one [163, 164, 168]. These studies are based on the so-called Layton model [163, 168].

Kwon *et al.* investigated the effects of different passivation layers on nAl produced by EEW. In their dataset, the passivation by air typically grants $85\% \leq C_{Al} \leq 90\%$ after production, while nAl passivated by other noninert or inert coatings are characterized by lower Al mass fractions. The fluoropolymer-passivated nAl characterized in [100], presents a $C_{Al} = 81\%$, while a nitrocellulose-coated formulation reaches 68%. After 12 months of storage in air with relative humidity <70%, ALEX^{TM} exhibits a reduction in C_{Al} lower than 5%, nitrocellulose-coated nAl exhibits $C_{Al} = 58\%$, while the fluoropolymer coating prevents powder active metal content reductions.

As a remarkable result, Kwon *et al.* demonstrated that stearic acid passivation does not prevent slow oxidation of metal particles with the formation of a thin alumina shell surrounding the core particle, thus reducing the C_{Al}. Kwok *et al.* [26] discussed reduced reactivity of APS nAl exposed to air contrasting different DSC–TGA data. In spite of this, their study is not focused on powder aging effects; thus, no detailed analyses were performed. Cerri *et al.* [168] conducted numerical and experimental investigations on aging of ALEX^{TM} powders and solid propellants loaded with nAl. In this work, reference formulations with μAl are compared to ALEX^{TM}-containing propellants. Thermochemical analyses, stress–strain tests, DMA, DSC, and impact and friction sensitivity studies were performed. Cerri *et al.* considered nAl-loaded propellants with maximum ALEX^{TM} mass fraction of 6% in order to limit viscosity increase during manufacturing (especially at the end-of-mix). In the presented results, after thermally simulated aging of 15 years, ALEX^{TM} shows a decrease in active Al content from the initial 88.2% to 4%. The lost active metal was consumed by the formation of $Al(OH)_3$. Bayerite formed by reaction with moisture accounts for the 84.2% of the aged ALEX^{TM} mass [168]. On the other hand, conventional μAl shows $C_{Al} = 99.6\%$ at the beginning of the aging process, while the final value reaches 72.5%. Also, in this case, the active Al reduction is caused by the reaction with moisture. This yields different thermochemical behavior of nonaged and aged μAl- and ALEX^{TM}-loaded propellants. By increasing aged nAl mass fraction in propellant formulation, an I_s detriment is achieved, while aged micron-sized powders still enhance performance, increasing their content into the formulations [168]. Nevertheless, it must be noted that Cerri *et al.* did not perform burning tests, and thus no data on the ballistic responses of the aged propellants are available. Under the investigated experimental conditions, friction sensitivity exhibits some enhancement by the aging process, while impact sensitivity seems affected by the filler particle sizes, but not from the additive aging [168].

12.4.6
Experimental Results: Data Processing and Discussions

Experimental results concerning mechanical properties of solid propellants/fuels loaded with nanopowders achieved at SPLab are hereby reported. Uniaxial tensile stress tests and DMA were performed.

In order to perform uniaxial tensile stress–strain tests, Instron Series 4302 with a 1000 N load cell is used. Samples are dog-bone shaped according to the standard DIN 53 504 S3A. All the tests are conducted up to the breakup of the specimen and recorded with a camera: the investigated polymeric materials, in fact, suffer from slippage and a refinement of data produced by the instrument is needed. Starting from the first movement, the distance between two marked points on the sample and the distance between the two separating clamps are monitored (Figure 12.28).

Data collected are plotted and the sample elongation (ΔL_{sample}) is expressed as a function of the clamp movement (ΔL_{clamp}), as reported in Equation 12.11.

$$\Delta L_{sample} = a_L \cdot \Delta L_{clamp}^{n_L} \tag{12.12}$$

Clamp displacement being the output of the instrument, it is possible to calculate linear strain $\varepsilon_{lin} = \Delta L_{sample}/L_{sample}$ and, from the applied load (F) and the initial section (A_0), the engineering stress $\sigma_{eng} = F/A_0$. Linear strain and engineering stress could well represent the material behavior for strain level lower than 5%, but for higher strain and beyond necking it is better to consider true stress and

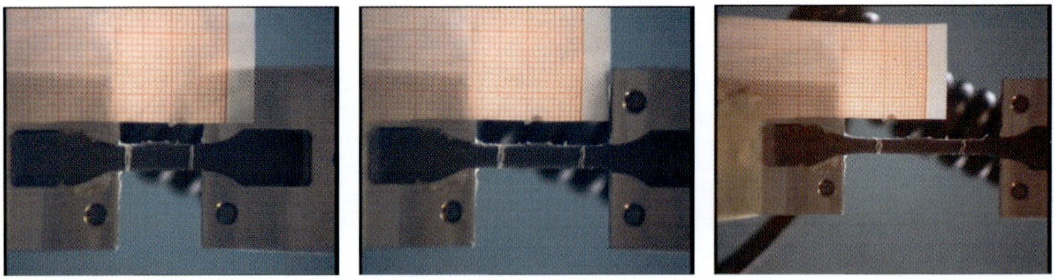

(a) (b) (c)

Figure 12.28 Uniaxial tensile stress–strain test at (a) beginning, (b) center, and (c) almost at the end of test. White markers on the sample enable slippage evaluation and correction.

logarithmic strain:

$$\sigma_{true} = \sigma_{eng}(1 + \varepsilon_{lin}) \tag{12.13}$$

$$\varepsilon_{log} = \ln(1 + \varepsilon_{lin}) \tag{12.14}$$

These parameters can be evaluated during the whole test and, in particular, when the sample breaks. Moreover, true stress versus logarithmic strain is plotted and the elastic modulus is evaluated at 10% of strain. The tangent modulus, E_{tan}, considers the slope of the curve tangent to the stress–strain one in the point at the chosen strain.

DMA are carried out using TA Instruments DMA 2980 with a dual-cantilever clamp. Specimens are parallelepipeds, whose size is defined considering the expected elastic modulus [134–137, 139]. Frequency sweep ranging from 0.3 and 100 Hz and strain sweep spanning from 0.3% to 1%, both in isothermal conditions at 313 and 323 K, are considered.

Viscosity of uncured formulations is measured using the Couette rheometer Rheotest 2.1. Tests are conducted varying the shear rate from 1 to 100 Hz in an isothermal condition at 293 K.

12.4.7
Tested Formulations

The effect of powder addition and quality of dispersion are evaluated considering samples filled with μAl or nAl, with mass fraction ranging from 1% to 10%. The used μAl has a mean size of 7–15 μm and it is mainly spherical. The considered nAl is ALEX™ with a nominal size of 100 nm. In order to maintain consistency with fuel formulations tested on SPLab 2D radial microburner, a small amount of carbon black (CB) is added. The CB used during this experimental investigation is nanosized and contains a small amount of carbon nanotubes.

All the compositions containing the same amount of filler shear the same composition, but they are obtained with different manufacturing techniques in order to improve the dispersion degree of nAl. Nanosized powders, in fact, are difficult to disperse homogeneously into a polymeric matrix: owing to van der Waals forces, very effective considering the high specific surface, very small elements tend to form clusters up to 100 μm in size.

In standard manufacturing procedure, the ingredients are added to liquid HTPB during mixing with a propeller at 50 rpm in vacuum. Two strategies have been selected to improve dispersion of nAl: enhancement of mixer velocity and solicitation with ultrasound devices. In particular, an ultrasonic bath (38 kHz) is used for dry powders before adding them to the compounds and an ultrasonic ring (24 kHz) is connected to the mixing setup after nAl joins the preparation. A summary of the formulations tested to find mechanical properties is presented in Table 12.19.

Rheological tests have been performed on uncured formulations of solid propellant. Analyses are carried out to see the effects of increasing solid fraction: first,

Table 12.19 Sum up of formulations investigated with tensile stress–strain tests and dynamic mechanical analyses.

Formulation	Matrix	Filler	Sonication time (min)		Mixer velocity (rpm)
			Bath	Ring	
1	HTPB	—	—	—	50
2	HTPB	2% CB	—	—	50
3	HTPB	1% μAl + 0.2% CB	—	—	50
4	HTPB	1% ALEX™ + 0.2% CB	—	—	50
5	HTPB	1% ALEX™ + 0.2% CB	15	30	50
6	HTPB	1% ALEX™ + 0.2% CB	15	30	100
7	HTPB	1% ALEX™ + 0.2% CB	15	30	150
8	HTPB	1% ALEX™ + 0.2% CB	15	60	100
9	HTPB	10% ALEX™ + 2% CB	—	—	50
10	HTPB	10% ALEX™ + 2% CB	30	—	50

liquid HTPB is tested and then 18% of μAl with a nominal size of 5–10 μm is added. Further samples are obtained adding to HTPB and μAl, AP with a nominal size of 50 μm in percentages spanning from 5% to 60%.

Addition of nanosized powders to heavily loaded formulations leads to a sharp increase in viscosity and problems connected to manufacture and casting [161]. A possible solution can be represented by coating powders, before the addition to the compound, with the same HTPB used in the propellant. Coated powders are obtained starting from ALEX™ (nominal size 100 nm) using acetylacetone as coupling agent and 1% of HTPB.

Specimens with partial substitution of micrometric aluminum with the nanometric variety were prepared keeping constant the relative fraction among ingredients: 50% AP + 32% HTPB + 18% Al. The 18% of aluminum is obtained with different combinations, μAl + nAl from 17% + 1% up to 5% + 13%. A comparison between coated powders and uncoated ALEX™ is carried out.

12.4.8
Uniaxial Tensile Stress–Strain Tests

First, results of the same composition containing 1% of ALEX™ obtained with different manufacturing techniques are presented and compared with the homologous formulation loaded with 1% of μAl. In Figure 12.29, maximum stress and maximum elongation are reported, while Figure 12.30a represents the tangent elastic modulus at 10% of strain.

The effect of filler addition to the HTPB matrix is then considered; moreover, a comparison between two specimens containing 10% of nAl is carried out: formulation 9 is obtained with standard manufacturing procedure, while

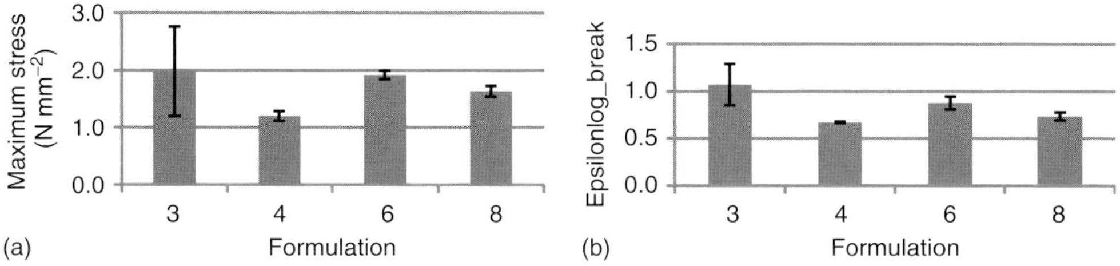

Figure 12.29 Comparison between formulations obtained with and without sonication: (a) maximum stress and (b) maximum elongation. Sonicated formulations bear higher stress and are more extensible.

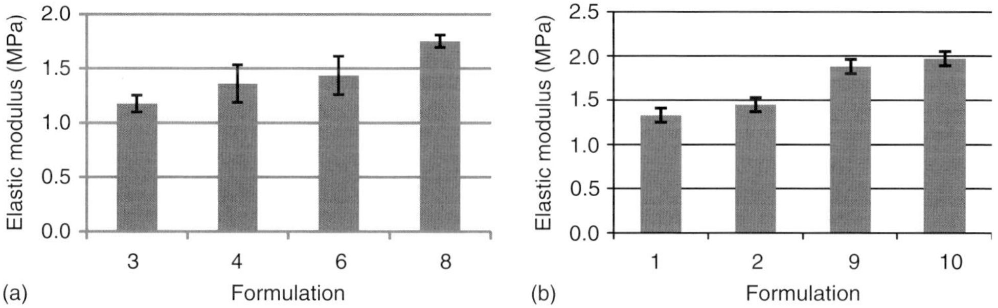

Figure 12.30 Elastic modulus: (a) sonication time effects and (b) filler content effects. Elastic modulus increase for increasing filler content and it is higher for sonicated formulations.

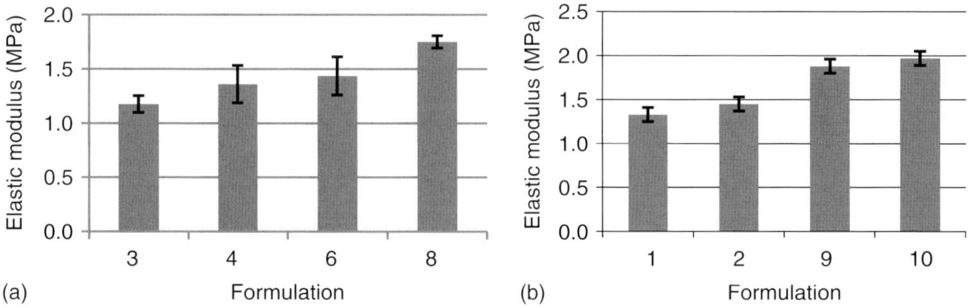

Figure 12.31 Increasing filler content effect and comparison between formulations obtained with and without sonication: (a) maximum stress and (b) maximum elongation. Sonicated formulations bear higher stress and are more extensible.

formulation 10 is treated with 30 min sonication. E_{tan} is reported in Figure 12.30b, while properties at sample failure are presented in Figure 12.31.

As can be seen in Figure 12.29, samples obtained ultrasonicating the compound (formulations 6 and 8) show higher extensibility and bare higher loading with respect to homologous formulation 4, prepared with the same nAl content but without any particular treatment to improve dispersion degree of the nanoadditive.

The same consideration can be done on 10% nAl-containing samples shown in Figure 12.30: maximum stress and elongation are higher for formulation 10 obtained with sonication compared with formulation 9 manufactured using the standard procedure. However, the highest values among formulations containing 1% of aluminum are shown by the samples containing μAl even if data for this formulation are quite scattered. Similar sonication effects are visible in Figure 12.30: elastic modulus increases with sonication time for formulation, presenting the same ingredients (4, 6, 8 and 9, 10). Moreover, preparation with μAl has a lower elastic modulus with respect to nAl-loaded formulations. Comparing unloaded HTPB with additives-containing formulations, it is possible to verify how the filler presence increases the elastic modulus of the polymer (Figure 12.30b), as well as properties at sample failure (Figure 12.31).

An explanation can be given considering that ultrasound treatment improves dispersion degree [147] and only well-dispersed powders can form a filler network, that is, those responsible for the increased properties of the system. On the other hand, cold cohesion leads to the formation of clusters, surrounded by a matrix that does not cooperate in improvement of mechanical properties. Moreover, by increasing filler content, the possibility of network formation also augments.

12.4.9
Dynamic Mechanical Analysis

In order to complete data from uniaxial tensile stress–strain tests, DMA are carried out. The comparison between μAl-loaded formulations and nAl-including ones obtained with different sonication times is shown in Figure 12.32 in terms of storage and loss moduli and tan(δ). Data presented in Figure 12.32 are related to strain sweep tests carried out at 313 K. It is possible to notice that for all the formulations considered, moduli remain constant in the whole strain interval considered. Moreover, as can be seen in Figure 12.32, the formulation loaded with μAl presents the lowest storage modulus among the investigated composition, while the loss modulus is comparable with the one shown by formulation 4, containing ALEX^{TM} but without any treatment.

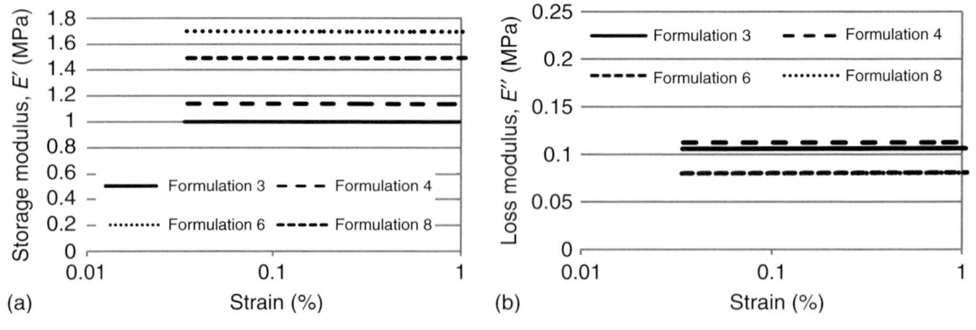

Figure 12.32 (a,b) Sonication time effects on E' and E''. Sonicated formulations have higher E' and lower E'' with respect to formulations obtained with standard procedure.

Ultrasound solicitations given to the preparation lead to higher storage moduli and lower E''. Loss modulus can be related with the energy dissipated: in case of cold cohesion, the solicitation given by the clamp during the test can break clusters and can improve interaction between particles; as a consequence, E'' results are higher with respect to the one shown by treated formulations. These results are consistent with the ones obtained from tensile stress–strain tests.

Another way to improve dispersion of nanoadditives in a polymeric matrix is connected to the mechanical solicitation given to the compound through the mixer propeller. Frequency sweep tests on formulations obtained with the same sonication time but changing mixer velocity are carried out and the results are shown in Figure 12.32.

In previews works [135], formulation 6 was verified to feature the best dispersion degree among the investigated formulations, probably because too high speeds can improve the possibility of particle collision and therefore of cluster formation. In Figure 12.33, one can see that formulation 6 shows also the highest E' and the lowest $\tan(\delta)$, while formulation 7 presents the lowest E' and the highest $\tan(\delta)$. Cluster presence reduces the possibilities of filler network formation and leads to the reduction of storage modulus. The $\tan(\delta)$ value shows that formulation 7 dissipates more energy with respect to the stored one, and this can be connected to the disruption of aggregates.

12.4.10
Rheological Tests

From the rheological point of view, heavily loaded suspensions show a different behavior with respect to dilute ones [125, 126, 138, 139, 147, 157]. As already discussed, compositions containing a small amount of solid particles (up to 20% in volume) exhibit moderate viscosity increases that are mainly due to the

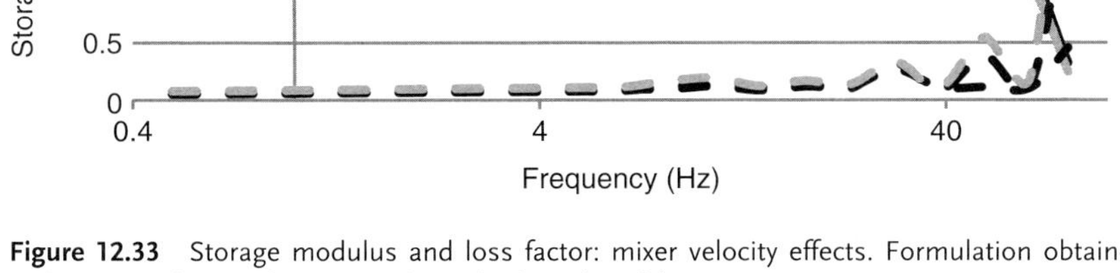

Figure 12.33 Storage modulus and loss factor: mixer velocity effects. Formulation obtained with mixer velocity of 100 rpm shows higher E' and lower $\tan(\delta)$.

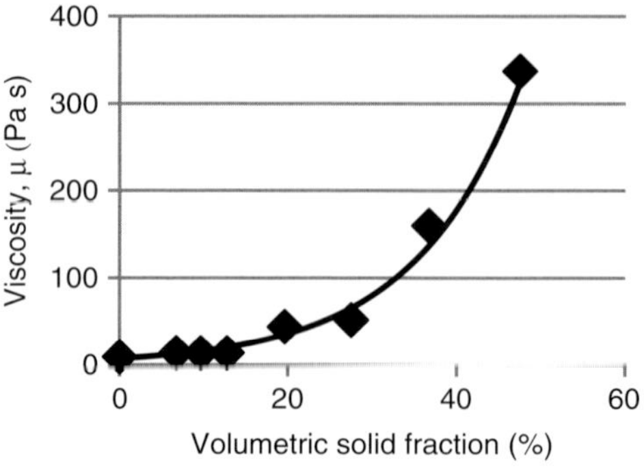

Figure 12.34 Viscosity as function of volumetric solid content of uncured propellant for $\dot{\gamma} = 1$ Hz.

hydrodynamic particle–liquid interactions. By increasing volumetric content over 20%, viscosity augments rapidly and reaches an asymptotic behavior for the maximum allowable volumetric content.

This behavior is verified also for uncured propellants. In Figure 12.34, viscosity as a function of volumetric content is presented. First, liquid HTPB is tested, showing a μ (293 K) = 9100 mPa s. Then, 18% by mass of μAl is added and further enhancement of volumetric content is obtained, adding AP up to 60% by mass.

A comparison between HTPB-coated and HTPB-uncoated ALEX^{TM} is carried out in formulations with partial substitution of μAl with nAl. Results increasing ALEX^{TM} fraction are shown in Figure 12.35 for two different shear rates. As can be seen in Figure 12.35, formulations containing nAl show lower viscosity with respect to uncoated ALEX^{TM} loaded ones for almost every amount. The only exception is visible in the plot obtained for a shear rate of 3 Hz, where the difference starts to become visible in the composition 3% ALEX^{TM} + 15% μAl. An explanation for this behavior can be given considering that coated particles should better disperse inside the liquid matrix. Clusters act like micrometric particles, increasing viscosity due to hydrodynamic forces.

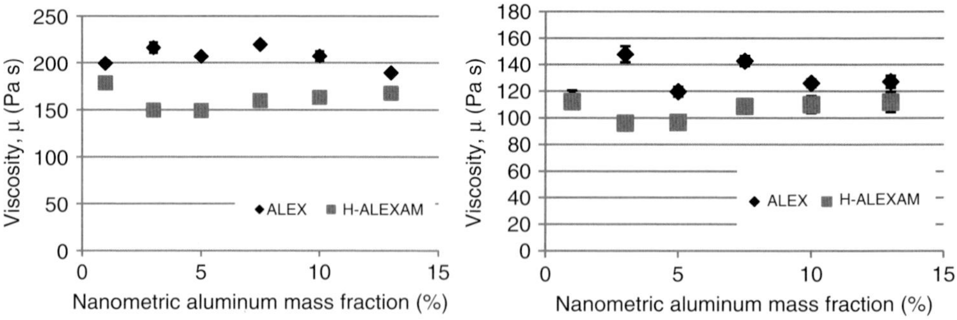

Figure 12.35 Comparison between HTPB-coated and HTPB-uncoated ALEX^{TM} for increasing nanosized aluminum fraction: (a) $\dot{\gamma} = 1$ Hz and (b) $\dot{\gamma} = 3$ Hz.

12.4.11 Concluding Remarks

The effects of filler content and the quality of its dispersion in mechanical and rheological properties of hybrid solid fuels and solid propellants are evaluated in this chapter. Uniaxial tensile stress–strain tests performed on hybrid solid fuels show an increase in the maximum bearable stress and in extensibility-enhancing filler content in the investigated range (up to 10%); in particular, the maximum values are shown by formulations treated with ultrasound in order to improve the dispersion degree of nAl inside HTPB.

Elastic modulus, evaluated as the slope of the tangent to 10% of the stress–strain curve, increases with the filler content and the highest value, for the same percentage of additive, is shown by ultrasonicated formulations.

In order to improve the dispersion degree of ALEX^{TM} into the HTPB matrix, various strategies are adopted; in particular, the effect of ultrasound and mixer velocity is evaluated. The best dispersed formulation shows the highest maximum stress and the highest extensibility together with the highest elastic modulus evaluated as the slope of the tangent to the 10% of stress–strain curve.

DMA confirm an increase of elastic modulus in its storage component with the homogeneity of dispersion; in particular, the highest value corresponds to the formulation mixed for 30 min with ultrasound and a mixer velocity of 100 rpm. The same formulation shows the lowest loss modulus, that is, the lowest amount of energy dissipated. This can be connected to the creation of a filler network, possible only for well-dispersed formulations, that contributes to increase E', while cohesion leads to the formation of particle clusters surrounded by the polymer, whose rupture, caused by the applied stresses, enhances the amount of energy dissipated.

Rheological tests on uncured propellants show an increase in relative viscosity with the volumetric filler content increasing AP fraction in AP + HTPB + μAl formulations. A gradual substitution of micrometric aluminum with the nanosized one allows to see the effect of an improved dispersion connected to an HTPB coating: relative viscosity is lower and less shear dependent for propellants containing coated powders, in particular for percentage ranging from 3% to 7.5%.

12.5 Combustion of Nanopowders in Solid Propellants and Fuels

From the energetic viewpoint, nanosized particles can be divided into coarse and fine, with a border size of about 10 nm. While for particles larger than 10 nm distinctive features are mainly determined by their specific surface, for the second group (made by particles containing tens of subparticles), an increase of energy takes place because of intermolecular interactions. In particular, the enthalpy of formation can reach [88] values of the order of megajoules per kilogram for Al particles of the second group. Thus, the current high interest in nanoparticles, in general, is motivated not only for internal ballistics reasons [170] but more generally also for both applied and fundamental reasons [171].

The ultimate goal for any energetic application is to assess the influence of Al particle size on the ballistic properties and flame structure of solid propellants or solid fuels for solid and hybrid rocket propulsion, respectively. The physical characterization of nAl powders and analyses of solid combustion residues were discussed previously [24, 172, 173] and further extended in Section 12.3 and Section 12.4. The companion study of agglomeration conducted for a variety of industrial and laboratory μAl propellants is briefly discussed in this work as a reference for nAl propellants.

12.5.1 Solid Rocket Propellants

In this section, the measured steady ballistic properties of many laboratory nAl formulations are presented and contrasted with the corresponding μAl ones. After detailing the tested solid propellant formulations, the distinctly different ballistic properties observed for μAl versus nAl samples are discussed. The achieved results, while in agreement with literature trends, are explained in terms of the propellant microstructure, that is the fine material distributed among the coarse oxidizer particles. In this respect, strikingly different features are exhibited by the tested μAl versus nAl formulations when examined by high-speed and high-resolution video camera recordings. Mastering these effects opens the way to innovative fine tuning of high-energy density fuels for a variety of applications [174].

12.5.1.1 Particle Clustering Phenomena

Cold Cohesion versus Hot Aggregation/Agglomeration Cold cohesion, a characteristic trait of the fluffy nanosized particles, takes place at room temperature (or close to) during storage, handling, and manufacture of the propellant. This kind of pulling together is responsible for microsized clusters reducing the specific surface of nanosized ingredients, and thus works against the major, although not unique, rationale of their enhanced reactivity. All precautions should be taken, including coating of the particles, to lessen the propensity of nAl to cold cohesion (i.e., clots or lumps of particles). Mechanical and rheological properties of matrices containing nanoparticles may be strongly affected by clustering. A possible approach to mitigate their propensity to cold cohesion is to coat the nanoparticles with specifically selected substances.

Hot agglomeration is a well-known trait of μAl occurring under high temperatures at/near the burning surface or in the gas phase of the combustion wave. Cold cohesion precedes and may influence hot aggregation/agglomeration effects. Both phenomena imply propulsive performance losses. The process of aggregation/agglomeration is discussed in detail hereunder.

Mitigation of Cold Cohesion and Hot Aggregation/Agglomeration Coating Al particles with high-melting metal films (e.g., Ni, Cu, Fe) decreases hot agglomeration [175] by making μAl ignition easier, thanks probably to highly exothermic intermetallic reactions at Al melting; a similar effect can also be reached by coating with organic substances [176]. Again, coatings or sonication can be used to mitigate the nAl propensity to cold cohesion. Hot aggregation is crucial for μAl, while cold cohesion is crucial for nAl.

Nomenclature for Cohesion, Aggregation, Agglomeration To distinguish between the different ways that the complex phenomena of particle clustering overall manifest, the following nomenclature is enforced: the word agglomeration is reserved for the formation of the spherical drops of liquid metal and oxide in combustion, while the word aggregation is reserved for the formation of the partially oxidized flakes of irregular shape seen at or near the burning surface (an intermediate step between the initial metal load and the possible final agglomerates). Agglomerates always imply a loss of the initial particle individuality; aggregates are agglomeration precursors and may keep some reminiscence of the initial particle individuality. As defined, both agglomeration and aggregation are processes typically manifested by Al burning. In addition, it is convenient to reserve the word cohesion for "a portion of a substance cleaving together in a thick nondescript mass" (Webster's Dictionary).

12.5.1.2 Propellant Volume Microstructure

Volume Microstructure and Pocket Composition Pockets are privileged zones for the occurrence of agglomeration processes thanks to the presence of favorable conditions, such as a local high concentration of metal particles [90, 91]. The pocket composition strictly depends on the propellant formulation, but by definition it includes only the fine fraction of AP (or whatever oxidizer is used) in addition to metal and binder. Thus, while AP coarse particles define the pocket size, the remaining ingredients identify the pocket composition and alter the microstructure formulation toward a fuel-rich mixture with respect to the nominal propellant formulation; for details, see [62], for example.

Volume Microstructure and Local Flame Temperature The local adiabatic flame temperature in the compound microstructure is computed assuming the combustion of a virtual propellant having the pocket formulation as discussed earlier. During combustion, the heat feedback to the burning surface causes AP to decompose into oxidizing gases and the binder to pyrolyze into fuel vapors. The fine AP fraction decomposes very fast because of its minute grain size. The two resulting gaseous streams mix together and react, generating a flame with a relatively low temperature, as a result of the fuel-rich local mixture. Standard thermochemical calculations allow obtaining the adiabatic flame temperature associated with the microstructure operating conditions. Only fine AP and HTPB are assumed to react in the pocket, producing a primary flame triggering a partial oxidation of the loaded metal and thus subsurface aggregation.

Surface Texture Microstructure Filling of the propellant surface microstructure (i.e., the fine material distributed among the coarse oxidizer particles) is conveniently examined by resorting to SEM visualization and energy-dispersive X-ray (EDX) microanalysis techniques. While coarse oxidizer grains are well visible under ordinary light illumination, EDX reveals the Al distribution and shows a uniform and well-distributed scattering of metal particles – among the coarse AP particles – only in the case of nAl.

12.5.1.3 Steady Combustion Mechanisms of AP/HTPB-Based Composite Propellants

Basic Flame Structure For the explored operating conditions, the well-known [177, 178] underlying unmetallized flame structure appears affected to a modest extent, as revealed by the minor changes of the burning rate pressure sensitivity. Al particle size affects burning according to the following trends:

1) For μAl propellants, the CCPs result from a competition between the homogeneous diffusion flames in the gas phase, yielding SOP and the heterogeneous reaction of liquid Al [179, 180]. The fraction of the metal participating in aggregation/agglomeration processes depends on the propellant properties [181–183]. The first stage of metal burning by agglomeration is carried out within the skeleton layer whose features determine [184] the burning properties. In either case (burning of metal with or without agglomeration), μAl particles feature a spacewise distributed burning overlapping the underlying unmetallized flame structure but extending much beyond the gas-phase flame thickness, and thus little affecting steady burning rates.
2) Metal powders in the nanometric range show a remarkable steady burning rate increase with decreasing nAl size and point out a primary effect of specific surface increase. Typically, under common operating conditions, by two orders of magnitude passing from μAl to nAl.
3) Qualitative and quantitative information was obtained comparing the growth mechanism of aggregation/agglomeration processes for propellants loaded with metals of different BET surfaces. Although following the same succession of events, the burning mechanism of nAl particles is plainly different, as discussed in the next sections.

Burning Surface of Metallized Formulations Contrasting burning processes at or near the combustion surface clearly indicates a stronger brightness of the region immediately above the burning surface for nAl versus μAl propellants. This fact can be associated with the rapid combustion of nAl in that region, which enhances near-surface heat release, thus increasing the burning rate. This remark is in qualitative agreement with the spectroscopic data by Weiser *et al.* [185] and explains the burning rate augmentation recorded by a number of investigators all around the world. In agreement with Trubert *et al.* at ONERA [186], further basic differences and a plausible explanation are discussed hereinafter.

Micrometric Al Powder: Agglomeration Controlled Following the pioneering work by Price [179, 180], during binder and oxidizer decomposition processes the metal particles, confined in the propellant microstructure among the coarse AP grains, warm up and reach the burning surface. Aluminum particles are naturally covered by a porous and refractive Al_2O_3 coating (unless of intentional provisions). The two materials have different thermal expansion coefficients and melting temperatures ($T_{melt\ Al_2O_3} = 2345$ K vs $T_{melt\ Al} = 933$ K); in addition, solid and liquid Al have different densities, yielding after heating a volumetric expansion of about 6%. All of these effects help the formation of clots of irregular shape in the skeleton layer of the burning propellant surface [14, 16, 187]. In particular, clusters of elongated shape (filigrees [188]) are observed accumulating on the burning surface, and protruding into the hotter regions of the gas phase, as typical of class B propellants (with the metal ignition temperature higher than the decomposition temperature of carbonaceous elements, yielding a skeleton layer made by the initial metal particles fastened among themselves). Then, a progressive but intense heating changes these protruding clots into reacting and hot aggregates, whose quick inflammation yields particle coalescence. The final result is a single drop in combustion with the characteristic round shape. Under the proper thermal and oxidizing conditions, ignition of aggregates and transition to agglomerates occurs [189, 190] when the protective oxide coating surrounding Al particles is cracked by fractures as a result of heating. Mainly the near-surface hot AP/binder flamelets, called *leading-edge flames* (*LEFs*), are suspected to be responsible for this phenomenon. After inflammation, a burning agglomerate sphere forms as chiefly a mixture of liquid Al and liquid Al_2O_3, with a solid lobe of Al_2O_3 (Figure 12.38a). This sphere of burning liquid Al keeps on growing until detachment [66]. The whole sequence is clearly illustrated in Figure 12.36, showing the details of the pre-inflammation process for a laboratory propellant burning at 2 bar, and Figure 12.37, showing the details of the post-inflammation transition from aggregate to agglomerate for an industrial propellant burning at 5 bar.

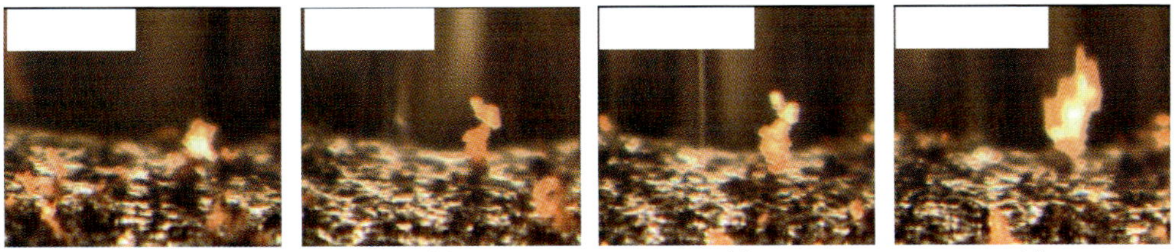

(a) (b) (c) (d)

Figure 12.36 Pre-inflammation sequence evidencing protruding clusters (filigrees) from the propellant burning surface just before agglomeration (with oxide cap) at 2 bar. (Video SPLab: 50 μm Al flakes, lab propellant, P'06.) (a) $t = 0$ ms, (b) $t = 6$ ms, (c) $t = 12$ ms, and (d) $t = 18$ms.

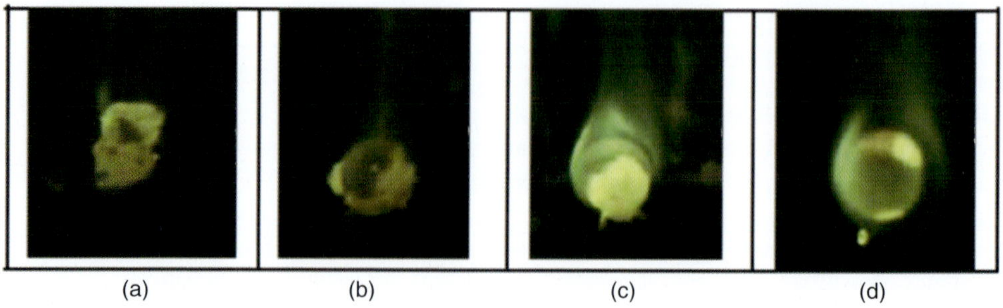

Figure 12.37 (a–d) Post-inflammation sequence capturing aggregate to agglomerate (with oxide cap) conversion at 5 bar. (Video SPLab: propulsion grade Al, reference industrial propellant A, 2000 fps.)

A vivid example of the agglomeration results in the particle-laden flame zone for a typical highly loaded Al propellant is shown in Figure 12.39a. For common industrial propellants, the observed [56, 189, 190] the incipient average agglomerate size is slightly above 250 μm at 5 bar and shrinks by almost 50% at 25 bar; see details in Section 12.2.6 and Section 12.3.6. For μAl, most of the heat release occurs far from the burning surface.

Nanometric Al Powder: Aggregation Controlled Specific attention is demanded for the combustion surface phenomena, probably the controlling zone for heat release of nAl propellant burning. The results in Figure 12.38b show the details of the near-burning-surface region for P˙01a propellant, revealing an unusual coral-like formation, reminiscent of the propellant volume microstructure, exuding from the sample. These pictures, validated for a variety of formulations and over a range of

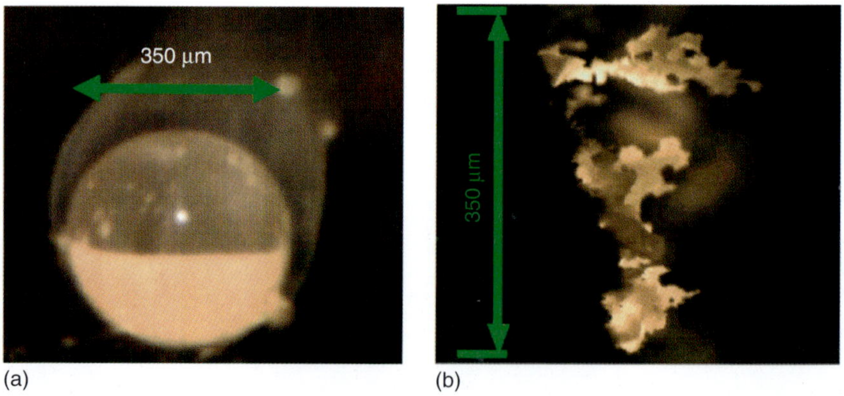

Figure 12.38 (a) Magnified view of a single spherical agglomerate of μAl in combustion *above* a propellant surface following inflammation of aggregates (filaments or filigrees) *at* the propellant surface, with loss of the initial particle individuality. (b) Oxidized metal flakes emerging from the burning surface of a 0.15 μm nAl propellant tested at 10 bar, with flakes replicating the 3D nAl distribution in the propellant matrix.

pressure, allow assigning the investigated propellants to class nano B propellants [16, 64]. Aggregation/agglomeration phenomena, although following the same succession of events, bluntly differ for Al and nAl particles. Chemical analyses of the friable exudation flakes reveal the presence of metal oxidation products together with some intermediate combustion products and possible residual unburned metal. Exudation is due to nAl aggregation probably starting in depth. Once the aggregation flakes have emerged at the burning surface, they can complete their oxidation/combustion while crumbling into CCP particles reminiscent of their original size (micrometric or submicrometric). Little agglomeration is observed following the exudation of aggregation flakes at the surface (Figure 12.39b). No significant inflammation (transition of burning into gas-phase mode) takes place, but the burning surface appears much more luminous thanks to the very large number of nAl flakes completing their oxidation (burning in heterogeneous mode) at/near the burning surface. Although agglomeration is possible for nAl as well, size and combustion times [191–193] of nAl particles are so small that the consequences are measurably reduced.

Moreover, for nAl, resealing after oxide layer crack is less likely to occur [15, 22]. An additional mechanism for fast reactions was recently identified as being due to melt aluminum dispersion under rapid heating conditions of small nAl particles [50–52].

Comparing Nanometric to Micrometric Al Powder The kind of image seen in Figure 12.39a for the particle-laden flame zone of μAl propellants is replaced by the blurred image of Figure 12.39b for nAl propellants. Aggregation, seen for μAl propellants only for short instants preceding agglomeration, is the main visible stage for nAl propellants. The distinctive feature of Al in the explored formulations is agglomeration, while the distinctive feature of nAl burning is aggregation. With respect to μAl particles, nAl particles feature the combined effects of earlier ignition

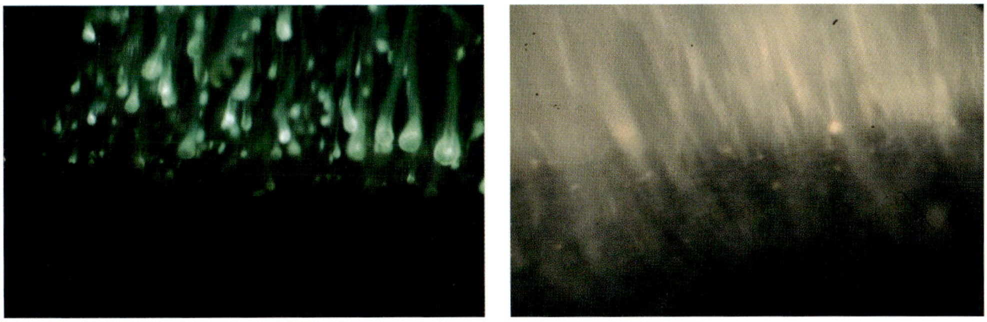

(a) (b)

Figure 12.39 (a) Particle-laden flame zone due to μAl combustion at some stand-off distance from the burning surface for industrial propellant *D* tested at 5 bar. (b) Fine aggregate emerging on the surface by exudation and reduced transition to spherical agglomerates at a short distance from the burning surface for nAl type 01a burning at 10 bar.

and quick premixed oxidation (possibly in depth), followed by oxidation completion at or very close to the burning surface.

Thus, at the propellant surface layer, combustion of nAl particles is almost concluded, whereas μAl particles are just starting to react. This, in turn, implies the absence of slag formation and also a probable increase of combustion inefficiency, as pointed out by thermochemical results [66] contrasting, under ideal operating conditions, the combustion temperature and product composition with the corresponding values achieved if only a fraction of the available fine AP reacts in the propellant microstructure, yielding a relatively low-temperature (due to the very fuel-rich mixture ratio of the local premixed flame) but prompt metal oxidation. Unless oxidation is fully completed downstream, combustion inefficiencies will arise. In exploratory tests with nAl type 01 burning under subatmospheric pressure, about 80% γ-Al_2O_3 and δ^*-Al_2O_3 were recovered, while up to 10% was unburned Al and the remaining CCP consisted of intermediate products, such as Al_2OC and Al_4C_3. Similar analyses were performed at ONERA [186]. Although the variety and complexities of the observed phenomena require further studies, the enhanced reactivity of nAl is well manifest. The average size of CCP is sharply reduced, as confirmed in other studies [194–196], with obvious advantages in terms of nozzle expansion efficiency. But the resulting overall specific impulse efficiency will depend on the combination of all of the above-mentioned factors.

Possible mechanisms responsible for the formation of the nAl flakes, for the considered class of propellants, include propellant manufacturing, propellant storage, and propellant heating during combustion (or a combination of all of them). The assumption that the contact between initial particles of nAl for propellants such as nanoB is related to technology factors was already discussed in [64]. At any rate, under any circumstance, the conductive thermal wave preceding the burning surface is able to warm up the nAl particles within the propellant microstructure to temperatures of the order of the ignition temperature, recalling the enhanced nAl reactivity (see Section 12.3.5). The intimate mixing with the surrounding decomposition products of the fine AP fractions and binder, favoring a local underoxidized premixed flame, promotes the formation of partially oxidized and intermediate metallized flakes (prompt metal oxidation). Depending on steady burning rate (i.e., pressure), type of binder, fine oxidizer size, and fraction, some unburned metal will also contribute to the flake formation.

12.5.1.4 **Transient Combustion Mechanisms**

The superior ballistic properties of nAl (in terms of burning rate) are typically ascribed to the more intense heat release rate near the burning surface compared to the corresponding μAl formulations. Although some evidence does exist in this respect, it is actually very difficult to get a convincing dataset proving this point beyond any reasonable doubt. To prove this, two solid rocket propellants, with the same composition but different aluminum particles size (μAl was replaced by nAl), were tested under transient combustion to assess their intrinsic resistance to strong but well-controlled dynamic disturbances. The go/no-go boundary between (continued) combustion and extinction was obtained in the same experimental apparatus by keeping the operating conditions as close as possible. If nAl formulations really imply a more vigorous energetic coupling at the burning surface in terms of increased heat feedback, this should clearly manifest in the compared experimental results of dynamic extinction boundary by fast depressurization. Moreover, the PDLs were compared. At last, the ignition delay (with ignition driven by an external heat source in the form of, mainly, a laser beam), was evaluated under a variety of operating conditions.

Fast Depressurization The depressurization tests were performed in an experimental rig realized at CNR-IENI Laboratory; detailed descriptions are reported in [197, 198]. Three main parts compose the rig: a windowed combustion bomb (1050 cm^3), a double-burst disk depressurization apparatus, and a discharge tank. For each test run, a different exhaust orifice was chosen to change the depressurization rate. The propellant samples were 5 mm × 5 mm section and 50 mm length strands. The data acquisition system recorded the pressure signal and the initial depressurization rate was used as indicative of the pressure disturbance; an automatic analysis of the video gives the burning surface regression during the steady burning period and the regression rate is calculated from the linear fitting of these data. Either extinction or continuous burning of the propellant sample gives the overall result of the pressure transient.

The go/no-go data are plotted in terms of initial depressurization rate versus initial pressure. A straight line was found to separate extinction from burning region in this plot. Two AP-based aluminized propellants were tested. Both had the same composition (AP 68% + HTPB 17% + Al 15%) and the same AP average particle size and size distribution. The only difference was the kind of aluminum powder: 50 mm nominal size μAl (flake type 06) and 0.150 mm nominal size nAl (uncoated type 01a) were used. The data previously collected [197] with an AP 80% + HTPB 20% nonmetallized propellant were kept as reference. The results of the steady-state burning rate measurement are summarized in Figure 12.40.

The nAl propellant exhibits a steady burning rate, higher compared to the μAl propellant, in the examined pressure interval: it is almost doubled at 2 MPa and an increase of about 50% at 5 MPa is noticed. The Vieille's law exponent decreases from 0.56 for the μAl propellant to 0.42 for the nAl propellant. On the contrary, adding μAl to a nonmetallized propellant does not modify significantly the ballistic properties of the propellant. The burning rate increases leads, considering constant thermal diffusivity of the condensed phase, to the decrease of the propellant thermal wave characteristic time (≈75% lower), thus suggesting a more prompt reaction of the propellant to any kind of disturbance. According to this, the significant effect of nAl particles is not limited to steady-state burning rates, but should be reflected also by the resistance to extinction of the propellant. The present experimental tests revealed a great increase in the depressurization rate needed to extinguish the propellant. To obtain extinction, in the experimental apparatus used in these tests, the nAl propellant requires an exhaust orifice area about four times the one required by the μAl propellant.

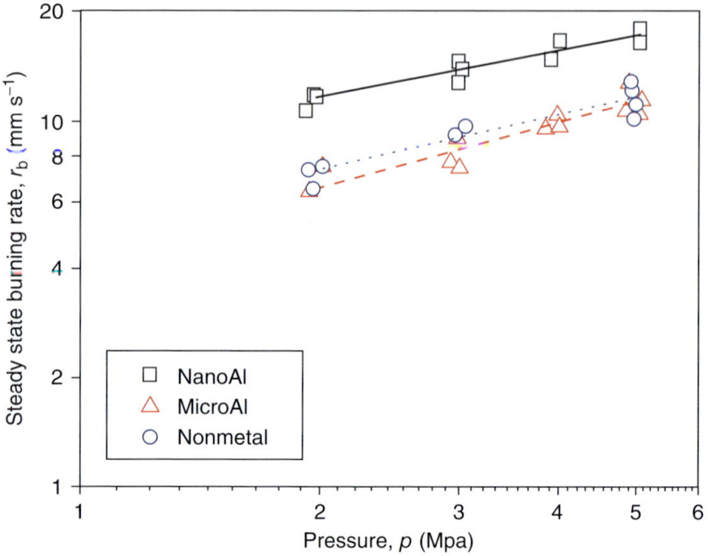

Figure 12.40 Steady-state burning rate comparison for the three tested formulations.

In Figure 12.41, the dynamic extinction boundaries of the different test propellants in terms of initial depressurization rate versus initial pressure are reported. Even in this case, the behavior of the μAl propellant and that of the reference nonmetallized propellant are comparable, with differences on the order at most of 20%. However, these differences are negligible if compared to the depressurization

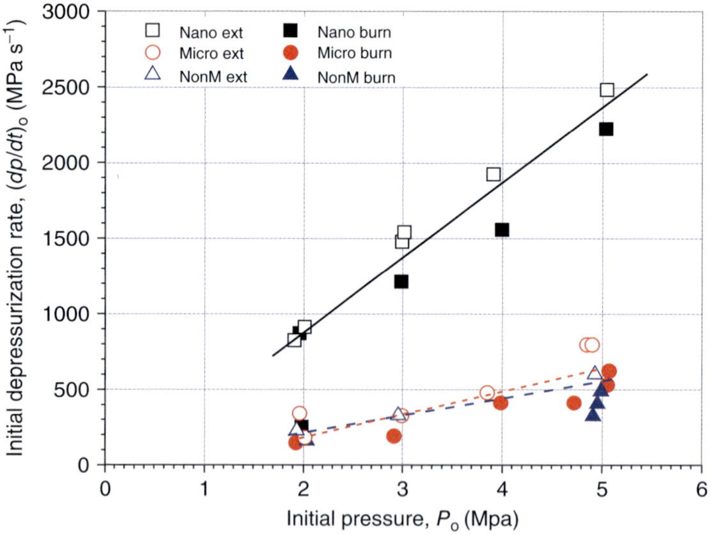

Figure 12.41 Experimental dynamic extinction boundaries by fast depressurization of the three tested formulations.

rate values associated with the nAl propellant. As a matter of fact, a 400% increase in the depressurization rate is needed for the nAl composition to achieve extinction.

Pressure Deflagration Limit (PDL) PDL phenomena are of interest for low-pressure operations in general, ignition, safe storage and transport, controlled extinction phenomena, and so on. For a given finite geometry sample, burning is no longer self-sustained for pressures below PDL. For a given formulation, the intrinsic value of PDL actually corresponds to a sample of infinite geometry, that is, suffering no border effects. Of course, this is only an ideal configuration. In practical terms, the importance of the border effects can be assessed by the shape factor, defined as the characteristic sample cross section ratio (area/perimeter). The experimental results obtained at SPLab point out a clearly lower PDL, and thus stronger resistance to static extinction, for propellants using nAl instead of μAl.

Tests on the PDL were performed in an experimental rig specifically realized for this task. The combustion bomb had a large internal volume of 40 l, necessary to damp out pressure oscillations typical of near-PDL burning behavior. Two photodiodes were faced on the optical access of the bomb, while the pressure signal was supplied by a transducer with a digital readout of 1 mbar. The chamber pressure was controlled by directly regulating a vacuum pump. Strand ignition was accomplished by a standard hot wire technique. Propellant samples were of square cross section and long enough (65 mm length strands) to allow applying a smooth pressure gradient (typically on the order of 1–0.5 mbar s^{-1}), thus avoiding fast depressurization effects. Strands were ignited at a pressure slightly higher than the expected PDL; the pressure was then smoothly reduced till complete extinction was reached. By direct observation of the signals of the photodiode and pressure transducer (Figure 12.42), the following trend can be recognized: the central part of the combustion is characterized by decreasing pressure, while the photodiodes

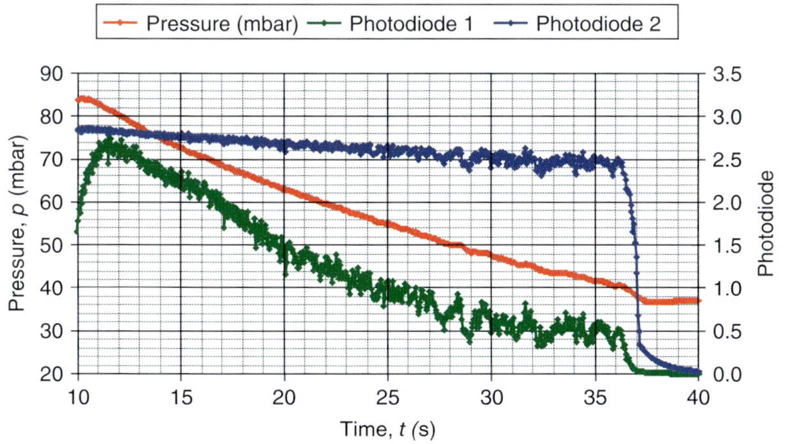

Figure 12.42 Photodiodes and pressure transducer signals during PDL testing.

show a reduction in frequency of the oscillating combustion and an increase in amplitude; when the flame extinction is reached, the signal from the photodiode falls to zero, while, at the same time, the pressure curve changes its inclination because there is no gas production. The pressure of the slope modification is taken as the wanted PDL value.

Experimental results indicate that the shape factor, defined as the area/perimeter ratio of the strand cross section, has a huge influence on the propellant PDL. The lowest PDL is reached when strands of large cross section area (that means large shape factor with samples of the described geometry) are used. The target of a test campaign is to produce a PDL versus shape factor plot for a given propellant composition. In this work, two different aluminum size propellants (both 68/18/14), with the same mass percentage and distribution of AP (0–71 μm), were tested. The results are reported in Figure 12.43: the propellant based on nAl has a lower PDL value compared to μAl at all shape factors.

Subatmospheric Burning A peculiar result indicating an increase in the steady burning rate but decrease in pressure sensitivity for propellants using nAl instead of μAl is shown in Figure 12.44. Two propellants having identical nominal composition, AP/Al/HTPB = 68/18/14 in mass%, respectively, with the same AP size (at most 71 μm) but different Al size (micrometric or nanometric), were tested. The nAl formulation is not only clearly faster than the μAl formulation but its pressure sensitivity is also different ($n = 0.52$ for nAl instead of 0.78 for μAl).

While the μAl formulation near PDL shows the typical self-sustained oscillatory burning [199], this effect is almost unperceivable in the nAl formulation. Moreover, the burning rate curves of the industrial AP5 propellant and the SPLab-made Al˙06 propellant, both using μAl, fully overlap over the explored pressure range.

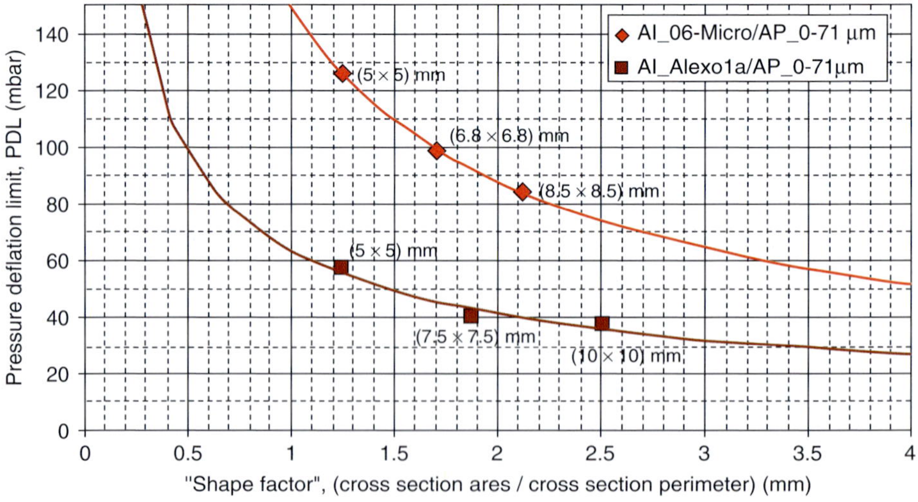

Figure 12.43 PDL versus shape factor for propellants loaded with μAl versus nAl.

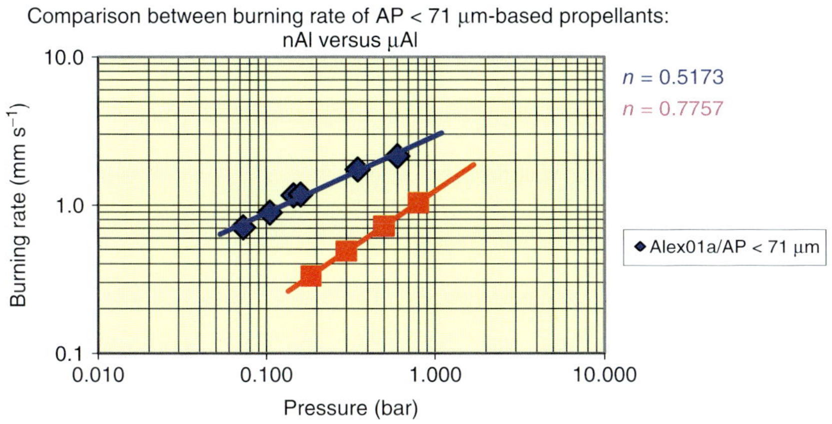

Figure 12.44 Subatmospheric burning rate versus pressure ($D_{AP} < 71\,\mu m$) showing faster combustion for nAl versus μAl propellants. (SPLab.)

12.5.1.5 **Concluding Remarks**

AP-/HTPB-based aluminized composite solid rocket propellants, containing nAl particles, show faster steady burning rates compared with the corresponding propellants containing Al particles. This increase is mainly determined by the intense energy released by nAl particle oxidation completion very close to/above the burning surface following a kind of prompt partial oxidation at/below the burning surface. The minute nAl particles particles manifest a strong reactivity mainly due to their increased specific surface, notwithstanding the simultaneous decrease of active Al content (typically, from 99.6% of 17 μm Al to 88.2% for Alex^{TM}). The number and speed of particles/agglomerates gradually increase with decreasing μAl size in the range 90–15 μm, while a remarkable decrease of flame heterogeneity is observed for nAl in the range 0.1–0.2 μm. Steady burning rates are not affected by particles in the micrometric range (say, $BET \leq 2\,m^2\,g^{-1}$) but strongly increased by decreasing size particles over the explored nanometric range; for very small particles, the appreciable decrease of active metal hinders a further increase of steady burning rate. Pressure exponents showed only minor changes within the limits of strand burner experiments conducted under standard operating conditions. Also, for ignition, an appreciable decrease is found for nAl particles in the range 0.1–0.2 μm, while little effect was observed for μAl.

The importance of the two different effects of metal agglomeration for μAl and metal aggregation (prompt metal oxidation) for nAl was shown. Smaller CCP sizes imply reduced specific impulse losses, and also reduced particulate damping especially at low pressures. However, the exact results depend on the many details of the propellant formulation: in particular, the pressure exponent of steady burning rates has been observed to follow different trends according to the specifics of the composition and operating conditions. This fact suggests that, in addition to the sheer increase of the specific surface, other subtle chemical and physical factors (intermetallic compounds, crystalline structure, etc.) may be important in nAl-enhanced reactivity. Furthermore, it is necessary to take into account that the way of inserting nAl in the propellant influences cold cohesion of particles.

12.5.2
Solid Rocket Fuels for Hybrid Propulsion

In this section, the ballistics of solid fuels tested by a laboratory-scale microburner is discussed. The investigated fuel formulations are introduced first, and then an overview of the test rig and of the data reduction technique is given. Finally, representative ballistic characterization results are presented.

12.5.2.1 Tested Ingredients and Solid Fuel Formulations

The baseline formulation for the relative ballistic grading of the metallized fuels is cured HTPB (see Table 12.10 for details). The nAl-loaded fuel formulations contain 10% of a metallic additive and 2% of CB (mass fraction). Details of the tested formulations are reported in Table 12.20. In the same table, the percentage gap between actual and maximum theoretical fuel density (due to porosity and other reasons) is defined as

$$\Delta\rho_{\%} = 100 \cdot \frac{\rho_{f,\text{Actual}} - \rho_{f,\text{TMD}}}{\rho_{f,\text{TMD}}} \tag{12.15}$$

Solid fuel manufacturing is conducted under controlled operating conditions. The compound is mixed under vacuum at 333 K (60 °C). In metallized fuels, after nanopowder addition, the compound is ultrasonicated all along the mixing phase [68, 114, 134–136].

Table 12.20 Density of investigated solid fuel.

Formulation	$\rho_{f,TMD}$ (kg m^{-3})	$\rho_{f,Actual}$ (kg m^{-3})	$\Delta\rho_{\%}$ (% $\rho_{f,TMD}$)
Baseline (HTPB)	920	919 ± 1	−0.1
HTPB + 100 nm ALEX^{TM}	992	991 ± 2	−0.1
HTPB + 50 nm L-ALEX$_{NPS}$	992	991 ± 2	−0.1
HTPB + 100 nm L-ALEX$_{NPS}$	992	990 ± 2	−0.2
HTPB + 100 nm L-ALEX$_{APS}$	992	1006 ± 7	1.4
HTPB + 50 nm P-ALEX$_{APS}$	992	1019 ± 6	3.0
HTPB + 100 nm P-ALEX$_{APS}$	992	995 ± 3	0.3
HTPB + 50 nm F-ALEX$_{APS,E}$	992	1014 ± 4	2.2
HTPB + 100 nm F-ALEX$_{APS,A}$	992	994 ± 1	0.2
HTPB + 50 nm VF-ALEX$_{APS,F}$	992	997 ± 1	0.5
HTPB + 50 nm VF-ALEX$_{APS,E}$	992	971 ± 3	−2.1
HTPB + 100 nm VF-ALEX$_{APS,F}$	992	934 ± 6	−5.8

Details about additive properties are reported in Section 12.3.

12.5.2.2 **Experimental Setup**

Experimental testing was conducted in a 2D radial microburner playing the role of a solid fuel strand burner (corresponding to the solid propellant strand burner widely used in solid rocket propulsion) [67, 68, 136, 200]. A schematic overview of the facility is reported in Figure 12.45. The main aim of the 2D radial microburner is a quick but relative ballistic grading of different fuel formulations under a variety of operating conditions. The test rig enables the combustion of cylindrical strands with a single central port perforation ($D_0 = 4$ mm, $L_p = 30$ mm, solid fuel web thickness of 7 mm). Thanks to its original design, the experimental rig enables a continuous visualization of the head-end section of the tested strand during the combustion, as shown in Figure 12.46. The r_f data are reduced by a continuous tracking of the central port diameter during combustion.

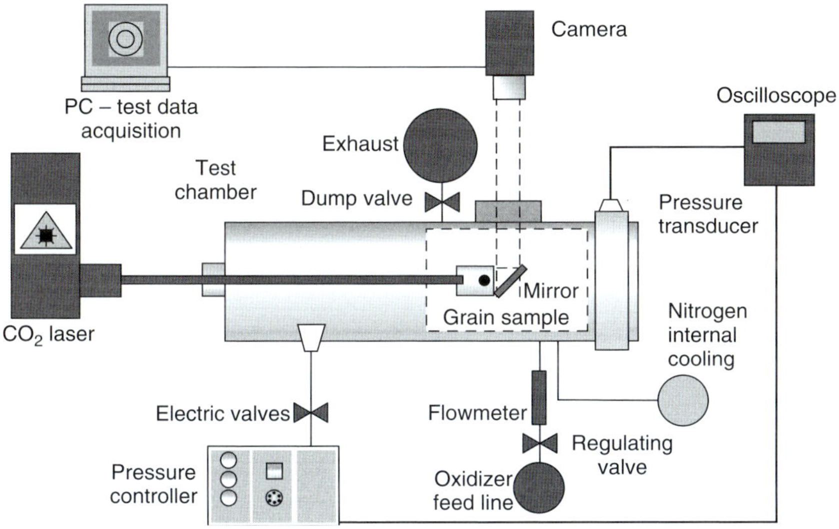

Figure 12.45 Overview of the 2D radial microburner; note the 45° mirror enabling strand head-end visualization [192]. (Reproduced with permission from [193] copyright © 1996 The American Institute of Aeronautics and Astronautics, Inc.)

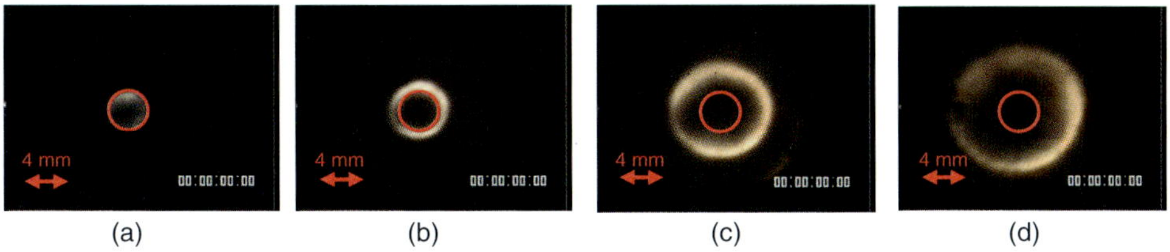

Figure 12.46 Strand head-end visualization for a typical 2D radial microburner test (HTPB burning in GOX, $p_c = 10$ bar, initial Gox = 140 kg m^{-2} s^{-1}). Red circle marks initial central port diameter. (a) $t = t_{ign}$, (b) $t = t_{ign} + 0.8$ s, (c) $t = t_{ign} + 4$ s, and (d) $t = t_{ign} + 7.8$ s.

Figure 12.47 The 2D radial microburner injector: (a) overall view and (b) close view of aft part; (1) oxygen inlet, (2) pre-injection chamber, (3) flanges for optical quartz windows accommodation, (4) injector head with screws for oxidizer injection control, (5) injector terminal for sample lodging, and (6) 45° mirror for head-end image reflection.

An image of the 2D radial microburner injector head is shown in Figure 12.47. The injector head consists of eight radial channels, enabling the oxidizer to flow from the pre-injection chamber to the injector terminal without visual obstacles. The cross sections of the eight injection channels are controlled by regulating stems. Thus, the characteristics of the flow reaching the sample can be tailored (stems completely/partially opened or closed). Thanks to this, standard/swirl flows can be achieved. The flanges limiting the pre-injection chamber volume lodge two quartz windows. The strand head-end view is thus reflected by the 45° mirror, as shown in Figure 12.45. This enables visualization of the strand head-end section from the lateral windows of the combustion chamber and the video recording by a high-speed digital camera operating with a frame rate ranging from 25 to 500 fps (depending on test operating conditions and combustion duration). A secondary, separated inlet delivers nitrogen to the combustion chamber in order to cool its internal volume and to prevent soot/combustion products to hinder the visualization.

The combustion chamber pressure, p_c, and the oxidizer mass flow rate, $\dot{m}_{ox}$, are measured and controlled independently. A digital flowmeter enables real-time measurement and control of $\dot{m}_{ox}$. Oxidizer mass flow rates up to $6 \cdot 10^{-3}\,kg\,s^{-1}$ can be achieved. The value of p_c is measured by a piezo-resistive transducer and controlled by a pressure regulator. The latter drives servo actuators acting on the exhaust gases pipeline (Figure 12.45). As a result, a quasi-steady p_c characterizes the performed tests, as shown by the $\dot{m}_{ox}$ and p_c histories in time shown in Figure 12.48. Strand ignition is achieved by a primer charge inserted in the central port at the head-end side of the strand. A laser beam impinging on the primer charge ignites it. The hot combustion products of the latter, in turn, ignite the strand. Primer charge size and composition are the same for the different fuel formulations in order to provide an identical initial energy input at ignition under given operating conditions.

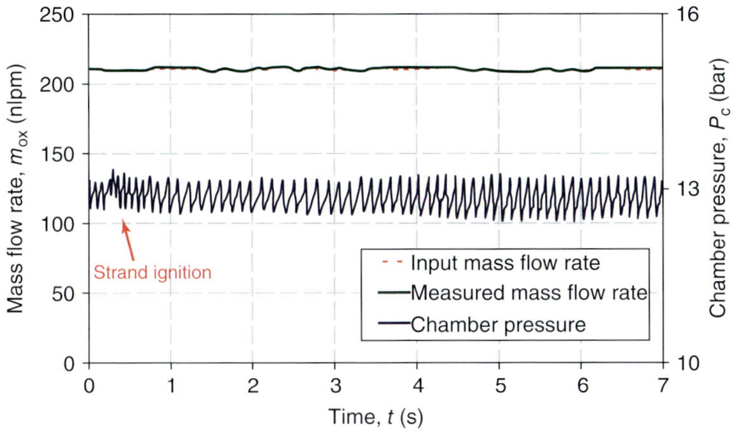

Figure 12.48 Typical histories of $p_c(t)$ and $\dot{m}_{ox}(t)$ of a ballistic test. Primer charge and strand ignitions are identified by a marked peak in $p_c(t)$; $\dot{m}_{ox}$ expressed in convenient units to improve readability.

12.5.2.3 Time-Resolved Regression Rate

An optical, nonintrusive time-resolved technique for r_f measurement was developed and validated at SPLab [67, 136, 201]. This data reduction technique is based on sampling the strand central port diameter in time from the recorded combustion video. Experimental investigations on hybrid fuel ballistics are usually conducted by thickness over time (TOT) techniques [202–204]. The TOT data reduction based on solid fuel grain size/weight measurement before and after the combustion run yields time-averaged information. As originally shown by Evans *et al.* [205], the time-resolved approach highlights ballistic responses not captured by the TOT data reduction and confirmed by numerical simulations [206].

For the SPLab time-resolved technique, data sampling starts from the instant when the central port diameter becomes fully visible (after primer charge ignition). Data sampling extends over a time interval depending on visualization quality. Sampling frequency is limited by the need for a clear visualization of the regressing surface. During combustion, flame flickering and head-end burning hinder high sampling frequencies of the port diameter. Under typical operating conditions, sampling is performed in the 5–10 Hz range for HTPB-based fuels and in the 10–100 Hz range for paraffin-based formulations. As extensively discussed in [69, 206], under the investigated conditions, the total average error for the regression rate determination is ~8%. During the combustion, for each given sampling time, t_i, central port diameter is sampled along two different radial directions (Figure 12.49). Usually, vertical and horizontal diameters are considered for the data reduction, although the final choice depends on combustion overall quality. If the measurements exhibit combustion anisotropies [136, 206], more than two radial directions are considered. The cross-sectional average diameter at time t_i can

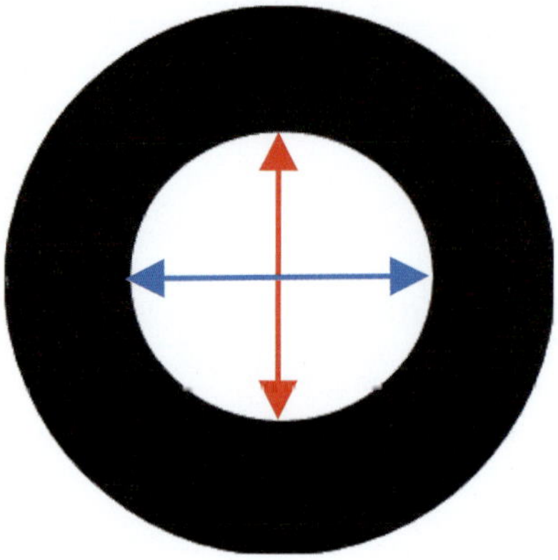

Figure 12.49 Definition of the sampled cross-section $\overline{D}_i$ at time t_i (see Equation 12.2): for each of the two radial directions, three local measurements are taken.

therefore be defined as

$$\overline{D}_i = \frac{\overline{D}_{i,1} + \overline{D}_{i,2}}{2} \tag{12.16}$$

The $\overline{D}_i$ sequence so collected represents a discrete information in time. Interpolating the $\overline{D}_i$ versus t_i series, a continuous history $\overline{D}(t)$ can be achieved as

$$\overline{D}(t) - D_0 = a_{\mathrm{D}} \cdot (t - t_0)^{n_{\mathrm{D}}}, \quad t \geq t_{\mathrm{ign}} > t_0 \tag{12.17}$$

The $\overline{D}(t) - D_0$ is considered valid starting from the ignition time t_{ign}. The latter parameter is *ad hoc* defined as the time delay maximizing the data fitting of Equation 12.17. For validation, the *ad hoc* defined $(t_{\mathrm{ign}} - t_0)$ can be compared to the ignition delay estimated under the hypothesis of exclusive convective heat transfer, $\Delta t_{\mathrm{ign,conv}}$. This specific ignition delay is evaluated by an equation originally proposed by Ohlemiller and Summerfield [207] as

$$\Delta t_{\mathrm{ign,conv}} = \frac{\pi}{4} \cdot \kappa_{\mathrm{f}} \cdot \left[\frac{\rho_{\mathrm{f}} \cdot C_{\mathrm{f}} \cdot \left(T_{\mathrm{s,fin}} - T_{\mathrm{s,in}}\right)}{h^*(T_{\mathrm{fl}} - T_{\mathrm{s,film}})} \right]^2 \tag{12.18}$$

In Equation 12.18, the film temperature of the solid fuel surface, $T_{\mathrm{s,film}} = (T_{\mathrm{s,fin}} + T_{\mathrm{s,in}})/2$ is introduced to account for the surface temperature changes during ignition. A detailed discussion of Equation 12.18 and a comparison between the *ad hoc* defined t_{ign} and the corresponding $t_{\mathrm{ign,conv}}$ is given in [68, 69, 136, 201, 202].

On the basis of Equation 12.17, the time-resolved ballistics of the burning strand can fully be assessed. In particular, $r_{\mathrm{f}}(t)$ is defined as

$$r_{\mathrm{f}}(t) = \frac{1}{2}\frac{d\overline{D}(t)}{dt} = \frac{1}{2} a_{\mathrm{D}} n_{\mathrm{D}} (t - t_0)^{n_{\mathrm{D}}-1}, \quad t \geq t_{\mathrm{ign}} > t_0 \tag{12.19}$$

The $G_{ox}(t)$, and other relevant parameters of interest such as $\dot{m}_f(t)$ and $O/F(t)$, can be evaluated for $t \geq t_{ign} > t_0$ as

$$G_{ox}(t) = \frac{\dot{m}_{ox}(t)}{\pi(\overline{D}(t)^2/4)} = \frac{\dot{m}_{ox}(t)}{\pi([D_0 + a_D(t - t_0)^{n_D}]^2/4)} \quad (12.20)$$

$$\dot{m}_f(t) = \rho_f A_f(t) r_f(t) = \rho_f \pi \overline{D}(t) L_p r_f(t) = \rho_f \pi [D_0 + a_D(t - t_0)^{n_D}] L_p r_f(t) \quad (12.21)$$

$$O/F(t) = \frac{\dot{m}_{ox(t)}}{\dot{m}_f(t)} = \frac{\dot{m}_{ox}(t)}{\rho_f \pi \overline{D}(t) L_p r_f(t)} = \frac{\dot{m}_{ox}(t)}{\rho_f \pi [D_0 + a_D(t - t_0)^{n_D}] L_p r_f(t)} \quad (12.22)$$

The instantaneous time-resolved data can be integrated in time to obtain the time-averaged ballistics. Thanks to this, time-resolved and TOT data can be compared. Thus, the consistency between different data reduction techniques can be evaluated according to the following conditions

$$r_f(t_{ign}) = \frac{1}{2} a_D n_D (t_{ign} - t_0)^{n_D - 1} \stackrel{?}{=} \frac{n_D}{2} \cdot \frac{\overline{D}(t_{ign}) - D_0}{t_{ign} - t_0} \quad (12.23)$$

$$< r_f(t_{fin}) >= \frac{1}{t_{fin} - t_{ign}} \int_{t_{ign}}^{t_{fin}} r_f(t) \cdot dt \stackrel{?}{=} \frac{1}{2} \cdot \frac{\overline{D}(t_{fin}) - \overline{D}(t_{ign})}{t_{fin} - t_{ign}} \quad (12.24)$$

$$< G_{ox}(t_{fin}) >= \frac{1}{t_{fin} - t_{ign}} \int_{t_{ign}}^{t_{fin}} G_{ox}(t) \cdot dt \stackrel{?}{=} \frac{\dot{m}_{ox}}{(\pi/4) \cdot \{[\overline{D}(t_{ign}) + \overline{D}(t_{fin})]/2\}^2} \quad (12.25)$$

With the time-resolved technique, a given fuel formulation can be characterized over a range of G_{ox} by a single test. Nevertheless, data scattering due to the diffusive nature of the hybrid flame and turbulence effects is possible. Therefore, the different $\overline{D}(t)$ collected for a certain fuel formulation, under the same operating conditions, are collapsed to an ensemble average curve of the kind of Equation 12.17, including all of the performed tests. For the latter, a power-law approximation of r_f versus G_{ox} is proposed as

$$r_f(G_{ox}) = a_r \cdot G_{ox}(t)^{n_r}, \quad t \geq t_{ign} > t_0 \quad (12.26)$$

The ensemble curve enables definition of proper error bars, defined in t and G_{ox} domains, respectively. In order to evaluate the error bars for a given ensemble, the evaluation of a proper t (or G_{ox}) range is required. The latter is defined considering the limits where the $\overline{D}(t)$ (or $r_f(G_{ox})$) curves of all the performed single tests are defined. Over this interval, the error bars are evaluated by confidence intervals centered on the mean value resulting from single test diameters at a given time (for Equation 12.17) or regression rates at a given G_{ox} (for Equation 12.19). Confidence intervals are evaluated with 95% accuracy.

The relative ballistic grading of the different tested fuel formulations is performed comparing $r_f(G_{ox})$ trends of loaded fuels with respect to the baseline data. In particular, for given fuel formulation and oxidizer mass flux, the following percent increase is defined

$$\Delta = 100 \cdot \frac{r_f - r_{f,\text{Baseline}}}{r_{f,\text{Baseline}}} \tag{12.27}$$

Defining Δ over different G_{ox}, an average percent increase can be defined (Δ_{av}). The standard deviation of this latter parameter, σ_Δ, is related to the oxidizer mass flux sensitivity of the considered fuel formulation. In particular, the higher the σ_Δ the higher the performance dependence on G_{ox}.

12.5.2.4 Ballistic Characterization: Analyses of the Results

The ballistic characterization of the tested fuel formulations is hereinafter discussed. The baseline data are discussed first in order to provide a detailed overview of the operating steps of the time-resolved technique. Fuel formulations loaded with nanopowders are then characterized on a relative grade with respect to the baseline.

At least three valid runs were performed for every fuel formulation. Ballistic tests were performed in GOX, with $p_c = 10\,\text{bar}$ and initial $G_{ox} \sim 390\,\text{kg}\,\text{m}^{-2}\,\text{s}^{-1}$.

Time-Resolved Ballistics of Baseline (HTPB) The $\overline{D}(t) - D_0$ for the baseline formulation is shown in Figure 12.50 for single tests and the resulting ensemble average. Under the investigated operating conditions, the $\overline{D}(t)$ exhibits a monotonically increasing behavior. As shown by the data reported in Table 12.21, Equation 12.17 properly fits the sampled diameter of the single tests as well as the ensemble average.

The ensemble average $\overline{D}(t) - D_0$ of the baseline formulation yields the r_f versus G_{ox} behavior reported in Figure 12.51. In the latter figure, time runs from right to

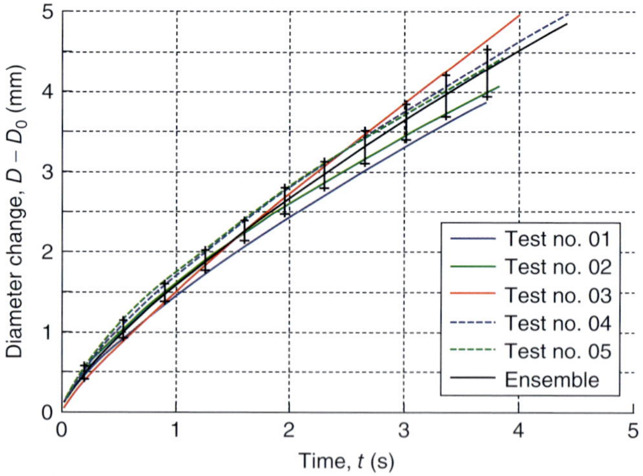

Figure 12.50 $\overline{D}(t) - D_0$ versus t for single tests and ensemble average for the baseline (pure HTPB fuel).

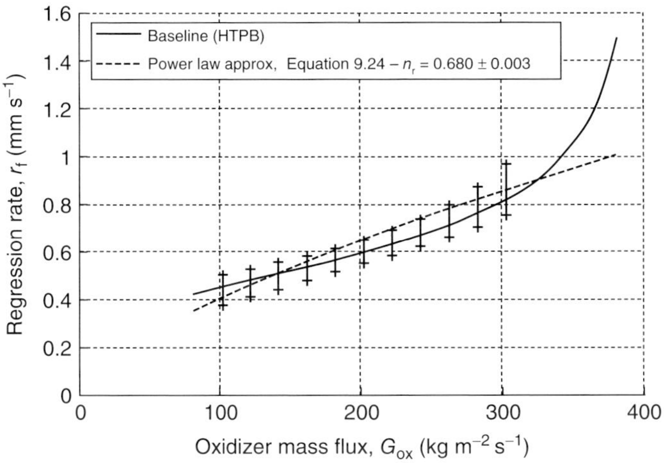

Figure 12.51 Ensemble average and power-law approximation of r_f versus G_{ox} for the baseline (pure HTPB fuel).

left, the earlier phases of the combustion being characterized by G_{ox} and r_f values larger than those of the final part. The marked, initial r_f exhibits a monotonic decrease during the run. This peculiar behavior of hybrid ballistics is captured by the time-resolved technique [67, 68, 205], but not completely by TOT [204]. The $\overline{D}(t)$ and $r_f(t)$ behaviors cause the performance shift characterizing hybrid rocket systems [208, 209].

In Figure 12.51, the power-law approximation of r_f versus G_{ox} is reported. The behavior of the time-resolved data is not captured by Equation 12.26, as testified by its poor data fitting reported in Table 12.22. In spite of this, the power-law approximation exhibits $n_r \sim 0.68$ in reasonable agreement with the value of 0.8 anticipated by the classical hybrid combustion model of Marxman and Gilbert [210, 211]. The difference found between the behavior identified by the commonly used Equation 12.26 and the time-resolved data is probably due to the overall O/F(t) evolution during burning. Thermal and transport properties of the flowing combustion

Table 12.21 Relevant parameters of Equation 12.17 for the baseline (HTPB).

Test no.	a_D (mm s$^{(-nD)}$)	n_D	R^2, Equation 12.21
01	2.104 ± 0.027	0.848 ± 0.010	0.998
02	1.715 ± 0.008	0.757 ± 0.004	0.999
03	1.881 ± 0.022	0.759 ± 0.009	0.998
04	1.448 ± 0.010	0.751 ± 0.007	0.999
05	1.613 ± 0.015	0.691 ± 0.006	0.999
Ensemble	1.600 ± 0.003	0.746 ± 0.002	0.984

Note the high quality data due to *ad hoc* definition of t_{ign}.

Table 12.22 Relevant parameters of Equation 12.26 for the baseline (HTPB).

Test no.	a_r (mm s^{-1}/[kg m^{-2}s^{-1})]^{n_r})	n_r	R^2, Equation 12.26
01	0.070 ± 0.001	0.483 ± 0.002	0.899
02	0.025 ± 0.001	0.633 ± 0.002	0.921
03	0.016 ± 0.001	0.735 ± 0.002	0.909
04	0.022 ± 0.001	0.609 ± 0.001	0.960
05	0.004 ± 0.001	0.971 ± 0.003	0.912
Ensemble	0.018 ± 0.001	0.680 ± 0.003	0.877

Relatively low data fitting due to marked initial r_f.

Table 12.23 Baseline (HTPB) data consistency checks.

Test no.	Equation 12.23 (%)[a]	Equation 12.24 (%)[a]	Equation 12.25 (%)[a]
01	0.1	0.2	−2.2
02	−0.1	1.8	−0.4
03	−0.2	−1.9	−8.1
04	0.1	0.3	−2.4
05	0.1	0.6	−0.4
06	0.5	−1.0	−3.0
07	1.0	−0.7	−1.1

Differences between data reduction techniques are due to solid fuel grain fragmentation and other effects of hybrid burning [68, 136].
[a] Percentage evaluated with respect to TOT values.

products mixture can be calculated by chemical equilibrium codes considering the detailed O/F history. The resulting data can be used in semi-empirical models of the regression rate able to capture the ballistic response even under high G_{ox} values [69, 136, 206]. Data reported in Table 12.23 shows the consistency checks for time-resolved data with respect to the conventional TOT technique. The reported results testify a general agreement between the two data sets.

Nanopowder Loaded Fuels: Ballistic Investigation A summary of the relative ballistic grading of the tested nAl-loaded fuel formulations is reported in terms of Δ in Table 12.24 and Table 12.25, while Table 12.26 reports data concerning $r_f(G_{ox})$ power-law approximations. In order to compare the ballistic response of all the tested formulations, $< r_f >$ evaluated between the initial oxidizer mass flux and the minimum common $< G_{ox} >$ value of 244 kg m^{-2} s^{-1} are reported in Table 12.26 and Figure 12.52 for the ensemble average.

The ballistic response of HTPB + 100 nm ALEX^{TM} is shown in Figure 12.53. Under the investigated conditions, this fuel formulation exhibits an impressive

Table 12.24 Regression rate percent increase with respect to the baseline for solid fuel formulations loaded with 100 nm powders.

Powder	Δ (%), Equation 12.27									Average parameters	
	G_{ox} (kg m^{-2} s^{-1}) 100	150	200	250	275	300	330	350	370	Δ_{av} (%)	σ_Δ (%)
ALEX^{TM}	−0.2	6	12	19	23	28	35	42	54	24	17.4
L-ALEX$_{APS}$	1	14	26	41	51	62	—	—	—	33	23.1
L-ALEX$_{NPS}$	—	−16	−6	7	14	24	40	—	—	11	20.3
P-ALEX$_{APS}$	—	—	3	10	13	18	25	31	41	20	13.1
F-ALEX$_{APS,A}$	6	17	28	41	49	58	74	89	—	45	28.1
VF-ALEX$_{APS,F}$	44	41	39	37	35	34	32	30	—	36	4.7

Table 12.25 Regression rate percent increase with respect to the baseline for solid fuel formulations loaded with 50 nm powders.

Powder	Δ (%), Equation 12.27									Average parameters	
	G_{ox} (kg m^{-2} s^{-1}) 100	150	200	250	275	300	330	350	370	Δ_{av} (%)	σ_Δ (%)
L-ALEX$_{NPS}$	—	7	16	26	33	40	52	64	83	40	25.3
P-ALEX$_{APS}$	17	19	22	24	25	27	29	31	—	24	4.8
F-ALEX$_{APS,E}$	36	35	34	33	32	32	30	29.5	28	32	2.7
VF-ALEX$_{APS,E}$	53	50	47	44	43	41	38	36	32.5	43	6.6
VF-ALEX$_{APS,F}$	15	19	18	20	21	22	24	26	28	22	4.0

initial r_f enhancement with respect to the baseline that vanishes as G_{ox} decreases. The $\Delta = 54\%$ for $G_{ox} = 370\,kg\,m^{-2}\,s^{-1}$ reduces to 42% at $350\,kg\,m^{-2}\,s^{-1}$ and nearly disappears below $130\,kg\,m^{-2}\,s^{-1}$. This behavior is testified by the $r_f(G_{ox})$ power-law approximation characterized by $n_r = 0.956 \pm 0.004$ (Table 12.26).

This behavior is due to a change in ALEX^{TM} effects on solid fuel ballistics as combustion proceeds. At the beginning of the combustion process, ALEX^{TM} is effective in producing r_f enhancement thanks to its reactivity and the high convective heat transfer.

When G_{ox} decreases, the convective heat transfer from the flame to the solid fuel reduces. Therefore, ALEX^{TM} starts behaving as a thermal sink, because of aggregation and agglomeration phenomena at/near the regressing surface [136, 212]. This behavior is suggested by experimental observations (under slightly different operating conditions) in a 2D slab microburner (Figure 12.54).

Under the investigated conditions, the r_f of all the tested formulations, but HTPB + 100 nm L-ALEX$_{NPS}$, is higher than that of the ALEX^{TM}-loaded fuel. The HTPB + 100 nm L-ALEX$_{NPS}$ provides $\Delta = 40\%$ for $G_{ox} = 330\,kg\,m^{-2}\,s^{-1}$. Nevertheless, owing to a high G_{ox} sensitivity, the r_f enhancement reduces

Table 12.26 Power-law approximations of r_f versus G_{ox} (see Equation 12.26) for the nAl-loaded formulations.

Nominal size	Powder additive	a_r	n_r	R^2	$<r_f>^a$ (mm s^{-1})
100 nm	ALEX™	0.005 ± 0.001	0.959 ± 0.022	0.853	0.855
	L-ALEX$_{APS}$	0.007 ± 0.001	0.905 ± 0.012	0.962	0.951
	L-ALEX$_{NPS}$	–[b, c]	1.299 ± 0.020	0.949	0.740
	P-ALEX$_{APS}$	–[b, c]	1.449 ± 0.033	0.895	0.784
	F-ALEX$_{APS,A}$	0.004 ± 0.001	1.011 ± 0.017	0.926	0.815
	VF-ALEX$_{APS,F}$	0.052 ± 0.003	0.535 ± 0.010	0.910	0.941
50 nm	L-ALEX$_{NPS}$	0.001 ± 0.001	1.181 ± 0.024	0.901	0.901
	P-ALEX$_{APS}$	0.022 ± 0.001	0.668 ± 0.011	0.924	0.853
	F-ALEX$_{APS,E}$	0.035 ± 0.002	0.607 ± 0.012	0.884	0.925
	VF-ALEX$_{APS,E}$	0.041 ± 0.003	0.595 ± 0.014	0.853	1.010
	VF-ALEX$_{APS,F}$	0.011 ± 0.001	0.804 ± 0.018	0.872	0.847

[a] Over the G_{ox} range from 387 to 244 kg m^{-2} s^{-1}, with the exception of 50 nm L-ALEX$_{NPS}$, $<r_{f,Baseline}>$ = 0.665 mm s^{-1}.
[b] Less than 0.001.
[c] Negligible.

progressively and a $\Delta = -16\%$ is achieved for an oxidizer mass flux of 150 kg m^{-2} s^{-1}. This behavior is captured by the high σ_Δ and n_r values achieved for this fuel formulation (see Table 12.25 and Table 12.26). The solid fuel formulation loaded with the 50 nm L-ALEX$_{NPS}$ offers better overall performance than its 100 nm counterpart. Considering G_{ox} of 330 and 150 kg m^{-2} s^{-1}, the achieved Δ are 52% and 7% respectively. The high G_{ox} sensitivity of HTPB + 100 nm L-ALEX$_{NPS}$ characterizes also the fuel loaded with the 50 nm powder for which $n_r = 1.181 \pm 0.024$. Considering the data reported in Figure 12.52, the percent $< r_f >$ increase of this formulation with respect to the baseline is 35%, while HTPB + 100 nm L-ALEX$_{NPS}$ achieves a 11% enhancement. This difference in the ballistic response of the L-ALEX™-loaded fuels can be related to the higher reactivity of the smaller powder particle size (Table 12.11 and Table 12.12). Fuel formulation loaded with 100 nm L-ALEX$_{APS}$ exhibits $\Delta = 62\%$ at 300 kg m^{-2} s^{-1}, and $\Delta = 1\%$ at 100 kg m^{-2} s^{-1}. The percent $< r_f >$ increase of HTPB + 100 nm L-ALEX$_{APS}$ between the initial oxidizer mass flux and G_{ox} = 244 kg m^{-2} s^{-1} is 43%. Performance enhancement of this latter formulation is therefore higher with respect to 100 nm L-ALEX$_{NPS}$ under the investigated conditions (Figure 12.55). This result can be related to the crucial role played by alumina shell surrounding the Al particle core of APS powders. Under rapid heating conditions (as those of solid fuel burning), the alumina shell and the small nAl particle sizes enable melt aluminum dispersion mechanism [50–52]. The latter originates fast reactions of the nanosized particles, thus enabling enhanced r_f.

A typical result for the ballistic characterization of HTPB loaded with fluoroelastomer-coated nAl is shown in Figure 12.56. Under the investigated

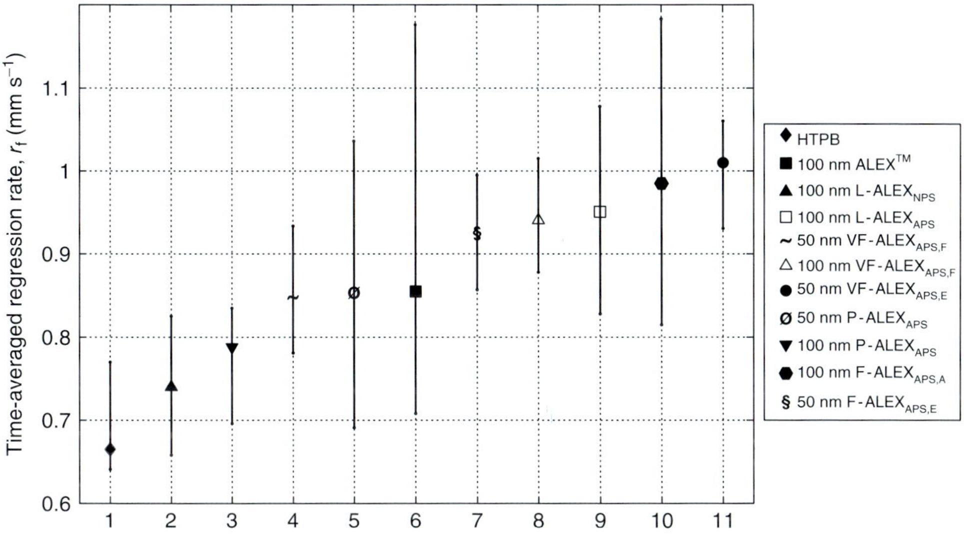

Figure 12.52 Overview of $< r_f >$ for $244\ \text{kg m}^{-2}\ \text{s}^{-1} \leq < G_{ox} > \leq 387\ \text{kg m}^{-2}\ \text{s}^{-1}$. Confidence intervals evaluated with 95% accuracy. Data concerning 50 nm L-ALEX$_{NPS}$ are not reported because of large data scattering.

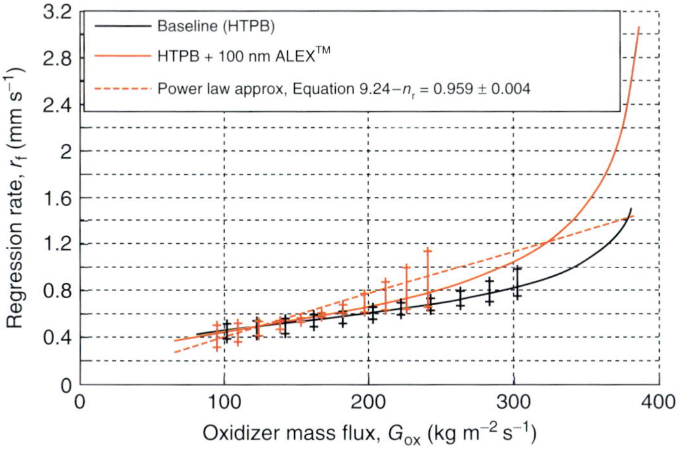

Figure 12.53 Ballistic characterization of HTPB + 100 nm ALEX^{TM}: the marked initial r_f enhancement monotonically decreases during combustion, yielding reduced values with respect to the baseline for $G_{ox} \leq 150\ \text{kg m}^{-2}\ \text{s}^{-1}$.

conditions, the VF-ALEX$_{APS}$ series provides r_f enhancement over the whole G_{ox} range. In spite of the absence of a strong r_f enhancement with respect to baseline in the earlier phases of the combustion, the percent $< r_f >$ increase for these fuel formulations ranges from 52% for 50 nm VF-ALEX$_{APS,E}$ to 27% for 50 nm VF-ALEX$_{APS,F}$ (Table 12.26). The absence of a definite initial r_f in the ballistic

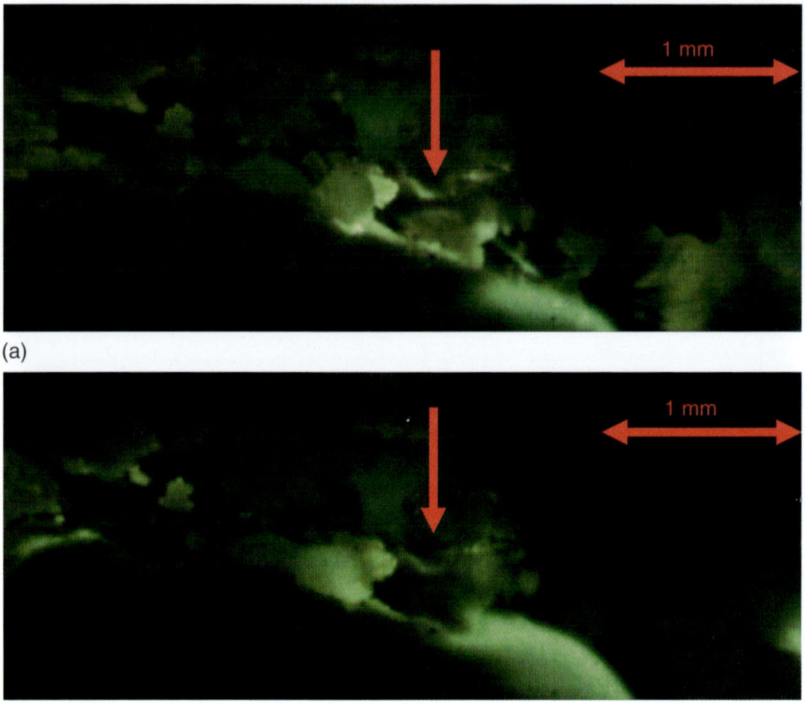

(a)

(b)

Figure 12.54 Merging of incipient CCP formation at the regressing surface of HTPB + 100 nm ALEX^{TM} (5% by mass) burning under 60% O_2 + 40% N_2 at 10 bar [212]. (a) $t = t_0$ and (b) $t = t_0 + 27.0$ ms.

response of VF-ALEX$_{APS}$-loaded fuels is probably related to the lower reactivity of fluoroelastomer-coated powders with respect to ALEX^{TM}, testified also by DSC–TGA tests (Table 12.15).

The main feature of the solid fuels loaded with Fluorel^{TM}-coated powders is the r_f enhancement for $G_{ox} < 250\ \mathrm{kg\,m^{-2}\,s^{-1}}$. At $G_{ox} = 100\ \mathrm{kg\,m^{-2}\,s^{-1}}$, $\Delta = 44\%$ and 53% are achieved by 100 nm VF-ALEX$_{APS,F}$ and 50 nm VF-ALEX$_{APS,E}$. Differences in the coating procedures, as the solvent used to solve the fluoroelastomer and the solvent removal technique (Table 12.10) could justify the observed behaviors between VF-ALEX powders (Table 12.25 and Table 12.26). With respect to the previous considered solid fuel formulations, fluoroelastomer-coated nAl led to a lower sensitivity to G_{ox} changes, and this resulted in an almost constant r_f increase over the whole investigated G_{ox} range. This is captured by σ_Δ and n_r values reported in Table 12.24 and Table 12.26. In particular, under the investigated conditions, power-law approximation of r_f versus G_{ox} for HTPB + 100 nm VF-ALEX$_{APS,F}$ yields $n_r = 0.535 \pm 0.010$, which is the lowest of the achieved values. The reduced G_{ox} effect on the ballistics of VF-ALEX$_{APS}$ series could be justified by different concurring phenomena:

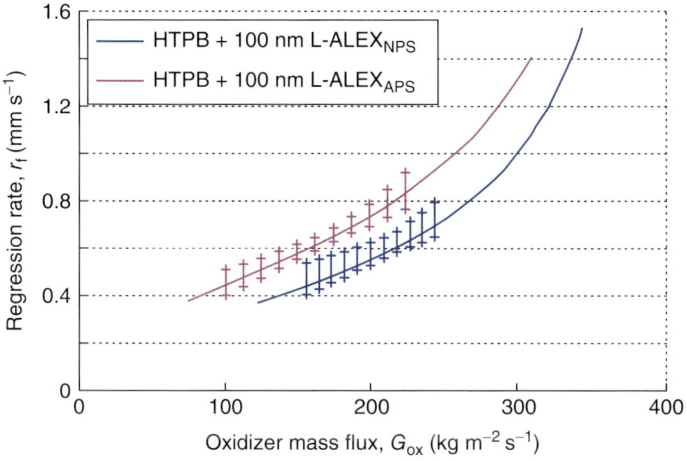

Figure 12.55 Ballistic characterizations of HTPB loaded with L-ALEX powders: the APS variant exhibit higher r_f over the whole G_{ox} range. Baseline not reported to improve readability.

- reduced reactivity with respect to ALEX^{TM} (Table 12.15), not only limiting the high flow r_f enhancement but also hindering possible surface/subsurface aggregation and agglomeration phenomena;
- presence of fluorine-containing species coming from the thermal decomposition of the fluoroelastomer [201] at/under the regressing surface and enhancing fuel ballistic response.

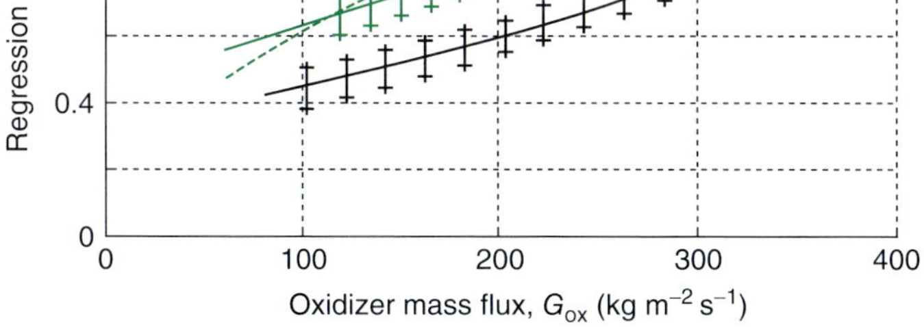

Figure 12.56 Ballistics of HTPB + 100 nm VF-ALEX$_{APS,F}$, r_f versus G_{ox}. Note the low G_{ox} sensitivity of the achieved r_f.

The best performance in terms of instantaneous r_f increase with respect to baseline at high G_{ox} is achieved by HTPB + 100 nm F-ALEX$_{APS,A}$. This fuel formulation achieves a $\Delta = 89\%$ for $G_{ox} = 350\,kg\,m^{-2}\,s^{-1}$. With $G_{ox} = 300\,kg\,m^{-2}\,s^{-1}$, this fuel formulation provides a two-fold r_f enhancement with respect to ALEX^{TM}. In spite of this, $n_r = 1.011 \pm 0.017$ and this strong G_{ox} sensitivity reduces the performance gain as combustion proceeds. The HTPB + 50 nm F-ALEX$_{APS,E}$ fuel exhibits the same behavior shown by VF-ALEX$_{APS}$ series with $\Delta = 30\%$ over all the investigated G_{ox}.

12.5.2.5 Concluding Remarks on Solid Fuel Burning

Experimental results achieved by testing HTPB-based solid fuel formulations loaded with different nAl powders were presented and discussed. Tested additives were produced by the EEW technique and characterized by nonisothermal oxidation before the burning tests (see Section 12.3.3). Powders differ because of the mean nominal particle sizes (100 and 50 nm), passivation methods, and possible presence of coatings. Industrial and laboratory-scale coatings of different natures were tested. Details of powders and coating characteristics are extensively discussed in Section 12.3. An overview of the tested fuel formulations is given in Table 12.20. Solid fuels were tested in a 2D radial microburner in GOX, with $p_c = 10\,bar$ and initial $G_{ox} \sim 390\,kg\,m^{-2}\,s^{-1}$. Data reduction was performed by a time-resolved technique for regression rate [67, 68, 114, 136]. A baseline formulation (cured HTPB) was considered for the relative ballistic grading of the investigated loaded fuels.

Under the investigated conditions, HTPB + 100 nm ALEX^{TM} solid fuel exhibits a significant initial r_f increase with respect to baseline. As combustion proceeds, this performance enhancement disappears. The overall percent $< r_f >$ increase of this fuel with respect to baseline is 28% (Table 12.24), although an instantaneous $\Delta = 54\%$ is achieved for $G_{ox} = 370\,kg\,m^{-2}\,s^{-1}$. In the absence of fluorinated coatings, performance of nAl is highly influenced by oxidizer mass flux changes: as G_{ox} decreases, so does the r_f enhancement (Figure 12.52). Fluorine-containing coating of the VF-ALEX$_{APS}$ series offers reduced sensitivity to oxidizer mass flux changes. Fluoroelastomer- and ester-coated nAl provides significant performance enhancement over the whole investigated G_{ox} range. With an oxidizer mass flux of $150\,kg\,m^{-2}\,s^{-1}$, HTPB + 100 VF-ALEX$_{APS,F}$ exhibits $\Delta = 41\%$. The overall percent $< r_f >$ increase of this fuel with respect to baseline is 42%. The low G_{ox} sensitivity of fluoroelastomer-coated nAl is possibly due to the release of oxidizing species from the thermal decomposition of coating and to the reduction of aggregation/agglomeration phenomena induced by its lower reactivity. Comparative tests conducted on APS, stearic acid-coated nAl and L-ALEX^{TM} revealed the importance of the alumina shell for the increase of the r_f (Figure 12.55). This behavior could be an indirect, experimental proof of the melt aluminum dispersion mechanism described by Levitas *et al.* [50–52]. Significant performance increases, with $\Delta = 89\%$ at $G_{ox} = 350\,kg\,m^{-2}\,s^{-1}$, is achieved by HTPB + 100 nm F-ALEX$_{APS,A}$. Comparing 100 nm powders with their 50 nm counterpart exhibited no definite trend of the ballistic response in terms of overall percent $< r_f >$ increase. No clear relationships can be identified between powder characteristics (S_{sp} and C_{Al}) and reactivity parameters from nonisothermal oxidation and nAl-loaded fuel ballistics.

The relative grading of the powders based on DSC–TGA data is not reflected in burning tests. On the basis of the data reported in Table 12.12, and considering the measured S_{sp}, 100 nm L-ALEX$_{APS}$ should be less effective than 50 nm L-ALEX$_{NPS}$ in enhancing the solid fuel r_f. In spite of this, ballistic characterization shows an opposite trend under the investigated conditions. Differences in operating conditions between nonisothermal oxidation and burning tests explain these different behaviors.

Further experimental work on energetic additives for solid fuel as well as studies of the thermophysical characteristics of nAl is recommended. The reactivity parameters from nonisothermal oxidation cannot be directly related to the solid fuel ballistic response, although they can drive research efforts toward promising new formulations. A better understanding of the complex phenomena involved in nAl combustion is required to properly relate physical analyses and ballistic data, especially for powders with organic coatings. Dedicated studies aimed at defining the possible interaction between coating decomposition products and the oxidation of nAl particles are needed in order to better understand the peculiar and promising behaviors of powders coated with fluorine-containing materials.

12.5.3
Chapter Summary

During the past nearly 15 years, nAl and other nanosized metals were extensively studied to augment performance of ES in terms of burning rate, ignition delay, and combustion efficiency. Experimental and theoretical investigations covered most energetic applications, including a variety of propulsion techniques, pyrotechnics, and explosives. Although attractive for fundamental studies and much used in laboratory experiments, no operational systems are reported in use for nAl powders. At least for rocket propulsion, loss of active metal, cold cohesion, and poor propellant castability globally overcome advantages such as increased burning rate (easily achievable by other ways) and reduced 2P flow I_s losses. The recommended approach to get the best of both worlds is to resort to dual metallic fuels, properly blending μAl and nAl. Another approach to exploit the potential of nAl is particle coating, the complexities of which, however, require careful manufacturing and do not lead to immediate solutions. Several approaches were also discussed so as to improve dispersion, and subsequently, mechanical properties of solid propellants or solid fuels containing nAl. A good control of particle size, metal content, and dispersion is a crucial requirement for applications of nanoingredients in propulsion.

Nomenclature

Acronyms and Abbreviations (Laboratory-Scale Coated nAl are Written in Italics)

2D	two-dimensional
ALEX^{TM}	aluminum exploded (produced by electrical explosion of wires)
ANFO	ammonium nitrate fuel oil
AP	ammonium perchlorate
APS	air-passivated
CB	carbon black
DMA	dynamic mechanical analysis
DOA	dioctyl adipate
DSC	differential scanning calorimetry
DSC–TGA	differential scanning calorimetry–thermogravimetric analysis
EEW	electrical explosion of wires
ES	energetic systems
F-ALEX$_{APS,A}$	FTOH 10 : 1 telomer (fluorotelomer) alcohol-coated ALEX^{TM}
F-ALEX$_{APS,E}$	fluorotelomer alcohol-derived ester-coated ALEX^{TM}
Fluorel^{TM}	fluoroelastomer:vinylidenefluoride hexafluoropropylene (70 : 30)
fps	frames per second
GAP	glycidyl azide polymer
GOX	gaseous oxygen
HTPB	hydroxyl-terminated polybutadiene
IPDI	isophorone diisocyanate
L-ALEX$_{APS}$	stearic acid-coated ALEX^{TM}
L-ALEX$_{NPS}$	stearic acid-passivated nAl (industrial version); more common: L-ALEX^{TM}
L-ALEX^{TM}	stearic acid-passivated nAl (industrial version); also denoted as L-ALEX$_{NPS}$
μAl	micron-sized aluminum powder
nAl	nanosized aluminum powder
Nlpm	normal-liter per minute
NPS	not air-passivated
P-ALEX$_{APS}$	palmitic acid-coated ALEX^{TM}
rpm	rounds per minute
SEM	scanning electron microscopy
SPLab	space propulsion laboratory
TEM	transmission electron microscopy
TED	transmission electron diffraction
TGA	thermogravimetric analysis
TMD	theoretical maximum density, kg m^{-3}
TOT	thickness over time
UDP-5	machine used for metal nanopowders production by EEW method

VF-$ALEX_{APS,F}$	fluoroelastomer and fluorotelomer alcohol-derived ester-coated ALEX™: final filtration and fast heating of suspension
VF-$ALEX_{APS,E}$	fluoroelastomer and fluorotelomer alcohol-derived ester-coated ALEX™: final evaporation under ambient conditions

Latin Symbols

A_0	initial sample cross section (mm^2)
a_L	pre-exponential factor (see Equation 12.6)
$\overline{A}$	space-average value (of parameter A)
$<A>$	time-average value (of parameter A)
A	multiplicative factor
a_s	mean surface particle diameter (nm)
C_{Al}	Al metal (or active Al) content, % mass
C_f	specific heat of the solid phase (kJ $(kg\ K)^{-1}$)
D	diameter, port diameter (mm)
D_i	i-th sampled port diameter (mm)
E'	storage tangent modulus (MPa)
E''	loss tangent modulus (MPa)
E^*	complex tangent modulus (MPa)
F	force (N)
G'	storage shear modulus (MPa)
G''	loss shear modulus (MPa)
G^*	complex elastic modulus (MPa)
G	mass flux ($kg\ m^{-2}\ s^{-1}$)
G_{ox}	oxidizer mass flux ($kg\ m^{-2}\ s^{-1}$)
h	convective heat transfer coefficient ($W\ m^{-2}\ K^{-1}$)
K	parameter (see Table 12.18)
k_E	Einstein coefficient (see Equation 12.11 and Table 12.18)
I_s, I_v	specific impulse (respectively gravimetric and volumetric)
L	length (mm)
L_{clamp}	clamp spacing (mm)
L_{sample}	sample length (mm)
L_p	solid fuel grain length (mm)
	mass flow rate ($kg\ s^{-1}$)
n	power-law exponent
O/F	oxidizer-to-fuel ratio
OH_{fun}	hydroxyl functionality
OH_v	hydroxyl value ($mequiv_{KOH}\ g^{-1}$)
p_c	chamber pressure (bar)
r_f	solid fuel regression rate ($mm\ s^{-1}$)
R^2	correlation index
S_{sp}	specific surface area ($m^2\ g^{-1}$)
t	time (s)

t_{ign}	ignition time (s)
T	temperature (K)
T_{onset}	oxidation onset temperature (K)
$T_{ox\ peak}$	oxidation peak temperature (K)
$T_{s,film}$	surface film temperature (K)
V_S	actual volume of the spheres in a typical aggregate
V_L	volume of suspending fluid entrapped within/on cluster surface
VoD	velocity of detonation ($m\,s^{-1}$)
X	molar fraction

Greek Symbols

α_{1000}	degree of transformation up to 1273 K
γ	shear
$\dot{\gamma}$	shear rate (Hz)
δ	phase shift angle (see Equation 12.8 and Equation 12.9)
Δ	difference between two quantities
$\Delta\%$	percent difference between two quantities (%)
$\Delta_{av.}$	average difference between two quantities
ΔH	variation of enthalpy ($J\,g^{-1}$)
ΔH_{ox}	heat release during oxidation ($J\,g^{-1}$)
Δm_1	initial mass loss in TGA traces (%)
Δm_{tot}	total mass gain in TGA traces (%)
$\Delta\rho\%$	percent difference between measured and theoretical maximum density (%)
Δt_{ign}	ignition delay ($t_{ign} - t_0$) (s)
ε	strain, nozzle area ratio
$\dot{\varepsilon}$	strain rate (Hz)
η	viscosity (mPa s)
$[\eta]$	exponent (see Table 12.18)
κ_S	thermal diffusivity (of solid fuel) ($m^2\,s^{-1}$)
ρ	density ($kg\,m^{-3}$)
ρ_f	solid fuel density ($kg\,m^{-3}$)
σ	normal stress (Pa)
σ_Δ	standard deviation of Δ_{av} (%)
$\sigma\Delta$	standard deviation of Δ with respect to $\Delta_{av.}$ (%)
τ	shear stress (Pa)
τ	mean residence time (ms)
φ	additive volumetric fraction (%)
φ_A	volume fraction of a cluster made up of spheres
ω	angular frequency ($rad\,s^{-1}$)
ξ	local mass fraction of condensed combustion products

Subscripts

APS	air-passivated
av	average
add	additive
cc	combustion chamber
conv	convective
D	diameter (Equation 12.2, Equation 12.3, and Equation 12.4)
eng	engineering (see Equation 12.13)
f	fuel
fin	final
fl	flame
i	*i*-th sampled value
ign	ignition
lin	linear
log	logarithmic
MAX	maximum
NPS	not air-passivated
ox	oxidizer
p	propellant, particle
r	regression rate
rad	radiation
rel	relative
s	surface
S	suspension
SM	suspending medium
t	throat section
true	real (see Equation 12.13)
tot	total

Superscripts

*	without blowing

Math Symbols

$\mathcal{M}$	molar mass ($kg\,mol^{-1}$)
T	thrust (N)
$-NCO/-OH$	curing level
$< \ldots >$	average value in time
$\overline{\overline{(\ldots)}}$	average value over composition

Chemical Nomenclature (IUPAC and Common Names)

Formula	IUPAC Common Name
$C_{16}H_{32}O_2$	hexadecanoic acid palmitic acid
$C_{18}H_{36}O_2$	octadecanoic acid stearic acid
$H(CF_2CF_2)_5CH_2OH$	(1,1,11) trihydroperfluoro-1-undecanol, or (2,2,3,3,4,4,5,5,6,6,7,7,8,8,9,9,10,10,11,11) eicosafluoro-1-undecanol FTOH 10 : 1 telomer alcohol
$C_4H_2O_3$	furan-2,5-dione maleic anhydride

References

1. Teipel, U. (2004) *Energetic Materials*, Wiley-VCH Verlag GmbH, Wineham.
2. Dreizin, E.L. (2009) Metal based reactive nanomaterials. *Prog. Energy Combust. Sci.*, **35** (2), 141–167.
3. Khabas, T.A. (2002) Solid-phase synthesis and sintering in highly disperse oxide-metal powder mixtures. *Glass Ceram.*, **59** (11–12), 404–408.
4. Oh, Y.H., Rhee, C.K., Kim, D.H., Lee, G.H., and Kim, W.W. (2006) Formation characteristics of an aluminum hydroxide fiber by a hydrolysis of aluminum nano powder. *J. Mater. Sci.*, **41** (13), 4191–4195.
5. Alinejad, B. and Mahmoodi, K. (2009) A novel method for generating hydrogen by hydrolysis of highly activated aluminum nanoparticles in pure water. *Int. J. Hydrogen Energy*, **34** (19), 7934–7938.
6. Gromov, A. and Vereshchagin, V. (2004) Study of aluminum nitride formation by superfine aluminum powder combustion in air. *J. Eur. Ceram. Soc.*, **24** (9), 2879–2884.
7. Ivanov, Y.F., Osmonoliev, M.N., Sedoi, V.S., Arkhipov, V.A., Bondarchuk, S.S., Vorozhtsov, A.B., Korotkikh, A.G., and Kuznetsov, V.T. (2003) Productions of ultra-fine powders and their use in high energetic compositions. *Propellants Explos. Pyrotech.*, **28** (6), 319–333.
8. Jayaraman, K., Anand, K.V., Chakravarthy, S.R., and Sarathi, R. (2009) Effect of nano-aluminium in plateau-burning and catalyzed composite solid propellant combustion. *Combust. Flame*, **156** (8), 1662–1673.
9. Risha, G.A., Evans, B.J., Boyer, E., and Kuo, K.K. (2007) in *Fundamentals of Hybrid Rocket Combustion and Propulsion, AIAA Progress in Aeronautics and Astronautics*, vol. 218 (eds M.J. Chiaverini and K.K. Kuo), American Institute of Aeronautics and Astronautics, Inc., pp. 413–456.
10. Lewis, W.K., Rumchik, C.G., Broughton, P.B., and Lindsay, C.M. (2012) Time-resolved spectroscopic studies of aluminized explosives: chemical dynamics and apparent temperatures. *J. Appl. Phys.*, **111** (1), 014903-1–014903-6.
11. Politzer, P., Lane, P., and Grice, M.E. (2001) Energetics of aluminum combustion. *J. Phys. Chem. A*, **105**, 7473–7480.
12. Duterque, J. (1997) Experimental studies of aluminum agglomeration in solid rocket motors. *Int. J. Energetic Mater. Chem. Propul.*, **4**, 693–705.
13. Price, E.W. and Sigman, R.K. (2000) *Progress In Astronautics And Aeronautics*, Solid Propellant Combustion and Motor Ballistic, vol. 185, AIAA Inc., Reston, VA, pp. 663–687.
14. Babuk, V.A., Vasilyev, V.A., and Sviridov, V.V. (2001) Propellant formulation factors and metal agglomeration in combustion of aluminized solid rocket propellant. *Combust. Sci. Technol.*, **163**, 261–289. doi: 10.1080/00102200108952159

15. Glotov, O.G., Zarko, V.E., Karasev, V.V. *et al.* (1998) Condensed Combustion Products of Metalized Propellants of Variable Formulations, AIAA Paper 98-0449.
16. Babuk, V.A., Vasilyev, V.A., and Malakhov, M.S. (1999) Condensed combustion products at the burning surface of aluminized solid propellant. *J. Propul. Power*, **15** (6), 783–793. doi: 10.2514/2.5497
17. Beckstead, M.W. (2005) Correlating aluminum burning times. *Combust. Explos. Shock Waves*, **41** (5), 533–546.
18. Ivanov, G.V. and Tepper, F. (1997) in *Fourth International Symposium on Special Topics in Chemical Propulsion, Challenges in Propellants and Combustion 100 Years after Nobel* (eds K.K. Kuo *et al.*), Begell House, pp. 636–645.
19. Mench, M., Yeh, C., and Kuo, K.K. (1998) *Energetic Materials Production, Processing and Characterization*, Institut für Chemische Technologie (ICT), Karlsruhe, pp. 30.1–30.15.
20. DeSena, J.T. and Kuo, K.K. (1999) Evaluation of stored energy in ultrafine aluminum powder produced by plasma explosion. *J. Propul. Power*, **15** (6), 794–800.
21. Cliff, M., Tepper, F., and Lisetsky, V. (2001) Ageing Characteristic of Alex Nanosize Aluminum, AIAA paper 2001-3287.
22. Dokhan, A., Price, E.W., Seitzman, J.M., and Sigman, R.K. (2002) Combustion Mechanisms of Bimodal and Ultra-fine Aluminum in AP Solid Propellant, AIAA 2002-4173.
23. Jones, D.E.G., Turcotte, R., Fouchard, R.C., Kwok, Q.S.M., Turcotte, A.M., and Abdel-Qader, Z. (2003) Hazard characterization of aluminum nanopowder compositions. *Propel. Explos. Pyrotech.*, **28**, 120–131.
24. Meda, L., Marra, G.L., Braglia, R., Abis, L., Gallo, R., Severini, F., Galfetti, L., and DeLuca, L.T. (2004) A wide characterization of aluminum powders for propellants, in *Ninth International Workshop on Combustion and Propulsion* (eds L.T. DeLuca, L. Galfetti, and R.A. Pesce-Rodriguez), Grafiche GSS, Bergamo.
25. Rietvield, H.M., Larson, A.C. and Von Dreele, R.B (1987) General Structure Analysis System (GSAS), Los Alamos National Laboratory, LA-UR-68-748.
26. Kwok, Q.S.M., Fouchard, R.C., Turcotte, A.M., Lighfoot, P.D., Bowes, R., and Jones, E.G. (2002) Characterization of aluminum nanopowder compositions. *Propell. Explos. Pyrotech.*, **27**, 229–240.
27. Brooks, K.P. and Beckstead, M.W. (1995) Dynamics of aluminum combustion. *J. Propul. Power*, **11**, 769–778.
28. Beckstead, M.W., Liang, Y., and Pudduppakkam, K.V. (2005) Numerical simulation of single aluminum Particle combustion. *Combust. Explos. Shock Waves*, **41**, 622–638.
29. Bucher, P., Yetter, R.A., Dryer, F.L., Vicenzi, E.P., Parr, T.P., and Hanson-Parr, D.M. (1999) Condensed-phase Species distributions about al particles reacting in various oxidizers. *Combust. Flame*, **117**, 351–361.
30. Sarou-Kanian, V., Rifflet, J.C., Millot, F., and Gokalp, I. (2006) Aluminum combustion in wet and dry CO_2: consequences for surface reactions. *Combust. Flame*, **145**, 220–230.
31. Bazyn, T., Krier, H., and Glumac, N. (2005) Oxidizer and pressure effects on the combustion of 10-μm aluminum particles. *J. Propul. Power*, **21**, 577–582.
32. Houim, R. and Kuo, K.K. (2008) in *Advancements in Energetic Materials and Chemical Propulsion* (eds K.K. Kuo and K. Hori), Begell House Inc., New York, pp. 454–475.
33. Goldschmidt, H. (1908) New thermite reactions. *Iron Age*, **82**, 232.
34. Wang, L.L., Munir, Z.A., and Maximov, Y.M. (1993) Review: thermite reactions: their utilization in the synthesis and processing of materials. *J. Mater. Sci.*, **28**, 3693.
35. Fischer, S.H. and Grubelich, M.C. (1998) Theoretical energy release of thermites, intermetallics, and combustible metals. Proceeding of the 24th International Pyrotechnics Seminar, Monterey, CA.
36. Granier, J.J. (2005) Combustion characteristics of al nanoparticles and

nanocomposite Al + MoO_3 thermites. PhD Dissertation in Mechanical Engineering, Texas Tech University.

37. Moin, E. (1988) The Current Status of Field Welding of Rail, Railway Track Struct.

38. Dixon, G.P., Martin, A.J., and Thompson, D. (1998) Lead-free Metastable Interstitial Composite (MIC) technology. US Patent 5,717,159.

39. Giles, J. (2004) Collateral damage. *Nature*, **427**, 580–581.

40. Pantoya, M.L. and Granier, J.J. (2005) Combustion behavior of highly energetic thermites: nano versus micron composite. *Propellants Explos. Pyrotech.*, **30** (1), 53–62.

41. Eckert, J., Holzer, J.C., Ahn, C.C., Fu, Z., and Johnson, W.L. (1993) Melting behavior of nanocrystalline aluminum powders. *Nanostruct. Mater.*, **2**, 407–413.

42. Qi, W.H. and Wang, M.P. (2004) Size and shape dependent melting temperature of metallic nanoparticles. *Mater. Chem. Phys.*, **88**, 280–284.

43. Trunov, M., Umbrajkar, S.M., Schoenitz, M., Mang, J.T., and Dreizin, E.L. (2006) Oxidation and melting of aluminum nanopowders. *J. Phys. Chem. B*, **110**, 13094–13099.

44. Moore, K., Pantoya, M.L., and Son, S.F. (2007) Combustion behaviors resulting from bimodal aluminum size distributions in thermites. *J. Propul. Power*, **23** (1), 181–185.

45. Popenko, E.M., Il'in, A.P., Gromov, A.M., Kondratyuk, S.K., Surgin, V.A., and Gromov, A.A. (2002) Combustion of mixtures of industrial aluminum powders with superfine powders of aluminum and aluminum oxide in air. *Combust. Explos. Shock Waves*, **38** (2), 157–162.

46. Granier, J.J. and Pantoya, M.L. (2004) Laser ignition of nanocomposite thermites. *Combust. Flame*, **138**, 373–383.

47. Sanders, V.E., Asay, B.W., Foley, T.J., Tappan, B.C., Pacheco, A.N., and Son, S.F. (2007) Reaction propagation of four nanoscale energetic composites (Al/MoO_3, Al/WO_3, Al/CuO, and Bi_2O_3). *J. Propul. Power*, **23** (4), 707–714.

48. Asay, B.W., Son, S.F., Busse, J.R., Jorgensen, B.S., Bockman, B., and Pantoya, M.L. (2004) Ignition characteristics of metastable intermolecular composites. *Propellants Explos. Pyrotech.*, **29** (4), 216–219.

49. Pantoya, M.L., Levitas, V.I., Granier, J.J., and Henderson, J.B. (2009) Effect of bulk density on reaction propagation in nanothermites and micron thermites. *J. Propul. Power*, **25** (2), 465–470.

50. Levitas, V.I., Asay, B.W., Son, S.F., and Pantoya, M.L. (2006) Melt dispersion mechanism for fast reaction of nanothermites. *Appl. Phys. Lett.*, **89** (7), 071–909.

51. Levitas, V.I., Asay, B.W., Son, S.F., and Pantoya, M.L. (2007) Mechanochemical mechanism for fast reaction of metastable intermolecular composites based on dispersion of liquid metal. *J. Appl. Phys.*, **101**, 8Paper 083524.

52. Levitas, V.I., Pantoya, M.L., and Dikici, B. (2008) Melt-dispersion versus diffusive oxidation mechanism for aluminum nanoparticles: critical experiments and controlling parameters. *Appl. Phys. Lett.*, **92**, 011921, Paper 011921.

53. Reshetov, A.A., Shneider, V.B., and Yavorovsky, N.A. (1984) Ultra dispersed aluminum's influence on the speed of detonation of hexogen. First All-Union Symposium on Macroscopic Kinetic and Chemical Gas-Dynamics, Chernogolovka, Russia, Vol. 1.

54. Miller, P.J., Bedford, C.D., and Davis, J.J. (1998) Effect of metal particle size on the detonation properties of various metallized explosives. Eleventh International Symposium on Detonation, Snowmass, CO, 30 August – 4 September, pp. 214–220.

55. Tulis, A.J., Sumida, W.K., Dillon, J., Comeyne, W., and Heberlein, D.C. (1998) Submicron aluminum particle size influence on detonation of dispersed fuel-oxidizer powders. *Arch. Comb.*, **18** (1-4), 157–164.

56. Beaudin, G., Lefrancois, A., Bergues, D., Bigot, J., and Champion, Y. (1998) Combustion of nanophase aluminum in the detonation products

of nitromethane. Eleventh International Symposium on Detonation, Snowmass, CO, 30 August-4 September, pp. 989-997.
57. Lefrancois, A. and LeGallic, C. (2001) Expertise of nanometric aluminum powder on the detonation efficiency of explosives. 32nd International Annual Conference of ICT, Karlsruhe, Germany, FRG, 3-6 July, pp. 36-1, 14.
58. Brousseau, P. and Cliff, M.D. (2001) The effect of ultrafine aluminum powder on the detonation properties of various explosives. 32nd International Annual Conference of ICT, Karlsruhe, Germany, FRG, 3-6 July, pp. 37.1–37.14.
59. Brousseau, P., Dorsett, H.E., Cliff, M.D., and Anderson, C.J. (2002) Detonation properties of explosives containing nanometric aluminum powder. 12th International Symposium on Detonation, San Diego, CA, 11-16 August.
60. Cooper, P.W. (1996) *Explosives Engineering*, John Wiley & Sons, Inc., New York, p. 292.
61. Dorsett, H.E., Brousseau, P., and Cliff, M.D. (2001) The Influence of ultrafine aluminum upon explosives detonation. 28th International Pyrotechnics Seminar, Adelaide, South Australia, 4-9 November.
62. DeLuca, L.T., Galfetti, L., Maggi, F., Colombo, G., Reina, A., Dossi, S., Consonni, D., and Brambilla, M. (2011) Innovative metallized formulations for solid or hybrid rocket propulsion. *Hanneng Cailiao/Chin. J. Energetic Mater.*, **20** (4), 465–474. doi: 10.3969/j.issn.1006-9941.2012.04.018Presented at the 2011 IASPEP (International Autumn Seminar on Propellants, Explosives and Pyrotechnics), China.
63. DeLuca, L.T. and Galfetti, L. (2008) Burning of metallized composite solid rocket propellants: from micrometric to nanometric aluminum size Proceedings AJCPP (3rd Asian Joint Conference on Propulsion and Power), The Korean Society of Propulsion Engineers, Gyeongju, Korea, 06-08 March.
64. Babuk, V.A., Dolotkazin, I., Gamsov, A., Glebov, A., DeLuca, L.T., and Galfetti, L. (2009) Nanoaluminum as a solid propellant fuel. *J. Propul. Power*, **25** (2), 482–489.
65. DeLuca L.T. *et al.* (2009) *Proceedings 2009 IASPEP (International Autumn Seminar on Propellants, Explosives and Pyrotechnics), China*, Ed. Li S., Wang Y., Cao F., Zhao S. and Zhou S., Science Press, 2009, pp. 258-270.
66. DeLuca, L.T., Galfetti, L., Colombo, G., Maggi, F., Bandera, A., Babuk, V.A., and Sinditskii, V.P. (2010) Microstructure effects in aluminized solid rocket propellants. *J. Propul. Power*, **26** (4), 724–733.
67. DeLuca, L.T., Galfetti, L., Colombo, G., Maggi, F., Bandera, A., Boiocchi, M., Gariani, G., Merotto, L., Paravan, C., and Reina, A. (2011) *Advances in Propulsion Physics*, vol. 2, Torus Press, Moscow, pp. 341–362.
68. Paravan, C., Reina, A., Sossi, A., Manzoni, M., Massini, G., Rambaldi, G., Duranti, E., Adami, A., Seletti, E., and DeLuca, L.T. (2013) in *Advances in Propulsion Physics*, vol. 4 (eds C. Bonnal, L.T. DeLuca, S.M. Frolov, and O.J. Haidn), Torus Press, Moscow, pp. 75–98.
69. Paravan, C., Manzoni, M., Rambaldi, G., and DeLuca, L.T. (2013) Analysis of Quasi-Steady and Transient Burning of Hybrid Fuels in a Laboratory-Scale Burner by an Optical Technique, *Int. J. of Energetic Mater. Chem. Propul.*, **12** (5), 385–410, doi:10.1615/IntJEnergeticMaterialsChemProp.20130
70. Galfetti, L., DeLuca, L.T., Maggi, F., Colombo, G., *et al.* (2012) Experimental investigation of transient combustion in a hybrid 2d lab-scale burner by a wire-cut transducer technique. 9-ISICP (9th International Symposium on Special Topics in Chemical Propulsion), Québec City, Canada, July 9–13, 2012.
71. Merotto, L., Boiocchi, M., Milova, P., Monzani, M.R., and Galfetti, L. (2012) Development of an experimental lab-scale set up for transient combustion investigation in hybrid rockets.

63rd IAC (International Astronautical Congress), Naples, Italy, October 1-5.
72. Gariani, G., Maggi, F., and Galfetti, L. (2011) Numerical simulation of HTPB combustion in a 2D hybrid slab combustor. *Acta Astronaut.*, **69**, 289–296.
73. Gariani, G., Maggi, F., and Galfetti, L. (2010) Simulation Code for Hybrid Rocket Combustion, AIAA Paper 2010-6872.
74. Maggi, F., Gariani, G., Galfetti, L., and DeLuca, L.T. (2012) Theoretical analysis of hydrides in solid and hybrid rocket propulsion. *Int. J. Hydrogen Energy*, **37**, 1760–1769.
75. Sutton, J.P. and Biblarz, O. (2001) *Rocket Propulsion Elements*, 7th edn, John Wiley & Sons, Inc., New York.
76. Sarner, S.F. (1966) *Propellant Chemistry*, Reinhold Publishing.
77. Glassman, I. and Sawyer, R.F. (1970) *The Performance of Chemical Propellants*, AGARDograph, p. 129.
78. Reydellet, D. (1986) A Computer Program for the Prediction of Solid Propellant Rocket Motor Performance, D. E. Coats, *et al.*, Technical Report AFRPL-TR-75-36, Air Force Rocket Propulsion Laboratories. Edwards, CA, USA.
79. Smith, E.B. (2004) *Basic Chemical Thermodynamics*, 5th edn, Imperial College Press.
80. Gordon, S. and McBride, B.J. *CEA: Chemical Equilibrium with Applications*, NASA, *http://www.grc.nasa.gov/WWW/CEAWeb* (accessed 26 September 2013).
81. Gordon, S. and McBride, B.J. (1994) *Computer Program for Calculation of Complex Chemical Equilibrium Compositions and Applications*, NASA Reference Publication RP-1311, NASA.
82. Chase M.W.J. (1998) NIST-JANAF thermochemical tables. *J. Phys. Chem. Ref. Data Monogr.* **9**, 1–1951.
83. Kubota, N. (2007) *Propellants and Explosives: Thermochemical Aspects Of Combustion*, 2nd edn, Wiley-VCH Verlag, Weinheim.
84. DeLuca, L.T., Marchesi, E., Spreafico, M., Reina, A., Maggi, F., Rossettini, L., Bandera, A., Colombo, G., and Kosowski, B.M. (2010) Aggregation versus agglomeration in metallized solid rocket propellants. *Int. J. Energetic Mater. Chem. Propul.*, **9**, 91–105.
85. Chan, M., Johnson, C.E., and DeLuca, L.T. ed. (2002) Evaluation of AlH for propellant applications. Proceedings of 8-IWCP, International Workshop on Rocket Propulsion: Present and Future.
86. Lide, D.R. (ed.) (2007) *Handbook of Chemistry and Physics*, 88th edn, CRC Press.
87. Altman, D. (2002) *Rocket Motors Hybrid*, vol. 14, Academic Press, pp. 303–321.
88. Yetter, R.A., Risha, G.A., and Son, S.F. (2009) Metal particle combustion and nanotechnology. *Proc. Combust. Inst.*, **32**, 1819–1838. doi: 10.1016/j.proci.2008.08.013
89. Dossi, S., Maggi, F., and DeLuca, L.T. (2012) Innovative metal fuels for solid rocket propulsion. *Int. J. Energetic Mater. Chem. Propul.*, **11** (4), 299–322.
90. Cohen, N.S. (1983) A pocket model for aluminum agglomeration in composite propellants. *AIAA J.*, **21** (5), 720–725.
91. Maggi, F., Bandera, A., Galfetti, L., DeLuca, L.T., and Jackson, T.L. (2010) *Acta Astronaut.*, **66**, 1563–1573.
92. Hermsen, R.W. (1981) Aluminum oxide particle size for solid rocket motor performance prediction. *J. Spacecraft Rockets*, **18**, 483–490.
93. Maggi, F., Dossi, S., and DeLuca, L.T. (2013) Combustion of metal agglomerates in a solid rocket core flow. *Acta Astronaut.*, **92**, 163–171.
94. NASA (1971) *Solid Rocket Motor Performance Analysis and Prediction*, NASA SP-8039, NASA, Washington, DC.
95. Kwon, Y., Jung, Y., Yavorovsky, N.A., Ilyin, A.P., and Kim, J. (2001) Ultrafine powder by wire explosion method. *Scr. Mater.*, **44** (8-9), 2247–2251.
96. Vollmer, B. and Janiak, C. (2011) Naked metal nanoparticles from metal carbonyls in ionic liquids: easy synthesis and stabilization. *Coord. Chem. Rev.*, **255**, 2039–2057.
97. Kwok, Q.S.M., Badeen, C., Armstrong, K., Turcotte, R., Jones, D.E.G., and Gertsman, V.Y. (2007) Hazard characterization of uncoated and coated

aluminum nanopowder compositions. *J. Propul. Power*, **23** (4), 659–668.
98. Gromov, A., Forther, B.U., and Teipel, U. (2006) aluminum nanopowders produced by electrical explosion of wires and passivated by non-inert coatings: characterization and reactivity with air and water. *Powder Technol.*, **164** (2), 111–115.
99. Kwon, Y.S., Gromov, A.A., Ilyin, A.P., and Rim, G.H. (2003) Passivation process for superfine aluminum powders obtained by electrical explosion of wires. *Appl. Surf. Sci.*, **211** (1-4), 57–67.
100. Kwon, Y., Gromov, A.A., and Strokova, J.I. (2007) Passivation of the surface of aluminum nanopowders by protective coatings of the different chemical origin. *Appl. Surf. Sci.*, **253** (12), 5558–5564.
101. Gromov, A., Ilyin, A., An, V., Faubert, F., de Izarra, C., Espagnacq, A., and Brunet, L. (2002) Characterization of aluminum powders: I. Parameters of reactivity of aluminum powders. *Propellants Explos. Pyrotech*, **27** (6), 361–364.
102. Gromov, A., Ilyin, A., Forter-Barth, U., and Teipel, U. (2006) Characterization of aluminum powders: II. Aluminum nanopowders passivated by non-inert coatings. *Propellants Explos. Pyrotech.*, **31** (5), 401–409.
103. Chung, S.W., Guliants, E.A., Bunker, C.E., Hammerstroem, D.W., Deng, Y., Burgers, M.A., Jelliss, P.A., and Buckner, S.W. (2009) Capping and passivation of aluminum nanoparticles using alkyl-substituted epoxides. *Langmuir Lett.*, **25** (16), 8883–8887.
104. Wang L.S. and Hong R.Y. (2011) in *Advances in Nanocomposites – Synthesis, Characterization and Industrial Application*, Boreddy S.R., In Tech pp. 289–322.
105. Advanced Powder Technology LLC, Tomsk, *www.nanosized-powders.com3*.
106. 3M Specialty Materials 3M Material Safety Data Sheet FC˙2175, Fluorel^{TM} Brand Fluoroelastomer, *www.machichemicals.com/pdf/3M˙FC-2175.pdf* (accessed 18 October 2013).
107. Hielscher, T. (2005) Ultrasonic production of nano-size dispersions and emulsions, ENS'05, Paris, France, 14-16 December.
108. Peralta-Videa, J.R., Zhao, L., Lopez-Moreno, M.L., de la Rosa, G., Hong, J., and Gardea Torresdey, J.L. (2011) Nanomaterials and the environment a review for the biennium 2008–2010. *J. Hazard. Mater.*, **186** (1), 1–15.
109. Mary, B., Dubois, C., Carreau, P.J., and Brousseau, P. (2006) Rheological properties of suspensions and polyethylene-coated aluminum nanoparticles. *Rheol. Acta*, **45**, 561–573.
110. Astankova, A.P., Godymchuk, A.Y., Gromov, A.A., and Il'in, A.P. (2008) The kinetics of self-heating in the reaction between aluminum nanopowder and liquid water. *Russ. J. Phys. Chem.*, **82** (11), 1913–1920.
111. Chen, L., Song, W.L., Wang, J.L.L., and Xie, C.S. (2009) Effect of heating rates on TG–DTA results of aluminum nanopowders prepared by laser heating evaporation. *J. Therm. Anal. Calorim.*, **96** (1), 141–145.
112. Mench, M.M., Kuo, K.K., Yeh, C.L., and Lu, Y.C. (1998) Comparison of thermal behavior of regular and ultra-fine aluminum powders (Alex) made from plasma explosion process. *Combust. Sci. Technol.*, **135** (1.6), 269–292. doi: 10.1080/00102209808924161
113. Sossi, A., Duranti, E., Paravan, C., DeLuca, L.T., Vorozhtsov, A.B., Gromov, A.A., Pautova, Y.I., Lerner, M.I., and Rodkevich, N.G. (2013) Non-isothermal oxidation of aluminum nanopowders coated by hydrocarbons and fluorohydrocarbons. *Appl. Surface Sci.*, **271** (2013), 337–343.
114. Sossi, A., Duranti, E., Manzoni, M., Paravan, C., DeLuca, L.T., Vorozhtsov, A.B., Lerner, M.I., Rodkevich, N.G., Gromov, A.A., and Savin, N. (2013) Combustion of HTPB-based solid fuels loaded with coated nanoaluminum. *Combust. Sci. Technol.*, **185** (1), 17–36. doi: 0.1080/00102202.2012.707261
115. Watson, K.W., Pantoya, M.L., and Levitas, V.I. (2008) Fast reactions with nano- and micrometer aluminum: a study on oxidation versus fluorination. *Combust. Flame*, **155**, 619–634.

116. Gromov, A., Ilyin, A., Foerter-Barth, U., and Teipel, U. (2006) Effect of the passivating coating type, particle size, and storage time on oxidation and nitridation of aluminum powders. *Combust. Explos. Shock Waves (Engl. Transl.)*, **42** (2), 177–184.

117. Cerri, S. (2006) Analisi sperimentale dei residui condensati di combustione solidi alluminizzati per propulsione spaziale. MSc Thesis, Politecnico di Milano.

118. Trunov, M.A., Schoenitz, M., Zhu, X., and Dreizin, E.L. (2005) Effect of polymorphic phase transformation in al_2o_3 film on oxidation kinetics of aluminum powders. *Combust. Flame*, **140**, 310–318.

119. Oliver, S.M. and Moylan, B.E. (1992) An Analytical Approach for the Prediction of Gamma-to-Alpha Phase Transformation of Aluminum Oxide (Al_2O_3) Particles in the Space Shuttle ASRM and RSRM Exhausts, AIAA-92-2915.

120. Kovalev, O.B. (2002) Prediction of the size of aluminum-oxide particles in exhaust plumes of solid rocket motors. *Combust. Explos. Shock Waves*, **38** (5), 535–546.

121. Glotov, O.G. (2000) Condensed combustion products of aluminized propellants. II. Evolution of particles with distance from the burning surface. *Combust. Explos. Shock Waves*, **36** (4), 476–487.

122. Galfetti, L., DeLuca, L.T., Severini, F., Meda, L., Marra, G., Lentini, L., Babuk, V.A., and Cerri, S. (2006) Experimental investigation and numerical modeling of the condensed combustion products of micro and nano-aluminized solid propellants, IAC-06-C4.3.03, 57th International Astronautical Congress.

123. Fabignon, Y., Orlandi, O., Trubert, J.F., Lambert, D., and Dupays J. (2003) Combustion of Aluminum Particles in Solid Rocket Motors, AIAA 2003-4807.

124. Dill, K.M., Reed, R.A., Calia, V.S., and Schulz, R.J. (1990) Analysis of crystalline phase aluminum oxide particles from solid propellant exhaust. *J. Propul. Power*, **5** (6), 668–671.

125. Nielsen, L.E. (1974) in *Mechanical Properties of Polymers and Composites*, vol. 1 (ed. L.E. Nielsen) Chapter 1, Marcel Dekker Inc., New York, pp. 1–37.

126. Teipel, U., Hordijk, A.C., Förter-Barth, U., Hoffman, D.M., Hübner, C., Valtsifer, V., and Newman, K.E. (2004) in *Energetic Materials* (ed. U. Teipel) Chapter 12, Wiley-VCH Verlag GmbH, Wenheim, pp. 433–508.

127. Klager, K. (1984) Polyurethanes, the Most Versatile Binder for Solid Rocket Propellants, AIAA Paper No. 84–1239.

128. Annovazzi, A. and DeLuca, L.T. (2009) in *Energetic Problems in Aerospace Propulsion*, 3rd edn – Notes for StudentsChapter 11 edn (ed. L.T. DeLuca), Premani, Pantigliate, Milano, pp. 1–104.

129. Humble, R.W. (2000) Fuel Performance Enhancements for Hybrid Rockets, AIAA Paper 2000–3437.

130. Larson, B.D., Boyer, E., Wachs, T., Kuo, K.K., *et al.* (2011) Characterization of the Performance of Paraffin/$LiAlH_4$ Solid Fuels in a Hybrid Rocket System, AIAA Paper 2011-5822.

131. Karabeyoglu, M.A., Altman, D., and Cantwell, B.J. (2002) Combustion of liquefying hybrid propellants: part 1, general theory. *J. Propul. Power*, **18** (3), 610–620.

132. Karabeyoglu, M.A. and Cantwell, B.J. (2002) Combustion of liquefying hybrid propellants: part 2, stability of liquid films. *J. Propul. Power*, **18** (3), 621–630.

133. Kim, S., Lee, J., Moon, H., Sung, H., and Kim, J. (2010) Effect of Paraffin-LDPE Blend in Hybrid Rocket Motor, AIAA Paper No. 2010-7031.

134. DeLuca L.T., Galfetti L., Maggi, F., Colombo, G., *et al.* (2012) Characterization of HTPB-based solid fuel formulations: performance, mechanical properties, and pollution, *Acta Astronaut.*, **92** (2), 150–162, doi. 10.1016/j.actaastro.2012.05.002.

135. Reina, A., Paravan, C., Maggi, F., Colombo, G., *et al.* (2011) Nanometric powders dispersion in a polymeric matrix. Presented at 4th EUCASS, Saint Petersburg, Russia, 04-08 July. ID No. 1214.

136. Paravan, C. (2012) Ballistics of innovative solid fuel formulations for hybrid rocket engines. PhD Dissertation. Aerospace Engineering Department, Politecnico di Milano.
137. Reina, A. (2013) Nano-metal fuels for hybrid and solid propulsion. PhD Dissertation. Aerospace Engineering Department, Politecnico di Milano.
138. Teipel, U. and Forther-Barth, U. (2001) Rheology of nano-scale aluminum suspensions. *Propellants Explos. Pyrotech.*, **26**, 268–272.
139. Nielsen, L.E. (1974) in *Mechanical Properties of Polymers and Composites*, vol. 2 (ed. L.E. Nielsen) Chapter 7, Marcel Dekker Inc., New York, pp. 379–452.
140. Smallwood, H.M. (1944) Limiting law of the reinforcement of rubber. *J. Appl. Phys.*, **15**, 758–766.
141. Fletcher, W.P. and Gent, A.N. (1953) Non-linearity in the dynamic properties of vulcanized rubber compounds. *Trans. Inst. Rubber Ind.*, **29**, 266–280.
142. Payne, A.R. (1962) The dynamic properties of carbon black-loaded natural rubber vulcanizates, part I. *J. Appl. Polym. Sci.*, **4** (19), 57–63.
143. Payne, A.R. (1962) The dynamic properties of carbon black-loaded natural rubber vulcanizates, part II. *J. Appl. Polym. Sci.*, **4** (21), 368–372.
144. Payne, A.R. (1965) Effect of dispersion on the dynamic properties of filler-loaded rubbers. *J. Appl. Polym. Sci.*, **9**, 2273–2284.
145. Wang, M.J. (1999) The role of filler networking in dynamic properties of filled. *Rubber Chem. Technol.*, **72** (2), 430–448.
146. Frolich, J., Niedermaier, W., and Luginsland, H.D. (2005) The effect of filler–filler and filler–elastomer interaction on rubber reinforcement. *Composites, Part A*, **36**, 449–460.
147. Russell, W.B. (1980) Review of colloidal forces in the rheology of suspensions. *J. Rheol.*, **24** (3), 287–317.
148. Risha, G.A., Evans, B.J., Boyer, E., Wehrman, R.B., and Kuo, K.K. (2003) Nano-sized Aluminum and Boron-Based Solid Fuel Characterization in a Hybrid Rocket Engine, AIAA Paper 2003-4593.
149. Lee, E.C. and Mielewski, D.F. (2005) Method for producing a well-exfoliated and dispersed polymer silicate nanocomposite by ultrasonication. US Patent US 2005/0122834 A1.
150. Gedanken, A. (2007) Doping nanoparticles into polymers and ceramics using ultrasound radiation. *Ultrason. Sonochem.*, **14**, 418–430.
151. Risha, G.A., Evans, B.J., Boyer, E., and Kuo, K.K. (2007) in *Fundamentals of Hybrid Rocket Combustion and Propulsion*, AIAA Progress in Astronautics and Aeronautics, vol. 218 (eds M.J. Chiaverini and K.K. Kuo) Chapter 10, AIAA Inc., Reston, VA, pp. 413–456.
152. Wang, Q., Xia, H., and Zhag, C. (2001) Preparation of polymer/inorganic nanoparticles composites through ultrasonic irradiation. *J. Appl. Polym. Sci.*, **80**, 1478–1488.
153. Caneba, G.T., Dutta, C., Agrawal, V., and Rao, M. (2010) Novel ultrasonic dispersion of carbon nanotubes. *J. Miner. Mater. Charact. Eng.*, **9** (3), 165–181.
154. Du, F., Scogna, R.C., Zhou, W., Brand, S., Fisher, J.E., and Winey, K.I. (2004) Nanotube networks in polymer nanocomposites: rheology and electrical conductivity. *Macromolecules*, **37**, 9048–9055.
155. Chen, Y., Liang, Y., Zheng, F., Zhou, R., and Feng, Z. (2001) The dispersion behavior of Si–C–N nanopowders in organic liquids. *Ceram. Int.*, **27**, 73–79.
156. Mackay, M.E., Tuteja, A., Duxbury, P.M., Hawker, C.J., Van Horn, B., Guan, Z., Chen, G., and Krishnan, R.S. (2006) General strategies for nanoparticle dispersion. *Science*, **311**, 1740–1743.
157. Metzner, A.B. (1985) Rheology of suspensions in polymeric liquids. *J. Rheol.*, **29**, 739–774.
158. Miller, R.R., Lee, E., and Powell, R.L. (1991) Rheology of solid propellant dispersions. *J. Rheol.*, **35** (5), 901–920.
159. Muthiah, R., Krishnamurthy, V.N., and Gupta, B.R. (1996) Rheology of HTPB propellant: development of generalized correlation and evaluation of pot life. *Propellants Explos. Pyrotech*, **21**, 186–192.

160. Wildemuth, C.R. and Williams, M.C. (1984) Viscosity of suspensions modeled with a shear-dependent maximum packing fraction. *Rheol. Acta*, **23**, 627–635.
161. Popenko, E.M., Gromov, A.A., Shamina, Y.Y., Ilyn, A.P., Sergienko, A.V., and Popok, N.I. (2007) Effect of the addition of ultrafine aluminum powders on the rheological properties and burning rate of energetic condensed systems. *Combust. Explos. Shock Waves*, **43** (1), 46–50.
162. Gromov, A.A., Forther-Barth, U., Popenko, E.M., and Teipel, U. (2006) Effect of nanoaluminum (ALEX^{TM}) additions on the rheology of energetic mixtures. Presented at HEM 2006, Belokhurika, Russia, 11-14 September.
163. Layton L.H. (1975) *Chemical Structural Aging Studies on an HTPB Propellant*, AD-A010 731, Thiokol Corporation, Brigham City, UT.
164. Akbas, A., Aksoy, S., and Hasirci, N. (1994) Effects of thermal ageing on the properties and lifetime prediction of hydroxyl–terminated polybutadiene. *Polym. J.*, **35** (12), 2568–2572.
165. Lillo, F., D'Andrea, B., Mercelli, G., and Sebasta, A. (2001) Long Term Aging of Aerospace and Tactical SRM. Experimental Study, AIAA Paper 2001-3284.
166. Kivity, M., Hartman, G., and Achlama, A.M. (2005) Aging of HTPB Propellant. AIAA Paper 2005-3802.
167. Shalom, A., Aped, H., Kivity, M., and Horowitz, D. (2005) The Effect of Nanosized Aluminum on Composite Propellant Properties, AIAA Paper 2005-3604.
168. Cerri, S., Bohn, M.A., Menke, K., and Galfetti, L. (2009) Ageing behavior of HTPB-based rocket propellant formulations. *Cent. Eur. J. Energ. Mater.*, **6** (2), 149–165.
169. Cliff, M., Tepper, F., and Lisetsky, V. (2001) Ageing Characteristics of ALEX^{TM} Nanosize Aluminum, AIAA Paper 2001-3287.
170. Arkhipov, V.A., Bondarchuka, S.S., Korotkikhc, A.G., Kuznetsova, V.T., Gromov, A.A., Volkov, S.A., and Revyagin, L.N. (2012) Influence of aluminum particle size on ignition and nonstationary combustion of heterogeneous condensed systems. *Combust. Explos. Shock Waves*, **48** (5), 625–635.
171. Babuk, V.A. and Salimullin, R.M. (2008) Nanoparticles as component of high-energy condensed systems. Proceedings of the Sixth All-Russia Conference Intrachamber Processes and Burning in Installations on Solid Propellants and in Artillery Systems, Izhevsk, Russia, September 2008, Institute of Applied Mechanics of the Russian Academy of Sciences, pp. 292–301.
172. Galfetti, L., Severini, F., DeLuca, L.T., Marra, G.L., and Meda, L. (2004) Ballistics and condensed combustion residues of aluminized solid rocket propellants, in *Ninth International Workshop on Combustion and Propulsion* (eds L.T. DeLuca, L. Galfetti, and R.A. Pesce-Rodriguez), Grafiche GSS, Bergamo, Italy, paper 18.
173. Galfetti, L., Severini, F., DeLuca, L.T., and Meda, L. (2004) Nano propellants for space propulsion. *4th International Spacecraft Propulsion Conference, CD Proceedings SP-555/556/557*, European Space Agency, Sardinia, Italy, June 04, ISBN: 92-9092-866-2/-867-0/-868-9.
174. DeLuca, L.T., Galfetti, L., Severini, F., Meda, L., Marra, G., Vorozhtsov, A.B., Sedoi, V.B., and Babuk, V.A. (2005) Burning of *n*Al composite rocket propellants. *Combust. Explos. Shock Waves*, **41** (6), 680–692.
175. Shafirovich, E., Escot, P., Chauveau, C., Gökalp, I., Rosenband, V., and Gany, A. (2004) in *Combustion of Single Aluminum Particles with Thin Nickel Coatings, Novel Energetic Materials and Applications* (eds L.T. DeLuca, L. Galfetti, and R.A. Pesce-Rodriguez), Grafiche GSS, Bergamo, pp. 47.1–47.6.
176. Risha, G.A., Boyer, E., Wehrman, R.B., and Kuo, K.K. (2002) Performance comparison of HTPB-based solid fuels containing nano-size energetic powder in a cylindrical hybrid rocket motor. 38th AIAA/ ASME/SAE/ASEE Joint Propulsion Conference and Exhibit, AIAA Paper 2002-3576.

177. Guirao, C. and Williams, F.A. (1971) A model for ammonium perchlorate deflagration between 20 and 100 atm. *AIAA J.*, **9** (7), 1345–1356. doi: 10.2514/3.6355

178. Babuk, V.A., Gamzov, A.V., Glebov, A.A., Dolotkazin, I.N., and Dosychev, A.V. (2006) Structure of composite metallized solid propellants and its role during burning. *Khim. Fizi. Mesoskopiya*, **8** (1), 33–45 (in Russian).

179. Price, E.W. (1979) Combustion of aluminum in solid propellant flames, AGARD PEP 53rd Meeting on Solid Rocket Motor Technology, Paper 14, AGARD, Paris, France.

180. Price, E.W. (1984) in *Combustion of Metallized Propellants, Fundamentals of Solid Propellants Combustion*, AIAA Progress in Aeronautics and Astronautics, vol. 90 (eds K.K. Kuo and M. Summerfield), Chapter 9, AIAA, New York.

181. Babuk, V.A., Glebov, A.A., and Dolotkazin, I.N. (2005) Burning mechanism of aluminized solid rocket propellants based on energetic binders. *Propellants Explos. Pyrotech.*, **30** (4), 281–290. doi: 10.1002/prep.200500012

182. Babuk, V.A., Vasilyev, V.A., Glebov, A.A., Dolotkazin, I.N., Galeotta, M., and DeLuca, L.T. (2004) Combustion mechanisms of AN-based aluminized solid rocket propellants, in *Ninth International Workshop on Combustion and Propulsion* (eds L.T. DeLuca, L. Galfetti, R.A. Pesce-Rodriguez, and G.S.S. Grafiche), Bergamo, Italy.

183. Babuk, V.A., Glebov, A., Dolotkazin, I.N., Conti, A., Galfetti, L., DeLuca, L.T., and Vorozhtsov, A. (2009) *Condensed Combustion Products from Burning of Nanoaluminum-Based Propellants: Properties and Formation Mechanism, Advances in Aerospace Sciences*, Torus Press, Moscow, pp. 3–16.

184. Babuk, V.A. (2009) Properties of a surface layer and regularities of burning metallized solid propellants. *Combust. Explos. Shock Waves*, **45** (4), 486–494.

185. Weiser, V., Eckl, W., Eisenreich, N., Kelzenberg, S., Plitzko, Y., Poller, S., and Roth, E. (2004) Burning behavior of aluminized composite propellants including nanoparticles, in *Ninth International Workshop on Combustion and Propulsion* (eds L.T. DeLuca, L. Galfetti, and R.A. Pesce-Rodriguez), Grafiche GSS, Bergamo.

186. Trubert, J.F., Hommel, J., Lambert, D., Fabignon, Y., and Orlandi, O. (2008) in *New HTPB/AP/Al Propellant Combustion Process in Presence of Aluminum Nano-Particles, Advancements in Energetic Materials and Chemical Propulsion* (eds K.K. Kuo and K. Hori), Begell House, New York, pp. 98–121.

187. Babuk, V.A., Vassiliev, V.A., and Sviridov, V.V. (2000) in *Solid Propellant Chemistry, Combustion, and Motor Interior Ballistics*, AIAA Progress in Astronautics and Aeronautics, vol. 185 (eds V. Yang, T.B. Brill, and W.Z. Ren), AIAA, Reston, VA, pp. 749–776.

188. Price, E.W., Sambamurthi, J.K., Sigman, R.K., and Panyam, R.R. (1986) Combustion of ammonium perchlorate: polymer sandwiches. *Combust. Flame*, **63** (3), 381–413. doi: 10.1016/0010-2180(86)90007-6

189. DeLuca, L.T. (2007) in *Theory and Practice of Energetic Materials*, vol. 7 (eds H. Ping, W. Yajun, and L. Shengcai), Science Press, Beijing, pp. 277–289.

190. DeLuca, L.T., Galfetti, L., Maggi, F., Colombo, G., *et al.* (2008) Burning of metallized composite solid rocket propellants: toward nanometric fuel size, Proceedings of ESA Space Propulsion 2008, ESA, Crete, Greece.

191. Belyaev, A.F., Korotkov, A.I., and Frolov, Y.V. (1968) Ignition and combustion of single particle of small-grained aluminum. *Combust. Explos. Shock Waves*, **4** (3), 323–330.

192. Olsen, S.E. and Beckstead, M.W. (1996) Burn time measurements of single aluminum particles in steam and carbon dioxide. *J. Propul. Power*, **12** (4), 662–671. doi: 10.2514/3.24087

193. Widener, J.F. and Beckstead, M.W. (1998) Aluminum Combustion Modeling in Solid Propellant Combustion Products, AIAA Paper 98-3824.

194. Galfetti, L., De Luca, L.T., Severini, F., Meda, L., Marra, G., Marchetti, M., Regi, M., and Bellucci, S.

(2006) Nanoparticles for solid rocket propulsion. *J. Phys. Condens. Matter*, **18**, S1991. doi: 10.1088/0953-8984/18/33/S15
195. Galfetti, L., De Luca, L.T., Severini, F., Colombo, G., Meda, L., and Marra, G. (2007) Pre and post-burning analysis of nano-aluminized solid rocket propellants. *Aerosp. Sci. Tech.*, **11** (2007), 26–32.
196. Cerri, S., Galfetti, L., DeLuca, L.T., D'Andrea, B., and Cianfanelli, S. (2007) Experimental Investigation of the Condensed Combustion Products of Micro Aluminized Solid Rocket Propellants, AIAA Paper 07-5766.
197. Dondè, R., Riva, G., and DeLuca, L.T. (1984) Experimental and theoretical extinction of solid rocket propellants by fast depressurization. *Acta Astronaut.*, **11** (9), 569–576.
198. Dondé, R., Guarnieri, C., Meda, L., Marra, G., *et al.* (2006) Experimental Investigations on transient burning of nano-aluminized solid rocket propellants. Changes in Aeronautical and Space Systems, Avignon, France, June 26-28, 2006, Paper 36-33-P.
199. Zanotti, C., Colombo, G., Grimaldi, C., Carretta, U., Riva, G., and De Luca, LT (1986), Oscillatory burning of composite propellants near pressure deflagration limit. XV International Symposium on Space Technology and Science (ISTS), Tokyo, Japan, May 18-24, 1986. ISTS Conference Proceedings, pp. 243–254.
200. DeLuca, L.T., Galfetti, L., Bosisio, F., Raina, H., Maggi, F., and Colombo, G. (2006) An Hybrid Micro-combustor for Regression Rate Measurements, IAC-06-C4.2.01.
201. DeLuca, L.T., Galfetti, L., Maggi, F., Colombo, G., *et al.* (2011) An Optical Time-Resolved Technique of Solid Fuels Burning for Hybrid Rocket Propulsion, AIAA Paper 2011-5753.
202. George, P., Krishnan, S., Varkey, P.M., Ravindran, M., and Ramachandran, L. (2001) Fuel regression rate in hydroxyl terminated polybutadiene/gaseous oxygen hybrid rocket motors. *J. Propul. Power*, **17** (1), 35–42.
203. Cauty, F., Seryn, N., and Gramer, D. (2007) *Fundamentals of Hybrid Rocket Combustion and Propulsion*, AIAA Progress in Astronautics and Aeronautics, vol. 218 Chapter 4, AIAA, pp. 167–205.
204. Houser, T.J. and Peck, M.V. (1964) *AIAA Progress in Astronautics and Aeronautics, Heterogeneous Combustion*, vol. 15, Academic Press, New York, pp. 559–581.
205. Evans, B., Boyer, E., Kuo, K.K., Risha, G., and Chiaverini, M.J. (2009) Hybrid Rocket Investigations at Penn State University's High Pressure Combustion Laboratory: Overview and Recent Results, AIAA Paper 2009-5349.
206. Fanton, L., Paravan, C., and DeLuca, L.T. (2012) Testing and modeling fuel regression rate in a miniature hybrid burner. *Int. J. Aerosp. Eng.*, **2012**, 673838. doi: 10.1155/2012/673838
207. Ohlemiller, T.J. and Summerfield, M. (1968) A critical analysis of arc image ignition of solid propellants. *AIAA J.*, **6** (5), 878–886.
208. Altman, D. and Holzman, A. (2007) *Fundamentals of Hybrid Rocket Combustion and Propulsion*, AIAA Progress in Astronautics and Aeronautics, vol. 218, Chapter 1, AIAA, pp. 1–36.
209. DeLuca, L.T. (2009) *Energetic Problems in Aerospace Propulsion*, 3rd – Notes for Students, Chapter 12 edn, Premani, Pantigliate, Milano, pp. 1–92.
210. Marxman, G.A. and Gilbert, M. (1963) *9th International Symposium on Combustion*, Academic Press, New York, pp. 371–383.
211. Marxman, G.A. (1967) *Proceedings of the 11th Symposium (Int.) on Combustion*, The Combustion Institute, Pittsburg, PA, pp. 269–289.
212. DeLuca, L.T., Paravan, C., Reina, A., Marchesi, E., Maggi, F., Bandera, A., Colombo, G., and Kosowski, B.M. (2010) Aggregation and Incipient Agglomeration in Metallized Solid Propellants and Solid Fuels for Rocket Propulsion, AIAA Paper 2010-6752.

Index

Metal Nanopowders: Production, Characterization, and Energetic Applications, First Edition.
Edited by Alexander Gromov and Ulrich Teipel.